TRANSPORT MODELS for Inland and Coastal Waters

Proceedings of the Symposium on Predictive Ability of Surface Water Flow and Transport Models, held at the Marriott Inn, Berkeley, California, August 18–20, 1980, under the sponsorship of the International Association for Hydraulic Research, the American Society of Civil Engineers, and the University of California, Berkeley, and with the financial assistance of the National Science Foundation and the Environmental Protection Agency.

TRANSPORT MODELS for Inland and Coastal Waters

Proceedings of a Symposium on Predictive Ability

Edited by
HUGO B. FISCHER
Department of Civil Engineering
University of California, Berkeley
Berkeley, California

1981
ACADEMIC PRESS
A Subsidiary of Harcourt Brace Jovanovich, Publishers
NEW YORK LONDON TORONTO SYDNEY SAN FRANCISCO

ACADEMIC PRESS, INC.
111 Fifth Avenue, New York, New York 10003

United Kingdom Edition published by
ACADEMIC PRESS, INC. (LONDON) LTD.
24/28 Oval Road, London NW1 7DX

Library of Congress Cataloging in Publication Data

Symposium on Predictive Ability (1980: Berkeley, Calif.)
Transport models for inland and coastal waters.

Sponsored by the International Association for Hydraulic Research, American Society of Civil Engineers, and University of California, Berkeley.
Includes bibliographical references and index.
1. Hydraulic models--Congresses. 2. Hydraulic engineering--Mathematical models--Congresses. I. Fischer, Hugo B. II. International Association for Hydraulic Research. III. American Society of Civil Engineers. IV. University of California, Berkeley. V. Title.
TC163.S95 1980 628.1'68 81-10990
ISBN 0-12-258152-0 AACR2

PRINTED IN THE UNITED STATES OF AMERICA

81 82 83 84 9 8 7 6 5 4 3 2 1

Contents

Contributors

Numbers in parentheses indicate the pages on which the authors' contributions begin.

M. B. Abbott (**222**), International Institute for Hydraulic and Environmental Engineering, Delft, The Netherlands

Gerrit Abraham (**1**), Delft Hydraulics Laboratory, Delft, The Netherlands

Keith W. Bedford (**172**), The Ohio State University, Columbus, Ohio 43210

François Boulot (**362**), Electricite de France, 78400 Chatou, France

M. A. M. de Ras (**408**), Rijkswaterstaat, The Hague, The Netherlands

G. Di Silvio (**112**), Instituto di Idraulica, The University of Padua, Padua, Italy

J. Dronkers (**451**), Rijkswaterstaat, The Hague, The Netherlands

G. Fiorillo (**112**), Instituto di Idraulica, The University of Padua, Padua, Italy

Jörg Imberger (**310**), The University of Western Australia, Nedlands, 6009, Western Australia

M. Karelse (**483**), Delft Hydraulics Laboratory, Dordrecht, The Netherlands

A. Langerak (**408**), Rijkswaterstat, The Hague, The Netherlands

J. J. Leendertse (**408, 451**), The Rand Corporation, Santa Monica, California 90406

A. McCowan (**222**), Danish Hydraulic Institute, Horsholm, Denmark

Nicholas V. M. Odd (**39**), Hydraulics Research Station, Wallingford, United Kingdom

John C. Patterson (**310**), The University of Western Australia, Nedlands, 6009, Western Australia

R. N. Pavlović (**63**), Sonderforschungsbereich 80 and Institute for Hydromechanics, University of Karlsruhe, Karlsruhe, Federal Republic of Germany

P. A. J. Perrels (**483**), Delft Hydraulics Laboratory, Emmeloord, The Netherlands

Donald W. Pritchard (**140**), Marine Sciences Research Center, State University of New York, Stony Brook, New York

W. Rodi (**63**), Sonderforschungsbereich 80 and Institute for Hydromechanics, University of Karlsruhe, Karlsruhe, Federal Republic of Germany

S. K. Srivatsa (**63**), Sonderforschungsbereich 80 and Institute for Hydromechanics, University of Karlsruhe, Karlsruhe, Federal Republic of Germany

Kim-Tai Tee (**284**), Atlantic Oceanographic Laboratory, Bedford Institute of Oceanography, Dartmouth, Nova Scotia, Canada

Adriaan G. van Os (**1**, **451**), Delft Hydraulics Laboratory, Delft, The Netherlands

Gerrit K. Verboom (**1**), Delft Hydraulics Laboratory, Delft, The Netherlands

I. R. Warren (**222**), Danish Hydraulic Institute, Horsholm, Denmark

Preface

The transport of dissolved substances through water bodies has been studied by means of hydraulic models for many years, and by means of numerical models since the early 1960s. Many symposia have been convened to discuss the building of such models. This book, however, gives the proceedings of a symposium convened to discuss whether models, once built, are truly predictive of natural events. The authors were asked to describe their models only to the extent needed for comprehension, and to concentrate on offering analytical discussion and prototype verification of predictive ability.

Chapter 1 sets forth a framework for evaluating predictive ability. As the spirited discussion shows, however, even the framework is not without question. The remaining chapters discuss models for pollutant transport in rivers, reservoirs, estuaries, and coastal areas. Among the results offered are examples from the Rhine River, near Karlsruhe, the Wellington Reservoir in Western Australia, the Lagoon of Venice, the French coast of the English Channel, and the Rhine Delta. Modeling of turbulence and techniques for acquiring field data are also discussed.

The authors of these papers were both self-selected, in that a call for proposals was issued, and then invited, in that a Scientific Committee appointed by the Committee on Fundamentals of the International Association for Hydraulic Research selected the papers from those proposed. The number of papers was strictly limited to allow each author one and a half hours for presentation and discussion. Thus the symposium could not hope to address all aspects of numerical and hydraulic modeling, and the subject matter is necessarily limited. In particular, none of the papers received by the Scientific Committeed made use of the finite element method, in spite of its popularity among many modelers. In the large laboratories, at least, the finite difference method seems at present the norm. It also seems likely, however, that predictive ability is not strongly dependent on the choice of numerical algorithm, since the limits we find on predictive ability are related primarily to gaps in physical understanding and ability to model detail, rather than the choice of numerical method.

The predictive ability of numerical and hydraulic models cannot be established by any one symposium or set of papers, nor can the state of predictive ability be easily summarized. These papers give many examples of prototype verification, yet they do not lead to the conclusion that all aspects of modeling are fully verified, or that either numerical or hydraulic models are fully predictive tools. As one important example, we find no adequate numerical models for stratified estuaries. Also, in spite of the remarkable advances in computational hardware, many of these papers mention limitations of computational ability and expense. Ten years ago it was widely believed that computer hardware limitations were a transient problem; now we find a broad realization that numerical models must make substantial concessions to the limitations of the computer. No model can reproduce all that is related to the external reality; thus all models must simplify, and the selection of the simplication, as Leendertse says in his discussion of Chapter 1, is an art. This book presents the skill and experience of twenty-four artists.

Acknowledgments

The appearance of this book is primarily a result of the dedication of Karen Earls and Peter Ray, who learned the operation of the Berkeley UNIX system and prepared all the manuscript in type-set form. I am also grateful to all those who assisted in the organization of the symposium, particularly the symposium coordinator, Dr. John McIlwrath. The papers were selected by a scientific committee whose membership was Dr. G. Abraham, Dr. F. Boulot, Dr. J. Imberger, Dr. H. Kobus, Dr. P. Ryan, Dr. T. Shaw, Dr. C. S. Yih, and myself. Financial support was received for the National Science Foundation, Grant ENG-7810818 and the Environmental Protection Agency, Grant R80642510.

MATHEMATICAL MODELING OF FLOWS AND TRANSPORT OF CONSERVATIVE SUBSTANCES: REQUIREMENTS FOR PREDICTIVE ABILITY

Gerrit Abraham
Adriaan G. van Os
Gerrit K. Verboom

Delft Hydraulics Laboratory, The Netherlands

1. INTRODUCTION

The purpose of this symposium is to further scientific understanding of the capabilities and limitations of mathematical models of flows and transport in natural receiving waters by means of presentations of complete modeling exercises. According to the instructions to the authors the end product should be a definite statement of what mathematical models can and, equally important, cannot do. To approximate such a definite statement the Scientific Committee asked authors presenting specific model applications to concentrate on giving an in-depth description of a specific example with enough detail to allow the audience to make an independent judgment on the value of the results. The authors of this general lecture were asked to assist the audience in making this judgment by providing a survey of the main factors which may lead to limited predictability. Consequently, the presentations on specific modeling exercises and this general lecture have opposite, though complimentary, functions within this symposium. The presentations on specific exercises have to concentrate on what can be accomplished under given well-defined circumstances. This general lecture has to present the main requirements for predictive ability.

In this lecture we discuss briefly the relationship between mathematical modeling, underlying research and engineering practice (Section 2). We give an analysis of the process of solving engineering problems by mathematical modeling (Section 3), followed by a detailed discussion of the different levels of approximation and the different feedback loops involved (Sections 4 through 10). Section 11 summarizes the findings insofar as relevant to the purpose of the lecture.

ISBN: 0-12-258152-0

2. MATHEMATICAL MODELING, UNDERLYING RESEARCH AND ENGINEERING PRACTICE

Mathematical surface water flow and transport models are applied to determine characteristics (such as magnitude of velocities and concentrations) relevant to the solution of design or management problems under consideration. In these engineering applications the modeling has to be predictive in the sense that it has to give an insight into relevant characteristics under conditions which do not exist as yet. In addition the predictions must be sufficiently accurate to allow the solution of the design or management problem for which they are made. Thus, predictive ability should mean more than the ability to produce numbers which are consistent with existing experimental data. *There must be prediction or calculation of the results of an experiment which has not been performed as yet and which becomes feasible only after the new conditions exist (after Saffman, 1977), while the predictions must be accurate enough to allow a sufficiently reliable answer to the engineering questions behind the mathematical model study.*

Finding which equations represent the development of a given physical phenomenon under given conditions requires research. Only after the research is completed does one have a quantitative insight into the performance of the equations to be selected in the predictive modeling effort. If not, quantitative insight into the performance of the selected aquations has to be obtained in one way or another in the predictive modeling process itself. *The predictive ability of a mathematical model is related to the state the research on these flow and transport phenomena which are important in the problem under consideration. In the ideal case the phenomena must be expressed in equations which are neither site-specific nor problem-specific.*

As explained in the subsequent text, at present several flow and transport processes can only be described by equations which are site-specific or problem-specific. It is important, then, to calibrate and verify the mathematical model. In this lecture calibration is defined as the process of making the output of the model agree with a limited number of known responses of the problem area to known excitations. In a negative sense this process can be referred to as the process of forcing agreement between the model output and the known responses. Verification is defined as the process of checking how well the calibrated mathematical model responds to excitations other than those utilized in the calibration, without changing the model. *A verified mathematical model may be used to interpolate within the range of known conditions covered in the calibration and the verification. It remains an intrinsic difficulty, however, to decide whether a verified mathematical model may be used to extrapolate beyond these known conditions. Yet, this may be required in the predictive application of the model.*

Engineering involves the use of several tools. These include field measurements, hydraulic scale models, empirical design graphs, mathematical models, etc. It requires engineering experience and engineering judgment to decide which tool to apply for a given problem, by itself or in combination with other tools, and to decide which value to attach to the extrapolations referred to at the end of the previous paragraph. *Experience in solving engineering problems with all available tools is a prerequisite for making appropriate use of a mathematical model as an engineering tool.*

A mathematical study may be research-oriented instead of prediction-oriented. For instance, a mathematical model may be applied as a tool for fundamental research to learn how to describe a given physical phenomenon in mathematical terms. In addition it may be used as a tool to investigate modeling procedures, such as methods to discretize the geometric features of the problem area. *The predictive ability which can be obtained when solving an engineering problem by mathematical modeling is the greater, the fewer elements of a research oriented character this mathematical modeling contains.*

3. MATHEMATICAL MODELING PROCESS

Figure 1 gives a schematic representation of the process of mathematical modeling for the solution of an engineering problem. The different levels of approximation and schematization contained in this figure and the corresponding feedback loops are explained as follows.

Solving engineering problems using mathematical models of flow and transport requires, to begin with, recognizing which characteristics or consequences of the flow and transport are to be known with which degree of accuracy to allow the solution of these problems. Thus, at the beginning of a mathematical modeling effort the relevant physical processes are to be distinguished (level 1 of approximation and schematization, Fig. 1). Also the design conditions are to be established. This initial engineering effort is decisive for the utility of a mathematical study for the solution of the engineering problem involved, and must be performed with care as only at the end of the mathematical model study it can be verified whether the information generated in the modeling effort is sufficiently accurate for the solution of the engineering problem under consideration (feedback loop 4, Fig. 1).

Solving a physical problem mathematically requires the use of equations which often have to be solved numerically. Thus in the mathematical solution of a physical problem two levels of approximation and schematization can be found: those needed to formulate the physical problem in mathematical terms (level 2 of Fig. 1) and those made in the numerical solution (level 3 of Fig. 1). In addition, the following feedback loops can be distinguished: a verification of the selected system of equations with boundary and initial conditions to verify whether the system may be expected to describe the physical phenomena (feedback loop 1, Fig. 1), a verification of the numerical solution procedure (feedback loop 2, Fig. 1) and a verification of the output of the computations (feedback loop 3, Fig. 1).

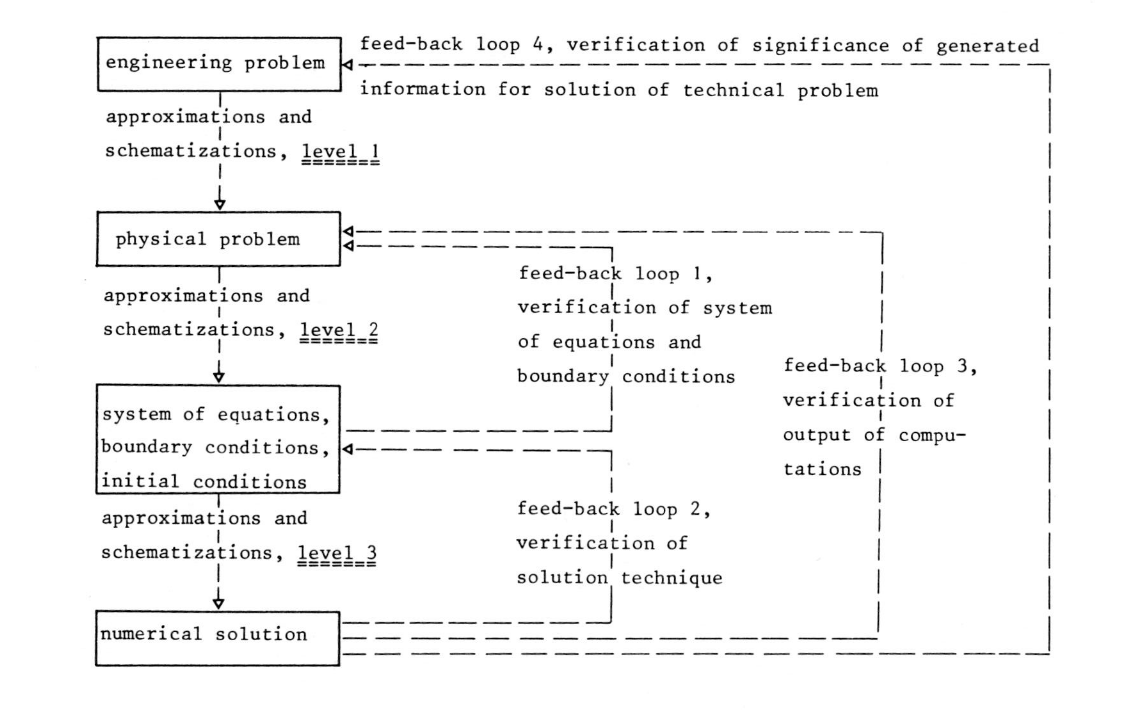

Figure 1. Approximations, schematizations and feed-backs in mathematical modeling for solution of engineering problem.

4. FROM ENGINEERING PROBLEM TO PHYSICAL PROBLEM (LEVEL 1 OF APPROXIMATION AND SCHEMATIZATION)

For the purpose of this general lecture it is important to distinguish between engineering problems whose solution requires knowledge of the spatial variation of the velocity and/or concentration of a constituent within the body of water, e.g. with depth or with the three space coordinates, and engineering problems whose solution can be derived from the variation of the water depth and the depth-mean values of the velocity and/or concentration with the two horizontal space coordinates. Problems requiring knowledge of the variation of the relevant parameters within the fluid include, for example, evaluating whether the three-dimensional temperature distribution in the mixing zone around a cooling-water outlet satisfies the temperature standards, and evaluating whether the salinity of the water flooding tidal flats in a partly mixed estuary remains below the tolerance limit set for a given ecological community.

Engineering problems which can often be solved on the basis of depth-mean parameters include, for example, evaluating the increase of salinity intrusion into a well mixed estuary because of an increased extraction of river water in the upstream range of the estuary, and evaluating the effects of dams on the tidal conditions of an estuary.

The engineering problems distinguished above lead to different types of physical problems. The problems requiring knowledge of the variation of the relevant parameters within the fluid lead to physical problems of the turbulent transport type as defined in Section 5.1 The engineering problems which can be solved on the basis of depth-mean parameters lead to physical problems of the dispersive transport type, as defined in Section 5.2, unless the effect of the dispersive transport turns out to be insignificant in the physical problem. If so, these engineering problems are of the wall shear type, as defined in Section 5.3.

5. EXPRESSING PHYSICAL PHENOMENA IN MATHEMATICAL TERMS (LEVEL 2 OF APPROXIMATION AND SCHEMATIZATION)

5.1 Physical Problems of Turbulent Transport Type

5.1.1 The closure problem of turbulence. The engineer's interest is often limited to the time-mean values of the relevant parameters, i.e., the values after turbulent fluctuations have been filtered out. The equations of motion and continuity governing the temporal and spatial variations of the time-mean values of velocity, pressure and concentration contain not only the time-mean values of these quantities as unknowns but also covariances of the turbulent fluctuations which represent the turbulent transport of momentum and mass. The covariances make the number of unknowns contained in these equations greater than the number of equations. The closure problem is the problem of making the number of equations equal to the number of unknowns by providing additional equations governing the covariances.

The engineering problems which ask for information on the variation of the relevant parameters within the fluid lead to a physical problem the solution of which can be obtained only by determining the spatial variation of the turbulent transports within the fluid. Physical problems requiring this information are referred to as physical problems of the turbulent transport type.

5.1.2 Gradient type of turbulent transport; eddy viscosity and eddy diffusivity models. Before the era of turbulence modeling it was the common practice to express the covariances in terms of the time-mean velocity and concentration by assuming a gradient type of turbulent transport. This approach to match the number of equations with the number of unknowns leads to the introduction of the coefficient of eddy viscosity ν_t, and eddy diffusivity K_t. The magnitude of these coefficients is not a property of the fluid, but depends upon the flow. If not derived from advanced turbulence models, the magnitude of the coefficients must be obtained from experiments. This empiricism limits the applicability of the gradient transport concept, because the needed experimental information can be provided only for flows with a single overall velocity scale and a single overall length scale. Tennekes and Lumley (1972, p. 57) indicate that the use of the gradient transport concept with given values of the coefficients ν_t and K_t should be limited to such flows.

The momentum jet and the buoyant plume in stagnant surrounding fluid are examples of flows with single overall length and single overall velocity scales. Tollmien (1926) and Schmidt (1941) solved these flows on the basis of the assumption of gradient transport with the magnitude of the coefficients ν_t and K_t as input data. These solutions, which showed the overall similarity of velocity and concentration profiles, were followed by the similarity solutions by Albertson, *et al.*, (1950) and Rouse, *et al.*, (1952), and later on by the entrainment solution by Morton, *et al.*, (1956). The entrainment concept had the advantage of being capable of dealing with jet type flows having different zones, each having a single overall length and single overall velocity scale (see, e.g., Abraham, 1978). By this concept it has been possible to solve three-dimensional buoyant jets in cross flows in an infinite space (see, e.g., Delvigne, 1979), though neglecting that the actual concentration distribution is horseshoe-shaped. In this way the similarity concept has been used to its limits. It has been pushed beyond its limits in the attempts to derive similarity solutions for buoyant surface jets (see, e.g., Policastro and Duyn, 1978). For such jets the similarity of velocity and concentration breaks down when buoyancy becomes important. For this reason Jirka, *et al.*, (1975), indicate that buoyant surface jet models should not be applied beyond a certain distance from the source. When buoyant surface jets are affected by the bottom and exhibit shoreline attachment, with and without circulation zones, the whole picture gets so complicated that empirical design graphs as presented by Naudascher and Schatzmann (1979) are the only simple model which makes sense.

Partly mixed estuaries are an example of flows with two different turbulence scaling possibilities (Fig. 2). In stably stratified fluids the processes producing turbulence and mixing must be examined carefully. According to Turner (1973) a distinction has to be made between "external" turbulence generated directly at a solid boundary and "internal" turbulence arising in the interior. This observation also applies to estuarine mixing. At the one extreme, in a well-mixed estuary the turbulence is primarily boundary generated. At the other extreme, in a highly stratified estuary with an arrested salt wedge the turbulence is primarily generated at the interface, i.e., in the interior of the stratified fluid. In a partly mixed estuary both types of turbulence occur (Abraham, 1980). For such estuaries it is common practice to express the turbulent transport of momentum and mass under stratified tidal conditions in terms of the eddy viscosity, eddy diffusivity or mixing length pertaining under neutral (homogeneous) conditions multiplied by a damping factor, the magnitude of which depends on the magnitude of the gradient Richardson number. This may be referred to as a well-mixed approach as the turbulence is assumed to be generated at the bottom, damping of turbulence by a vertical density gradient being the only internal effect which is considered explicitly. Conceptually, the well-mixed approach should not be applied above a certain level of stratification, as internal turbulence dominates in the limiting case of an arrested salt wedge. Then the well-mixed approach is based on a source of turbulence which is of secondary importance. From a practical point of view this implies that the damping function does not depend uniquely on the gradient Richardson number.

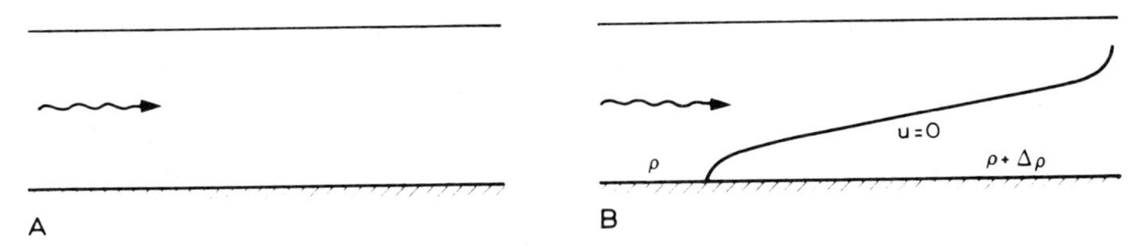

Figure 2. Homogeneous flow (A) with turbulence generated at the bottom. Arrested saltwedge (B) with turbulence generated at the interface as lower layer is at rest and the bottom shear is equal to zero. Both types of turbulence occur in partially mixed estuary.

5.1.3 Turbulence theory and turbulence modeling. Turbulence modeling, i.e., research aiming at expressing turbulence in as generally valid equations as possible and reducing the empiricism, has been initiated during the last decade. The principle of turbulence modeling is to work at a higher level of precision than with given coefficients of eddy viscosity and eddy diffusivity by introducing equations which govern the temporal and spatial variation of the covariances. In principle this leads to equations containing triple correlations as unknowns. By working at a higher level of precision it is hoped that eventually generally valid empirical relationships are found. Considerable progress in this direction has been made, as can be seen from state-of-the-art surveys given by Reynolds (1976), Saffman (1977) and Rodi (1980).

A certain group of turbulence models is referred to as one-equation models. In this type of model the coefficients of eddy viscosity and eddy diffusivity are expressed as functions of the locally available turbulent kinetic energy (k) and a length scale (L). The system of equations is closed by introducing an equation governing the temporal and spatial variation of k and by specifying the variation of L. This becomes difficult in flows with more than one single overall length scale. For this reason both the gradient transport concept and the one-equation models should not be applied to flows whose scaling laws are not known beforehand. This explains the present trend away from one-equation models for complex flows (Rodi, 1980). One equation models of partly mixed estuaries have been presented by Smith and Dyer (1979) and Liu and Leendertse (1978), without indicating how to take into account that estuarine turbulence can be generated both externally and internally.

Two-equation models are another group of turbulence models. In these turbulence models the coefficients of eddy viscosity and eddy diffusivity are expressed as a function of the locally available turbulent energy (k) and the local dissipation rate of turbulent energy (ϵ). The system of equations is closed by introducing equations governing the temporal and spatial variation of k and ϵ. In his state-of-the-art review Rodi (1980) indicates that the two-equation k-ϵ model has a fair degree of universality and can be applied with some confidence to new problems. This statement is supported by a summary of a fairly large variety of basic hydraulic flow situations which could be predicted by the k-ϵ model.

The main conclusion of Rodi's (1980) state of the art review reads that the available turbulence models (two-equation models and other types of advanced turbulence models) are suitable for hydraulic problems. Since the application of advanced turbulence models to hydraulic problems has only started recently and was restricted to fairly idealized problems, Rodi indicates that further extensive testing is needed to find out how well the models work for more realistic situations and where the limitations are. Research aimed at this objective will be reported at this symposium by Rodi (see Chapter 3).

5.2 *Physical Problems of Dispersive Transport Type*

5.2.1 Reduction of number of spatial dimensions by spatial averaging. When the engineer's interest is limited to the depth-mean values of the relevant parameters the physical problem may be described by depth-averaged equations, i.e., equations obtained by integrating the governing equations over the depth, provided that this approach gives sufficiently accurate information. Similar arguments apply when the engineer's interest is limited to cross-sectional mean values of the relevant parameters. Then, if sufficiently accurate, the physical problem may be described by equations integrated over the cross-section.

The integrated equations contain the depth- or cross-sectional mean values of the horizontal velocity and the concentration as unknowns. In the integrating process, the turbulent transport of momentum leads to the wall shear as an unknown. The non-linear terms lead to the dispersive transport of momentum and of a constituent as unknowns (see, e.g., Fischer, 1967 and Flokstra, 1977). Thus, engineering problems which ask for depth mean or cross-sectional mean values of the relevant parameters lead to a physical problem in the solution of which the closure problem involves expressing the magnitude of the dispersive transport and the wall shear in the depth mean or cross-sectional mean values of velocity and concentration. Physical problems requiring the solution of these two closure problems are referred to as physical problems of the dispersive transport type.

5.2.2 Dispersive transport of constituent. After integration over the cross section of the water course, the continuity equation for a constituent contains the dispersive transport of the constituent as an unknown. This dispersive transport can be expressed in local one-dimensional parameters (water depth, values of velocity and concentration averaged over the cross section, and derivatives of these quantities in the longitudinal direction), provided that the conditions are such that the classical assumptions made by Taylor (1953) are satisfied (see, e.g., Fischer, 1967).

The continuity equation which governs the variation of the concentration over the profile may be expressed as

$$\underset{(a)}{\frac{\partial \bar{\bar{c}}}{\partial \tau}} + \underset{(b)}{\frac{\partial c'}{\partial \tau}} + \underset{(c)}{u' \frac{\partial \bar{\bar{c}}}{\partial \xi}} + \underset{(d)}{u' \frac{\partial c'}{\partial \xi}} + \underset{(e)}{v \frac{\partial c'}{\partial y}}$$

$$+ \underset{(f)}{w \frac{\partial c'}{\partial z}} + \underset{(g)}{\frac{\partial T_\xi}{\partial \xi}} + \underset{(h)}{\frac{\partial T_y}{\partial y}} + \underset{(i)}{\frac{\partial T_z}{\partial z}} = 0 \tag{1}$$

where ξ is the longitudinal coordinate in coordinate system moving at velocity $\bar{\bar{u}}$, τ is the time within this coordinate system, y, z are the transverse and vertical coordinates, u, v, w are the velocities in ξ, y and z-directions, c is concentration, T_ξ, T_y, T_z are the turbulent transport of mass in direction of index, $=$ is the superscript referring to profile mean value, and $'$ is the superscript referring to local deviation from profile mean value.

When the assumptions made by Taylor are satisfied it is justified to neglect terms (d) and (g) containing $\partial c'/\partial \xi$. Then, treating $\partial \bar{\bar{c}}/\partial \xi$ as a parameter the variation of c' over the profile can be determined without integration in the longitudinal direction. This implies that the dispersive transport can be expressed in local one-dimensional parameters. When the Taylor assumptions are not satisfied, determining the variation of c' over the profile requires an integration in the longitudinal direction. Then the dispersion coefficient depends on the geometric features of the water course, and consequently becomes a site-specific parameter.

The concentration has to be distributed sufficiently uniformly over the cross section for the Taylor conditions to be satisfied. This is certainly not the case for the salinity distribution in partly mixed estuaries. Therefore, the dispersion coefficient is a site-specific parameter in one-dimensional salinity intrusion modeling. This explains the finding of the Delft Hydraulics Laboratory (1980) that the dispersion coefficient formulation proposed by Harlemann and Thatcher (1974) is not valid for the Rotterdam Waterway Estuary. Odd (see Chapter 2) elaborates further on this limitation of predictive ability.

5.2.3 Dispersive transport of momentum (effective shear stress). After integration over the depth, the equation of motion contains the dispersive transport of momentum and the depth-integrated turbulent transport of momentum as unknowns. Flokstra (1977) refers to the sum of these unknowns as the effective shear stress.

On theoretical grounds Flokstra (1977) argues that the solution of the depth-integrated equations of motion can produce a steady circulating flow driven by the main flow only if (1) the advective acceleration terms are included, (2) vorticity is introduced into the flow by a no-slip boundary condition and (3) the effective shear stress term is included. Lean and Weare (1979) confirm these conclusions on the basis of a series of numerical experiments, emphasizing that the numerical accuracy of the computational scheme used must be sufficiently high so that the effects of spurious numerical dispersion are small compared with the physical shear stresses to be modeled.

The above arguments make it necessary to express the effective shear stress in terms of depth-mean quantities in order to produce recirculating flows correctly in depth-averaged numerical computations. This is a difficult closure problem since it can be shown that generally the dispersive transport of momentum transfers energy out of the circulating flow, leaving only the turbulent stresses as a mechanism transferring energy to the recirculating flow. So both the turbulent shear (turbulent transport of momentum) and the dispersive transport of momentum have to be taken into account. This closure problem has not been resolved as yet (Flokstra, 1977, Lean and Weare, 1979). Consequently circulating flows are difficult to reproduce correctly in two-dimensional computations. It must be mentioned, however, that studying the near field of side discharges into open channel flow, McGuirk and Rodi (1978) found that the contribution of the dispersive transport of momentum to the effective shear stress could be neglected.

5.3 Physical Problems of Wall Shear Type

As long as the streamlines are more or less straight the effect of the dispersive transport of momentum may be neglected. When the concentration is approximately uniformly distributed over the depth and the phase effect prevails (Fischer, 1976), the influence of the dispersive transport of a constituent may also be neglected. This applies to wide, irregularly shaped well-mixed estuaries with ebb channels which are different from flood channels with tidal flats, etc. (Fig. 3). In such estuaries the mixing is governed by large scale advective transport processes, which depend on the large scale current pattern. At this symposium these arguments are substantiated by Dronkers, *et al.*, (see Chapter 12).

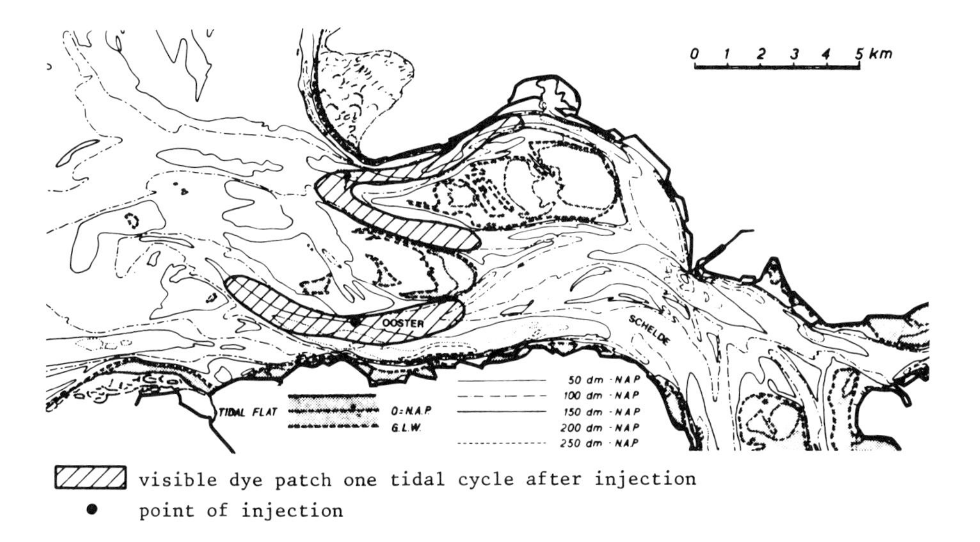

Figure 3. Concentration distribution primarily controlled by large scale advection induced by combined influence of tidal flow and bathometry; dye patch separated into two parts. DHL Project M 1603.

In physical problems of the dispersive transport type expressing the wall shear in the mean velocities becomes the only closure problem when the dispersive transport turns out to be insignificant. Physical problems to which this applies are referred to as physical problems of the wall shear type.

5.4 Boundary Conditions

The value of the boundary conditions cannot be obtained from the equations used to describe the physical phenomena, but must be inserted on the basis of other information.

5.4.1 Solid boundaries. In the direct vicinity of the bottom the velocity varies much more rapidly with depth than in the remaining part of the vertical. Consequently in physical problems of the turbulent transport type the flow near the bottom requires a much finer computational grid than the remaining part of the flow. A common method to avoid this fine grid is to relate the variation of the velocity over the lowest vertical grid with the velocity at one grid step above the bottom by a "law of the wall".

In physical problems of the dispersive transport type and in those of the wall shear type, the wall shear has to be related to the mean velocity and to the bottom roughness by the empirical wall shear coefficient. This closure problem can be solved on the basis of the experimental information available on the magnitude of the wall shear coefficient, obtained from uniform flow experiments in rectangular channels. It is not known precisely how to apply this experimental information to non-uniform tidal flows in irregularly shaped channels (Dronkers, 1969). In addition, when dealing with large tidal areas it may be that prototype bottom roughness is not known precisely either. This means that adjusting the wall shear coefficient and the schematization of the geometry (Sec. 7.1) are interconnected items, which have to be taken care of simultaneously on an ad hoc basis in feedback loop 3. The same applies to the effect of non-uniform density on the wall shear coefficient. At this symposium the schematization of the geometry in conjunction with the adjustment of the wall shear coefficient is included in the contribution of Leendertse, *et al.*, (see Chapter 11).

5.4.2 Open boundaries. The conditions at the boundaries of the area reproduced in a mathematical model are to be prescribed. These boundary conditions may be derived from field measurements only when they are not affected by the technical measures, whose effect the model will study, i.e., when the waves generated by these technical measures can leave the problem area without coming back (e.g., due to reflections) as an incoming wave. Otherwise predicting the effect of the technical measures on the boundary conditions becomes part of the study.

Deriving the boundary conditions for mathematical models of large areas with open boundaries from field measurements is a problem in itself. A mathematical model of the net residual flow through the Dover Strait, for example, requires mean sea levels and amplitudes of the tide at its southern boundary (e.g., a line across the Channel) and at its northern boundary (e.g., a line across the North Sea). Using a mathematical model as a tool for interpretation of field data Prandle (1978) concluded that about one quarter of the net residual flow is due to a sea level difference of 6 cm between the boundaries, which are 300 km apart in the mathematical model. This illustrates the difficulties of determining the net residual flows from mathematical models of large areas with open boundaries (see Chapter 10).

5.4.3 Mathematical considerations for open boundaries. From the mathematical point of view the initial conditions, and the number and the kind of the boundary conditions must be such as to make the mathematical problem well-posed. In the sense of Hadamard (1923) the main aspect of well-posedness is that in the interior of the problem area the solution only changes a little with small changes of the boundary conditions. When an incorrect number of boundary conditions or boundary conditions of the incorrect type are prescribed the problem becomes ill-posed. Then for given combinations of parameter values the solution of the system of continuous equations may show that small changes of the boundary conditions lead to large changes in the interior of the problem area, often only after the simulation period exceeds a given length of time, while this would not be found if the problem were well-posed. A vast amount of literature on this subject deals with problems such as linear and non-linear wave propagations, the diffusion equation with a source term proportional to the concentration squared and the inverse diffusion problem (Payne, 1975). No such studies are known in connection with problems of flows and transports.

The number of boundary conditions to be given to make a problem well-posed depends on the type of equations and the actual flow conditions, as can be seen from Table 1, for the system of shallow water equations, i.e., the system of the continuity equation and two equations of motion (in horizontal direction) which gives the water depth and the two depth-averaged horizontal velocity components as functions of time and the horizontal space coordinates. This system of equations is hyperbolic without including the effective shear stress (see Section 5.2.3) in the depth-averaged equations of motion. The system is incompletely parabolic when the effective shear stress is included as a gradient type of transport of momentum, as then the two equations of motion become of a parabolic type, while the continuity equation remains of a hyperbolic type. The actual flow conditions are important insofar as distinction has to be made between boundaries with sub-critical flow and with super-critical flow and between boundaries with outflow and with inflow.

Table 1.
Number of Boundary Conditions for Two Types of Shallow Water Equations

Hyperbolic Equations			*Incomplete Parabolic Equations*		
Solid Boundary	*Open Boundary with Inflow*	*Outflow*	*Solid Boundary*	*Open Boundary with Inflow*	*Outflow*
Sub-critical flow					
1	2	1	2	3	2
Super-critical flow					
1	3	0	2	3	2

In addition to being correct in number, the boundary conditions have to be of the correct type in order to make a problem well-posed. In recent years the kind of boundary conditions required to make a problem well-posed has been studied intensively, especially by Kreiss and his co-workers (see Oliger and Sundstrom, 1978).

For the hyperbolic system the sub-critical inflow and outflow boundary condition may be given by a combination of Riemann invariants (characteristic variables), which represent a combination of incoming and outgoing waves. For the inflow boundaries a normal procedure in numerical practice is, however, to prescribe the water level and the velocity component parallel to the boundary. Only recently it has been shown that this procedure leads to a well-posed problem (Verboom *et al.,* 1981), though the numerically formulated problem is ill-posed if the leap-frog scheme is used. The latter observation has been made by Kreiss on the basis of normal mode analysis (Oliger and Sundström, 1978).

As yet, the conditions for the well-posedness available for the incomplete parabolic system are difficult to be understood physically or to be applied in numerical practice (Gustafsson and Sundström, 1978).

6. VERIFICATION OF SYSTEM OF EQUATIONS AND BOUNDARY CONDITIONS (FEEDBACK LOOP 1)

6.1 Physical Considerations

Screening the physical concepts behind the mathematical formulation for a given physical phenomenon is a prerequisite in mathematical modeling. The screening provides a first answer to the question whether a given mathematical formulation is generally valid, problem-specific, or site-specific. It also gives a first answer to whether the research behind a given mathematical formulation is completed or whether there are still open questions. By answering these questions the screening gives indications of the predictive ability which may be expected in a given modeling exercise. The predictive ability is the greater, the more generally valid are the mathematical formulations and the smaller is the number of open research questions. Consequently, *generally speaking, because of the difference in difficulty of the closure problems involved, for physical problems of the wall shear type the predictive ability tends to be greater than for physical problems of the turbulent transport type and of the dispersive transport type.*

The following example serves the purpose of illustrating that in the screening of the physical concepts the arguments to be considered vary with the physical phenomena to be studied and with the method selected to describe these phenomena in mathematical terms.

In one-dimensional long wave theory the vertically integrated conservation equations for mass, momentum and energy are to be expressed as

$$\frac{\partial F(x,t)}{\partial t} + \frac{\partial G(x,t)}{\partial x} = K(x,t) \tag{2}$$

where t is time, x is the horizontal coordinate, and F, G, K are functions of x and t. At a long-wave discontinuity Abbott (1979, Chapter 5) applies the formula

$$-c\,\Delta F + \Delta G = 0 \tag{3}$$

where c is the velocity of propagation of discontinuity, ΔF is the change in value of F over discontinuity, and ΔG is the change in value of G over discontinuity. Introducing a coordinate system (ξ, τ) moving with the velocity of propagation of the discontinuity Eq. (2) reads

$$\underset{(a)}{\frac{\partial F}{\partial \tau}} - \underset{(b)}{c\frac{\partial F}{\partial \xi}} + \underset{(c)}{\frac{\partial G}{\partial \xi}} = \underset{(d)}{K} \tag{4}$$

where $\xi = x - \int_0^t c(t)\,dt$ is the horizontal coordinate in the coordinate system moving with velocity c, and $\tau = t$ is time in the moving coordinate system. Term (a) represents the variation of F with time observed when moving with the discontinuity. Term (b) gives the variation of F with time observed when the (frozen) discontinuity passes a stationary observer. At the discontinuity term (a) is much smaller than term (b). Hence term (a) may be neglected in Eq. (4). Thus, if Eq. (2) is valid at the discontinuity integrating Eq. (4) with respect to ξ from a section downstream of the discontinuity (indicated by index $-$) to a section upstream of the discontinuity (indicated by index $+$) gives

$$-c\Delta F + \Delta G = \int_{\xi_-}^{\xi_+} K \, d\xi \tag{5}$$

Therefore, Eq. (3) is based implicitly on the assumption that in Eq. (5) the integral is either zero or sufficiently small to be neglected. For long surface waves this assumption is satisfied by the conservation equations for mass and momentum, which contain enough information to determine the velocity of propagation of the long surface wave (c_{surf}) as a function of wave height and water depth. Determining the velocity of propagation of the long internal wave in stratified flow (c_{int}) as a function of wave height, total depth and lower layer thickness it is necessary to introduce an empirical energy loss coefficient in the solution for c_{int} (Kranenburg, (1978). This is due to the fact that the assumption of the intergal of Eq. (5) being zero or sufficiently small to be neglected only applies to the conservation equations for mass for the upper layer and for the lower layer separately and to that for momentum for both layers combined (Abraham and Vreugdenhil, 1971).

The above arguments illustrate the wide range of issues to be considered in the screening of the physical concepts.

The tidal computations made in the study of the feasibility to generate tidal power in the Bay of Fundy illustrate the issues to be considered when screening the boundary conditions. In this particular instance, answering the question whether the boundary conditions at the open boundary were affected by the technical measures studied was a topic of main interest because the resonance characteristics of the Bay were found to be modified by the proposed measures, and relatively small modifications could have a relatively large effect. Within this context the open boundary problem was studied by Garrett and Greenberg (1977), who show how to correct for, or at least estimate, the effect of the technical measures on the conditions at the open boundary, using output from the tidal computation and estimates of the impedance of the exterior ocean. The latter impedance cannot be determined precisely as it requires a numerical model of the whole exterior ocean region. Thus, an estimate of the impedance is needed for an estimate of changes in the tidal regime at the open boundary.

6.2 Mathematical Considerations

In most instances the system of equations has to be solved by numerical methods. This makes it important to have insight into the general behavior of the system of equations, as well as to have the boundary conditions formulated properly.

Insight into the general behavior of the system of equations can be obtained from analytical solutions for simplified conditions and from studying the dispersion relation, i.e., the relation between the wave length and the wave frequency of periodic small disturbances superimposed on a background flow. The dispersion relation contains sufficient information to determine under which circumstances the disturbance is stable and under which circumstances the disturbance is unstable. From the numerical point of view it is important to know under which conditions unstable solutions are to be expected.

Two-dimensional tidal problems can be solved analytically for simple problem areas of constant depth. Analytical solutions for this simplified geometry may reveal essential features for a more complex geometry (see, e.g., van der Kuur and Verboom, 1975).

Determining the conditions for well-posedness and the implications of ill-posedness is a matter of ongoing fundamental mathematical research, which for systems of equations of the incomplete parabolic type has not led to proper guidelines for the numerical practice as yet. Generally speaking these fundamental studies are too broad to be included in mathematical modeling studies for the solution to an engineering problem. Nevertheless, in practical studies it must be verified in some way and to some extent that problems due to ill-posedness do not occur and that the solution represents the essential features of the problem to be solved.

7. NUMERICAL SOLUTION TECHNIQUE (LEVEL 3 OF SCHEMATIZATION AND APPROXIMATION)

This chapter presents some general arguments on the numerical solution process. It does not discuss numerical solution techniques as such. The arguments presented apply both to the finite difference method and to the finite element method, though the literature cited primarily deals with the finite difference method.

7.1 Discretizing of Mathematical Formulae, Physical Considerations

In the numerical solution of a system of equations the dependent variables are determined at grid points only. In the numerical schemes functions which are continuous in space and time are approximated by grid functions, that is, by functions which are defined at the grid points only. These functions must represent the continuous functions as accurately as required for the solution of the engineering problem behind the mathematical model study. Assuming the required accuracy to be prescribed, the grid intervals in time (Δt) and space ($\vec{\Delta x}$) must be selected such as to allow the smallest time and length scales (τ and $\vec{L}$) in the relevant phenomena to be represented in the numerical solution with the prescribed accuracy. An interconnected requirement is that the geometric features of the problem area can be represented sufficiently accurately at grid scale.

The minimum time and length scales in the physical problem have to be known and the accuracy up to which they have to be reproduced in the numerical solution has to be prescribed. Otherwise there is no yardstick available to judge the predictive ability of a mathematical model.

For hyperbolic problems with real waves the smallest time and length scales to be reproduced (τ and $\vec{L}$) can be related to the period and the length of the shortest wave reproduction of which is deemed relevant. For parabolic problems without real waves τ and $\vec{L}$ have to be determined otherwise. Then $\vec{L}$ may be related to a characteristic length scale of the spatial gradients of, for example, concentration, while τ may be related to a characteristic time scale of the diffusion and/or advection process. However, these length and time scales vary with time, and so do τ and $\vec{L}$ determined by this procedure (Daily and Harleman, 1972, Reynolds, *et al.*, 1973).

Depending on the numerical method used the grid intervals Δt and $\vec{\Delta x}$ must be smaller or much smaller than τ and $\vec{L}$ to warrant reproduction of the relevant small scale phenomena in the numerical solution. If, for instance, for economic reasons $\vec{\Delta x}$ cannot be chosen much smaller than $\vec{L}$, the relevant small scale phenomena become subgrid scale, and can only be reproduced as such.

The grid interval $(\vec{\Delta x})$ is related not only to the smallest relevant length scale $(\vec{L})$ in the considered physical phenomena, but is also related to the smallest relevant length scale in the geometry of the problem area because of the potential importance of subgrid scale geometric features.

Subgrid scale geometric features induce subgrid scale deterministic processes which influence the spatial variation of velocity and concentration at subgrid scale and generate subgrid scale turbulence. These subgrid scale processes cannot be included in the computational grid. This makes it necessary to determine whether their combined net effect is important enough to have it represented in the computational results. For instance, in two-dimensional depth-averaged tidal and water quality computations, channels with subgrid width may be among the subgrid scale geometric features. If so, in the modeling process it is necessary to determine whether the net effect of the subgrid scale advective transport processes induced by these channels must be reproduced. This may lead to the question of how to represent (schematize) the actual geometry at grid scale without losing the net effect of the subgrid scale channels and possibly other subgrid scale geometric features. Unfortunately there is no straightforward recipe to answer questions of this type other than by reducing the grid intervals. In general the influence of the subgrid scale geometric features is unknown. Their influence is problem, site and numerical method specific.

In cooling-water problems the near-field processes may slip through the computational grid used for modeling far-field processes. If so, this leads to the question of how to introduce the net result of the near-field processes into the far-field modeling.

Zimmerman (1978) shows that subgrid scale bottom topography may cause tide-induced residual recirculations, presenting quantitative estimates of their magnitude.

Though outside the scope of this general lecture, the grid intervals of the field data to be used for calibration and verification of the mathematical model are also to be related to τ and $\vec{L}$ (Reynolds, *et al.*, 1973).

7.2 Discretizing of Mathematical Formulae, Mathematical Considerations

Numerical schemes must be stable, consistent and convergent. A scheme is consistent if for steps in time and space approaching zero the differential equation coincides with the difference equation. A scheme is convergent if the solutions also coincide. For linear differential equations, having constant coefficients, it has been proven that a scheme being both stable and consistent implies it to be convergent (Lax's equivalence theorem, Richtmeyer and Morton, 1967). For more complicated equations this has not been proven. Therefore the subsequent text concentrates on the conditions for stability and methods to verify the accuracy of the numerical solution.

7.2.1 Stability analysis. Stability of the numerical method used is a prerequisite for accurate results. Stability can be defined as: (a) the numerical solution remains bounded in a finite period of time with the grid intervals (Δt and $\vec{\Delta x}$) going to be zero; or, (b) the numerical solution remains bounded in an infinite period of time with fixed values of Δt and $\vec{\Delta x}$. The latter definition is the most important for the numerical practice, e.g., for water quality studies with long simulation periods.

Stability analysis involves all grid functions as well as the discretized initial and boundary conditions. Stability analysis is related to the problem of determining the conditions for well-posedness of the boundary and initial conditions to be given with a system of continuous equations. Both problems can be studied using techniques of the same type, i.e., by proving that some positive definite norm of the grid functions is bounded and also is a continuous function of the discretized initial and boundary conditions. For many problems the stability has been investigated in this way. However, generally speaking, the mathematical work involved is too substantial to be included in the numerical practice.

In order of decreasing stringency and increasing frequency of application, the following methods of stability analysis are available in connection with the numerical solution of the shallow water equations.

The energy method (Richtmeyer and Morton, 1967) is the most stringent method available. It studies the conditions for stability using the aforementioned procedure based on the bounded positive norm. In the numerical practice the stability is only proven occasionally in this way; for the shallow water equation see Elvius and Sundström (1973) and Marchuk, *et al.*, (1973).

In the *normal mode method* of stability analysis the equations are linearized. Then stability can be investigated by analyzing the eigen frequencies or normal modes. When the considered problem is treated as an initial value problem, i.e., when boundary conditions are omitted, the normal modes in space are the Fourier components and those in time give the amplification matrix, just as in the Von Neumann analysis. The conditions found are necessary, not sufficient for stability. The same technique can be applied to the initial boundary value problem. In this case only necessary conditions are derived (Kreiss, 1962 and Wirz, *et al.*, 1977).

The *Fourier method* developed by Von Neumann (Richtmeyer and Morton, 1967) is only applicable within the inner part of the area of computation (i.e., far away from its boundaries) for differential equations having constant coefficients and coincides with the Von Neumann analysis referred to in the previous paragraph.

The *modified equation approach* (Hirt, 1968, Warming and Hyett, 1974) which gives indications about the stability, though it does not give definite (i.e., mathematically proven) answers. In this method the truncation error is used to derive stability conditions, one of which being that the coefficient of the lowest even order space derivative should be a positive one.

Because of the effort involved in the mathematical modeling of flows and transport for the solution of engineering problems it is only common practice to apply the Fourier method and the modified equation approach of stability analysis.

However, the necessary conditions found by these methods leave room for unstable behavior, linear or non-linear. If the growth rate of the instability is small the results may be rather accurate for small period simulations, whereas the results may become unbounded in a longer period simulation (Fig. 4). This is especially important when applying a mathematical model, calibrated and verified in short term simulations (Stelling, 1980). Not knowing the stability properties of the numerical method used it remains a delicate problem to guarantee that a longer period simulation leads to accurate and meaningful results.

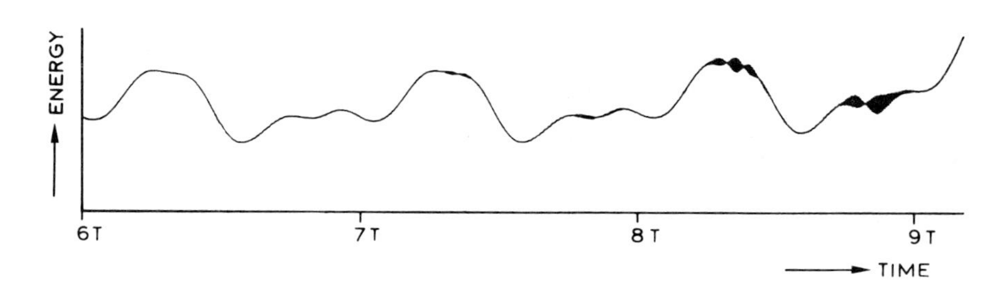

Figure 4. Weak instability apparent after a number of tidal periods (T). DHL Project S 333 (Stelling, 1980).

The more stringent the information available on the stability characteristics of the numerical scheme used, the better it can be excluded that the predictive ability of a mathematical model is hampered by the numerical solution technique.

7.2.2 Analysis of computational accuracy. It is often possible to show that the difference between the continuous solution and the continuation of the discretized solution in some suitable norm is bounded and proportional to the grid interval to some power. This proves convergence, i.e., accuracy for $\Delta t \rightarrow 0$ and $\vec{\Delta x} \rightarrow 0$. This only gives global information on the accuracy of the solution for finite grid intervals.

When checking the conservation of certain integral properties such as mass, momentum, energy, vorticity, and so on, the numerical solutions are integrated over the spatial coordinates. Thus, these checks leave the question of the local accuracy unanswered. They provide a necessary but not a sufficient condition for local accuracy.

More local and less complete checks on the accuracy include: (a) simulation of simple cases for which analytical solutions are known; (b) comparing the normal modes of the continuous and the discretized problem; (c) refining the grid by reducing Δt and $\vec{\Delta x}$; (d) modified equation approach; and (e) comparison with other schemes.

Simple cases for which analytical solutions are known are often not decisive for the real problem to be solved. This applies, for instance, to Orzag's (1971) analytical test for numerical dispersion induced by the advective terms. Orzag's test case is that of a cloud of scalar quantity being displaced without diffusion in a divergence-free flow field of constant depth. Diffusion being absent, in this flow field the cloud should keep its original shape. Considerations presented by Piacsek and Williams (1970) imply that this test case is not as stringent as a test case including variations of depth with location and time (non-divergence free flows).

For the hyperbolic shallow water equations the wave propagation factor (Leendertse, 1967) gives the ratio of the normal modes for the continuous and discretized pure initial value problem. Values of this ratio in the order one are a necessary condition for accuracy. If this analysis were performed for the initial boundary value problem a better indication for the accuracy would have been obtained, though still incomplete, as the equations are linearized in the analysis.

For non-linear systems of equations decreasing the grid intervals Δt and $\vec{\Delta x}$ is the only way to see how the numerical solution procedure behaves, though in essence this method gives indications for convergence and not for accuracy. In practice, however, the choice of Δt and $\vec{\Delta x}$ is usually a compromise between accuracy and economy. Dealing with the shallow water equations and keeping the ratio of $\Delta t/\Delta x_i$ constant, the costs of a computation are proportional to the grid size to the third power. Consequently checking the behavior of the solution procedure by a reduction of the grid interval by a factor two may eventually become a costly proposition. Nevertheless, there are instances with strong reasons to do so, such as when strong non-linear effects in the physical phenomenon occur.

The modified equation approach can be used to compare the magnitude of the numerical viscosity and diffusivity with the magnitude of their physical counterparts.

Comparing the results obtained making use of a given scheme with those obtained making use of another scheme gives information on the accuracy of the former scheme only insofar as the accuracy of the latter scheme is known.

7.2.3 Non-uniform grids. In practical problems the information needed for some parts of the problem area may be more detailed than the information needed for other parts. This may lead to non-uniform grids or to nested models. The accuracy of the numerical method depends on the grid interval, and so do the wave (error) propagation properties. Special artificial, dissipative interfaces (Abbott, 1979, chapter 4) may be needed if the grid changes too abruptly, whereas their influence on the overall accuracy is insufficiently known. It is likely that issues of this type are to be considered both in connection with finite difference methods and with finite element methods. When using nested models the accuracy of the models representing a small area may be limited because of inaccuracies in the boundary conditions to be derived from the models representing a large area (see Chapter 10).

8. VERIFICATION OF NUMERICAL SOLUTION TECHNIQUE (FEEDBACK LOOP 2)

Whether the numerical scheme adopted produces solutions with sufficiently close agreement to the actual solution of the considered system of equations is the question to be answered in feedback loop 2. Answering this question leads to questions of the following types: (a) Is the choice of the grid intervals Δt and $\vec{\Delta x}$ based on information related to stability, accuracy, and minimal time and length scales of the physical problem, or is the choice also based on economic arguments and, if so, to what extent? (b) What is known of the stability and the accuracy of the numerical solution technique as a function of Δt, $\vec{\Delta x}$ and the duration of the simulation period? (c) What is the effect on the solution of the choice and the treatment of the boundary conditions, i.e., the number and kind, and the situation of open boundaries? (d) What are the previous applications?

Generally speaking, in mathematical modeling for the solution of engineering problems, the questions on stability, accuracy and well-posedness, for example, are not answered in as much detail as possible given the state-of-the-art of numerical modeling techniques. It is common practice to derive insight into the capabilities of a numerical scheme primarily or to a large extent from experience. However, by itself experience cannot provide stringent answers to the questions referred to at the beginning of this section. For engineering-oriented mathematical modeling, it seems appropriate to rely on experience when it is done with precaution and care, and when it includes a check on the convergence of the solution by decreasing Δt and $\vec{\Delta x}$. However, to apply a mathematical model as a tool for fundamental research (see Section 9.2) there must be a thorough understanding of the capabilities of the numerical scheme used.

Because of the costs and effort involved, engineering oriented mathematical modeling of complex flows and transports often does not include studies providing stringent answers in connection with the stability and accuracy of the numerical scheme used and the well-posedness of the mathematical problem solved. Instead, it is common practice to rely on experience, which does not give stringent answers in connection with stability, accuracy and well-posedness.

9. VERIFICATION OF OUTPUT OF COMPUTATIONS (FEEDBACK LOOP 3)

9.1 Feedback Loops 1 and 2 Combined

The quality of the output of the computations depends on the quality of the mathematical expressions for the physical phenomena, and the mathematical solution techniques. There is limited possibility to determine the quality of these items separately. Often the quality of both these items combined has to be checked by comparing the output of the computations with experimental data.

This involves feedback loops 1 and 2 simultaneously. Hence, if the experimental data is assumed to be correct, disagreement between the output of computations and the experimental data may be due to deficiencies in the mathematical expressions for the physical phenomena and/or in the solution technique.

The early literature on recirculating flows shows how both these items may be interwoven. As indicated in Section 5.2.3, reproduction of these flows requires the effective shear stress terms to be included in the governing depth-averaged equations. Yet, reasonably correct circulating flows were calculated without the effective shear stress terms, this disagreement between theory and calculations being due to the numerical smoothing procedure adapted in the computations to suppress instability (see Fig. 5).

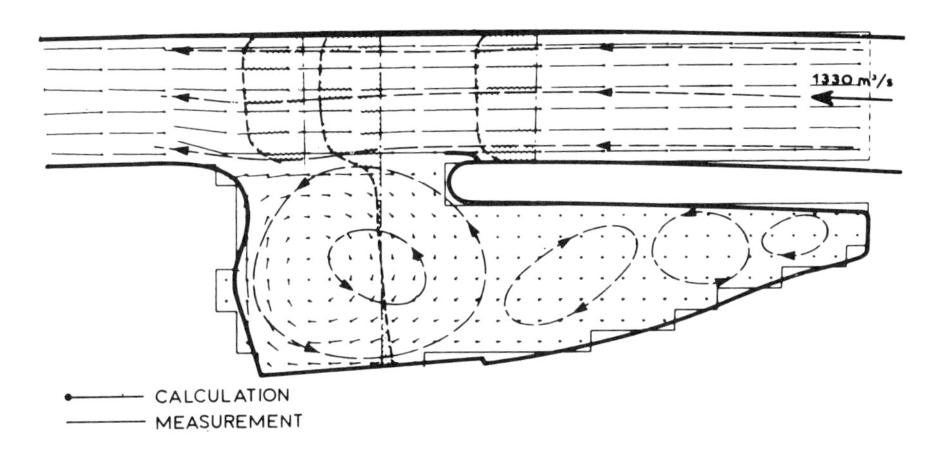

Figure 5. Example of numerical smoothing procedure selected so as to produce effective shear stresses of correct order of magnitude. DHL Project S 163.

9.2 *Mathematical Model as a Tool for Research*

A mathematical model is a tool which gives the integrated effect of separate physical phenomena. Consequently, learning how to describe separate physical phenomena in mathematical terms (e.g., the turbulent transports of momentum and mass in given flow conditions) may be the objective of a research oriented mathematical model study. Then, one looks for an objective method to describe the separate physical phenomena in mathematical terms, which makes the output of computations agree with experimental data on the integrated effect of these phenomena. Then, in essence, feedback loop 3 is the final feedback loop in a research effort to understand the separate physical phenomena involved, and the mathematical model is applied as a tool for fundamental research. Perrels and

Karelse present elements of this type of fundamental research at this symposium. (See Chapter 13.)

In turbulence modeling given methods to describe the turbulence properties have been found to work for specific flow conditions. Another type of fundamental research is the research made to determine the range of applicability of these methods to flows of a different nature. Fundamental research of this type is summarized by Reynolds (1976), Saffman (1977) and Rodi (1980).

The output of a mathematical model varies with the mathematical modeling procedures used in the computations. Consequently feedback loop 3 also may be the final feedback loop in research efforts to develop specific mathematical modeling procedures. Examples of subjects which can be treated in this context are the following.

Schematizing the geometry in two-dimensional tidal models has become largely a matter of experience. Developing formal methods to do so may be a subject for research in itself. At this symposium the subject of schematizing the geometry is included in the contribution of Leendertse, *et al*., (Chapter 11).

There is a tendency to apply different types of mathematical models for different zones within the near-field of cooling-water discharges. Different chains of models are being tested for this purpose (Rodi, Chapter 3). Providing formal procedures to select the appropriate chain for a given application is another type of research of mathematical modeling procedures.

Mathematical modeling for the solution of an engineering problem is to be based on quantitative insight into both the performance of a given mathematical formulation to describe a given physical phenomenon and the performance of a given mathematical modeling procedure under given circumstances. If not available beforehand this insight is to be obtained by applying the mathematical model as a tool for research in feedback loop 3.

9.3 End Product of Feedback Loop 3 in Mathematical Modeling for Solution of Engineering Problem

In mathematical modeling for the solution of engineering problems the end product has to be a verified mathematical model plus a statement on the model's accuracy and range of applicability.

The statement on the model's accuracy requires questions of the following type to be answered: (a) how well does the verified mathematical model reproduce all available experimental information? and (b) can the reproduction be improved while remaining within the structure of the model, e.g., by adjusting the magnitude of numerical coefficients contained in the model?

These questions are often answered qualitatively. This is reflected by the fact that in the literature the agreement between a mathematical model and measurements often is described as being "good", "fairly good", "promising", etc. (see Fig. 6). Yet, qualitative information may be sufficient as input into the solution of the engineering problem.

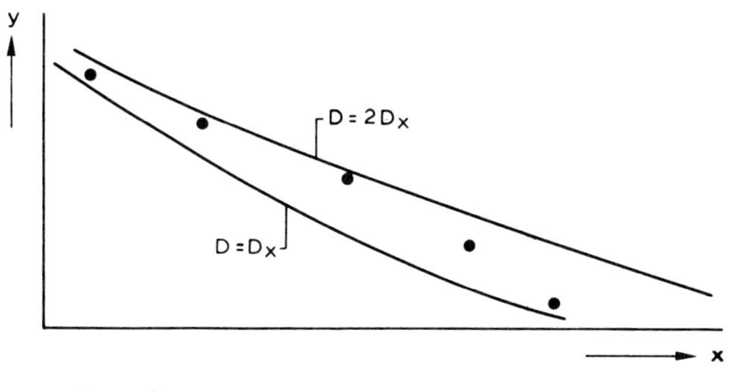

● Experiments

D_X: value of D according to X's theory

Figure 6. Computations give order of magnitude correctly; thus "fairly good" agreement. They give incorrect dependency of y on x (concave versus convex); still "fairly good" agreement?

As a rule a verified mathematical model may be used for interpolation within the range of experimental conditions covered in the calibration and verification of the model. If asked for by the engineering problem under consideration the statement on the model's range of applicability has to indicate whether the mathematical model may be used for extrapolations outside these conditions. Then it has to indicate to what extent the accuracy of the model as found in the verification may be expected to be retained when making the extrapolations.

Evaluating whether extrapolations may be made requires general questions of the following type to be answered. (a) Has the agreement between the verified mathematical model and available experimental data been obtained by generally valid or by site-specific or problem-specific mathematical formulations of the relevant physical phenomena? (b) In how many previous applications has the model been proved to work satisfactorily under which circumstances? (c) To what extent do the geometric conditions for which predictions are to be made deviate from those of the calibration and the verification?

In addition it is necessary to consider which factors are the critical ones in a given extrapolation. For instance, making two-dimensional tidal computations to study the effect of a reduction of the tidal amplitude induced by a dam it is relevant to ask whether the schematization of the geometry adapted for the tidal flats remains valid when water depth above the tidal flats is greatly reduced.

The answer to the question whether the accuracy of the model will be retained in the extrapolation is likely to be positive when the closure problem involved has been solved over a wide range of conditions, when it is not necessary to extrapolate the boundary conditions, and when the computational grid is fine in comparison with the relevant geometric features of the problem area. These conditions may be satisfied for physical problems of the wall shear type. When not satisfied the question how the extrapolation affects the accuracy can often only be answer subjectively on the basis of experience.

In several instances the question whether the accuracy of the verified model will be retained in the extrapolations is unanswerable except by experience. However, then the judgment on this matter often has to remain a subjective one. For instance, this applies when the closure problem is still the subject of research or has been solved in a site-specific or problem-specific fashion as may be true for complex problems of the turbulent transport type and of the dispersive transport type.

10. ENGINEERING SIGNIFICANCE OF GENERATED INFORMATION (FEEDBACK LOOP 4)

To evaluate the significance of the generated information for the solution of the engineering problem behind the mathematical model study, the accuracy and the range of applicability offered by the mathematical model have to be known and have to be compared with the accuracy and range of applicability asked for to solve the engineering problem. In this context distinction has to be made between engineering problems with accuracy requirements imposed from outside and those with accuracy requirements imposed from within.

10.1 Accuracy Requirements Imposed from Within

The quality of the available information is one of the factors which affect the safety margin to be adopted in the solution of the engineering problem. Thus the information generated for the solution of the engineering problem is relevant when the benefits in the form of a reduced safety margin derived from the information exceeds the costs of generating the information. This criterion contains implicitly an accuracy requirement as the reduction of the safety margin can be the greater, the more accurate the generated information is. For instance, when determining the location of a new cooling water intake with respect to the location of a new cooling water outlet the distance between both may be taken the shorter, the more accurate predictions of the temperature field around the outlet are available.

10.2 Accuracy Requirements Imposed from Outside

Environmental standards are allowable limits on certain parameters with the purpose of constraining the impact of effluent discharges to acceptable limits. These standards lead to accuracy requirements imposed from outside. These accuracy requirements are linked with the standard of non-compliance, i.e., the amount by which the allowable limit has to be exceeded during a period of time of given length to cause a violation of the standard. When stringent, the standard of non-compliance may be small in comparison with the accuracy of the best available method to predict the effect of the effluent discharge. If so, this may have the same result as reducing the allowable limit by an amount equal to the difference between the accuracy which can be provided and the standard of non-compliance.

The criterion of Section 10.1 remains valid for engineering problems with accuracy requirements from outside.

11. SUMMARY AND CONCLUSIONS

This general lecture is a reaction to the Scientific Committee's request for a survey of the main factors which may lead to limited predictability and for a survey of the main requirements for predictive ability. In addition it attempts to contribute to the end product of the symposium which should be a definite statement on what mathematical models can and cannot do. This final section of the lecture summarizes the finding within these perspectives.

11.1 Main Requirements for Predictive Ability

The main requirements for predictive ability are that, at the end of feedback loop 3, a verified mathematical model has to be available plus a statement on the model's accuracy and range of applicability. The statement on the range of applicability has to indicate whether the accuracy of the verified mathematical model may be expected to be retained when extrapolating outside the range of experimental conditions covered in the calibration and verification of the model.

Drafting the statement on the model's accuracy and range of applicability makes it necessary to, amongst other things: (a) screen the physical concepts behind the mathematical formulation for the relevant physical phenomena; (b) distinguish between the type of physical problem to be considered, as the type of the problem gives indications for the general state of the underlying research; (c) screen the boundary conditions; and (d) determine whether the stability and accuracy of the numerical scheme used are known in sufficient detail.

Considerations of this type have to be made explicit in the engineering decision that the mathematical model may be applied for given interpolations or extrapolations. This is in essence what is asked for in the statement on the accuracy and range of applicability of the verified mathematical model.

11.2 Main Factors Leading to Limited Predictive Ability

The main factors leading to limited predictive ability given in the preceding text were found in an analysis of the process of solving engineering problems through mathematical modeling as represented in Fig. 1. In this analysis the three levels of approximation and schematization and the four feedback loops distinguished in Fig. 1 were discussed separately.

Summarized in a nutshell the findings of the analysis indicate that predictive ability is limited whenever research underlying the mathematical model is not completed. In addition limited predictive ability may be due to uncertainty with respect to the (open) boundary conditions, or may be caused by an excessively coarse grid size for the geometric representation of the problem area.

The state of the research often is such that the mathematical formulation of the physical phenomena, which are relevant in the considered problem, has to be either site-specific or problem-specific. This is a factor leading to limited predictive ability sufficiently broad to be mentioned separately.

The preceding text lists several factors for potentially limited predictive ability. Therefore, as a general rule, when making predictions by means of a mathematical model it has to be investigated whether or not one or more of these factors are valid in this specific instance. Finding one or more of these factors to be valid does not by necessity imply that no predictions may be made. One possibility is that the effect of these factors can be shown to be of minor importance. Another possibility is that the mathematical model remains the most appropriate engineering tool in comparison with the other available tools, notwithstanding possible restrictions. This is due to the fact that the other engineering tools have their own capabilities and limitations.

11.3 Contribution to End Product of Symposium

Within its restricted scope and function, as defined in the beginning of this paper, this general lecture attempts to contribute to the statement on what mathematical models can and cannot do, the end product this symposium aims at. It does so in the form of the general statements on the relationship between mathematical modeling, underlying research and engineering practice given in Section 2, and further in the general statements on the state of the modeling art given in Sections 6.1, 8 and 9.3.

In connection with the main predictive requirements at the one hand this general lecture indicates that the reasons why a given mathematical model is deemed applicable in a given engineering context have to be made explicit (Section 11.1). On the other hand it acknowledges that often subjective judgment must play a role in answering the questions involved (Section 9.3). Therefore mathematical models should not be applied as an engineering tool without input from engineering.

It is a good thing to realize that given its function within the symposium this general lecture concentrates on what may limit the predictive ability of mathematical models of flows and transports. As such it concentrates on one side of the medal. The total picture has to be obtained from the in-depth descriptions of specific model applications. These descriptions are the backbone of the symposium. They show how far the predictive requirements can be satisfied under specific circumstances, and what can be accomplished under these specific circumstances. Their specific information has to provide the main reference material for the end product of the symposium.

REFERENCES

Abbott, M. B. (1979). "Computational Hydraulics," Pitman, London.

Abraham, G. (1978). Entrainment solutions for jet discharge into deep water. *In* "Thermal Effluent Disposal from Power Generation" (Z. P. Zaric, ed.), 11-44, Hemisphere, Washington, D.C.

Abraham, G. (1980). On internally generated estuarine turbulence. *Proc. Int. Symp. Stratified Flows, 2nd Trondheim, Norway* **1**, 344-353.

Abraham, G., and Vreugdenhil, C. B. (1971). Discontinuities in stratified flows. *J. Hydr. Res.*, **9**, No. 3, 293-308.

Albertson, M. L., Dai, Y. B., Jensen, R. A. and Rouse, H. (1950). Diffusion of submerged jets. *Trans. Am. Soc. Civ. Eng.* **115**, Paper No. 2409, 639-697.

Daily, J. E., and Harleman, D. R. L. (1972). Numerical model for the prediction of transient water quality in estuary networks. Tech. Rep. 158, Ralph M. Parsons Laboratory, Department of Civil Engineering, Massachusetts Institute of Technology.

Delft Hydraulics Laboratory, 1980. Application of one-dimensional dispersion concept to salinity intrusion in Rotterdam Waterway Estuary. Report M 896-42 (Dutch text).

Delvigne, G. A. L. (1979). Round buoyant jet with three-dimensional trajectory in ambient flow. *Proc. Congr. Int. Assoc. Hydraul. Res. 18th, Cagliari, Italy* **3**, 193-201.

Dronkers, J. J. (1969). Tidal computations for rivers, coastal areas and seas. *J. Hydraul. Div. Proc. Am. Soc. Civ. Eng.* **95**, No. HY1, 29-77 (in particular, p. 35).

Elvius, T., and Sundström, A. (1973). Computationally efficient schemes and boundary conditions for a fine-mesh barotropic model based on the shallow-water equations. Tellus, **24**, 132-157.

Fischer, H. B. (1967). The mechanics of dispersion in natural streams. *J. Hydraul. Div. Proc. Am. Soc. Civ. Eng.* **93**, No. HY6, 187-216.

Fischer, H. B. (1976). Mixing and dispersion in estuaries. *Ann. Rev. Fluid Mech.* **8**, 107-133.

Flokstra, G. (1977). The closure problem for depth-averaged two-dimensional flow. *Proc. Congr. Int. Assoc. Hydraul. Res. 17th, Baden-Baden, Germany* **2**, 247-256.

Garrett, C., and Greenberg, D. (1977). Predicting changes in tidal regime: the open boundary problem. *J. Phys. Oceanogr.* **7**, 171-181.

Gustafsson, B., and Sundström, A. (1976). Incompletely parabolic problems in fluid dynamics. Department of Computer Science, Uppsala University.

Hadamard, J. (1923). Lectures on Cauchy's problem in linear partial differential equations. Yale University Press.

Harleman, D.R.F., and Thatcher, M. L (1974). Longitudinal dispersion and unsteady salinity intrusion in estuaries. *La Houille Blanche,* **29**, No. 1/2, 25-33.

Hirt, C. W. (1968). Heuristic stability theory for finite difference equations. *J. Comp. Phys.* **2**, 339-355.

Jirka, G. H., Abraham, G., and Harleman, D.R.F. (1975). An assessment of techniques for hydrothermal prediction. Tech. Rep. 203. Ralph M. Parsons Laboratory, Department of Civil Engineering. Massachusetts Institute of Technology.

Kranenburg, C. (1978). On internal fronts in a two-layer flow. *J. Hydraul. Div. Proc. Am. Soc. Civ. Eng.,* **104**, No. HY10, 1449-1453.

Kreiss, H. O. (1962). Über die Stabilitätsdefinition für Differenzgleichungen die partielle Differentialgleichungen approximieren. *Nordisk Tidskr. Informations Behandlung,* **2**, 153.

Kuur, P. van der, and Verboom, G. K. (1975). Computational analysis for optimal boundary control of two dimensional tidal model. Publication No. 148, Delft Hydraulic Laboratory.

Lean, G. H., and Weare, T. J. (1979). Modelling two-dimensional flow. *J. Hydraul. Div. Proc. Am. Soc. Civ. Eng.* **105**, No. HY1, 17-26.

Leendertse, J. J. (1967). A water quality simulation model for well-mixed estuaries and coastal seas. Vol. I-VI, Rand Corporation, Santa Monica, California.

Liu, S. K., and Leendertse, J. J. (1978). Multi-dimensional numerical modelling of estuaries and coastal seas. *In* "Advances in Hydroscience" (V. T. Chow, ed.), **11,** 95-164.

McGuirk, J. J. and Rodi, W. (1978). A depth-averaged mathematical model for the near field of side discharges into open-channel flow. *J. Fluid Mech.* **86,** Part 4, 761-781.

Marchuk, G. I., Gordeev, R. G., Rivkind, V. Ya., and Kagan, B. A. (1973). A numerical method for the solution of tidal dynamics equations and the results of its applications. *J. Comp. Phys.*, **13,** 15-35.

Morton, B. R., Taylor, G. I., and Turner, J. S. (1956). Turbulent gravitational convection from maintained and instantaneous sources. *Proc. Roy. Soc. London, Ser. A* **1196,** 1-23.

Naudascher, E., and Schatzmann, M. (1979). Teil A: Einfluss der Einleitungsparameter auf die Fernfeldausbreitung. *In* "Das Ausbreitungsverhalten von Abwärme- und Abwassereinleitungen in Gewässern" (E. Naudascher, ed), Erich Schmidt Verlag. (See also Schatzmann and Naudascher, 1980).

Oliger, J., and Sundström, A. (1978). Theoretical and practical aspects of some initial-boundary value problems in fluid dynamics. SIAM **35,** 419-446.

Orszag, S. A. (1971). Numerical simulation of incompressible flows within simple boundaries: accuracy. *J. Fluid Mech.* **49,** 75-112.

Payne, L. E. (1975). Improperly posed problems in partial differential equations. Regional Conference Series in Applied Mathematics, SIAM, Philadelphia.

Piacsek, S. A., and Williams, A. P. (1970). Conservation properties of convection difference schemes. *J. Comp. Phys.* **6,** 392-405.

Policastro, A. J., and Dunn, W. E. (1978). Evaluation of integral and phenomenological models for heated surface plumes with field data. *In* "Thermal Effluent Disposal from Power Generation" (Z. P. Zaric, ed.), Hemisphere, Washington, D. C., 45-59.

Prandle, D. (1978). Residual flows and elevations in the southern North Sea. *Proc. Roy. Soc. London, Ser. A* **359,** 189-228.

Reynolds, S. D., Roth, P. M., and Seinfeld, J. H. (1973). Mathematical modeling of photo chemical air pollution--I. *Atm. Env.* **7,** 1033-1061.

Reynolds, W. C. (1976). Computation of turbulent flows. *Ann. Rev. Fluid Mech.* **8,** 183-208.

Richtmeijer, R. D., and Morton, K. W. (1967). "Difference Methods for Initial-value Problems." Interscience Publications, New York.

Rodi, W. (1980). "Turbulence Models and their Application in Hydraulics--a State of the Art Review." Edited by *Int. Assoc. Hydraul. Res.*

Rouse, H., Yih, C.S., and Humphreys, H. W. (1952). Gravitational convection from boundary source. *Tellus,* **4,** No. 3, 201-210.

Saffman, P. G. (1977). Problems and progress in the theory of turbulence. Proceedings of Structure and Mechanics of Turbulence II, Springer-Verlag, Berlin, 273-306.

Schatzmann, M. and Naudascher, E. (1980). Design criteria for cooling-water outlet structures. *J. Hydraul. Div. Proc. Am. Soc. Civ. Eng.* **106,** No. HY3, 397-408.

Schmidt, W. (1941). Turbulente Ausbreitung eines Stromes erhitzter Luft. *Zeitschrift für Angewandte Mathematik und Mechanik,* Band 21, 265-278; 351-363.

Smith, T. J., and Dyer, K. R. (1979). Mathematical modelling of circulation and mixing in estuaries. In "Mathematical Modelling of Turbulent Diffusion in the Environment," pp. 301-341. (C. J. Harris, ed.), Academic Press, London.

Stelling, G. S. (1980). Improved stability of Dronker's tidal schemes. *J. Hydraul. Div. Proc. Am. Soc. Civ. Eng.* **106,** No. HY8, 1365-1379.

Taylor, G. I. (1953). Dispersion of soluble matter in solvent flowing slowly through a tube. *Proc. Roy. Soc. London, Ser. A* **219,** 186-203.

Tennekes, H., and Lumley, J. L. (1972). "A First Course in Turbulence." M.I.T. Press, Cambridge, Massachusetts.

Tollmien, W. (1926). Strahlverbreiterung. *Zeitschrift für Angewandte Mathematik und Mechanik,* Band 6, 468-478.

Turner, J. S. (1973). "Buoyancy Effects in Fluids." Cambridge University Press, Section 4.3.

Verboom, G. K., Stelling, G. S., and Officier, M. J. (1981). Boundary Conditions for the Shallow Water Equations. "Computational Hydraulics: Homage to Alexandre Preissmann", (M. B. Abbott, and J. A. Cunge, editors). Pitman, London.

Warming, R. F., and Hyett, B. J. (1974). The modified equation approach to the stability and accuracy of finite difference methods. *J. Comp. Phys.* **14,** 159-179.

Wirz, H. J., Schutter, F. de, and Turi, A. (1977). An implicit, compact, finite difference method to solve hyperbolic equations. *Math. Comp. Sim.* **19,** 241-261.

Zimmerman, J. T. F. (1978). Topographic generation of residual circulation by oscillatory (tidal) currents. *Geophys. Astroph. Fluid Dynamics* **11,** 35-47.

DISCUSSION

Jan J. Leendertse

The Rand Corporation, Santa Monica, California, U.S.A.

Reading the general lecture of this symposium on the requirements of predictive ability of mathematical models, it struck me how diverse the views on this subject can be. My views, which are, I believe, the views of an experienced modeler, are quite different.

On the third day of this symposium, three papers were presented concerning the Delta Works. In this project--one of the largest engineering construction jobs in the world--a large prefabricated storm surge barrier will be constructed. Prefabricated parts weighing up to 8000 tons will be transported through tidal waters and placed on the specially prepared and protected seabed. Naturally, good predictions of current magnitudes and current reversals are required for the placement. Inaccurate predictions would seriously endanger the whole construction job.

For these important flow predictions a very simple mathematical model is used which relates observed currents with observed water levels and the predicted water levels with the observed ones. On the basis of this information, current predictions can be made. The effect of the gradual restriction will be accounted for in the predictions on the basis of previous rates of change in the current observations.

This simple mathematical model does not satisfy at all the requirements set forth in the general lecture. Nevertheless, it was chosen on the basis of the proven capability of the modelers to make accurate predictions during a previous closure. It was chosen even though physical models, one- and two-dimensional models were available.

In the general lecture an attempt is made to make a scientific analysis of the process of mathematical modeling of flow and transports of conservative substances and derive from it the requirements for prediction ability. This is the problem: the process by which a model in general is derived can best be described as an intuitive art, and creativity of the modeler is *the* important ingredient for a successful model investigation; creativity cannot be replaced by scientific knowledge.

Modeling is certainly not a scientific endeavor, even though it is customary to report it as such. Successful modeling is nearly always reported as a more or less logical reconstruction of occurrences. Derivations are presented in a logical sequence which has little relationship to the manner in which the modeling effort progressed. Generally many attempts were made to produce results and only the one finally chosen was reported.

The reason that we do this is that the modeler wants to make his work acceptable to the scientific community. Also, it provides a convenient, acceptable frame for reporting. As a result, an unrealistic view is presented as to what modeling actually is and how it is done.

The modelers of surface water flows and transports are really modelers of systems, in this case physical systems. In the last thirty years, with the advent of computers, modeling and system simulation have been applied to many fields such as space flight, economics, management, marketing, politics, social sciences, transportation, and many others. As a result of this we now have well-established concepts, meanings and experiences which are generally applicable, but not widely known in the scientific hydraulic research community. To lay the groundwork for proper discussion I will mention a few important ones.

A model is a representation of an object, system, or idea in some form other than the entity itself.

An important usage of models is for making predictions and comparisons, the latter generally to find preferences between different courses of action. These are not the only uses; modeling provides a systematic and often efficient way for experts and decision makers to focus their judgment and intuition. A model can serve as an effective means of communication and certainly as an aid to thought. Modeling goes hand-in-hand with progress in science and technology.

Forester, in his book *Industrial Dynamics* (1961), gives a good overview of how models are used in different disciplines and what their basis is. He writes, for example, "In engineering systems, models have been built upward from available knowledge about separate components. Designing a system model upward from identifiable and observable pieces is a sound procedure with a history of success." He writes also, "Physics is a foundation of principle to explain underlying phenomena but not a substitute for invention, perception and skill in applying the principles." How true that is for modeling of flows and transports of substances!

It is difficult to describe the function which models fulfill; we are able to list at least the following five functions: (1) aid to thought; (2) aid to communication; (3) aid in training and instruction; (4) instrument of prediction; and (5) aid to experimentation.

All of us easily see these applications from our own experience. Models can be classified in many different ways, for example: (1) iconic versus analog versus mathematical; (2) static versus dynamic; (3) deterministic versus stochastic; (4) discrete versus continuous. It is noted here that mathematical models can be static or dynamic, deterministic or stochastic, and discrete or continuous.

As I indicated earlier, the way a modeler derives a model for the system he is studying can best be described as an intuitive art. No fixed rule can be given. The modeler must have the ability to analyze a problem, abstract its *essential* features, select and modify assumptions that characterize the system, and subsequently extend and enrich it until a useful approximation is found. One of the more important steps in the process appears to be the separation of the system into simple parts--parts which are relevant to the solution of the problem and/or the processes which govern the system. Simplification can be done by making variables constant, eliminating variables, assuming linearity, and restricting boundary conditions and assumptions.

Modeling is an evolutionary process; its progress is dependent upon the flexibility of the model or modeler to make changes and on the relationship of the model builder and the ultimate model user (or user of the model results). The act of modeling can only be mastered by those who are resourceful, have ingenuity, and have insight into the processes they are trying to model *and* into the problems which the ultimate model user faces in his decision making. In other words, the goals and objectives of the model should be clear to the modeler.

A management scientist once wrote that a good simulation model should be: simple to understand by the user; goal and purpose directed; robust, not giving absurd answers; easy for the user to control and manipulate; complete on important issues; adaptive, thus easily modifiable, e.g., include more complicated processes; and evolutionary. You will note that the simple mathematical model mentioned in my introduction completely satisfies the first five criteria.

Before discussing in more detail the general lecture, let me bring the general context of this paper in perspective. Karplus, in his keynote address at the 1972 Computer Simulation Conference in San Diego, noted that all technological advances go through a similar irreversible evolution which is characterized by different types of individuals who play a key role. They are: (1) the inventor or innovator; (2) the engineer; (3) the theoretician; (4) the ultimate user. From my own experience with two-dimensional models, I found that the first two phases ran until about 1975. By then it became quite established that this modeling was possible. For the one-dimensional models these two phases ran until about 1970.

In these phases the first simulation systems were built. This is naturally still continuing, but now the theorists are the most obvious in the literature. We see publications about boundary conditions, "improved" solution methods, stability aspects, and with this conference an attempt will even be made to focus on what the requirements of mathematical models should be.

Verification has been of much concern in modeling. Many models are being used effectively which have never been verified. In accordance with Shannon (1975), we can rather speak of validation. In his book he comes to the conclusion that validation is the process of bringing to an acceptable level of the user's confidence that any reference about a system derived from the simulation is correct. He writes that it is impossible to prove that any simulation is a correct or "true" model of the real system and that we are fortunately seldom concerned with proving the "truth" of a model. Instead, modelers are mostly concerned with validating the insights which are gained or will be gained from the simulation. Thus it is the operational utility of the model and not the truth of its structure that usually concerns us.

A useful model is one from which some manager or decision maker can gain insight into the value of possible decision alternatives. He may need predictions or merely comparisons. He will judge a model by its applicability. The model should be *relevant, usable, valid, and cost-effective.*

The scientist looks differently at a model. He equates quality with non-triviality. He wants it to be elegant and powerful. In our context it should simulate all processes so he is able to get an insight on all details.

It is noted here that the intent of this conference on predictive modeling was to obtain contributions from modeling studies for the solution of *real* problems. Also in these studies verification should have been made with field data.

The committee organizing this symposium, however, apparently looked at it from a scientific point of view. I have the impression that many of the studies which will be presented do not fulfill the function of being instruments of prediction for the solution of real problems. With their presentation they will certainly be an aid to communication.

With this background I want to return to the paper of Abraham, van Os, and Verboom. This paper refers to a very small subset of mathematical models of flow and transport. In engineering practice in this field we are using statistical models for prediction of water levels due to storm surges, and correlation models for prediction of tides and currents. We use steady-state models of flow over weirs and under sluice gates, and correlation models between river flow and salinity, locking frequencies and salinity intrusion, to mention a few. This type of mathematical modeling is much more common than the deterministic mathematical simulations which the authors have in mind in their paper. Mathematical simulation is naturally a process we should avoid if we can. It is expensive and difficult, and can be done only by those who really have the skills.

On this subject the authors and the speaker agree. I do not agree with the concepts of approximations, schematizations, and feedbacks which are the basis of their paper (Fig. 1), and have reasons to question the process of modeling as presented by the authors. I have described experiences of successful modelers in other fields, and I see no reason why it should be different in this case. More importantly, I have been involved in modeling for many years and my experience has been different and in line with those in other fields.

As an example, Dr. Liu and I have developed a three-dimensional model. This was directly developed from our understanding of the physics into a finite difference model. Partly our understanding of the physics was translated directly in FORTRAN program statements, and only after the basic model was working did we write down the partial finite difference equations.

Man's thought processes are turned to real concepts which we can visualize. Taking the limits and working with differentials is certainly a respectable computation method, invented before we had computers, but this is not now necessarily the point of departure.

We can build directly in finite differences and then investigate the effects of how we took the differences. This approach makes conservation visible from the outset. Dr. Liu and I followed Forester without specifically realizing it. The model was built from identifiable and observable pieces, and finite difference expressions were used which were proven to work well in other models.

All the steps and requirements set forth in the general lecture are certainly too extensive to be done when solving practical problems. You will never finish studying all aspects of each step, and the problem will be solved by others or by other means while you are still in loops. It is the modeler who has to cut through all options there are and select the relevant processes and formulate his model. He will always be open to criticism from those who are on the sidelines and are not concerned with cost effectiveness. Predictive ability is directly related to cost. More detailed programs, including more refined representations of the processes, will cost more to built, operate and validate. Better resolution may considerably increase the predictive capabilities.

In modeling there is much more involved than the steps indicated. It is too simple to presume that we have known excitations or inputs to a model. All measurements which are made in the prototype are state-estimates; the model also generates state estimates. It is with this degree of uncertainty that we work and the processing of data and the assessment of their worth are an essential part of any model investigation. Also with this uncertainty it is somewhat naive to define calibration as the process of making the output of the model agree with a limited number of known responses of the problem area to known excitations. The predictive capabilities of a model are directly related to available data. In many instances data is at least as important for prediction as the model which is used and the skill and capabilities of the modeler who makes the investigation.

Much of the discussion in the general lecture about the selection of the appropriate representation of flow in partial differential equations and difference equations for a particular problem will be irrelevant in many practical applications. The development cost of a new program for a particular problem is generally so large that it is not cost-effective to do so. Consequently, model investigations will be done by use of program systems designed for a certain group of problems. By making use of such a system, the practical predictive capability for a particular problem will be much greater than when starting from scratch.

As to the technical content of the paper, I want to make only a few remarks. The authors certainly give evidence of a thorough understanding of present knowledge of the physical processes of flows and transports and the mathematical description thereof. It is with their rigid view on modeling that I disagree. Under "stability analysis" the energy method is mentioned as the most stringent method available, but it assures no stability. One of the major problems in my experience has been instabilities caused by vortices which could not be properly represented on the grid. Conservation of the squared vorticity was the predominant stability requirement.

Furthermore, the turbulence model Dr. Liu and I developed was discussed. Contrary to the statement in the general lecture, our description of this turbulence model contains externally and internally generated turbulence. As the models we are building are adaptive and evolutionary, we have now taken into account the work required to mix heavier layers with lighter layers, which dampens the turbulence.

The authors of the general lecture propose that mathematical model reports should contain a statement on the model's accuracy and range of application. The statement should not be limited to information about the model, but should also contain information about the data which is used for the investigation.

In reporting all model studies it seems at least appropriate to make comparisons of state estimates obtained by model and prototype based upon a statistical basis rather than making subjective comparisons such as "good" and "excellent." Moreover, as the modeler is at least as important as the model which is used it seems appropriate to put the name of the modeler on the report rather than to bury it somewhere in the introduction as is customary in several institutions.

I do think that this symposium will contribute to the understanding of present capabilities of mathematical models of flows and transports in natural waters used for the solution of real problems. However, predictive ability is only one the factors to be considered in an investigation. Predictive ability depends on the modeler, the model, and the available data. Whether research underlying the mathematical model is completed or not seems to have only relative importance in solving practical problems.

In solving practical problems we are seldom concerned with the "truth" of a model and whether or not the model is verified. We expect that the model is validated. It is the operational utility and applicability of the model upon which the model and its modeler will be judged in engineering practice.

REFERENCES

Forrester, J. W. (1961). "Industrial Dynamics." John Wiley and Sons, New York.

Shannon, Robert E. (1975). "Systems Simulation, the Art and Science." Prentice Hall, Inc., Englewood Cliffs, New Jersey.

REPLY

In defining predictive ability (Section 2 of the lecture), we observe that "the predictions must be accurate enough to allow a sufficiently reliable answer to the engineering questions behind the mathematical model study". Leendertse expects a model to be validated and adopts Shannon's conclusion that "validation is the process of bringing to an acceptable level of the user's confidence that any reference about a system derived from the simulation is correct." Thus, in essence we and Leendertse expect the model to fulfill the same function.

Though our perception of the function of a model coincides with his, Leendertse makes it appear that we are primarily interested in the "truth" of a model, that we equate quality with non-triviality--in short, that our perception of modeling is unrealistic. In this way he creates differences where in reality there are none. The Delft Hydraulics Laboratory is an institute for applied research. There we develop and apply mathematical models as one of the available engineering tools. Leendertse's discussion and our general lecture are written from the same engineering background. Therefore at several instances we feel that our observations coincide with his, and vice versa. In addition, we agree with several of his observations which have no counterpart in our lecture: for instance, his observations on cost effectiveness, on the operational utility, and on the availability of data, a subject we knew to be dealt with in a separate lecture later on in the symposium. We also agree with his observations on the significance of the modeler's experience, except where we feel he is exaggerating. It is true that creativity cannot and should not be replaced by scientific knowledge. Neither can scientific knowledge (and factual knowledge, as for instance that accumulated in engineering experience) be replaced by creativity. These three ingredients must complement one another.

Comparing our arguments with those of Leendertse, one has to look through a difference in semantics. That, in Leendertse's terminology, the modeler has to abstract the essential features of a problem is equivalent to our first level of schematization and approximation. His observation that the modeler has to have insight into the problem which the ultimate model user faces in his decision-making is contained in our feed-back loop 4. The same applies to his observation that a complicated mathematical model has to be avoided as an engineering tool, where a simple tool is appropriate.

Leendertse does not agree with the concept of approximations, schematizations, and feed-back loops of Figure 1, his argument being that it is not in line with his experience. Nevertheless, his description of how modeling proceeds resembles a description of the figures in words: "The modeler must have the ability to analyze a problem, to abstract its essential features [level 1], select and modify assumptions that characterize the system [level 2], and subsequently extend and enrich it [level 2] until a useful approximation is found [feed-back loops 1 and 4]". And later on, "We can build directly in finite differences [levels 2 and 3] and then investigate the effects of how we took the differences [feed-back loop 3]". That a modeler must have insight into the ultimate decision problem confirms that he has to begin at level 1 and to end with feed-back loop 4. And so on.

The scheme of Figure 1 remains valid whether the approximations at level 2 and level 3 are made as successive steps (equations first and numerical scheme later) or quasi-simultaneously (equations directly in Fortran). For instance, in both cases one has to solve the same closure problem when dealing with turbulent transport type problems. In addition, as indicated in the preceding paragraph, one has to investigate the effects of how the differences are taken.

To bring the general context of our lecture into perspective, Leendertse refers to a keynote address delivered by Karplus in 1972. Karplus (1976) adds to this perspective, examining the history of mathematical modeling in the various application disciplines. After a discipline reaches a certain maturity, he notes, efforts to improve the predictive validity of mathematical models can be expected to be relatively unproductive. Then the important point is that the ultimate use of a model must confirm its validity. Only when both the designers of models and their customers appreciate the ultimate limitations of the models, Karplus observes, can full effort be brought to bear upon optimizing the models in terms of their usefulness. Within this context we introduced the statement on accuracy and range of applicability to fulfill the function explained in the final paragraph of Section 11.1.

Karplus (1976) brings the concepts derived from other application areas into perspective by introducing the concept of the model's "shade of gray" as an inference based upon the experience of modeling physical and other systems. Karplus introduces this parameter as a measure for the model's reliability and validity. He relates this parameter with the amount of factual and scientific knowledge available for a given application area. He relates the ultimate mode of application of a model with its shade of gray. The darker the shade of gray, the more the modeling coincides with black box modeling and the more the model assumes more qualitative and less quantitative roles. In Figure (A) Karplus compares the shade of gray of different application areas, including water quality. For the models presented at this symposium going through feed-back loops of Figure 1 leads to insight into their shade of gray. This information can be given rather objectively.

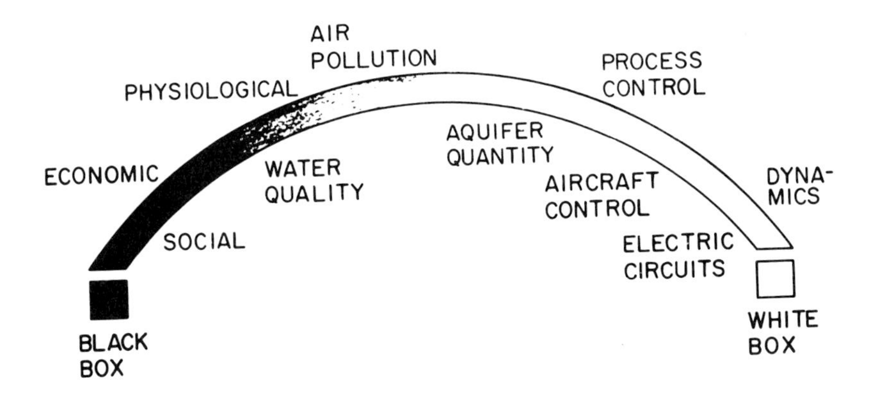

Figure A. Karplus' (1976) comparison of different application areas with respect to their "shade of gray".

In engineering one has to know the accuracy and range of applicability of the engineering tool one uses. This is nothing new. It applies to a hydraulic scale model and to an empirical design graph as well as to a mathematical model. If not, one is not able to ascertain whether or not the output of the engineering tool is accurate enough to allow sufficiently reliable answers to the relevant engineering questions (our terminology), or to bring to an acceptable level the user's confidence that references about the system derived from the model are correct (Leendertse's terminology). Because of this, we introduced the statement on the model's accuracy and range of applicability as the main predictive requirement. We used the scheme of Figure 1 as a means to explain which type of information has to be made explicit in the statement. In this respect one has to collect information needed to appraise such items as: whether or not subjective arguments have to enter into the statement (Sec. 9.3); whether or not there are data base limitations; the model's shade of gray; and the implications of these factors for the permitted mode of model application.

Asking the modeler at the end of his efforts to make explicit the information needed to appraise items of this type by going through the feed-back loops of Figure 1 is not an excessive requirement. It is reasonable to expect that at the end of his efforts the modeler will re-examine his final results and the way they were produced. It is as reasonable that by then he will make his final considerations on the accuracy and range of applicability explicit as input into the decision process. This is the essence of what we asked for in Section 11.1 We trust that Leendertse does not question this conclusion--though this is not clear from his text--and that he agrees that for a conscientious modeler it is a normal thing to provide this input.

In Section 1 we explained the aim of the lecture to be two-fold: to elaborate upon the factors which may limit predictive ability and to derive predictive requirements. We further explained that we left the subject of how to model to the presentations on specific model applications. Therefore, we see no evidence in our lecture to support Leendertse's inference that we presented a rigid view on modeling. That, nevertheless, he comes to this conclusion is due, we believe, to an incorrect perception by him of the subject of our lecture. This becomes evident in his statement that "this simple mathematical model does not satisfy *at all* the requirements set forth in the general lecture." Apparently the predictions which can be obtained using this simple model mentioned in Leendertse's introduction are accurate enough to allow a sufficiently reliable answer to the engineering questions behind the mathematical model study. So it satisfies the requirements put forward in Sec. 2. In addition its statement on accuracy and range of applicability seems to explain explicitly why sufficiently reliable extrapolations can be made. So the main predictive requirement of Sec. 11.1 is satisfied. Our requirements are met when the statement on accuracy and range of applicability takes due note of the factors which may lead to limited predictive ability. Then there is a guarantee that the mode of application of a model is in accordance with its capabilities and limitations. Because of this, we were explicit in explaining what may limit predictive ability. That the mode of model application is in accordance with its capabilities and limitations is the modeler's responsibility, and hence a requirement for predictive ability he is judged upon.

Whether or not the turbulence model of Liu and Leendertse (1978) contains externally and internally generated turbulence is not what we had in mind in commenting on this model. What we were referring to was that the parameterization of the length scale L contained in the model does not reflect that turbulence can be generated both externally and internally. Our consideration of this matter can be found in Section 5.1.2 of the lecture and in Abraham (1980) as referenced in the lecture.

REFERENCES

Karplus, W. J. (1976). The future of mathematical models of water resources systems. *In* "System Simulation in Water Resources," pp. 11-18 (G. C. van Steenkiste, ed.), North Holland Publishing Company.

THE PREDICTIVE ABILITY OF ONE-DIMENSIONAL ESTUARY MODELS

Nicholas V. M. Odd

Hydraulics Research Station
Wallingford, United Kingdom

1. INTRODUCTION

One-dimensional numerical models of estuaries are commonly used by consultants and public organizations responsible for planning engineering works. They are often employed in conjunction with large-scale physical models and multi-dimensional numerical models. The one-dimensional model is still one of the most widely used methods for predicting the effects of engineering works on tidal propagation, tidal flooding, evacuation of fluvial floods, saline intrusion, siltation, thermal balance and water quality in tidal channel systems.

The term numerical model is used in the context of estuaries to describe the systematic calculation of the motion of bodies of matter, such as water or dissolved solutes throughout the estuary system at successive time intervals. The main features of the tidal motion are determinate and respond to basically simple laws of physics. That is to say, the imposition of an external force upon a body of water must result in its acceleration, and the net amount of water that enters or leaves a length of channel equals the amount stored or lost by it. Similarly, laws governing the conservation of matter may be applied to the motion of dissolved salt and suspended sediment, but certain aspects of their motion can, as yet, only be defined by empirical relationships.

Models based on the laws of physics are termed deterministic models, whereas models based on empirical relationships are termed functional models. Most one-dimensional estuary models are hybrid since although they are based on the laws of Newtonian physics they also contain empirical relationships which have an important influence on the solution.

The predictive ability of hybrid models is limited by the range of conditions for which their empirical functions are known to hold true.

An ideal test of the predictive ability of a model requires one to forecast a given event, without prior knowledge of the outcome, and to obtain precise observations under conditions that match those used in the forecast. Nor should the boundary conditions have an overriding influence on the results as in the case of

TRANSPORT MODELS FOR
INLAND AND COASTAL WATERS

ISBN: 0-12-25852-0

the simulation of conditions in a very short reach of an estuary. Such conditions are rarely satisfied in practice. Instead, at best, models are proved against one set of observations and then validated by hindcasting one or more other events and comparing the results with independent sets of field data. The definition of the word "hindcast" in this context is a numerical simulation of a natural event, which has already taken place. Proving the model is a process in which coefficients are adjusted until the results agree with observations. Validation is a process in which the model is used to hindcast an event that is significantly different from the proving test, without changing the values of coefficients.

In practice, models are often used to predict the effects of engineering works under a series of critical design conditions, such as extreme tides, which are unlikely to be observed in a follow-up study. Predictions often take the form of comparative tests with different amplitudes for the forcing motion (tidal or fluvial), in which case one is more concerned with the incremental changes, Δ_r, than the absolute values. It is argued that the accuracy of these incremental changes, (true Δ_r − model Δ_r)/true Δ_r, is about the same as the accuracy of the model as defined in proving or validation tests, ϵ_r/r_{obs}, as illustrated in Fig. 1. This argument cannot be applied to the accuracy of predictions arising from large-scale changes in the geometry of a tidal system.

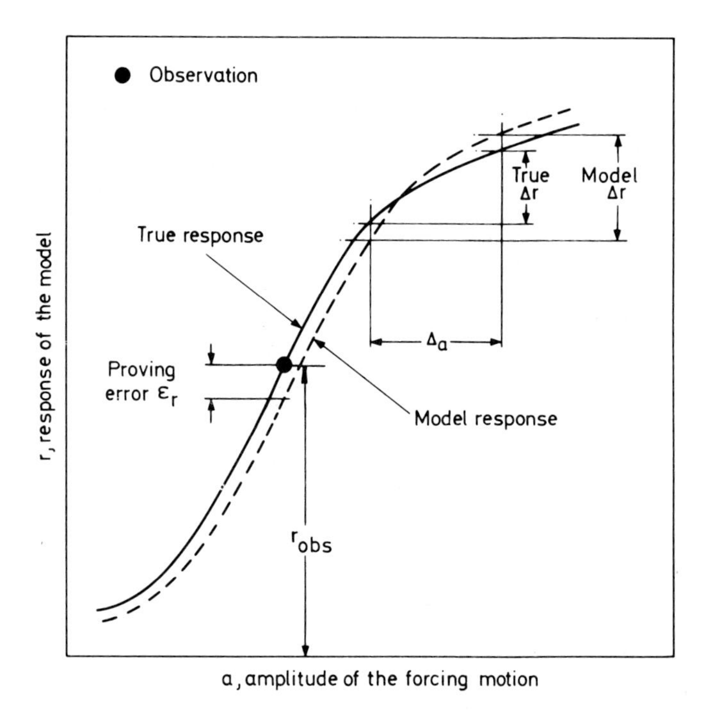

Figure 1. The accuracy of incremental changes.

If one-dimensional estuary models are set up without giving due attention to detail or predictive tests are ill-conceived and the results are analyzed or interpreted without care they can give rise to false predictions and badly designed works. This paper points out some of the pitfalls in applying the technique and various ways of analyzing and presenting the results. It also indicates the weaknesses and limitations of one-dimensional estuary models.

2. BASIC THEORY AND ASSUMPTIONS

In a one-dimensional estuary model the effective flow-carrying cross section of the waterway is considered to be a streamtube of variable dimensions, Fig. 2, and the flow in it is described by the mean velocity over the section. The process of averaging over a cross section gives rise to additional terms or coefficients whose value is dependent upon the shape of the velocity and solute distributions over the cross section, Fig. 3.

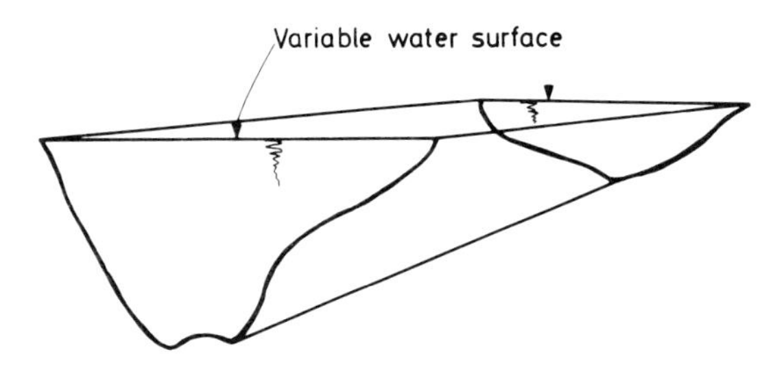

Figure 2. A tidal channel as a stream tube of variable dimensions.

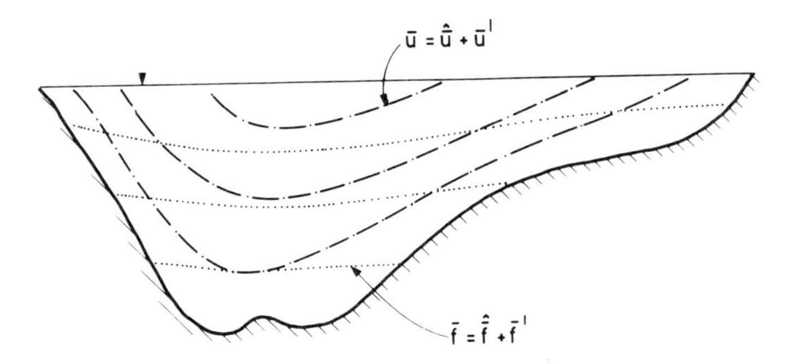

Figure 3. Instantaneous tidal velocity and solute distributions over a cross section.

Consider a streamtube of cross-sectional area A. In order to take into account the cross-sectional distribution of a variable f, with turbulent fluctuations smoothed out, it is convenient to express the value at a point as the sum of the cross-sectional mean value $\bar{f}$ and the variation from the mean, f', as follows:

$$f = \bar{f} + f' \tag{1}$$

where

$$\bar{f} = \frac{1}{A} \int_A f dA \tag{2}$$

and, by definition

$$\overline{f'} = 0 \tag{3}$$

the cross-sectional average value of the advective term can be expressed as follows

$$\frac{\overline{\partial (fu)}}{\partial x} = \frac{1}{A} \left\{ \frac{\partial}{\partial x} (A\bar{f}\bar{u}) + \frac{\partial}{\partial x} (A\overline{f'u'}) \right\} \tag{4}$$

The term $\frac{1}{A} \frac{\partial}{\partial x} (A\overline{f'u'})$ can be treated in the form of a coefficient α attached to the advective term, where

$$\alpha = \frac{1}{A\bar{f}\bar{u}} \int_A uf dA \tag{5}$$

as in the case of the convective acceleration term in the equation for the conservation of momentum, (in this case α may have a value as high as 1.4). It may also be treated as a longitudinal dispersion term so that

$$\overline{f'u'} = -D_T \frac{\partial \bar{f}}{\partial x} \tag{6}$$

as in the case of the equation for the conservation of solutes, where D_T is the effective coefficient of longitudinal dispersion (m^2/s). One-dimensional models are most accurate for cross-sectionally well-mixed conditions when f' is small compared to $\bar{f}$.

The process of averaging the longitudinal pressure forces over a streamtube approximates the longitudinal density gradient term.

Consider the streamtube in which the density variations over the cross section are negligible. The longitudinal pressure gradient can be expressed as follows

$$\frac{\partial p}{\partial x} = \rho_o g \frac{\partial \eta}{\partial x} + g(\eta - z) \frac{\partial \rho}{\partial x} \tag{7}$$

Since η is assumed to be constant across the width of the channel it follows that for a single streamtube

$$\int_A \frac{\partial p}{\partial x} dA = Ag\rho_o \frac{\partial \eta}{\partial x} + AX_o g \frac{\partial \rho}{\partial x} \tag{8}$$

where

$$X_o = \frac{1}{A} \int_A (\eta - z)\, dA \tag{9}$$

X_o, the distance to the center of gravity of the section is often assumed to be half the hydraulic radius.

The equivalent incremental rise in mean tide level at the head of an estuary, $\Delta\eta$, due to longitudinal salinity variations is approximately as follows

$$\Delta\eta = \frac{d\Delta\rho}{2\rho_o} \tag{10}$$

$\Delta\eta$ often has a value of about 0.3 m in deep estuaries.

The basic equations describing one-dimensional tidal motion are as follows

$$w \frac{\partial \eta}{\partial t} + \frac{\partial Q}{\partial x} = 0 \tag{11}$$

$$\frac{\partial Q}{\partial t} + gA\frac{\partial \eta}{\partial t} + \alpha_1 \frac{\partial}{\partial x} \left(\frac{Q^2}{A}\right) + \frac{AR_g}{2\rho_o} \frac{\partial \rho}{\partial x} + \frac{\tau_o P}{\rho_o} = 0 \tag{12}$$

where

$$\frac{\tau_o P}{\rho_o} = \alpha_2 \quad \frac{f|Q|Q}{8RA} \tag{13}$$

and, w is the water surface width (m); η is the elevation of the water surface to a common datum; Q is the tidal discharge (m^3/s); A is the effective cross-sectional area of flow (m^2); τ_o is the bed stress (N/m^2); f is the friction factor of the channel; R is the hydraulic radius (m) (A/P); P is the wetted perimeter of the channel (m); ρ is the density of water (kg/m^3); ρ_o is average density ($\simeq$ 1010 kg/m^3); and, g is the gravitational constant (9.81 m/s^2).

There are many ways of representing frictional resistance terms depending on the assumptions made about the lateral distribution of the roughness and the depth mean velocity. If the channel is wide compared to its depth ($w/R > 20$) and if the bed of the channel can be assumed to consist of rigid immobile roughness elements, the friction factor, f, is a known function of the relative roughness of the channel and the flow Reynolds number as prescribed by the Colebrook-White transition law

$$\frac{1}{\sqrt{f}} = -2\log_{10}\left\{\frac{k_s}{14.8R} + \frac{2.51\nu}{4UR\sqrt{f}}\right\} \tag{14}$$

where: k_s is the equivalent spherical grain size roughness (m); ν is the kinematic viscosity of water (m^2/s); and, $U = Q/A$ is the area mean tidal velocity (m/s).

The value of k_s varies from zero for a smooth mud bed to about 0.2 meters for a bed of rippled sand. The value of f is not very sensitive to the estimated value of k_s.

As the water levels vary in a shallow tidal channel, the area of the banks submerged in the flow and the relative roughness and hence the friction factor of both the bed and the banks may change appreciably. If the sloping sides of a channel have a different roughness from the bed, i.e., they are in smooth mud and the bed is rippled sand, a composite friction factor may be calculated as follows, provided the depth mean velocity does not vary significantly across the channel.

$$f \text{ composite} = \frac{(f\text{bed}\times P\text{bed}) + (f\text{bank}\times P\text{bank})}{P\text{bed} + P\text{bank}} \tag{15}$$

Another method of handling a cross section with wide variations in depth is to assume that the friction factor is constant and that the local depth mean velocity, $u(y)$, is proportional to the square root of the local depth, $d(y)$, as occurs in some steady flows. One can then define a modified hydraulic radius as follows

$$R' = \left[\frac{1}{A}\int_{o}^{w} d(y)^{3/2}\, dy\right]^2 \tag{16}$$

The modified hydraulic radius increases monotonically with rising water levels, which is not necessarily the case with the normal definition of $R = A/P$, if there are shallow berms at the sides of the channel. In predicting tidal levels the most important correction factors and coefficients are those in the dissipation term. Errors arising in averaging the temporal acceleration term are not usually of any practical importance. The magnitude of the Froude number, which is generally low in an estuary, gives a good idea of the importance of the convective acceleration term and hence the coefficient attached to it. The main problem with the term is that its magnitude varies with the ability of the model to resolve longitudinal variations in the cross-sectional area and hence the tidal velocity. For example, any given reach of an estuary, represented quite reasonably by one flow element in a model, may contain many changes of cross section.

As regards the convective diffusion equation for solutes and suspended matter the most critical coefficient arising from averaging over the stream tube is the so-called coefficient of effective longitudinal dispersion, D_x, whose magnitude depends on the variation of the velocity and the concentrations from the mean values over the section. The predictive ability for a one-dimensional model depends strongly on the manner in which correction factors or coefficients are used to counteract the errors introduced by averaging over a cross section.

3. SCHEMATIZATION OF AN ESTUARY SYSTEM

Schematization of an estuary system is a method whereby the various reaches are divided into a number of short elements. The average geometric properties of the streamtube in each element, such as width, effective flow area and wetted perimeter are defined in the model as a function of level in the form of sets of tables. The first step in the process of schematization is to identify the network of channels, stagnant storage zones, ocean outfalls, junctions and fluvial inputs to be represented in the model. The manner in which a tidal channel system is schematized, which is to a great extent a matter of personal judgment, can have an overriding influence on the ability of the model to simulate natural events.

The engineer has to define and model the paths of all the main flow-carrying channels and the extent of the intertidal storage zones. In many tropical deltas the tidal prisms stored in these zones, which are often mangrove swamps, account for a significant and often major part of the tidal prism of the estuary at its mouth. The problem is further complicated by the fact that the mangrove swamps prevent one surveying the area and they make aerial photography difficult to interpret. In such cases it is advisable to make tidal discharge observations in the channels leading to the larger swamps as a means of defining their tidal volume and hence their surface area.

As regards the identification of channels in large deltas, which can only be readily done from satellite images, one has to take into account the resolution of the images. One can represent the most important aspects of the effects of intertidal storage by assuming them to be stagnant zones and it is not usually necessary to model the detailed drainage systems in each tidal flat.

A branched or looped network of one-dimensional elements may be used to simulate tidal processes in quite complicated natural bays and harbors which often consist of storage zones linked together by a system of relatively narrow channels. The storage properties of intertidal flats may be defined from a combination of plans and cross sections. The geometry of the channels in a system is generally determined from cross sections. Measuring cross sections is expensive and slow especially when they have to be related to a common datum in what is often a previously unsurveyed area. A minimum requirement is a cross section at the center and end of each flow or continuity element. If these cross sections are taken from navigation charts they are related to chart datum which varies along an estuary. The spacing of the cross sections and the length of the model elements is a matter of judgment between the ideal and the practical. In practice, we have found the average spacing may vary from as little as 100 m on a small estuary to as much as 8 km in a large tropical delta. A plot of the cross-sectional area as a function of distance along a channel at a standard uniform level is a useful way of defining whether the chosen model element length will be able to resolve the major features of the longitudinal variations in the geometry of the channel, Fig. 4. In any event, the element length should generally be at least several times the width of the deep flow-carrying channel in a one-dimensional model.

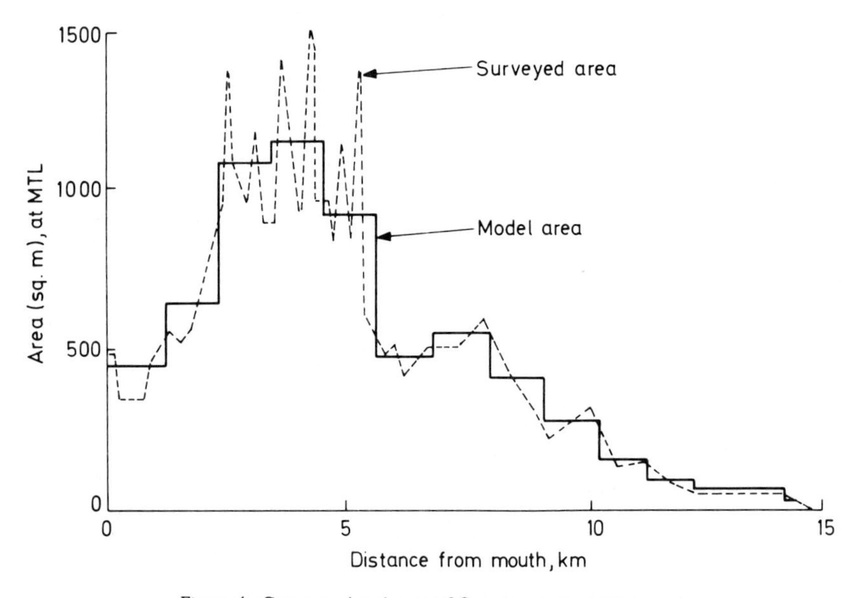

Figure 4. Cross-sectional area of flow in a typical tidal creek.

One inevitably smooths out some of the variations in cross-sectional areas. However, one usually has only one or two cross sections per element. In these cases, the accuracy of the storage properties of the model can be increased by adjusting the value of the mean water surface width using a map or chart to define its variations between cross sections. For most practical tidal calculations, one usually averages each geometrical property such as the cross-sectional area at a given level. Some modelers arrange to have their elements start and end at surveyed cross sections believing that this makes best use of the field data. However, this may give rise to unacceptably rapid variation in the length of elements simply due to the irregular spacing of surveyed cross section. Rapidly varying element lengths reduce the accuracy of the numerical solution. In some applications investigators prefer to average a derived property like a modified hydraulic radius rather than some simple geometric property of the cross section such as the wetted perimeter.

The vertical spacing between tabulated values should be able to resolve sudden changes in water surface width as a function of level which often occurs in some estuaries. It should be noted that in these cases the numerical scheme should be time-centered to avoid errors in the conservation equation if it is based on the water surface width, as is often the case for tidal calculations. It is also important to define the level at which water will spill over natural levees or artificial embankments, especially in the case of tidal surge or fluvial flood calculations. The water level at the head of a tidal river or delta in the wet-season may rise as much as 10 or even 20 m above dry-season conditions, giving rise to a dramatic variation in the storage and flow-carrying capacity of the channel, Fig. 5. The resonance characteristics of estuaries depends strongly on their depth and length, and the inaccurate representation of the geometry can give rise to unrealistic results from one-dimensional models, which is another good reason for taking care over the collection and schematization of geometric data.

4. FRICTIONAL RESISTANCE OF TIDAL CHANNELS

Investigators generally do not take enough care over the representation of frictional losses in tidal calculations. Frictional resistance caused by channel roughness usually has an important influence on tidal flow in estuaries. It controls the peak velocities and generally slows, distorts and dampens the propagation of the tide. The importance of the effect of frictional resistance on the propagation of the tidal flow in a particular estuary system can be gauged by the value of the non-dimensional dissipation factor, D_s, defined by Harder (Ref. 1)

$$D_s = \frac{fT\ U_{\max}}{64R} \tag{17}$$

where, T is the tidal period (s) and $U_{\max}$ is the maximum tidal velocity (m/s).

The dissipation factor is proportional to the ratio between the amount of energy that is dissipated by frictional resistance and the total energy in the tidal motion. A value of D_s of less than 1 indicates that frictional effects are relatively unimportant. A value of D_s greater than about 5 indicates that frictional effects have an important influence on the propagation of the tide. In modeling a tidal regime with a high dissipation factor special attention must be paid to the manner in which it is represented in the calculation.

In many shallow estuaries the flow on the late part of the incoming tide is often controlled by its inertia (momentum), whereas it is almost entirely friction controlled on the longer ebb tide, when the estuary gradually drains out.

The form of the energy loss term is invariably based on the assumption that the instantaneous velocity distribution is the same as the logarithmic profile of steady flow. This assumption is usually valid in fast flowing, well-mixed estuaries. In deep or stratified estuaries when the velocity profiles are often S-shaped, the assumption is obviously not valid, but it seldom matters because the exact form of

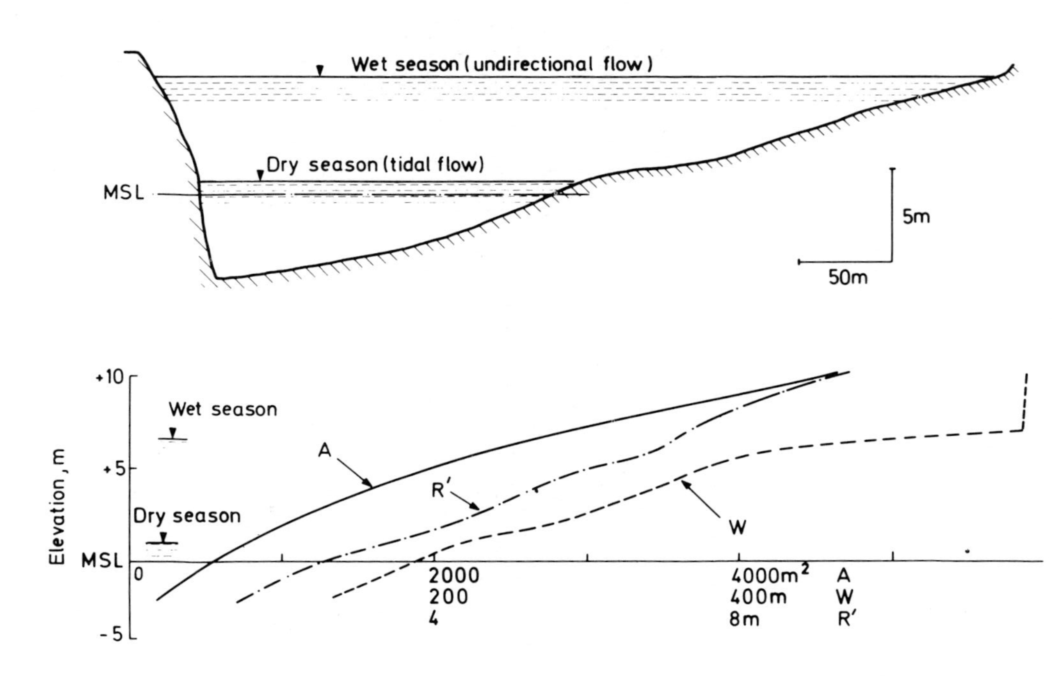

Figure 5. Hydraulic variables as a function of elevation in a tropical tidal delta channel.

the energy loss terms is relatively unimportant in deep slow moving flows.

The simplest empirical resistance functions, such as the Chezy formula, provide an inadequate description of frictional resistance in shallow estuaries with an appreciable tidal range. In order to prove mathematical models based on these types of resistance laws, investigators have often been forced to make the friction coefficient vary in an apparently arbitrary manner with depth, or even with the direction of flow. The situation becomes even worse if at the same time the model is based on an approximate channel geometry, such as an equivalent rectangular section, because the optimum friction coefficient may vary with the plan of schematization. The proving stages of such models are long and tedious and open to the criticism of forcing a solution from an inadequate model.

For tidal channels with immobile or effectively rigid beds, it is advisable to assign separate roughness values for the bed and bank perimeters of each model element, based on the concept of equivalent spherical grain-size roughness, k_s. Separate roughnesses are given because in the lower parts of many estuaries the banks are of relatively smooth mud while the bed is rougher, whereas in the landward reaches the bed is often smooth mud and the banks are rough tufted grass slopes. The D'arcy Weisbach friction, f, is evaluated separately for bed and banks at every time step using the Colebrook-White transition law for open channels, together with the prevailing water depths and tidal velocities. Then a composite friction factor for the whole channel cross section is evaluated by weighting the relative lengths of the wetted perimeter of the bed and the submerged parts of the banks. Experience has shown that the composite friction factor of a typical estuary channel with a rigid bed may vary by as much as thirty percent during the tidal cycle, although the surface roughness remains constant. The change is often in excess of that predicted by relative roughness effects and is attributed to the relatively large areas of bank that become submerged at high water.

The friction factor of a sandy river channel often falls dramatically during the passage of a fluvial flood. The great reduction in channel roughness during high flows can usually be attributed primarily to sediment transport processes smoothing out ripples, dunes and bars on the river bed, while the damping effect of the suspended sediment is of secondary importance. The same phenomenon is thought to occur in sandy tidal channels during the passage of the incoming and outgoing tide.

The roughness of a sandy tidal channel falls as the tidal currents increase until the bed becomes effectively smooth. The small-scale ripples are the source of the roughness. Their effective roughness, k_s, is solely dependent on a mobility number based on the mean depth velocity,

$$k_s = \text{function}\left[\frac{U^2}{gD(S-1)}\right] \tag{18}$$

where S is the specific gravity of sand, provided the sand grain size D is between 0.15 and 0.30 mm, the Froude number is less than 0.3, the depth of flow is very large compared to the grain size, and the changes in the rate of flow are slow compared to the rate of growth or decay of the ripples.

This appears to be the case in most estuaries except during periods of flow reversal. Ripples adapt themselves fairly quickly to changes in tidal velocity. The relationship between the effective roughness of the ripples and the mobility of the sand can be used to evaluate the resistance to flow in a sandy tidal channel, taking into account the gradual changes in the tidal velocity and depth of flow. This method will overestimate the change in the friction factor which occurs during brief reductions in the flow velocity at slack water, when the bed forms are probably frozen.

The presence of dunes in a tidal channel is easily detected on normal echo-sounding traces. Their occurrence is unlikely in channels with beds of very fine sand ($D_{50} < 0.15$ mm), which usually only exhibit ripples and are therefore more amenable to the aforementioned method. Dunes that occur in tidal channels are usually fairly gentle undulations of the bed that remain almost constant in height and length throughout the tidal cycle. They may give rise to appreciable resistance if the water depths become comparable to their height (0.1 - 0.8 m). The computed velocity for an element in a numerical model is often much lower than the depth-mean velocity in the mid-region of the channel. If the mobile bed resistance function obviously overestimates the bed form roughness effect, a coefficient can be introduced into the calculation. This coefficient is the ratio between the mean velocity over the sand bed in the mid-region of the channel and that over the whole cross section. The value of this coefficient may be determined from field observations or calculations, or its value may be optimized to produce the best simulation for a range of tide and river flow conditions. It is unlikely to exceed the value of 1.5 in regularly shaped channels.

The trial and error process of proving a tidal model by adjusting the roughness values in each element until the correct pattern has been reproduced can be lengthy. An estimate of the value of the composite friction factor and the equivalent surface roughness in each element can be readily determined by carrying out a cubature and associated calculations based on the equations of continuity and motion using the averaged geometric functions of the schematic model. For good results the coverage of the tidal level observations has to be close enough to allow accurate spatial interpolation of water levels at each model mesh point. Surface roughness values determined in such a manner have been used to prove a model of the Yorkshire Ouse system of tidal estuaries, with over 150 km of shallow channels, in days rather than weeks or months.

The method has an additional advantage that significant mobile bed roughness effects should show up in the way the calculated composite friction factor varies during the tidal cycle.

Bend losses in even very tortuous estuaries are much less important than boundary roughness effects provided the Froude number is low. This ceases to be the case in the upper reaches of estuaries during large fluvial floods. The bend losses in the upper reaches of the Brisbane tidal river probably account for about twenty-five percent of the friction losses during peak flood stages.

5. OCEAN TIDES AND FLUVIAL DISCHARGES

The predictive ability of a tidal model depends to a great extent on the accuracy of the predicted tide curve at the seaward boundary of the model. The ideal position of such a boundary is at a point where the narrow channel suddenly widens into the open ocean. However, even in these cases there is sometimes a sand bar on a constriction which causes a local head loss significantly altering the amplitude of the tide over a short distance at the mouth of the estuary. Ideally, harmonic constants determined from several years of continuous tidal records are necessary. In practice, a minimum of 29 days continuous tidal level observations will suffice. Certain tidal processes such as saline intrusion are sensitive to the level of high water and there may be significant differences between hindcast tidal levels and observed tidal levels at the mouths of estuaries, Fig. 6, especially if there are a large number of tidal constituents. Any tidal level record contains variation in mean tide level caused by meteorological effects. These effects are sometimes so strongly seasonal that they can be represented by equivalent annual and bi-annual tidal constituents, as in the case of the Suez Canal. If the seaward limit of the model is not located on the edge of a large body of water, engineering works within the tidal system may influence the tide boundary. In this case it may be necessary to construct a two-dimensional model of a larger area to evaluate the changes at the boundary of the inner model. In the case of tidal surge predictions it is necessary to define the surge residual at the mouth of the estuary which may be done from observations or again may be derived from two-dimensional storm surge models.

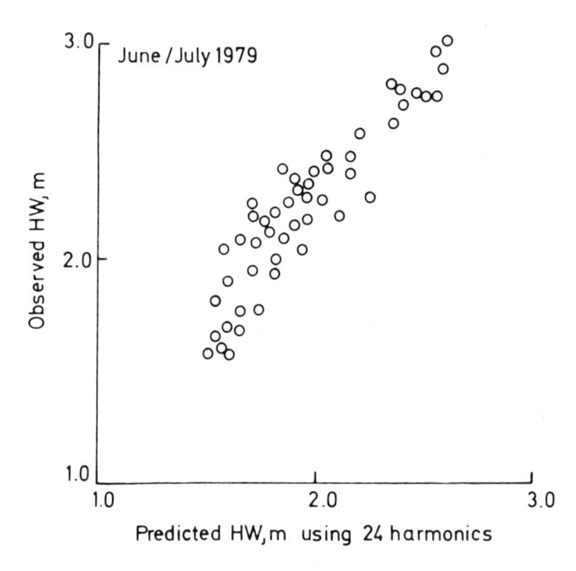

Figure 6. Comparison between observed and predicted HW levels at the mouth of an estuary.

The characteristics of the fluvial hydrographs play an important role on the problem of interactions between the saline and fresh water in a tidal channel system. If the hydrograph is flashy with rapidly varying flows, flood levels may be sensitive to the amount of off-channel storage available in the system. However, one can often treat the low fluvial flows as steady for the purpose of tidal calculations. Ideally many years of river discharge data are needed and it is always important to establish river gauging stations early in an investigation of a previously ungauged estuary.

6. PRACTICAL CONSIDERATIONS AFFECTING THE CHOICE OF NUMERICAL METHODS

The work of university departments, research establishments and consulting engineers throughout the world demonstrates that there is a multitude of different numerical solutions to what appears to be the same set of equations. Several attempts have been made to compare the performance of different methods in terms of the accuracy, economy, adaptability and general usefulness. These comparisons, useful as they are, have not discovered any one numerical method that has outstanding advantages in all situations. In fact, it is probably true to say that in an estuary where the tidal range is small relative to the mean depth almost any second order accurate method will give a reasonable answer in terms of water levels. The engineer should be open-minded in this matter., even if practical considerations, such as the need to make use of a costly suite of programs, limit him to the same method most of the time. Although in each new investigation the basic equations appear to be similar, the relative importance of the various terms may be quite different.

In recent years the estuary division at HRS have used an implicit 6-point scheme almost exclusively in their one-dimensional tidal calculations, (Ref. 2). The scheme essentially uses a sparse grid in space with overlapping continuity and conveyance elements and overlapping solutions in time to provide good time-centering of the water levels for determining the water surface width, which often varies rapidly as a function of level in British estuaries. The improved time-centering of an overlapping scheme is generally considered to compensate quite adequately for the increased computational time. One of the solutions is treated as a secondary predictor and it is repeatedly corrected by averaging from the main solution of the end of each main time step. An alternative method would be to iterate, but in this case a closing error has to be defined, which may necessitate several iterations to achieve an accurate solution in one element of the model. The fluvial hydraulics division at HRS use 4-point schemes almost exclusively. The 6-point scheme has a considerable advantage over 4-point schemes for tidal calculations in the tri-diagonal matrices which are easily solved by simple recurrence relationships even in the case of complex channel systems can be formulated and the 6-point scheme is well-adapted to solving the advective diffusion equation as regards space and time-centering. It also has simple boundary conditions, namely, a tidal level at seaward boundaries and a discharge at the landward boundaries. The main disadvantages of the 6-point scheme as compared to a 4-point scheme are the difficulty of representing the convective acceleration term, the peculiar shape of continuity elements at junctions and the greater difficulty of including hydraulic structures placed across the channel.

The extremely large tidal range and shallowness of British estuaries give rise to tidal bore effects which severely test all types of numerical schemes. The problem of calculating flows in complex looped and branched systems such as tidal deltas is handled by setting up recurrence relationships for the individual reaches and solving for the tidal levels at the junction separately with a junction matrix. The unknown levels at each of the junctions is related to the unknown levels at adjacent junctions via the recurrence relationships for each reach. In the case of large deltas most of the computational effort goes into the solution of the junction matrix.

7. JUDGING THE QUALITY OF RESULTS FROM AN ESTUARY MODEL

AT HRS the methods of proving and validating one-dimensional estuary models as regards tidal propagation during low fluvial flows have gradually changed. Firstly, the best possible schematic representation of the geometry of the channel system is attempted and, secondly, the correct mechanisms of frictional dissipation for the particular estuary in question is modeled. As far as possible the bed roughness values are predetermined in each reach of the system from a knowledge of the local bed sediments, k_s being zero for smooth mud, variable for fine sand and constant with a value of about 0.2 m for a wide range of rigid gravel and rough beds. The k_s value may also be determined by a cubature or by other means.

Frictional effects are most important during spring tides. A repeating spring tide cycle of 12.5 or 25 hours, depending on the relative magnitude of the diurnal and semi-diurnal tidal constituent, is simulated and the model results are compared with tidal level observations. These types of runs are short and the k_s values are adjusted to get the best results in terms of the phase and level of high and low water throughout the tidal system under investigation.

For the next part of the proving process, the model is run to a state of dynamic equilibrium with a full neap-spring tidal cycle lasting 15 days. The fluvial flows are kept constant and the model is run until the effects of any incorrectly prescribed initial conditions have disappeared. A species analysis is then applied to the computed tidal levels, tidal velocities and sometimes tidal discharges.

The phase and amplitude of the main tidal constituents simulated in the model are plotted out as a function of distance along the estuary and compared with observed values at particular points, Fig. 7. The method does require continuous, but not necessarily simultaneous, tidal observations at several points throughout the tidal system. However, this type of observation is easier to mount and supervise than simultaneous manual observations during particular tides.

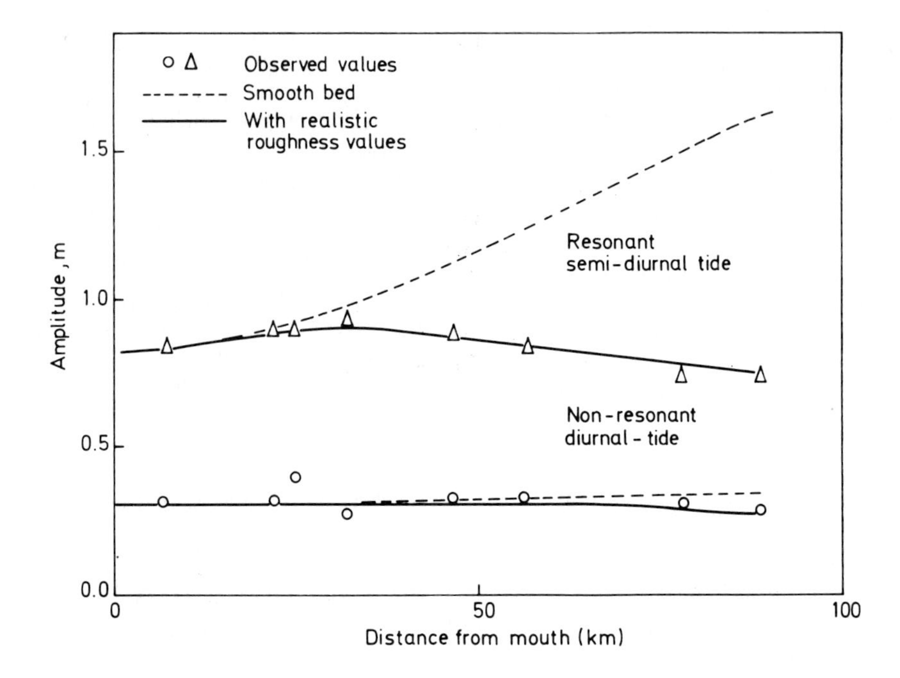

Figure 7. Effect of setting realistic roughness values on simulated tidal motion.

The advantage of this method is that it shows whether the model simulates the important features of the tidal motion, which are not necessarily obvious from a comparison of local tide curves, the traditional method of proving a one-dimensional model. The quality of model results can appear to be quite poor when comparing local tide curves. This often occurs due to poor tidal level observations and imperfectly prescribed boundary conditions, both common problems, as well as imprecise representation of the geometry of the estuary in the model.

However, when a model simulates the important features of the tidal regime as regards the propagation of all the main tidal constituents it is a useful predictive tool. In addition to proving the model the method of species analysis allows a good understanding of the tidal motion in the estuary, normally considered to be the main advantage of analytical models, in terms of resonance, shallow water, frictional and water density effects during periods of steady fluvial flows.

The aforementioned method is not suited to proving models set up to simulate and predict the interaction of tides and fluvial floods in the landward reaches of tidal channel systems. In this case, it is still better to simulate several fluvial flood events in detail to ensure that the model has the correct conveyance and storage characteristics.

A tidal estuary model that takes into account all the important processes that are taking place in the particular estuary and simulates the main features of the tidal motion can be trusted. It is sometimes argued that adjusting the frictional coefficients forces the model to give any desired answer. However, in practice, this is not the case if the geometry of the channel system is known and if the frictional resistance is calculated on a solidly based physical theory. The structure of the equations of motion and continuity determine the form of the solution. The main problem lies in getting the correct distribution and type of roughness where it varies along an estuary. The frictional resistance of tidal channels is still a subject of physical research, and it is true to say that investigators have sometimes in the past been forced to make the friction factor vary with the height of the tide or the direction and magnitude of the tidal current in an apparently arbitrary way to prove a model. The necessary physical research is being done and the frictional resistance of channels can now be prescribed more readily in terms of physical processes such as sand transport or the effect of bends or stratification.

One-dimensional models are used to predict the effect of engineering works at a feasibility study stage of a project. In these cases the field data is usually very sparse, especially if the estuary is in a developing country. One-dimensional estuary models are often the only practical means of planning and designing engineering works and they are no worse and possibly better than many of the techniques used to predict stresses in structures such as buildings and dams. For example, the most sophisticated structural analysis may assume Hooke's law for concrete and make gross assumptions about the transfer of forces at joints and the effect of cladding and internal walls. One of the big differences between structural and hydraulic calculations is that in the latter case it impractical to design works with safety factors as large as those often used in structural calculations.

The quality and predictive ability of one-dimensional estuary models as regards tidal motion depends on the degree of complexity of the cross sections of the channel and the intertidal regions and the accuracy with which the flow and storage properties of the channel are represented as a function of distance and water level. Special care should be taken over the representation of frictional losses in estuaries with a high dissipation factor. The distortion of the tide curve due to shallow water effects, quantified by the quarter diurnal tidal constituents, is one of the most difficult processes to simulate in a model.

8. USES OF ONE-DIMENSIONAL MODELS TO SOLVE ENGINEERING PROBLEMS IN ESTUARIES

In recent years the Hydraulics Research Station at Wallingford has used one-dimensional tidal models to determine the effect of large power stations abstractions, flood control schemes, tidal surges, proposed weirs, sluices and barriers, dredging and the reconstruction of existing channels and extension to existing channel systems on tidal levels and overbank flooding, (Refs. 3-11 and 16).

Design and feasibility studies of engineering works now tend to involve both mathematical and physical models in complementary roles. The mathematical model may cover a relatively large network of channels surrounding the region of main interest such as a bridge crossing or a tidal sluice. The focal region may in turn be represented in a steady state physical model or in another more detailed multi-dimensional mathematical model, whose boundary conditions are supplied by the one-dimensional model.

It is easy to simulate the performance of common control structures, such as weirs and sluices, in a computer program which enables one to optimize their main dimensions by a series of tests.

One of the problems of analyzing the output of numerous test runs is the tremendous volume of results which may be derived from unsteady flow calculations even in the case of one-dimensional models. The use of species analysis of results eliminates the time element and produces more simple and readily understood diagrams, such as Fig. 8. Unlike analytical solutions, which usually give the trend of effects resulting from a change in conditions, numerical calculations usually require a series of comparative predictive tests before one can determine the effect of a change such as an increase in channel depth. However, the results from a series of numerical predictions can usually be analyzed and presented in a monograph. Besides predicting the change in water levels that is likely to occur as a result of engineering works it is also important to be able to explain and demonstrate the cause for the changed conditions. There is considerable scope in devising means of analyzing the results from numerical models towards this end.

9. SIMULATION OF TRANSPORT PROCESSES IN ONE-DIMENSIONAL ESTUARY MODELS

The satisfactory planning and design of engineering works in estuaries often requires a prediction of changes on the auxiliary tidal processes of saline intrusion, thermal or oxygen balance and sediment transport, (Refs. 12-15). One-dimensional models simulate more or less correctly mixing processes caused by the joining and parting of flows at junctions at different stages of the tide and the temporary storage of water on stagnant intertidal flats. However, the correct representation of both these processes depends on the accuracy of the schematization of the plan and geometry of a given channel system. The total effect of all processes causing longitudinal dispersion are generally represented in such models using the concept of a coefficient of effective longitudinal dispersion. This gross assumption has an overriding influence on the predictive ability of one-dimensional estuary transport process models. The resulting advection diffusion equation is as follows

$$\frac{\partial}{\partial t}(AC) + \frac{\partial}{\partial x}(QC) - \frac{\partial}{\partial x}(AD_T \frac{\partial c}{\partial x}) + KAC = 0 \tag{19}$$

There is no generally accepted theory or empirical formulation for prescribing the value of the effective coefficient of longitudinal dispersion, D_T, in terms of

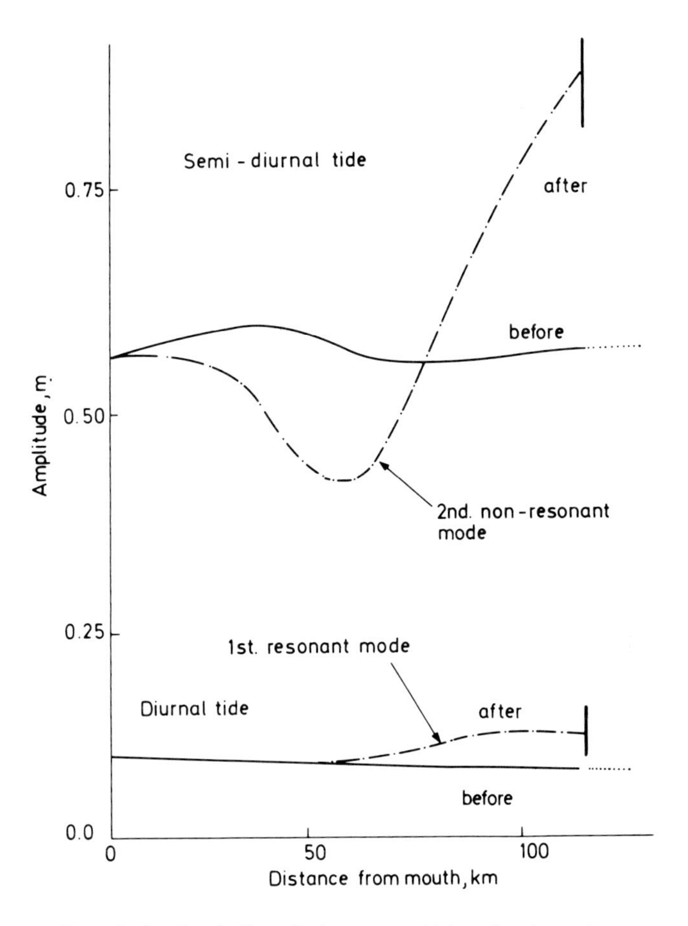

Figure 8. Predicted effect of a barrage on tidal motion down stream.

local area averaged parameters. The value of the coefficient is more predictable when the solute is uniformly mixed over the cross section. This condition is approached in relatively narrow turbulent estuaries, especially during periods of low fluvial flows, which are of most interest to the engineer.

The practice in the estuary division at HRS at present is to prescribe the value of D by the following type of function, which is loosely based on the work of Sanmuganathan, (Ref. 17)

$$D_T(x) = D_1 \frac{A(x,t)\ U(x,t)}{A(o)\ U(o)} + D_2 \frac{dS}{dx} \cdot \frac{L}{So} \tag{20}$$

The coefficients D_1 and D_2 are adjusted to give a best-fit between model simulation and observations, Fig. 9. If observations are available during a period of steady fluvial flows, it is sometimes possible to pre-determine a mean tide value for D_1, which helps define the right range of values for D_1 and D_2. The fact that these are two unknown coefficients means that even with this formulation there is a wide range of combinations which could give approximately the same answer as regards the salinity at a given point in the upper estuary. This is best illustrated from a consideration of steady state tide averaged conditions at which the salinity at a point is given by the following integral

$$S(x) = S(o)\exp\left[-\int_0^x \frac{Qf\,dx}{A(x)D_T(x)}\right] \tag{21}$$

Both the shape of the longitudinal salinity distribution as well as the limit of saline intrusion has to be considered (Fig. 10).

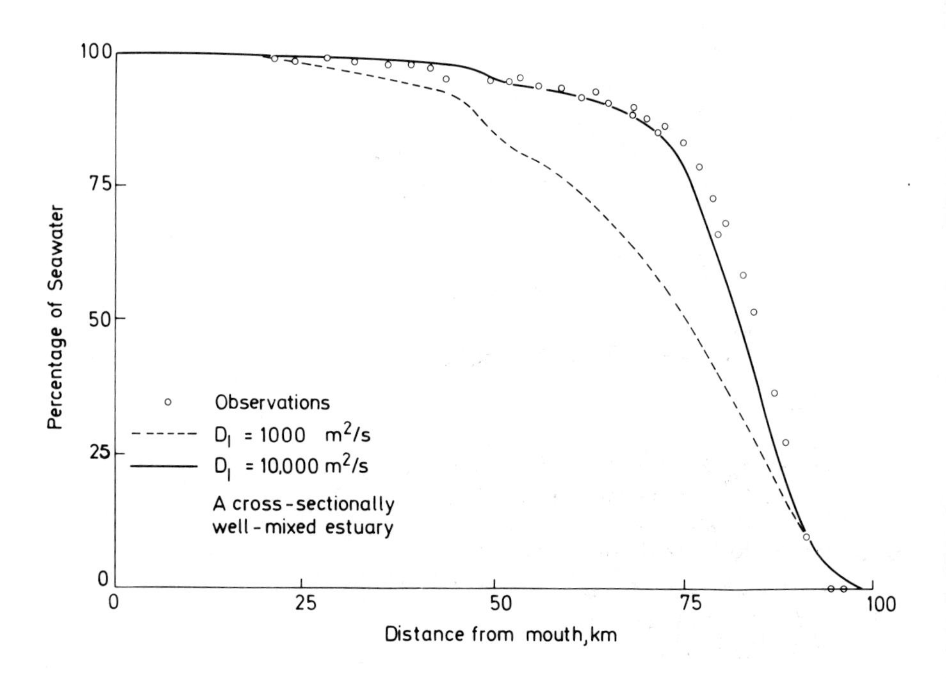

Figure 9. Sensitivity of mean tide salinity distribution in an estuary to the value of D_1.

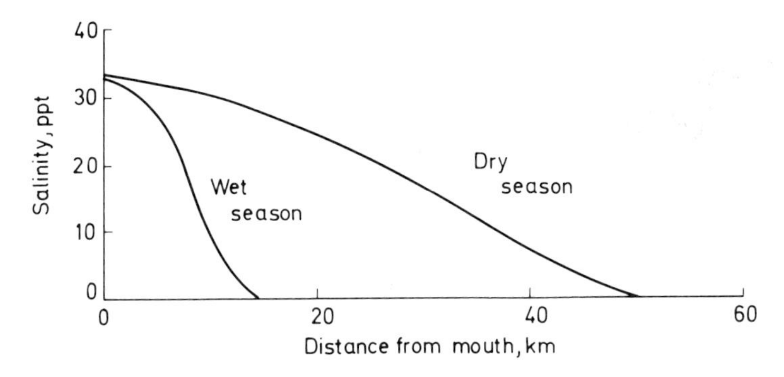

Figure 10. Typical wet and dry season salinity distributions in a tropical estuary.

The transport of other solutes, such as dissolved oxygen or suspended mud, are generally assumed to be governed by the same diffusion process with the same value of a diffusion coefficient as that required for the correct simulation of saline intrusion. This is obviously a gross assumption if they are distributed differently over the cross section compared to salinity.

The other major problem arising in the simulation of transport processes in one-dimensional estuary models is the so-called numerical dispersion that can arise from an inaccurate numerical solution of the advective term. This problem becomes most acute in the case of estuaries with a large tidal excursion and for conditions where there are high longitudinal gradients in the area mean concentrations of the solute. The longitudinal salinity gradients in even cross-sectionally well-mixed estuaries can be extremely steep, for example, salinity may fall from full sea water strength to almost fresh water in a few kilometers in a small estuary. In such conditions the chlorinity, which is the critical factor as regards pollution of intakes, varies from as little as 20 ppm in river water to about 20,000 in sea water, Fig. 11. The critical value for irrigation purposes is about 300 ppm. One is therefore often trying to simulate the movement and shape of the landward toe of the salinity distribution in an estuary. Longitudinal variations in dissolved oxygen and pollutants and even suspended mud are generally much smaller and therefore the prediction of their motion is less sensitive to numerical dispersion problems.

The predictive ability of one-dimensional estuary models as regards the transport of dissolved and suspended matter is therefore strongly dependent on the reliability of the function prescribing the value of coefficient of longitudinal dispersion and in some cases on the accuracy of the numerical solution, especially as regards the advective term.

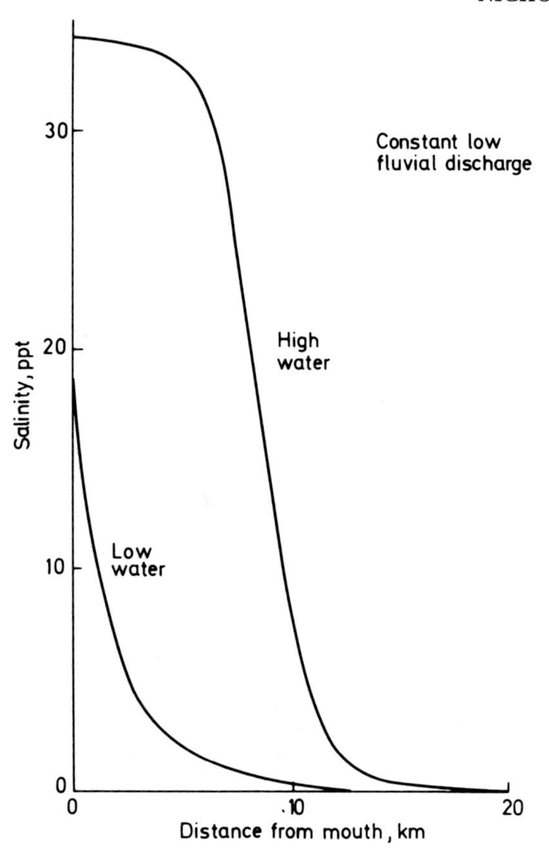

Figure 11. Typical HW and LW salinity distributions in a shallow estuary with a large tidal range.

9.1 Sediment Transport

There are not many conditions for which one-dimensional estuary models can be used to simulate sediment transport. However, HRS has used such models to simulate and predict siltation in the landward reaches of shallow British estuaries during periods of negligible fluvial flows. In the case of the Great Ouse Estuary which has an artificial channel with a bed of very mobile fine sand decreasing in size in the landward direction, the flux of each sand fraction was calculated as a function of the local area mean velocity, and the proportion of the bed surface covered by that sand fraction. This fraction varied during the course of the calculation. The main uncertainty centered on the ratio between the effective depth-mean velocity over the flat mobile bed and the area mean velocity over the whole cross section, because the sand transport function was proportional to a high power of the tidal velocity.

The bed of the channel, defined as the cross-sectional area between the bed surface and inerodible stratum below, was divided into layers containing differing sand mixtures. The proportion of the bed surface available for erosion, was assumed to be related to the volume of each sediment fraction contained in the exposed horizon. Erosion of the sub-surface horizons were not allowed to take place until all the sediment of whatever type had been removed from the surface horizon. This allowed a thin layer of coarse sediment or cohesive mud to protect or armor an underlying layer of easily erodible material. The model solved the following equation governing the conservation of sediment volume in the surface horizon

$$W(\eta_{\text{bed}})\left\{\frac{\partial \eta_{\text{bed}}}{\partial t} - \frac{G}{\gamma D_{\text{mud}}}\right\} + \frac{1}{\gamma s}\sum_{i=1}^{N}\frac{\partial q_{s_i}}{\partial x} = 0 \tag{22}$$

where, G is the vertical flux of mud to and from the bed surface caused by deposition or erosion; γD is the dry density of a sediment deposit in the bed; q_{s_i} is the horizontal flux of a sand fraction; N is the total number of fractions.

In the absence of a significant amount of mud transport it was found possible to accelerate the sand transport calculation with respect to the flow and thereby simulate long periods of siltation in a relatively few tidal cycles. The main problem with the numerical solution of Eq. (22) is the growth of alternatively high and low bed levels in successive elements. This was caused by the use of spatially averaged velocities in the sand transport function and problems with sand passing over inerodible reaches of the estuary and the rapidly changing availability of a given sand fraction in the bed surface. It was solved in an *ad hoc* manner by determining what sediment was available in the up-flow element. It was not possible to use a dissipative scheme such as the Lax-Wendroff scheme because of the variability of the surface sediments. The use of information from an up-flow element has the unwelcome effect of making the net transport (i.e., the difference between flood and ebb) in a rapidly narrowing tidal channel a function of the model element length. The predictive ability of one-dimensional estuary models as regards sand transport is therefore strongly dependent on the physical representation of transport processes and the numerical solution of the conservation equation.

10. CONCLUSIONS

The main factors affecting the accuracy and predictive ability of one-dimensional estuary models are: (a) The schematic representation of the geometry of the tidal channel system; (b) The representation of frictional dissipation in the case of estuaries with a high dissipation factor; (c) The empirical representation of the physical processes causing longitudinal dispersion of solutes; (d) The accuracy of the numerical treatment of the advective terms especially in the case of steep longitudinal gradients; (e) The methods of proving and using such models especially the formulation of comparative predictive tests; and (f) The method of analyzing the results, which should lead to a good understanding of the causes of any changes predicted by the model.

Acknowledgements

This paper is published with the permission of the Director of Hydraulics Research Station, Wallingford, United Kingdom.

REFERENCES

1. Harder, J.A. (1963). The Prediction of Tidal Flows in Canals and Estuaries, Part II. University of California, Berkeley, California.
2. Maskell, J.M. (1975). A One-Dimensional Implicit Model Incorporating Species Analysis. HRS Report INT 145.
3. HRS (1969). Derwent Tidal Sluice: The Effect of the Construction of the Sluice on Peak Water Levels in the Yorkshire Ouse Tidal System. Report 428.
4. HRS (1969). Port Rashid, Dubai: Dubai Creek Tidal Calculation. Report EX 453.
5. HRS (1970). M27 Tipner Lake Bridge Tidal Calculation. Report EX 482.
6. HRS (1970). River Boyne Drainage Scheme: Mathematical Model Studies. Report EX 500.
7. HRS (1974). Simulation of Tidal Processes in the Great Ouse Estuary. Report DE 18.
8. HRS (1975). Orfordness Nuclear Power Station: Numerical Studies of Ore/Alde Estuary. Report EX 682.
9. HRS (1976). Suez Canal Feasibility Study: The Effect of Deepening and Widening the Canal on Tidal Currents. Report EX 731.
10. HRS (1977). Gambia Barrage Study: Effect of the Barrage on the Tidal Regime Downstream. Report EX 795.
11. HRS (1978). Port of Brisbane Siltation Study: Tidal Propagation in the Brisbane River in the Dry Season. Report EX 829.
12. HRS (1977). Sha Tin New Town: Mathematical Model Studies of Pollution in Shing Mun Tidal Creek. Report EX 802.
13. HRS (1979). Great Ouse Estuary: The Effects of the Reduction and Redistribution of Fluvial Flows on Siltation, Flood Evacuation and Land Drainage. Report EX 865.
14. HRS (1979). Shannon Estuary Thermal Study: The Effect of a Cooling Water Discharge on the Thermal Balance of the Estuary. Report EX 868.
15. HRS (1979). Brunei Water Resources Study: The Effect of Increased Freshwater Abstractions on Saline Intrusion in the Tutong and Belait Estuaries. Report 881.
16. HRS (1980). Lower Burma Paddyland Development Project: Mathematical Model Studies of the Irrawaddy Delta. EX 929 (first report)
17. Sanmuganathan, K. (1975). One-Dimensional Analysis of Salinity Intrusion in Estuaries. HRS Report OD/2.

PREDICTION OF FLOW AND POLLUTANT SPREADING IN RIVERS

W. Rodi, R. N. Pavlović and S. K. Srivatsa

Sonderforschungsbereich 80 and Institute for Hydromechanics
University of Karlsruhe, Karlsruhe, FRG

1. INTRODUCTION

1.1 The Problem Considered

The discharge of thermally, chemically or biologically polluted waste water into rivers causes ecologically endangered zones with relatively high pollutant concentration stretching along the discharge bank for many kilometers. Work is in progress at the Sonderforschungsbereich 80 of the University of Karlsruhe on the development of mathematical models for simulating the pollutant spreading and consequently the effect of the pollutant discharge on the water quality. The models should be able to predict, for given river and emission conditions, the distribution of pollutant concentration with an accuracy sufficient for practical purposes. The distribution of pollutant concentration is governed by two mechanisms: heat as well as chemical substances can be transported from one point in the river to another by the mean motion of the flow (convective transport) as well as by the turbulent motion, and the latter process is also called turbulent diffusion. Both processes must be simulated correctly by a mathematical model, which means that the mean-flow field must either be known or must in general also be determined by the mathematical model.

The discussion in this paper is restricted to side-discharges, the only type of discharge occurring in Germany, but similar phenomena may be found when the waste water is discharged from the bottom of the river. The side-discharge situation is sketched in Fig. 1, which shows that the discharge jet is bent over by the river flow, which in turn is bent away from the bank by the jet. The discharge jet entrains water, which leads to the formation of a recirculation zone behind the jet because, near the bank, the entrainment water has to be provided from downstream. In the immediate vicinity of the discharge, a very complex flow field is therefore set up. Outside the relatively small recirculation zone, however, a predominant flow direction prevails, and this fact is of great importance for the development of economical mathematical models. Apart from the occurrence of reverse flow, the flow in the vicinity of the discharge is usually complicated further

ISBN: 0-12-25852-0

by three-dimensional effects which are due either to the discharge geometry or to buoyancy effects arising from a density difference between the discharged waste water and the river water. Especially in the case of cooling-water discharges, a significant stratification can be set up near the outlet as is indicated in Fig. 1; the cross-sectional isotherms illustrate how the discharged warm water rises to the surface and spreads out laterally. As the temperature can vary significantly with depth in such a situation, the problem is strongly three-dimensional. Usually, however, the stratification is eroded rather quickly by the turbulence in the river which evens out vertical non-uniformities by way of vertical mixing. This soon leads to approximately constant temperature with depth, as is indicated by the vertical isotherms in Fig. 1. In such well-mixed regions only the horizontal distribution of heat and other pollutants is of interest so that the problem can be treated as two-dimensional.

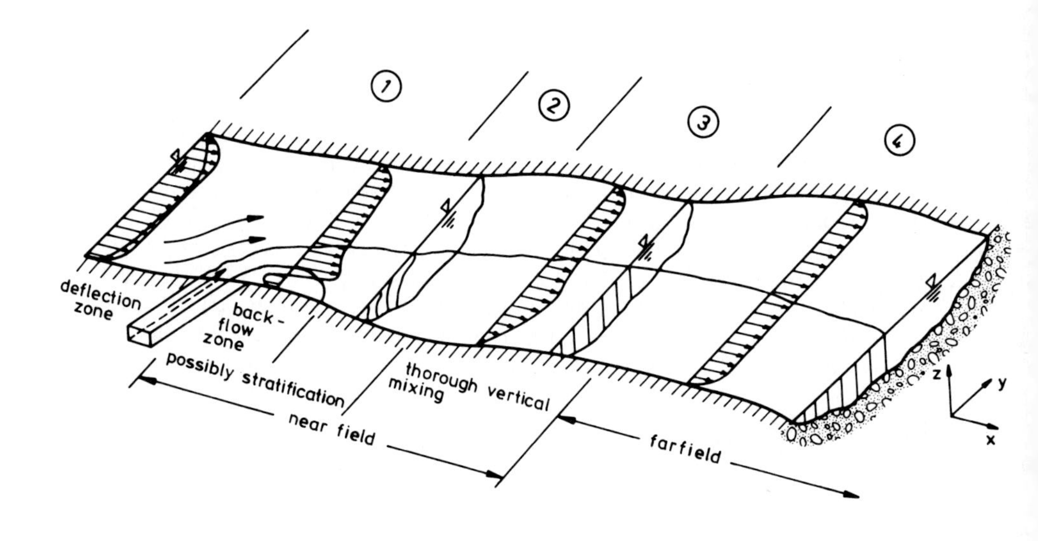

Figure 1. Typical discharge into a river with characteristic flow regions.

The discharged jet influences the flow field in the river only within a certain distance downstream from the outlet. This region is called the near-field, and here the turbulence stems partly from generation at the river bed but also from generation within shear layers induced by the discharge, and hence the turbulence behavior in the near-field is particularly difficult to simulate in a mathematical model. From a certain distance, the flow in the river is no longer influenced by the discharge, and this region is called the far-field. Here the turbulence is governed entirely by the generation at the river bed.

Special situations may occur in both the near- and the far-field which were not included in the sketch of Fig. 1. In the further discussion, however, only rivers without groynes and bends are considered, and unsteady situations are also excluded.

1.2 The Basic Modeling Concept

The aim of this research is the development of a mathematical model to simulate all the physical phenomena described above, with special focus on the near-field phenomena. In this section, the model approach followed is outlined briefly. Because of the complexity of the near-field phenomena, integral methods based on profile assumptions and entrainment laws are not generally suitable, and diffusion methods, which assume the velocity field to be known, are only applicable to the far-field. Hence the method adopted is the so-called field-method which is based on the time-averaged partial-differential equations governing the mean-flow quantities, including those for the velocity field. The turbulent transport terms occurring in these equations have to be determined with the aid of a turbulence model as discussed below.

Considering the complete region of interest, the discharge problem is generally three-dimensional and the governing equations are elliptic in nature. The numerical solution of three-dimensional elliptic equations requires much computer storage and time because the flow field must be discretized in all three directions (leading to large storage requirements) and the elliptic equations require an iterative solution procedure (leading to large computing times). Therefore, only rather coarse numerical grids can be used with present computers, which leads to a poor resolution of the flow field. The presently achievable resolution is by far not sufficient to simulate the various phenomena in the river section sketched in Fig. 1. The solution of the general three-dimensional elliptic equations for the whole river stretch of interest is therefore not feasible at the present time.

In many river zones, certain physical phenomena dominate over others so that the equations can be simplified accordingly. Two simplifications can be particularly significant. Firstly, the dimensions of the problem can be reduced and secondly, in certain regions, the elliptic equations can be simplified to parabolic ones, which are amenable to the much more efficient noniterative solution procedure of marching-forward integration. A short description of these simplifications will be given here; a more detailed account is given in Rodi (1980a).

Except in relatively small recirculation regions, the flow in the river has a predominant direction and turbulent transport in the main-flow direction is negligible compared with that in the cross-sectional plane. Thus, in most parts of the river, influences cannot be transmitted upstream by the turbulent diffusion nor by the mean flow. When the transmission of pressure effects from downstream to upstream points is also unimportant, the equations can be made parabolic in the main-flow direction (x-direction in Fig. 1). The solution of the parabolic equations at a certain river cross section does not depend on the solution at cross sections located downstream nor on the conditions at the outflow-boundary.

Therefore, the equations can be integrated in a numerical scheme by marching from one cross section to the next, starting with given initial conditions at the cross section furthest upstream. The marching-forward procedure is particularly economical because the flow field is covered only once during the solution procedure and all the variables have to be stored only at grid nodes in one cross section, so that only two-dimensional storage is required. A three-dimensional parabolic finite-difference procedure of this type has been developed by Patankar and Spalding (1972) and has been applied to open-channel-flow situations by McGuirk and Spalding (1972) and by Rastogi and Rodi (1978). As part of the simplification to parabolic equations, the pressure gradient with respect to x (main-flow direction) appearing in the longitudinal momentum-equation is assumed to be constant over each cross section. Further, the corresponding width-average surface elevation has to be known in subcritical flows from a one-dimensional backwater profile calculation.

Close to discharges, the upstream transmission of influences due to the action of pressure is often important. It leads, for example, to the bending of the river flow upstream of the discharge, as sketched in Fig. 1. Such influences can be accounted for in a so-called partially-parabolic procedure, in which the pressure is treated elliptically, but all other flow variables parabolically, as before. In this case, the pressure is stored three-dimensionally. Starting with an estimated pressure-field, the marching-integration is carried out several times, each time improving the pressure-field until convergence is achieved. The main advantage of the partially-parabolic procedure over a fully-elliptic procedure is that three-dimensional storage is required only for the pressure variable but not for the other variables, which require only two-dimensional storage. Such a procedure was developed by Pratap and Spalding (1978) and was applied by Leschziner and Rodi (1979) to strongly curved open-channel flow, where the upstream transmission of pressure effects is also very important. The basic procedure is restricted to flows without any reverse flow and was therefore extended by Rodi and Srivatsa (1980a) to a locally-elliptic procedure in which flow-reversal may occur in a small predefined part of the flow field. In this small embedded region, all variables are now stored three-dimensionally.

Simplification of the three-dimensional equations to two-dimensional ones is possible under well-mixed situations when temperature or pollutant concentration vary little over the river depth. Two-dimensional depth-averaged equations can be derived formally by integrating the three-dimensional equations over the depth. In general, the depth-averaged equations are elliptic and a number of well-tested numerical procedures are available for the solution of these equations. In this case, the computational effort is already reduced significantly compared to the 3D-elliptic procedure so that further simplification to a partially-parabolic or locally-elliptic method is not necessary. One example of a 2D-elliptic finite-difference procedure is that of Gosman and Pun (1974) which was used in the calculations described in this paper. Although existing 2D-elliptic procedures are quite economical, they would again lead to unacceptably high computation expenses when applied to long river stretches. However, downstream of the recirculation zone, the elliptic equations can be simplified to 2D-parabolic equations, which can again be solved with efficient marching-forward procedures, in this case requiring only one-dimensional storage. One such procedure is that of Patankar and Spalding (1970), which will be described in Section 3.3 as it is used in the present calculations. The 2D-parabolic models can be applied without excessive computing costs also to the far-field, but there diffusion models can be used also (e.g. Yotsokura and Sayre, 1975) when the velocity field is known from measurements or can be estimated with sufficient accuracy. The 2D-parabolic models have, however, the advantage that they account automatically for effects of changes in the river cross section.

The various models can be combined to form a model chain with which the whole field of interest can be covered. Figure 2 gives two possible model chains, one for situations with strongly three-dimensional behavior in the vicinity of the discharge and the other for situations with weak three-dimensionality even near the discharge. In the former case, the calculation would be started with the 3D partially-parabolic/locally-elliptic model covering Region 1 of Fig. 1. In Region 2, where the three-dimensional effects (e.g. stratification) are still significant but the transmission of influences from downstream to upstream locations is not, the 3D-parabolic model would be used. The calculation would switch to the 2D depth-average parabolic model as soon as the 3D-calculations indicate nearly uniform profiles of pollutant concentration over the depth (Region 3), and this calculation can be carried on into the far-field (Region 4). Alternatively the far-field can be calculated with a diffusion method, using a measured or estimated velocity distribution as input. For each model component, the conditions at the inflow-boundary (initial conditions) are provided by the calculations with the model component for the adjacent upstream-region. Of course, for the first member of the model chain the initial conditions must be prescribed at a cross section upstream of the waste-water discharge. The second model chain given in Fig. 2 is suitable for situations with weak three-dimensional effects everywhere so that the calculation can be started with a depth-averaged model. In this case, Region 1 of Fig. 1 would be calculated with the 2D-elliptic model. Downstream of the recirculation zone, the calculation would be continued with the 2D-parabolic model as described for the other model chain above. In the well-mixed case, there is no difference between Regions 2 and 3. The great advantage of the model-chain concept is that, for each flow region, the simplest possible mathematical model is used so that the computational effort is minimized.

The turbulent transport terms in the mean-flow equations, representing the transport of momentum, heat and mass by the turbulent fluctuations, are additional unknowns in the mean-flow equations. Hence, approximate hypotheses must be introduced to obtain a closed set of equations. Such hypotheses are called a "turbulence model" and a review on existing turbulence models has recently been given by Rodi (1980b). Usually, the eddy-viscosity concept is introduced, which relates the turbulent momentum transport to mean-velocity gradients. Analogously, the eddy-diffusivity concept is used relating the turbulent heat or mass transport to the gradients of temperature or species concentration. Models are available, however, which do not use the eddy-viscosity/diffusivity concept but employ differential transport equations for the individual turbulent transport

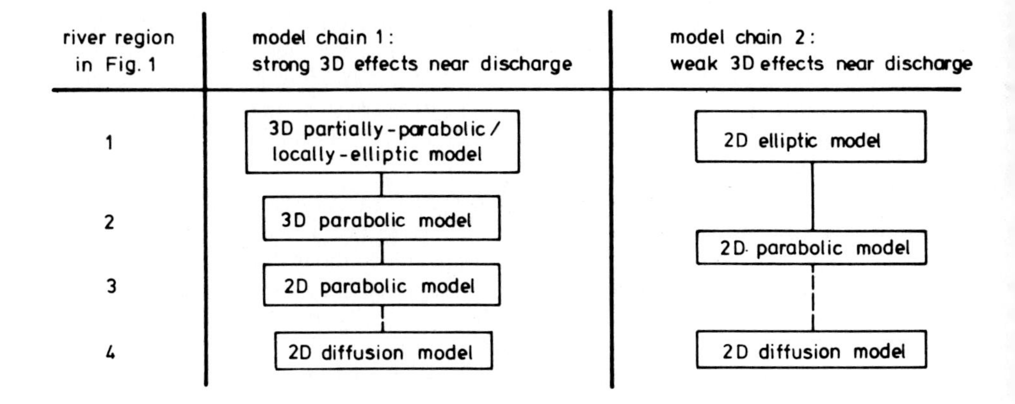

Figure 2. Model chains for calculating the flow and pollutant spreading in rivers.

quantities (like the turbulent stresses and heat or mass fluxes), but these models are rather complex and, at their present state of development, they are not very suitable for practical calculations, as is discussed in greater detail by Rodi (1980b). As the eddy-viscosity and diffusivity are not fluid properties but depend on the turbulence structure, they may vary strongly from one point in the flow to another and from one flow situation to another. Hence, the introduction of the eddy-viscosity/diffusivity concept alone is not sufficient to determine the turbulent transport quantities, but a model must also be introduced to determine the distribution of the eddy viscosity/diffusivity over the flow field. In many depth-average calculations of the far-field, the use of a constant eddy-diffusivity (the eddy viscosity is not very important in such calculations) may be sufficient for practical purposes. However, the complex near-field phenomena cannot be simulated with a constant eddy viscosity and diffusivity. There the distribution of the eddy viscosity and diffusivity over the flow-field has to be determined. The approach taken herein is to characterize the local state of turbulence by two parameters, namely the kinetic energy k and the rate of its dissipation ϵ, and to relate the eddy viscosity and diffusivity at each point to these parameters. The kinetic energy k is a measure of the intensity of the turbulent fluctuations and ϵ is a measure of the length-scale L of the fluctuations since one can assume that $\epsilon \propto k^{3/2}/L$. The distribution of the turbulence parameters k and ϵ over the flow field is determined from semi-empirical transport equations for these two quantities. This so-called $k-\epsilon$-turbulence model has been found to work well in many different flow situations with the same set of empirical constants, both in hydraulics and in other areas of fluid mechanics. A special form of the $k-\epsilon$-model suitable for depth-average calculations is introduced in detail in Section 2.3.

1.3 Present Contribution

In many rivers the vertical mixing is strong enough for the depth-average approach to be sufficiently accurate for practical purposes, even when there are some three-dimensional effects in the direct vicinity of the waste-water outlet. This paper describes the experience gained so far with the two-dimensional model chain applied to real-life situations, and shows how well the 2D-model chain can describe the pollutant spreading and the velocity field near existing waste-water discharges. The model chain consists of a 2D-elliptic model for the direct vicinity of the discharge, where the transmission of influences in the upstream direction is important and which includes the recirculation region, followed by a parabolic model for the downstream region where all influences in the upstream direction are negligible. The model chain is first described in detail in Section 2 where the mean-flow equations are given in their elliptic and parabolic forms respectively, the turbulence model is introduced and the main features of the two numerical schemes for solving the elliptic and parabolic equations are described. Section 3 presents the application of the model chain to the discharge from the Stuttgart sewage plant into the river Neckar and to the cooling-water discharge from the Karlsruhe power station into the river Rhine. Section 4 closes the paper with an assessment of the performance of the 2D model chain.

2. THE MATHEMATICAL MODEL

In this section, a composite depth-average model is described which consists of an elliptic model for the vicinity of the discharge and of a parabolic model for the downstream region where the flow has boundary-layer character. First, the mean-flow equations are introduced in the more general elliptic form, and the simplification to parabolic equations is then discussed. The treatment of the bed shear and the turbulent transport quantities via a turbulence model are basically the same for the elliptic and the parabolic model components and will therefore be discussed together in Sections 2.2 and 2.3 respectively. On the other hand, the numerical schemes for solving the elliptic and parabolic equations are very different so that they will be discussed separately in Sections 2.4 and 2.5.

2.1 Mean-Flow Equations

The task for a depth-average model is to calculate the horizontal distribution of the depth-averaged velocity components $\overline{U}$ and $\overline{V}$, the local water-depth h, and the depth-averaged scalar quantity $\overline{\Phi}$. These depth-averaged quantities are defined by the following relations.

$$\overline{U} = \frac{1}{h}\int_{z_B}^{z_B+h} U dz \qquad \overline{V} = \frac{1}{h}\int_{z_B}^{z_B+h} V dz \qquad \overline{\Phi} = \frac{1}{h}\int_{z_B}^{z_B+h} \Phi dz \tag{1}$$

where z is the vertical co-ordinate and z_B represents the channel bottom as sketched in Fig. 3. The scalar quantity $\overline{\Phi}$ may stand for either temperature of species concentration. In a depth-averaged model, the velocity field is independent of the $\overline{\Phi}$-distribution as buoyancy forces acting directly on the vertical motion cannot be accounted for in such a model. For steady flow situations, to which attention is restricted in the paper, the equations governing the distribution of $\overline{U}, \overline{V}$ and h are as follows (for a derivation see Kuipers and Vreugdenhil, 1973).

Continuity equation:

$$\frac{\partial}{\partial x}(h\overline{U}) + \frac{\partial}{\partial y}(h\overline{V}) = 0 \tag{2}$$

x-momentum equation

$$\overline{U}\frac{\partial \overline{U}}{\partial x} + \overline{V}\frac{\partial \overline{U}}{\partial y} = -g\frac{\partial}{\partial x}(h + z_B) + \frac{1}{\rho h}\frac{\partial(h\overline{\tau}_{xx})}{\partial x} + \frac{1}{\rho h}\frac{\partial(h\overline{\tau}_{xy})}{\partial y} - \frac{\tau_{bx}}{\rho h}$$

$$\underbrace{+ \frac{1}{\rho h}\frac{\partial}{\partial x}\int_{z_B}^{z_B+h}\rho(U-\overline{U})^2 dz + \frac{1}{\rho h}\frac{\partial}{\partial y}\int_{z_B}^{z_B+h}\rho(U-\overline{U})(V-\overline{V})\,dz}_{\text{dispersion}} \tag{3}$$

y-momentum equation

$$\overline{U}\frac{\partial \overline{V}}{\partial x} + \overline{V}\frac{\partial \overline{V}}{\partial y} = -g\frac{\partial}{\partial y}(h+z_B) + \frac{1}{\rho h}\frac{\partial(h\overline{\tau}_{xy})}{\partial x} + \frac{1}{\rho h}\frac{\partial(h\overline{\tau}_{yy})}{\partial y} - \frac{\tau_{by}}{\rho h}$$

$$\underbrace{+ \frac{1}{\rho h}\frac{\partial}{\partial x}\int_{z_B}^{z_B+h}\rho(U-\overline{U})(V-\overline{V})\,dz + \frac{1}{\rho h}\frac{\partial}{\partial y}\int_{z_B}^{z_B+h}\rho(V-\overline{V})^2 dz}_{\text{dispersion}} \tag{4}$$

Similarly, the equation governing the distribution of the scalar $\overline{\Phi}$ (concentration or temperature) is

$$\overline{U}\frac{\partial\overline{\Phi}}{\partial x} + \overline{V}\frac{\partial\overline{\Phi}}{\partial y} = \frac{1}{\rho h}\frac{\partial(h\overline{J}_x)}{\partial x} + \frac{1}{\rho h}\frac{\partial(h\overline{J}_y)}{\partial y}$$

$$\underbrace{+ \frac{1}{\rho h}\frac{\partial}{\partial x}\int_{z_B}^{z_B+h}\rho(U-\overline{U})(\Phi-\overline{\Phi})\,dz + \frac{1}{\rho h}\frac{\partial}{\partial y}\int_{z_B}^{z_B+h}\rho(V-\overline{V})(\Phi-\overline{\Phi})\,dz}_{\text{dispersion}} \tag{5}$$

In the momentum equations (3) and (4), the pressure gradients have been replaced by the gradients of water-depth h via the hydrostatic pressure assumption (g is the gravitational acceleration). τ_{bx} and τ_{by} are the bottom shear-stresses in x- and y-direction respectively (forces per projected area). Wind-shear stresses at the surface are neglected in the present paper. The heat flux through the surface appearing as source/sink term in Eq. (5) when $\overline{\Phi}$ stands for temperature is also neglected in this paper because calculations are carried out only over relatively short river stretches where the heat exchange with the atmosphere is unimportant. It would, of course, not be difficult to include the heat flux using one of the heat-transfer models available in the literature.

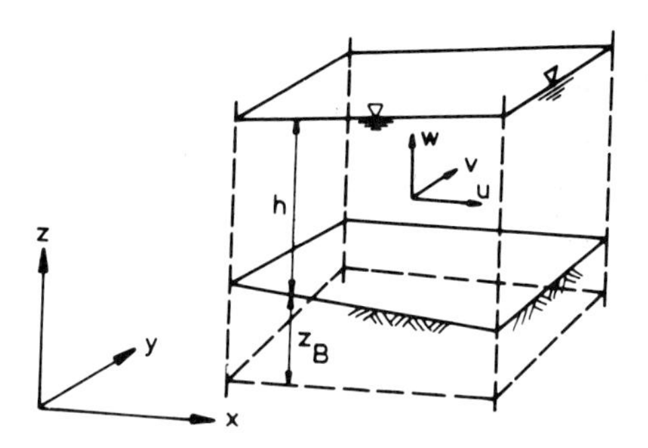

The momentum equations (3) and (4) contain the depth-averaged turbulent stresses $\overline{\tau}_{xx}$, $\overline{\tau}_{yy}$ and $\overline{\tau}_{xy}$ and the scalar transport Eq. (5) contains the depth-averaged turbulent heat or mass fluxes $\overline{J}_x$ and $\overline{J}_y$. A turbulence model described in Section 2.3 is needed to determine the horizontal distribution of these stresses and fluxes over the flow domain.

Equations (3) to (5) contain so-called dispersion terms accounting for vertical non-uniformity of the U, V and Φ profiles. These terms arise from splitting local quantities into depth-averaged values and deviations from these values and then carrying out the depth-averaging of the equations. The physical meaning of the dispersion terms is similar to that of the turbulent stress and flux terms in that both represent gradients of the transport of momentum (effective stresses) and of heat or mass. From measurements of depth-averaged quantities it is usually not possible to distinguish between the turbulent and the dispersion contribution to the transport and thus to the spreading, and consequently an eddy diffusivity chosen to simulate the experimental results partly accounts also for dispersion. For steady effluent discharge and steady flow conditions as considered here, it is in line with the depth-average approach to neglect the dispersion terms in the $\overline{\Phi}$ equation (5) since this approach is based on the assumption of a nearly uniform distribution of temperature or concentration over the depth, and Rastogi and Rodi (1978) have confirmed by way of a three-dimensional calculation that the dispersion terms are indeed negligible in the case of a full-depth, co-axial steady discharge into developed channel flow. It should, however, always be kept in mind that the eddy-diffusivity used to simulate the turbulent heat or mass flux $\overline{J}$ may account also for dispersive transport. The reader is also reminded that the situation is quite different under unsteady conditions (e.g. spill of a quantity of pollutant) where the longitudinal dispersion is known to influence greatly the distribution of depth-average concentration. In contrast to the temperature/concentration distribution, the velocity distribution can never be entirely uniform because the velocity has to go to zero at the river bed. Hence, there will always be some vertical non-uniformity of the velocity profiles, and the dispersion terms in the momentum equations (3) and (4) may be of some importance. Rastogi and Rodi (1978) have calculated these dispersion terms with the aid of a three-dimensional model and have found that they are negligible in the boundary-layer type flow considered by them. As was pointed out by Flokstra (1976), the dispersion terms may be of greater significance in depth-average calculations of recirculating flows. The work of McGuirk and Rodi (1978) has shown, however, that the characteristics of the recirculation region developing behind a side discharge can be described well without including the dispersion terms. Hence, these terms are also neglected in the present model.

2.1.1 Elliptic equations used. In addition to the neglect of surface shear stresses, surface heat transfer and of the dispersion terms, the elliptic equations are simplified further by the introduction of the rigid-lid approximation which neglects the variation of surface elevation in the continuity equation (2) and in the diffusion terms of (3) to (5). In effect, the local water-depth h is written as the sum of a water-depth h_s that would prevail under stagnant conditions and a super-elevation Δh due to the motion of the water

$$h = h_s + \Delta h \tag{6}$$

The rigid-lid approximation involves the replacement of the local water depth h by the static water-depth h_s in the continuity equation (2) and in all the terms of Eqs. (3) to (5) except for the gradients of the super-elevation which are retained in the momentum equations (3) and (4). With this approximation, the equations read as follows

$$\frac{\partial(h_s \overline{U})}{\partial x} + \frac{\partial(h_s \overline{V})}{\partial y} = 0 \tag{7}$$

$$\overline{U}\frac{\partial \overline{U}}{\partial x} + \overline{V}\frac{\partial \overline{U}}{\partial y} = -g\frac{\partial \Delta h}{\partial x} + \frac{1}{\rho h_s}\frac{\partial}{\partial x}(h_s \overline{\tau}_{xx}) + \frac{1}{\rho h_s}(h_s \overline{\tau}_{xy}) - \frac{\tau_{bx}}{\rho h_s} \tag{8}$$

$$\bar{U}\frac{\partial \bar{V}}{\partial x} + \bar{V}\frac{\partial \bar{V}}{\partial y} = -g\frac{\partial \Delta h}{\partial y} + \frac{1}{\rho h_s}\frac{\partial}{\partial x}(h_s \bar{\tau}_{xy}) + \frac{1}{\rho h_s}(h_s \bar{\tau}_{yy}) - \frac{\tau_{by}}{\rho h_s} \tag{9}$$

$$\bar{U}\frac{\partial \bar{\Phi}}{\partial x} + \bar{V}\frac{\partial \bar{\Phi}}{\partial y} = \frac{1}{\rho h_s}\frac{\partial}{\partial x}(h_s \bar{J}_x) + \frac{1}{\rho h_s}\frac{\partial}{\partial y}(h_s \bar{J}_y) \tag{10}$$

These are the mean-flow equations used in the elliptic model. It should be emphasized that the static depth h_s is not unknown but is specified by the river geometry. Also, owing to the use of the rigid-lid approximation, the elliptic model cannot be applied to long stretches of gradually varied flow, where the water depth can vary significantly in the streamwise direction. The elliptic model is intended only for use in the direct vicinity of the discharge, however, and downstream of the recirculation zone the parabolic model, which can account for the streamwise variation of h, is used.

2.1.2 Parabolic equations. The elliptic equations will now be simplified to parabolic ones suitable for river stretches where downstream events cannot influence the flow upstream apart from an influence on the water level. In such situations the following boundary-layer approximations are satisfied

$$\frac{\partial}{\partial x} << \frac{\partial}{\partial y}, \quad \bar{V} << \bar{U} \tag{11}$$

Owing to the first condition, the turbulent transport in the longitudinal direction is negligible compared to that in the transverse direction so that the transport terms containing gradients with respect to x in Eqs. (3) to (5) are negligible. Introducing conditions (11) into the lateral momentum equation (4) yields that the water level above datum $H = h + z_B$ is approximately constant over the width of the river. Therefore, in the longitudinal momentum equation (3), the gradient of the surface elevation $\partial H/\partial x$ is replaced by dH/dx, and the equations read

$$\frac{\partial h\bar{U}}{\partial x} + \frac{\partial h\bar{V}}{\partial y} = 0 \tag{12}$$

$$\bar{U}\frac{\partial \bar{U}}{\partial x} + \bar{V}\frac{\partial \bar{U}}{\partial y} = -g\frac{dH}{dx} + \frac{1}{\rho h}\frac{\partial}{\partial y}(h\bar{\tau}_{xy}) - \frac{\tau_{bx}}{\rho h} \tag{13}$$

$$\bar{U}\frac{\partial \bar{\Phi}}{\partial x} + \bar{V}\frac{\partial \bar{\Phi}}{\partial y} = \frac{1}{\rho h}\frac{\partial}{\partial y}(h\bar{J}_y) \tag{14}$$

The flow field is described in this case solely by the x-momentum equation (13) and by the continuity equation (12). The streamwise variation of the water-level H above datum has to be known *a priori* and must in general be calculated with a one-dimensional backwater-profile method (see, for example, Rastogi, *et al.*, 1975). With H known, the local water depth $h = H - z_B$ can then be calculated from the known topography of the river bed.

2.2 Determination of the Bed-Shear Stress τ_b

A model assumption has to be introduced for the bed-shear-stress components τ_{bx} and τ_{by} appearing respectively in the momentum equations (3) and (4). Here these shear stresses are related to the depth-averaged velocity components by the usual quadratic friction law

$$\tau_{bx} = \frac{1}{\cos\phi} c_f \rho \overline{U} (\overline{U}^2 + \overline{V}^2)^{1/2} \quad \tau_{by} = \frac{1}{\cos\phi} c_f \rho \overline{V} (\overline{U}^2 + \overline{V}^2)^{1/2} \tag{15}$$

where τ_{bx} and τ_{by} are forces per projected horizontal area. The friction coefficient c_f, on the other hand, relates the actual stress, which is the force per actual (inclined) area, to the depth-averaged velocities, and hence the ratio of projected horizontal area to the actual area has to appear in relations (15). Here only the inclination of the river bed in the lateral direction (angle ϕ defined in Fig. 4) is considered because the bed slope in the streamwise direction is usually very small; $\cos\phi$ appears in Eq. (15) as the ratio of horizontal projection to inclined area.

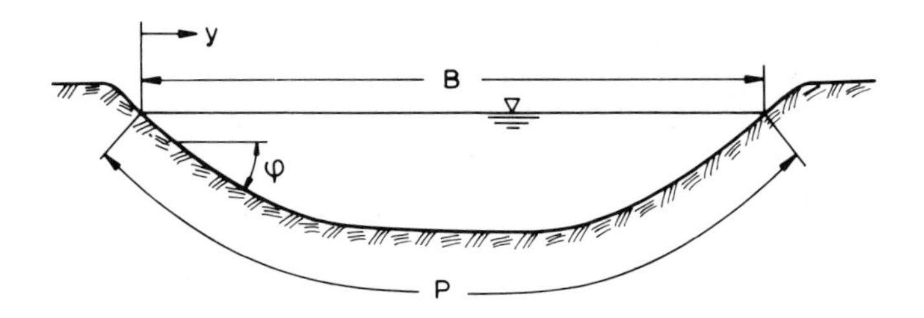

Figure 4. Typical river cross-section.

In rectangular channels, where $\cos\phi$ is equal to unity at the bottom, a laterally constant value of c_f can be used, which is directly related to the Manning roughness factor n. In channels with sloping banks however, c_f has been found to vary from one edge of the channel to the other; in particular, c_f increases towards the edge of the river where the water becomes shallower. This behavior is well known from measurements in channels of trapezoidal and elliptic cross section. A preliminary study was carried out to find a functional relationship between c_f and a suitable flow parameter from three-dimensional calculations as well as experimental findings. The increase in c_f at sloping river banks is not well understood at present, but it appears to be associated with increased turbulence production at the banks. The increased turbulence production also leads to a higher turbulent (or eddy) viscosity ν_t and the preliminary three-dimensional calculations have shown that the relative eddy viscosity

$$f = \frac{\tilde{\nu}_t}{\overline{U}h} \tag{16}$$

is a suitable parameter with which to correlate c_f. Here $\tilde{\nu}_t$ is the depth-average eddy viscosity whose distribution will be calculated with the aid of a turbulence model as explained in the following subsection. The three-dimensional calculations suggested the following functional relationship between c_f and f

$$c_f = c_{fo} + (f - f_o)^{1.5} \tag{17}$$

where f_o is the value of f obtained in the center portion of the river with horizontal bottom. c_{fo} is the friction factor pertaining to this center portion; its actual value is determined such that the total shear force on the river bed is equal to that determined from the one-dimensional backwater profile calculation

$$\int_o^B c_f \rho \overline{U}^2 dy = c_{fo}\rho \int_o^B \frac{\overline{U}^2}{\cos\phi} dy + \rho \int_o^B (f - f_o)^{1.5} \frac{\overline{U}^2}{\cos\phi} dy = \tau_{b,1D} P_w \tag{18}$$

where $\tau_{b,1D}$ is the shear stress determined from the one-dimensional calculation and P_w is the wetted perimeter as defined in Fig. 4. The c_f function (17) was tested for various trapezoidal and triangular channel flows, and the calculated depth-averaged velocity profiles agree favorably with the measurements of Holly (1975), Lau and Krishnappan (1978) and Powell and Possey (1959) for such channels. The authors will report the study of the behavior of c_f and the test calculations elsewhere.

In the parabolic model, the variation of c_f was determined from (17) and (18) for each cross section while in the elliptic-model calculations the cross section was taken as constant and the c_f-distribution was determined only once. c_f was further assumed not to be influenced by the discharge.

2.3 Turbulence Model

In this subsection, a model is given for determining the depth-averaged turbulent stresses $\bar{\tau}_{xx}$, $\bar{\tau}_{xy}$ and $\bar{\tau}_{yy}$ in the momentum equations (3) and (4) and the depth-averaged heat or mass fluxes $\bar{J}_x$ and $\bar{J}_y$ in the scalar-transport equation (5), using a modification of the k-ϵ turbulence model. A detailed description of the k-ϵ model was given by Launder and Spalding (1974) and by Rodi (1980b), and these papers also present numerous application examples.

Rastogi and Rodi (1978) have adapted the model for use in depth-averaged open-channel flow calculations, and their version is employed here. The k-ϵ model uses the eddy-viscosity/diffusivity concept, which relates the turbulent stresses and heat or mass fluxes respectively to the gradients of the depth-averaged velocity components and the depth-averaged temperature or concentration

$$\frac{\bar{\tau}_{ij}}{\rho} = \tilde{\nu}_t\left(\frac{\partial \bar{U}_i}{\partial x_j} + \frac{\partial \bar{U}_j}{\partial x_i}\right) - \frac{2}{3}\tilde{k}\delta_{ij}\,, \qquad \frac{\bar{J}_{\Phi i}}{\rho} = \tilde{\Gamma}_t\frac{\partial\bar{\Phi}}{\partial x_i} \tag{19}$$

In analogy with the original k-ϵ model, Rastogi and Rodi (1978) assumed that the local depth-averaged state of turbulence can be characterized by two parameters, namely the turbulence energy and dissipation parameters $\tilde{k}$ and $\tilde{\epsilon}$. The eddy viscosity and diffusivity can then be related to these parameters by dimensional analysis

$$\tilde{\nu}_t = c_\mu\,\frac{\tilde{k}^2}{\tilde{\epsilon}}\,, \qquad \tilde{\Gamma}_t = \frac{\tilde{\nu}_t}{\sigma_t} \tag{20}$$

where c_μ and σ_t are empirical constants, the latter one being called turbulent Prandtl or Schmidt-number. The eddy viscosity and diffusivity $\tilde{\nu}_t$ and $\tilde{\Gamma}_t$ are not true depth-averaged quantities in the sense of the mathematical definition of (1); rather (19) defines the turbulent viscosity and diffusivity such that, when they are multiplied by the relevant gradient of the transported quantity, the depth-averaged turbulent shear stress or heat flux is obtained. The variation of the turbulence parameters $\tilde{k}$ and $\tilde{\epsilon}$ is determined from the transport equations

$$\bar{U}\,\frac{\partial\tilde{k}}{\partial x} + \bar{V}\,\frac{\partial\tilde{k}}{\partial y} = \underline{\frac{\partial}{\partial x}\left(\frac{\tilde{\nu}_t}{\sigma_k}\,\frac{\partial\tilde{k}}{\partial x}\right)} + \frac{\partial}{\partial y}\left(\frac{\tilde{\nu}_t}{\sigma_k}\,\frac{\partial\tilde{k}}{\partial y}\right) + G + P_{kv} - \tilde{\epsilon} \tag{21}$$

$$\bar{U}\,\frac{\partial\tilde{\epsilon}}{\partial x} + \bar{V}\,\frac{\partial\tilde{\epsilon}}{\partial y} = \underline{\frac{\partial}{\partial x}\left(\frac{\tilde{\nu}_t}{\sigma_\epsilon}\,\frac{\partial\tilde{\epsilon}}{\partial x}\right)} + \frac{\partial}{\partial y}\left(\frac{\tilde{\nu}_t}{\sigma_\epsilon}\,\frac{\partial\tilde{\epsilon}}{\partial y}\right)$$

$$+\, c_1\,\frac{\tilde{\epsilon}}{\tilde{k}}\,G + P_{\epsilon v} - c_2\,\frac{\tilde{\epsilon}^2}{\tilde{k}} \tag{22}$$

where

$$G = \tilde{\nu}_t \left[2\left(\frac{\partial \bar{U}}{\partial x}\right)^2 + 2\left(\frac{\partial \bar{V}}{\partial y}\right)^2 + \left(\frac{\partial \bar{U}}{\partial y} + \frac{\partial \bar{V}}{\partial x}\right)^2 \right] \quad (23)$$

is the production of turbulent kinetic energy k due to interaction of turbulent stresses with horizontal mean velocity gradients, and c_1, c_2, σ_k and σ_ϵ are further empirical constants. Equations (21) and (22) can be considered as depth-averaged forms of the three-dimensional k and ϵ equations (Rodi, 1980b) when all terms originating from non-uniformity of vertical profiles are assumed to be absorbed in the source terms P_{kv} and $P_{\epsilon v}$. The main contribution to these terms stems from significant vertical velocity gradients near the bottom which, by interaction with the relatively large turbulent shear stresses in this region, produce turbulence energy. This production is in addition to the production G due to horizontal velocity gradients, and depends strongly on the bottom roughness. Because it is governed by the near-bottom region, the additional vertical production terms are related to the resultant bottom shear-stress τ_b via the friction velocity U_* with the following result

$$P_{kv} = c_k \frac{U_*^3}{h}, \qquad P_{\epsilon v} = c_\epsilon \frac{U_*^4}{h^2} \quad (24)$$

where the friction velocity is $U_* = [c_f(\bar{U}^2 + \bar{V}^2)/\cos\phi]^{1/2}$. The empirical constants c_k and c_ϵ were determined by Rastogi and Rodi (1978) from the rate of energy dissipation (related to the energy slope) and the dye spreading in undisturbed normal channel flow. They obtained

$$c_k = \frac{1}{\sqrt{c_f}}, \qquad c_\epsilon = c_{\epsilon\Gamma} \frac{c_2}{c_f^{3/4}} \sqrt{c_\mu} \quad (25)$$

where the coefficient $c_{\epsilon\Gamma}$ depends on the dimensionless diffusivity $e^* = \tilde{\nu}_t/(U_* h \sigma_t)$. When the value of $e^* \approx 0.15$ recommended by Fischer, *et al.*, (1979) as average of experimental results for wide laboratory flumes is adopted, there results $c_{\epsilon\Gamma} = 3.6$, the value used by Rastogi and Rodi (1978) and McGuirk and Rodi (1978) in their calculations of laboratory situations. Dye-spreading measurements in rivers have shown however (see Fischer, *et al.*, 1979) that the dimensionless diffusivity e^* usually has values well above 0.15 measured in laboratory flumes. The reasons for this are not yet entirely clear, but it should be recalled that a diffusivity determined from dye-spreading measurements accounts not only for turbulent transport but also for dispersive transport due to vertical non-uniformities of scalar quantities and velocity components. When significant secondary motions in cross-sectional planes are present, a relatively small non-uniformity of temperature or concentration may cause relatively large dispersion contributions in the $\overline{\Phi}$-equation (5). In natural rivers, such secondary motions may arise from large-scale irregularities in the river bed and, of course, also from river bends. The secondary motion set up in river bends decays only slowly in straight sections after the bend so that there will usually be a remnant secondary motion even in relatively straight sections between the bends. In the application of the present model to river-flow situations in Section 3, reasonable results can, of course, be expected only when a realistic diffusivity is used as empirical input for determining c_ϵ according to (25). Hence, in the calculations to be presented in Section 3, a value of $e^* = 0.6$ is used as this seems to be a typical value for many river situations (see Yotsukura, Fischer and Sayre, 1970; Glover, 1964) and is recommended by Fischer, *et al.*, (1979) for practical purposes. With this value for e^*, there results a value of 2 for the coefficient $c_{\epsilon\Gamma}$ in the c_ϵ-relation (25). The values for the remaining empirical constants in the turbulence model were simply adopted from Launder and Spalding (1974) and are given in Table 1. It should be emphasized that none of the empirical constants was adjusted to suit the problems considered here but that they are all taken from the literature where most of them were determined from experiments on quite different flows.

Table 1. *Empirical Constants in Turbulence Model*

c_μ	c_1	c_2	σ_k	σ_ϵ	σ_t
0.09	1.43	1.92	1.0	1.3	0.5

2.3.1 Parabolic form of the model. The turbulence-model relations introduced so far are suitable for use in the elliptic model. As can be seen from Eqs. (13) and (14), only the lateral turbulent momentum transport (shear stress $\overline{\tau}_{xy}$) and the lateral heat or mass flux $\overline{J}_y$ have to be determined, and for these the eddy viscosity/diffusivity relation (19) reads

$$\overline{\tau}_{xy} = \rho\tilde{\nu}_t \frac{\partial \overline{U}}{\partial y}, \qquad \overline{J}_y = \rho\tilde{\Gamma}_t \frac{\partial \overline{\Phi}}{\partial y} \tag{26}$$

while Eq. (20) remains the same. The boundary-layer approximation (11) also applies to the $\tilde{k}$- and $\tilde{\epsilon}$-equations (21) and (22) and reduces these to parabolic forms. In the parabolic model, the underlined terms in Eqs. (21) to (23) are omitted accordingly. Otherwise the same model relations are used as described above.

2.4 Finite-Difference Procedure for Elliptic Equations

The elliptic equations (7-10) describing the mean flow and (21) and (22) governing k and ϵ are solved numerically with a modified version of the computer program for 2-D elliptic flows by Gosman and Pun (1974), which incorporates the solution algorithm of Patankar and Spalding (1972). The finite-difference procedure uses a rectangular numerical grid, and the dependent variables are the primitive variables $\overline{U}$, $\overline{V}$ and P (as opposed to stream-function and vorticity used in some methods). A staggered grid is used so that different variables are stored at different places, as shown in Fig. 5. Also shown in this figure are micro-control volumes around the individual grid points, which are of course also staggered for the various variables. The finite-difference equations are derived by integrating the differential equations over these control volumes, making suitable assumptions about the variation of the dependent variables between the grid points. These assumptions will be discussed in detail below. Alternatively, one may consider the finite-difference equations as expressing the balance between transport by mean-flow (convection) or turbulent motion (diffusion) of a certain quantity into and out of the control volume and sources or sinks (e.g. in the k and ϵ equations), and in the case of momentum equations also pressure forces acting on the faces of the control volume. Of course the spatial variation of depth has to be considered when a balance is constructed, and Fig. 6 gives the control volume used in deriving the finite-difference equations for the depth-average model. The faces of the control volume are located midway between the grid points and are labeled by the lower case letters n, s, e, and w for the control volume around the grid point P, while the neighboring points are denoted by capital letters N, S, E, W. The continuity equation reads in finite-difference form

$$h_e \overline{U}_e \Delta y - h_w \overline{U}_w \Delta y + h_n \overline{V}_n \Delta x - h_s \overline{V}_s \Delta x = 0 \tag{27}$$

and the finite-difference equation for the scalar quantity $\overline{\Phi}$ is

$$\begin{aligned} h_e \overline{U}_e \overline{\Phi}_e \Delta y - h_w \overline{U}_w \overline{\Phi}_w \Delta y + h_n \overline{V}_n \overline{\Phi}_n \Delta x - h_s \overline{V}_s \overline{\Phi}_s \Delta x \\ = \tilde{\Gamma}_{\phi e} \left(\frac{\partial \overline{\Phi}}{\partial x} \right)_e h_e \Delta y - \tilde{\Gamma}_{\Phi w} \left(\frac{\partial \overline{\Phi}}{\partial x} \right)_w h_w \Delta y \\ + \tilde{\Gamma}_{\Phi n} \left(\frac{\partial \overline{\Phi}}{\partial y} \right)_n h_n \Delta x - \tilde{\Gamma}_{\Phi s} \left(\frac{\partial \overline{\Phi}}{\partial y} \right) h_s \Delta x \end{aligned} \tag{28}$$

The left-hand side expresses the difference between the flow into and out of the control volume of the quantity $\overline{\Phi}$ and the right-hand side the corresponding difference between the turbulent transport into and out of the volume. The eddy-diffusivity assumption has been introduced so that the gradients of $\overline{\Phi}$ at the control volume faces appear. This way of deriving the finite-difference equations ensures overall conservation of all quantities. The h's appearing in (27) and (28) are

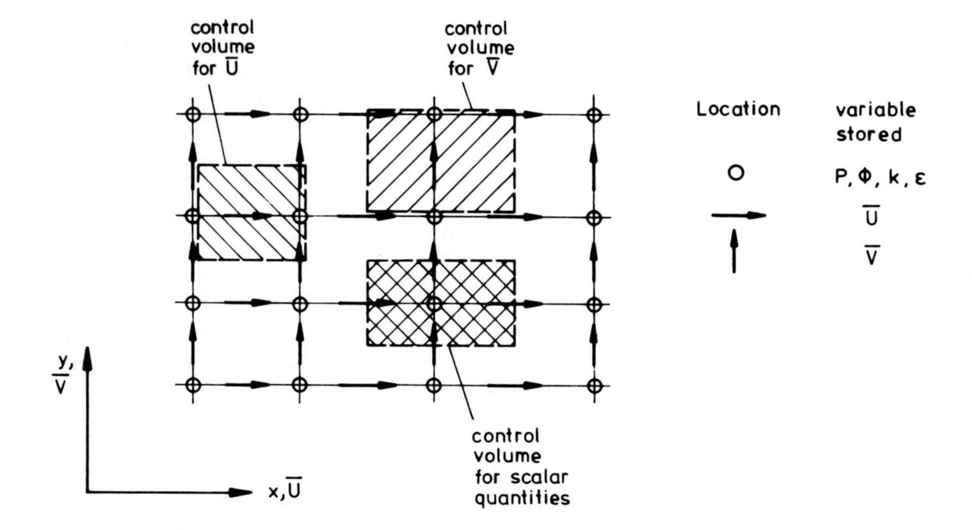

Figure 5. Staggered grid.

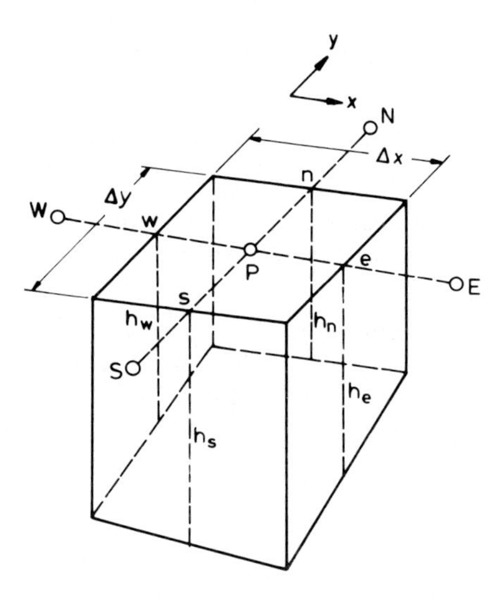

Figure 6. Control column for elliptic flow calculations.

prescribed values of the static depth, stored at the grid points, and obtained at the control volume faces by linear interpolation. In the calculations reported in Section 3, however, the static depth h_s varies only with y as the river cross-section was assumed to be constant for the relatively short river stretch for which elliptic calculations were carried out.

The values of the dependent variables at the faces of the control volume and also of their derivatives must be expressed in terms of values at the grid points. Here a hybrid difference scheme is used; this employs central differences when the grid Peclet-number* $Pe \leqslant 2$ so that $\bar{\Phi}_w = \frac{1}{2}(\bar{\Phi}_W + \bar{\Phi}_P)$. The gradient is expressed as $(\partial\bar{\Phi}/\partial x)_w = (\bar{\Phi}_P - \bar{\Phi}_W)/\Delta x$. This discretization scheme is second-order accurate. When, however, the convective transport through the faces is large compared with the diffusive one, this scheme leads to numerical stability problems. Hence for $Pe > 2$, an upwind-difference scheme is used so that $\bar{\Phi}_w = \bar{\Phi}_W$ when $\bar{U} > 0$ and $\bar{\Phi}_w = \bar{\Phi}_P$ when $\bar{U} < 0$, and the diffusion flux through the w-face is omitted. This scheme is only first-order accurate and introduces numerical diffusion, which is a particular problem in recirculating flows where the streamlines are at an angle to the numerical grid. For the side-discharge situation this problem was investigated by McGuirk and Rodi (1978) and these authors concluded that numerical diffusion is of course present and is strongest in the shear layer bordering the recirculation zone, but that it is always small compared with the dominant terms in the equations; for example, it is small compared with the turbulent diffusion in the shear layer. It should be noted that these conclusions are of course based on calculations with a relatively fine grid, and care has always to be taken that the grid is sufficiently fine.

With the relations between the values at the control-volume faces and the grid points introduced above and the linearization of the convection terms (i.e. assuming the $\bar{U}$'s in Eq. (28) to be known from the previous iteration) the finite-difference equations can be written as follows

$$A_P\phi_P = A_N\phi_N + A_S\phi_S + A_W\phi_W + A_E\phi_E + B_P \tag{29}$$

where ϕ can stand for any of the dependent variables. This equation relates the value of ϕ at grid-point P to the ϕ-values at the neighboring points. The coefficients A and B_P contain only quantities that are known or assumed to be known from the previous iteration. B_P contains the pressure gradients; how these are treated will be discussed shortly.

At the river edges, the control volumes are triangular in the cross-sectional plane. Hence, if the edge is at the w-face, this face reduces to a line with $h_w = 0$. The same finite-difference equations are used as for control volumes away from the edge, except that for the boundary control volume there is no connection to a grid-point W, so that the coefficient A_W in Eq. (29) is 0. For boundary control volumes, the convection and diffusion flux through the w-face (or rather line) is

*For example for direction x the grid Peclet-number at face w is defined as $Pe = \bar{U}_w\Delta x/\tilde{\Gamma}_{\Phi w}$.

0. The treatment of boundaries at the discharge itself will be discussed in the application section.

One of the most important features of the method is the determination of the pressure field, in the present case the determination of the superelevation Δh. A guess-and-correct procedure is used whereby the momentum equations are solved with a guessed pressure field, which is taken from the previous iteration. The resulting velocity field does not necessarily satisfy the continuity equation because the guessed pressure field was not quite correct. This leads to mass sources for each control volume, and corrections to the pressure and velocities are now introduced in order to annihilate the mass sources. Combination of the continuity and momentum equations leads to a Poisson-equation for the pressure correction. By an iterative procedure, the pressure and velocity fields are improved until the mass sources are below a prespecified value. Here the criterion of convergence used was that the sum of the mass source at all grid nodes arising from non-satisfaction of the continuity equation be less than half a percent of the total mass flow of the river.

The set of finite-difference equations at the various grid points is brought into tri-diagonal-matrix form by allowing unknowns only along one grid-line, assuming the neighboring values off this line to be known from the previous iteration. The solution of the finite-difference equations can then be carried out by efficient recursion formulae known as the tri-diagonal-matrix algorithm.

At the inflow boundary (upstream of the discharge), the velocity and temperature conditions for the elliptic calculations are specified from experiments whenever possible and are guessed otherwise. k- and ϵ-distributions are taken that correspond to normal flow (see Rastogi and Rodi, 1978). In any case, the inflow boundary should be sufficiently upstream of the discharge that the exact conditions have little influence on the calculations in the region of interest. The outflow boundary should be placed at a cross section where the flow has become parabolic so that no influences are transmitted upstream and the details of the boundary condition have little influence on the solution. The actual boundary condition chosen is that of zero-gradient with respect to x of all variables. To ensure that this assumption has no significant influence on the predictions in the vicinity of the recirculation zone, it was found necessary to place the downstream boundary at least one recirculation-length distance from the reattachment point (see McGuirk and Rodi, 1978).

Details on the numerical grids used and on the convergence are given in the application section.

2.5 Finite-Difference Procedure for Parabolic Equations

In the parabolic flow region, the equations are solved with a modified version of the 2D-boundary-layer procedure of Patankar and Spalding (1970). As the solution at a cross section x_p does not depend on the solution for $x > x_p$ (see Fig. 7), the equations can be solved by a marching-forward (or rather downstream) procedure. Starting from the prescribed initial conditions at a line along the inflow boundary (here these conditions are provided by the elliptic procedure), the solution is obtained for the next downstream line by solving finite-difference equations, and from there on one proceeds again to the next line and so on. In contrast to the elliptic procedure, the parabolic one does not require iterations. Also, while the elliptic procedure requires 2D-storage, the parabolic method requires only 1D-storage along cross-sectional lines. When the solution moves to the next line, the old values can be overwritten. In parabolic calculations, conditions at the outflow boundary need not be specified at all.

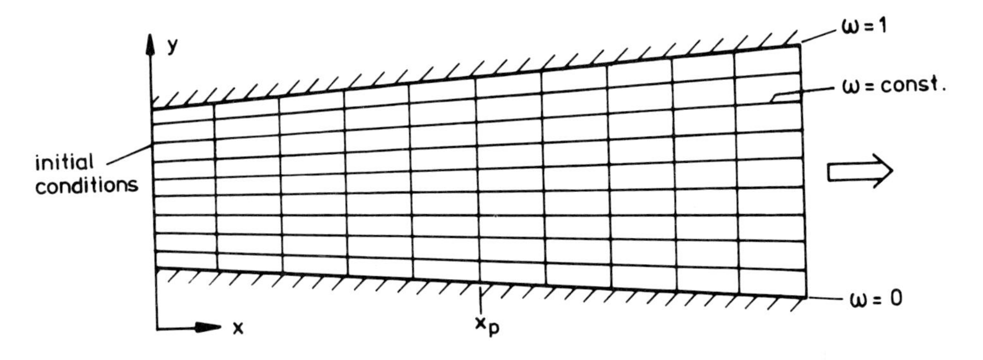

Figure 7. Self-adjusting grid in parabolic procedure.

When applied to channel flow, the method of Patankar and Spalding (1970) uses a streamline coordinate system. The lateral coordinate y is replaced by the dimensionless stream function ω defined by

$$\omega = \frac{\Psi}{\rho Q} = \frac{\int_0^y \rho \bar{U} h dy}{\int_0^B \rho \bar{U} h dy} \tag{30}$$

where Ψ is the stream function, which is the mass flow between the bank with $y = 0$ and a certain distance y from this bank, and Q is the total volume flow rate. Lines of constant ω are used as grid lines (see Fig. 7), causing the grid to expand and contract automatically with the river width B. The lateral coordinate ω always has values between 0 at one bank and 1 at the other. The original program of Patankar and Spalding (1970) is intended for more general applications, including shear layers with free boundaries such as wall boundary layers and jets. In these cases, fluid is entrained into the calculation domain (if this coincides with the shear-layer region) so that the volume flow Q in the layer grows and the lines of constant ω are not streamlines.

The finite-difference equations are derived again by considering a control volume and by constructing the balance for the processes acting on this control volume. A typical control volume is shown in Fig. 8. The derivation of the finite-difference equation for the U-momentum will now be discussed briefly. The V-momentum equation need not be solved in parabolic flows. In a streamline-coordinate system the n and s-faces of the control volume coincide with streamlines so that there is no net flow across these faces and hence there is convection only through the upstream and downstream faces. This convection is equal to the mass flux ρQ times the convected quantity. The finite-difference equation for U-momentum reads

$$(\Psi_n - \Psi_s)(\bar{U}_{P,D} - \bar{U}_{P,U}) = (\bar{\tau}_n h_n - \bar{\tau}_{s} h_s)\Delta x - \tau_b \Delta x \Delta y - \rho g \Delta H h \Delta y \tag{31}$$

where the convection on the left-hand side is simply the mass flux times the difference of the velocities at the upstream and downstream faces. The convection is balanced by shear stress acting on the n- and s-faces, by the bottom shear and by pressure forces on the upstream and downstream faces which are expressed here by the change in surface elevation ΔH. Because of the use of the streamline coordinate system, the continuity equation is satisfied automatically and does not appear any longer. Also, since there is no convection through the n- and s-faces, assumptions are not necessary for expressing the values at these faces in terms of grid-point values. Hence, upwind differencing is not introduced and numerical diffusion is absent. According to Eq. (30), $\Delta\Psi$ can be expressed very simply by $Q\rho\Delta\omega$, where ω is specified by the grid. The corresponding Δy is not prespecified but depends on the velocity distribution. When the shear stresses in Eq. (31) are expressed by the eddy-viscosity relation

$$\bar{\tau} = \tilde{\nu}_t \rho \frac{\partial \bar{U}}{\partial y} = \tilde{\nu}_t \rho \frac{\bar{U} h}{Q} \frac{\partial \bar{U}}{\partial \omega} \tag{32}$$

and the velocity gradients then appearing at the n- and s-faces are expressed by the neighboring values, the following finite-difference equation is obtained

$$\begin{aligned} \rho Q(\omega_n - \omega_s)\left(\frac{\bar{U}_{P,D} - \bar{U}_{P,U}}{\Delta x}\right) &= (\tilde{\nu}_t \rho \bar{U})_{n,U} \frac{h_n^2}{Q} \frac{\bar{U}_{N,D} - \bar{U}_{P,D}}{\omega_n - \omega_p} \\ &\quad - (\tilde{\nu}_t \rho \bar{U})_{s,U} \frac{h_s^2}{Q} \frac{\bar{U}_{P,D} - \bar{U}_{S,D}}{\omega_p - \omega_s} - \tau_b \Delta y - \rho g \frac{\Delta H}{\Delta x} h \Delta y \end{aligned} \tag{33}$$

The velocities for calculating the gradients at the n- and s-faces are taken at the downstream cross section, so that the procedure is fully implicit. Some linearization has to be introduced however by taking the terms in the parentheses involving the eddy viscosity at the upstream cross section. The finite-difference equation for the scalar quantity $\bar{\Phi}$ is very similar. It can be obtained by omitting the last two terms in Eq. (33) and by replacing the $\bar{U}$'s by $\bar{\Phi}$'s except in the eddy-viscosity term. It can be seen that the local water depth h appears in various places, and this is calculated from the water level H (obtained from the one-dimensional back-water profile calculation) and the known bed geometry. The original program of Patankar and Spalding (1970) had to be modified only a little because it was set up for axisymmetric flow calculations, and in such flows the local radius r appears in the same way in the equations as does the water depth h in the depth-averaged equations.

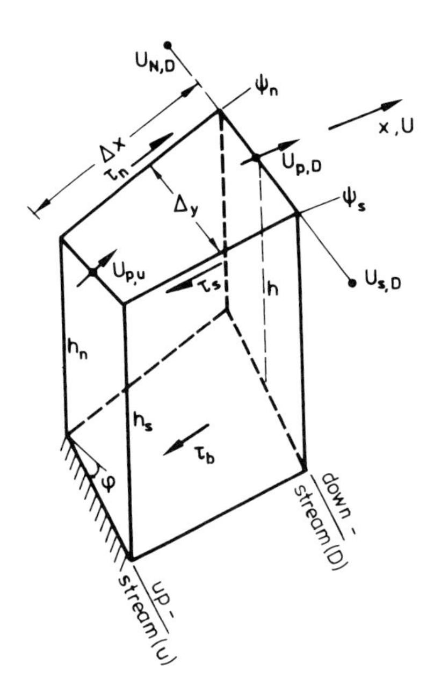

Figure 8. Control volume for parabolic flow calculations.

The finite-difference equations can be expressed in the following form

$$A_p\phi_p = A_N\phi_N + A_S\phi_S + B_p \tag{34}$$

where again the coefficients A and B_p are known. In this case they contain only upstream quantities. Equation (34) relates three neighboring unknown values in each finite-difference equation, and the whole set of equations is of the tri-diagonal-matrix form, which can be solved efficiently with the tri-diagonal-matrix algorithm already mentioned. As in the case of the elliptic method, triangular control volumes are used at the edge of the river, and there is no transport across the n-face (or rather line) when the boundary coincides with the n-face. Hence, there is no connection to a grid-point N and the coefficient $A_N = 0$ in Eq. (34).

The values of ω are specified across the river, while y for a given ω depends on the velocity distribution and must be calculated at each downstream station from

$$\Delta\omega = \frac{\overline{U}h}{Q}\Delta y \tag{35}$$

assuming linear distributions of $\overline{U}$ and h between grid points. Of course this calculation should yield $y = B$ for $\omega = 1$ at the opposite bank. When the surface elevation gradient in Eq. (33) is simply taken from the backwater-profile calculation, this is not necessarily the case because the velocity distribution calculated in the 2D-procedure may not always be entirely compatible with that assumed in the 1D-calculation. Hence, the surface-elevation gradient from the 1-D calculation is slightly corrected to make sure that $y = B$ for $\omega = 1$. This procedure is discussed in detail by Patankar and Spalding (1970).

Because of the linerization of certain terms in the finite-difference equations, the accuracy of the solution depends on the forward step. Values of about 0.02 times the river width were found suitable. The lateral grid distribution is as in the elliptic procedure, so that the values at the outflow boundary of the elliptic region are simply taken over and no interpolation is required.

3. APPLICATION OF THE MATHEMATICAL MODEL

This section describes the application of the mathematical model to the Stuttgart sewage-plant discharge into the river Neckar and the Karlsruhe power station cooling-water discharge into the river Rhine and compares the predictions with field measurements. In each case, the direct vicinity of the discharge was calculated with the elliptic model component while the parabolic model was used for river zones extending further downstream.

Here it is in place to mention briefly the tests to which the two model components have been subjected previously. McGuirk and Rodi (1978) have applied the elliptic model to the near field of side discharges into a rectangular laboratory flume and compared their calculations with the measurements of Carter (1969). The investigations were carried out for ratios of discharge to channel velocities varying from 2 to 10. The size of the recirculation zone forming behind the discharge was calculated correctly as a function of the ratio of discharge to channel momentum. The jet trajectory was also predicted in agreement with the measurements for various velocity ratios, and the calculated isotherms are in general agreement with measurements, although there are differences in details, partly because some buoyancy effects were present in the experiments. The elliptic model was also applied by McGuirk (see Rodi, 1980b) to a combined intake/discharge problem. Of particular interest was the contamination of the intake water when the intake is fairly close to the discharge. The amount of discharged contaminant reaching the intake as a function of intake/discharge flow rate to river flow rate could be predicted correctly. As a last example, McGuirk and Rodi (1979) have calculated unsteady mass exchange between a dead-water zone (e.g. bay, harbor) with a main stream passing by this zone. In this case, the "washing out" of the contaminant from the dead-water zone was calculated correctly by the elliptic model.

The parabolic model was tested by calculating the depth-averaged velocity profile in a rectangular channel under developed normal flow conditions. The channel had a width-to-depth ratio of 30 and a Manning roughness of $n = 0.029$, and the calculated velocity profile was found to agree well with laboratory data (see Rodi, 1980b). The model was also tested for gradually varied flow in rectangular channels (Pavlović and Rodi, 1979). Rastogi and Rodi, (1978) applied the parabolic model to coaxial discharges into rectangular channel flow (the discharge was located at the channel center) and compared the calculations with those obtained with a three-dimensional model. Of particular interest were the conditions under which buoyancy effects due to heated-water discharge are significant. These effects, which may lead to an increased pollutant spreading, can be accounted for in the 3D-calculation but not in the 2D-model. For high densimetric Froude numbers the 3D- and 2D-calculations were virtually the same but for low values of $F = U_d/\sqrt{gh\Delta\rho/\rho}$ a significant increase in pollutant spreading due to the secondary motion induced by buoyancy effects was observed in the 3D-model when the channel bed was smooth. For a rough channel bed however (Manning roughness $n = 0.025$), little influence was found down to values of $F = 5$ so that nearly the same results were obtained with the 2D- and 3D-model. Hence, it appears that the depth-average model is good enough in many practical circumstances.

3.1 Discharge from Stuttgart Sewage Plant into the the River Neckar

3.1.1 Problem specification. The aim of this case study is to see how well the mathematical model can simulate the pollutant-concentration field induced by the discharge from the Stuttgart sewage plant into the river Neckar. The study site is shown in Fig. 9. The discharge is located between the locks at Hofen and Aldingen. The river is fairly straight between the discharge and the downstream lock, but there is a slight bend just upstream of the discharge which may induce some secondary motion and lead to increased pollutant spreading. For the river stretch covered by the calculations, details of the river geometry can be seen from Fig. 11, where the variation of width and mean as well as maximum depth is given together with the bed profiles at the locations where they have been measured. The river widens and deepens towards the downstream lock. The cross section is approximately trapezoidal and does not change significantly in shape over the river stretch considered. A situation is examined when the river discharge was $24 m^3/s$, corresponding to a relatively low cross-sectional average velocity of about 0.15 m/s.

The discharge from the sewage plant was $2.4 m^3/s$ perpendicular to the river bank, and the discharge velocity was 0.75 m/s. The geometrical details of the

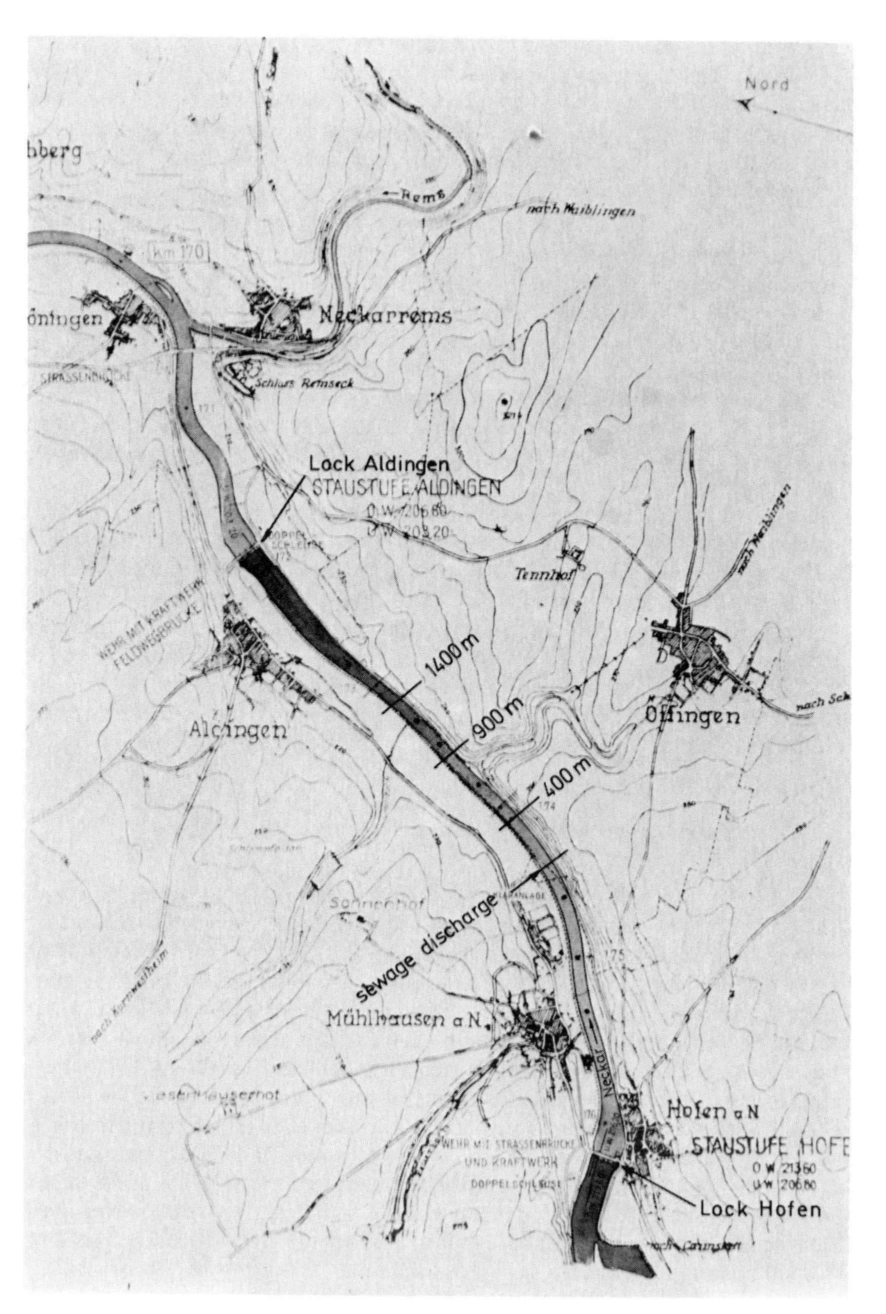

Figure 9. River Neckar between the locks at Hofen and Aldingen and location of discharge from Stuttgart Sewage Plant.

discharge are given in Fig. 10. The discharge channel was cut into the sloping bank and was divided in two equal sections. The surface discharge extended over about 1/3 of the river depth.

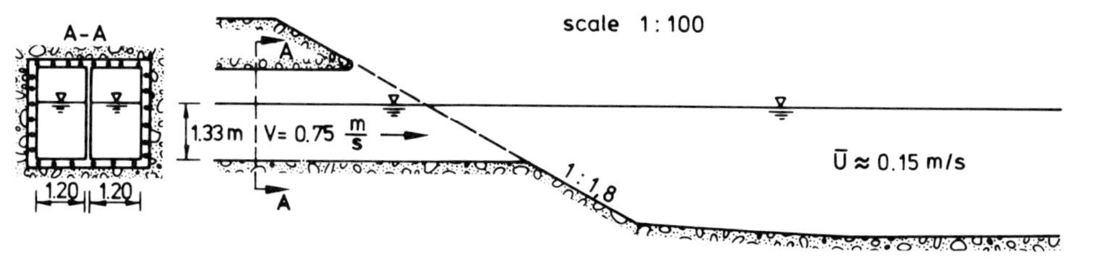

Figure 10. Configuration of discharge into the River Neckar.

3.1.2 Field measurements. The above discharge conditions relate to a period when measurements were taken as reported in detail by Käser, *et al.*, (1980). During this period there was no ship traffic and no lock operation so that the flow conditions were reasonably steady. The sewage discharge was seeded with Rhodamine B, and the concentration profiles were measured at one depth from a boat at various cross sections (preliminary measurements indicated little variation with depth). The velocity profiles were also measured with propellers, in this case also at one cross section upstream of the discharge, and the local water depth was determined with a Sonar. The actual measurement points are shown in Fig. 13. They had to be corrected because they are in conflict with the laws of conservation of mass and species. The corrected profiles satisfying these laws are also shown in Fig. 13. The need to correct the actual measurements is due to certain measurement errors and the fact that Rhodamine B attaches itself to suspended particles and settles with these to the river bed.

3.1.3 One-dimensional backwater profile calculations. The field measurements indicated that the water level was approximately horizontal so that H is constant. The 1D-backwater profile calculation method of Rastogi, *et al.*, (1975) was used to determine the variation of the cross-sectional bed friction $\tau_{b,1D}$ which enters the calculation of the friction coefficient c_f employed in the 2D-model. The measured river-bed profiles were used in the 1D-calculation, with a linear interpolation between the measurement cross sections. Starting with the given water elevation at the downstream end of the calculation domain ($x = 1400$ m), the backwater profile was calculated with various values of the Manning friction coefficient n. A value of $n = 0.035$ gave a horizontal water level in agreement with the observations (the backwater profile is not very sensitive to n because the flow velocity is so low). The resulting distribution of $\tau_{b,1D}$ and also of the hydraulic radius R and the cross-sectional velocity are presented in Fig 11.

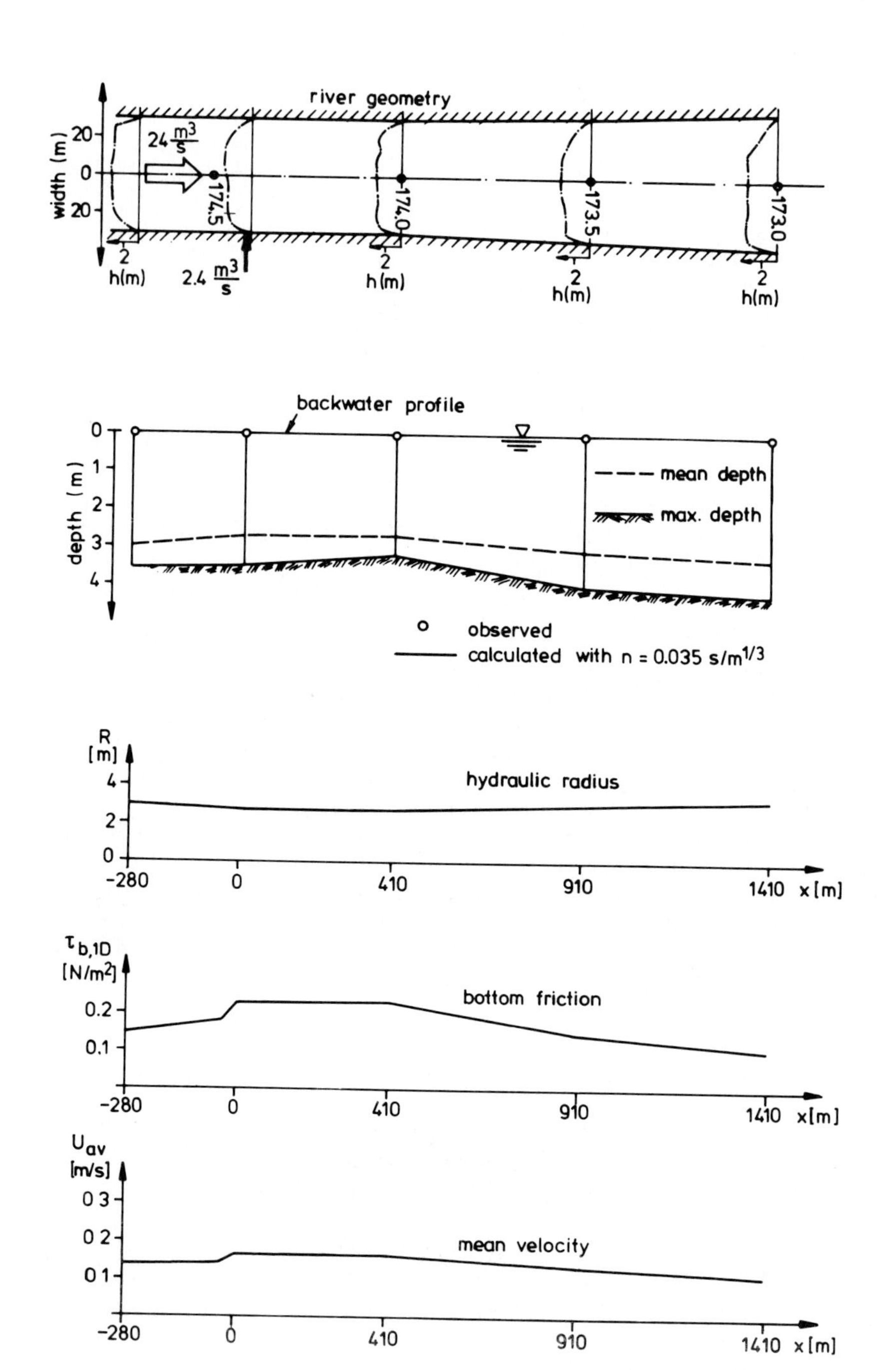

Figure 11. One-dimensional backwater profile calculations for the River Neckar.

3.1.4 Computational details of the 2D-calculation. As indicated in Fig. 13, the elliptic model component was used for a river stretch extending from 53 m upstream of the discharge to 165 m downstream, which was sufficient for the flow to be parabolic at this cross section. From there onward the parabolic model was employed down to $x =$ 1400m. Over the relatively short stretch covered by the elliptic calculation, the constant cross section shown at $x = 10$ m was assumed. The numerical grid used for the elliptic calculation is indicated in Fig. 14, showing 34 grid lines in the x-direction and 31 in the y-direction, with a grid-point concentration in the discharge and recirculation regions. The y-distribution of grid 'ines was retained for the parabolic model, although of course in this model the grid expands as the river widens. In the parabolic model the actual river profiles were taken, with a linear interpolation between the measurement cross sections. The forward steps of the marching forward procedure were 0.015 times the local river width B, except for the first 50 integration steps where the forward step was 0.005 B.

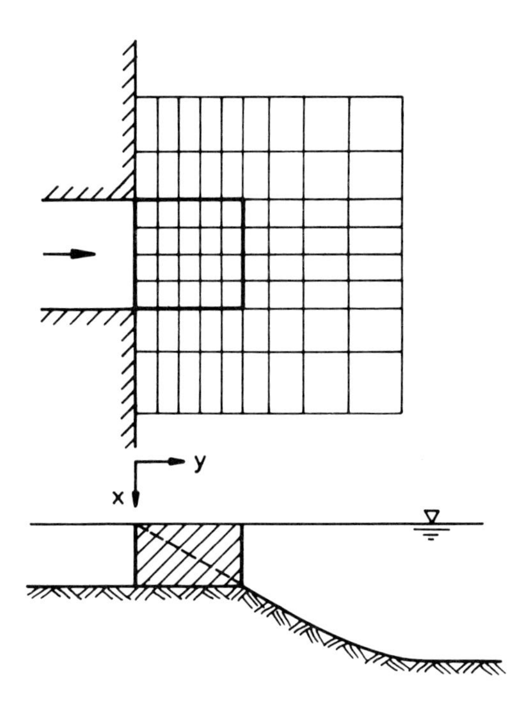

Figure 12. Representation of discharge in mathematical model.

The velocity profile at the inflow boundary ($x = -53$ m) was taken to be the same as that measured 200 m upstream of the discharge (the corrected profile was taken). The inflow concentration is, of course, zero. The discharge is treated so that for the shaded grid zones in Fig. 12 the water depth assumes the value of the discharge channel depth (h_s = 1.33 m), while outside these cells the depth is given by the river geometry. For the first row of these boundary cells the lateral inflow through the left faces of the control volumes are specified by the discharge velocity and a concentration of unity (only the concentration relative to the discharge concentration is of interest).

The initial conditions for the parabolic calculation starting at x = 165 m, provided by the elliptic model at that cross section, are given in Fig. 13.

The elliptic calculation took 239 iterations to reach the convergence criterion specified in Section 2.4, computing time was 17 minutes on a UNIVAC 1108. The computing time for the parabolic calculation covering a much longer river stretch was only 2.5 minutes.

3.1.5 Comparison of 2D-predictions with field measurements. The predicted profiles of velocity and concentration are shown for the measurement stations in Fig. 13 where they are compared with the experimental profiles. The comparison should be carried out with the corrected measurements. Results of two calculations are included for the concentration profiles. When e^* is taken as 0.15 as measured in wide and straight laboratory flumes, the dye-spreading in the far-field can be seen to be considerably smaller than that observed. When a value of e^* = 0.6 is adopted as measured in various rivers (e.g. Glover 1964, Yotsukura, *et al.*, 1970), the dye-spreading is in good agreement with the corrected field measurements. The higher value for the diffusion coefficient may be necessary because the river is bent upstream of the discharge. In the near-field (say for x < 300 m) the calculations are however not sensitive to the value of the diffusion coefficient as empirical input because there the turbulence is more generated by the discharge jet and the concentration field is more influenced by convection processes. Further, the change in the diffusion coefficient had little effect on the velocity profiles. In the upper part of Fig. 13 these can be seen to be very well predicted by the model, in particular near the discharge, where the velocity is reduced near the bank (no measurements were taken in the reverse-flow region) and is increased outside the recirculation zone. The calculated streamlines in the elliptic near-filed region are shown in Fig. 14. They indicate that a recirculation zone develops which is approximately 50 m long and 13 m wide. Detailed measurements were not taken in this region, but a recirculation zone of about the same dimensions was observed.

3.2 Cooling-Water Discharge from Karlsruhe Power Station into the Rhine

3.2.1 Problem specification. In this section, the model is tested further by application to the cooling-water discharge from the 600 MW Karlsruhe power station (Rheindampfkraftwerk Karlsruhe). The course of the Rhine in the vicinity of Karlsruhe is given in the map of Fig. 15. A close-up of the discharge area is presented in Fig. 16. In the river section downstream of the discharge the river is only very slightly curved, but there is a more severe bend some 5 km upstream.

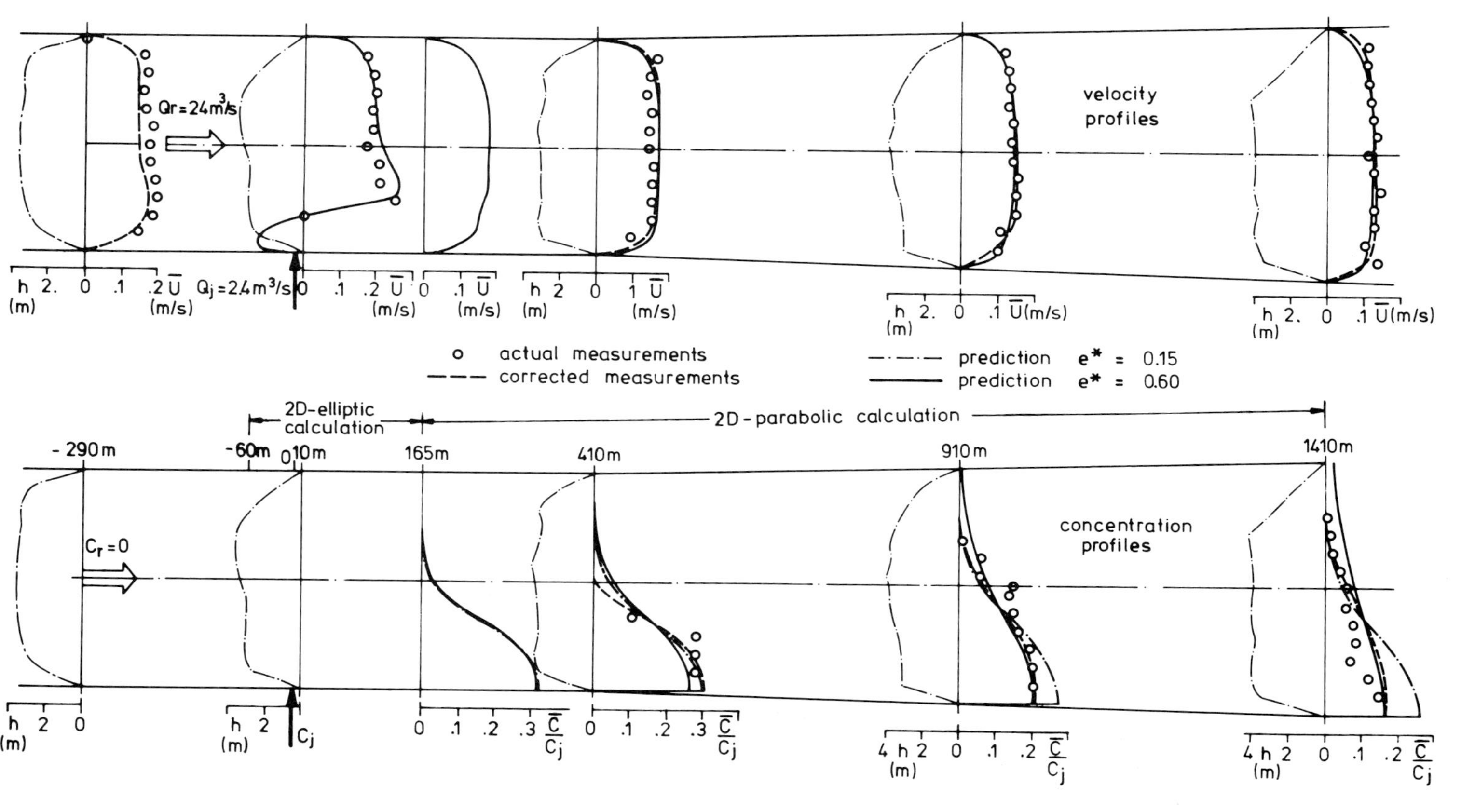

Figure 13. Depth-averaged velocity and concentration profiles for the River Neckar.

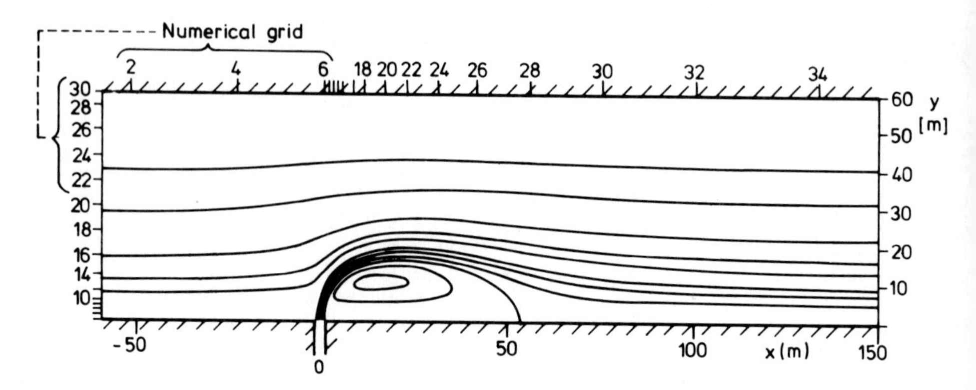

Figure 14. Predicted streamlines in the vicinity of the discharge into the Neckar.

Groynes are located at the bank opposite the discharge, and the groyne line meanders causing also a meandering of the flow. As a consequence, the main flow shifts from the right bank upstream of the discharge to the left bank 900 m downstream, as can be seen from the bed profiles and the velocity profiles presented in Fig. 20. One feature that makes this particular discharge situation not so ideal for model verification studies is the fact that 300 m downstream of the discharge the entrance to the Karlsruhe harbor interrupts the bank line (see Fig. 16). Measurements have indicated, however, that there is little exchange of flow and heat between the almost stagnant water body in the harbor and the river passing by.

The discharge geometry is sketched in Fig. 17. The discharge is at a right angle to the bank, and the discharge channel forks into three outlet channels. The cooling water is rejected at the surface only at very low river discharges. For all cases considered here, the discharged jet was submerged. Figure 17 shows the water-levels for the two river discharges mainly considered in this study, $Q \approx 840 m^3/s$ and $Q \approx 1400 m^3/s$. Calculations were carried out for three different conditions, as summarized in Table 2.

Table 2. *River and Discharge Situations Simulated*

Case	Date	River Discharge (m^3/s)	Cooling Water Discharge (m^3/s)	Excess Temperature of Cooling Water (° C)
1	Dec. 18, 1975	835	16.2	7.0
2	Nov. 20, 1975	1370	16.0	8.3
3	May 22, 1980	1420	10.64	5.38

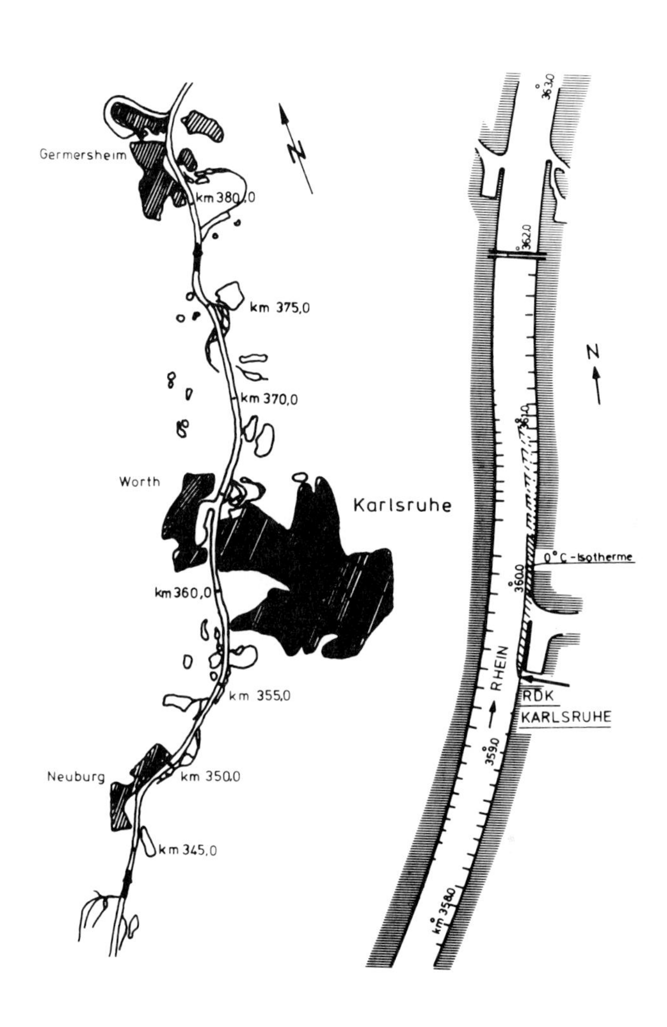

Figure 15. River Rhine in the vicinity of the discharge.

3.2.2 Field-measurements. Grimm-Strele carried out field-measurements on the dates shown in Table 2 to determine the flow and temperature field in the Rhine near the discharge from the Karlsruhe power station. For Cases 1 and 2 he used a boat to measure the velocity distribution with propellers and the

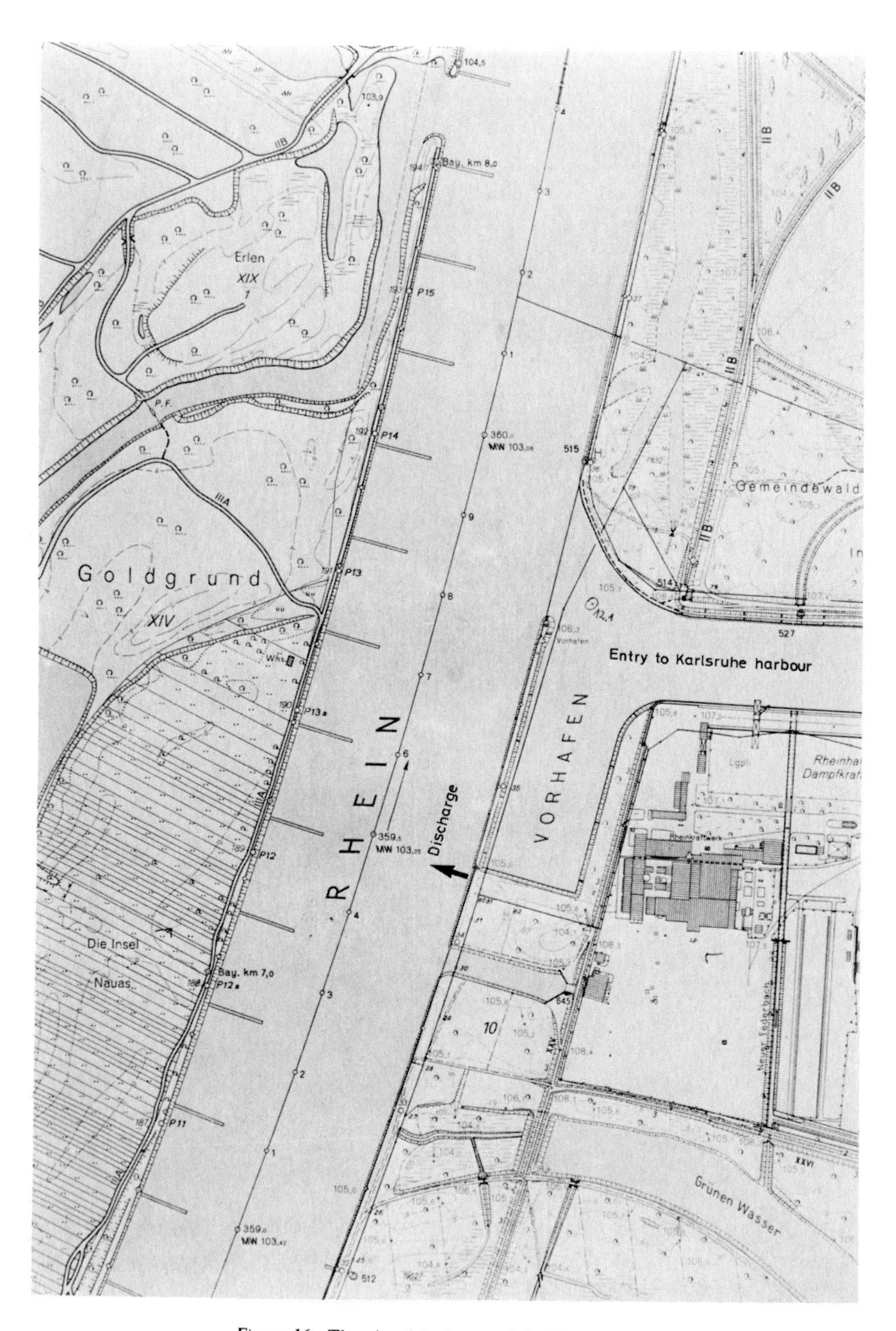

Figure 16. The simulated part of the River Rhine.

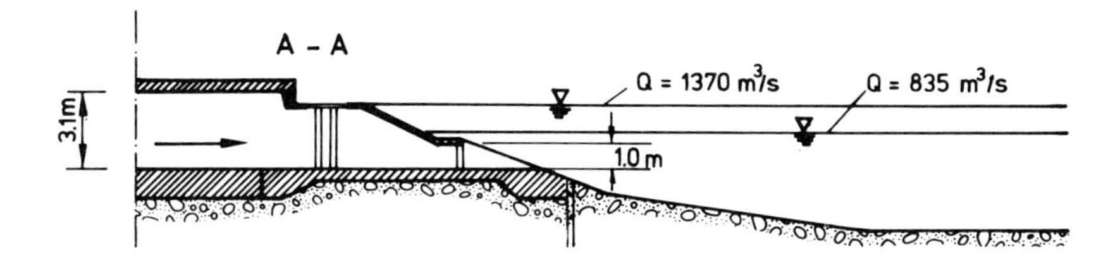

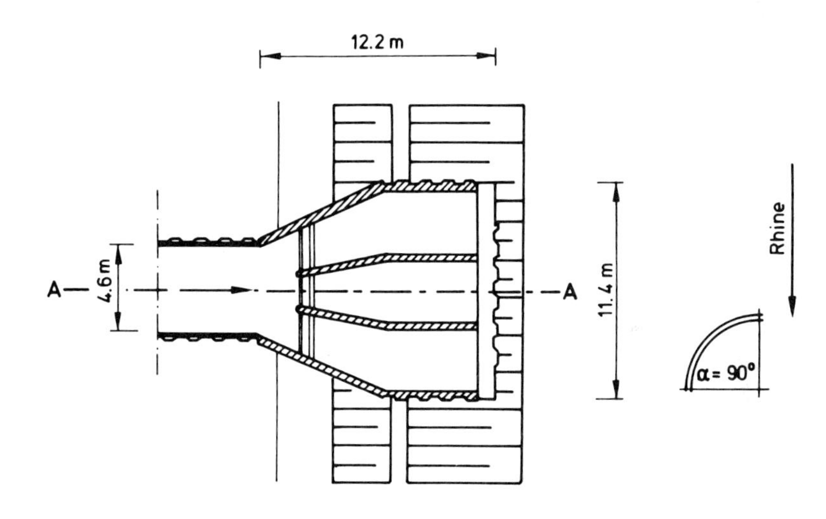

Figure 17. Configuration of cooling water discharge into the Rhine River.

temperature distribution with thermistors (Grimm-Strele, 1976). He measured both temperature and velocity at various depths so that depth-averaged values could be determined. At the measurement cross sections only weak vertical non-uniformities of the temperature were observed. The measured depth-averaged profiles of velocity and concentration are shown in Fig. 20. Within the experimental accuracy, the vertical and horizontal velocity distribution for different discharges was found to be the same. The vertical profile in the undisturbed river region is nearly logarithmic. A shift in the main flow from the right bank to the left bank in the region of interest was observed, as discussed already. For Case 1 with the lower river discharge, the conservation laws are satisfied reasonably well. For Case 2 with the high discharge, the heat flux determined from the measured temperature and velocity profiles varies considerably with x and is only 50 percent of the rejected heat at $x = 100$ and $x = 200$ (see Grimm-Strele, 1976). Hence, the temperature profiles were corrected such that the heat is conserved. For approximately the same river discharge, the near-field measurements for Case 3 shown in Fig. 21 indicate significantly more spreading. Based on this finding, the temperature profiles for Case 1 were corrected such that they are wider than the measured ones.

The measurements reported in Grimm-Strele (1976) do not cover the region very close to the discharge where a recirculation zone may occur. Hence, this author has recently carried out additional measurements just in the near-field region, for conditions listed as Case 3 in Table 2. Only the temperature distribution was measured by placing a thermistor in 40 cm depth on a 18 m long aluminum bar hinched at the river bank and supported by floats on the river. During these measurements, the river discharge was very high, and so was the water level. The cooling water jet was therefore well submerged and surfaced only several meters away from the bank. Between this zone and the bank, river water entered the region behind the discharge suppressing any tendency for reversed flow, which in this case was not observed. Instead, a zone of low velocity developed behind the discharge.

3.2.3 One-dimensional backwater profile calculations. For yet a different measurement campaign (November 28, 1975, Q=900m^3/s), Grimm-Strele (1976) gives detailed information on the backwater profile. This case is therefore used to determine the Manning roughness factor n prevailing in the calculation domain. One-dimensional backwater profile calculations were carried out with the river-bed profiles as given in Fig. 18, the groyne line representing the opposite bank. The entry to the Karlsruhe harbor was ignored, that is the bed profile in the entry region was simply taken by interpolating between the profiles at $x = 300$ m and $x = 700$ m. The influence of the value of n was examined. Good agreement with the observed backwater profile was obtained with $n = 0.0225$. In this case, the backwater profile was quite sensitive to the value of n because of the relatively high velocity of between 1.5 to 2 m/s. One-dimensional calculations were then carried out for the Cases 1 and 2 specified in Table 2 with the chosen n. The results for the backwater profile, the bed shear $\tau_{b,1D}$, the hydraulic radius R and the cross section average velocity are given in Fig. 18.

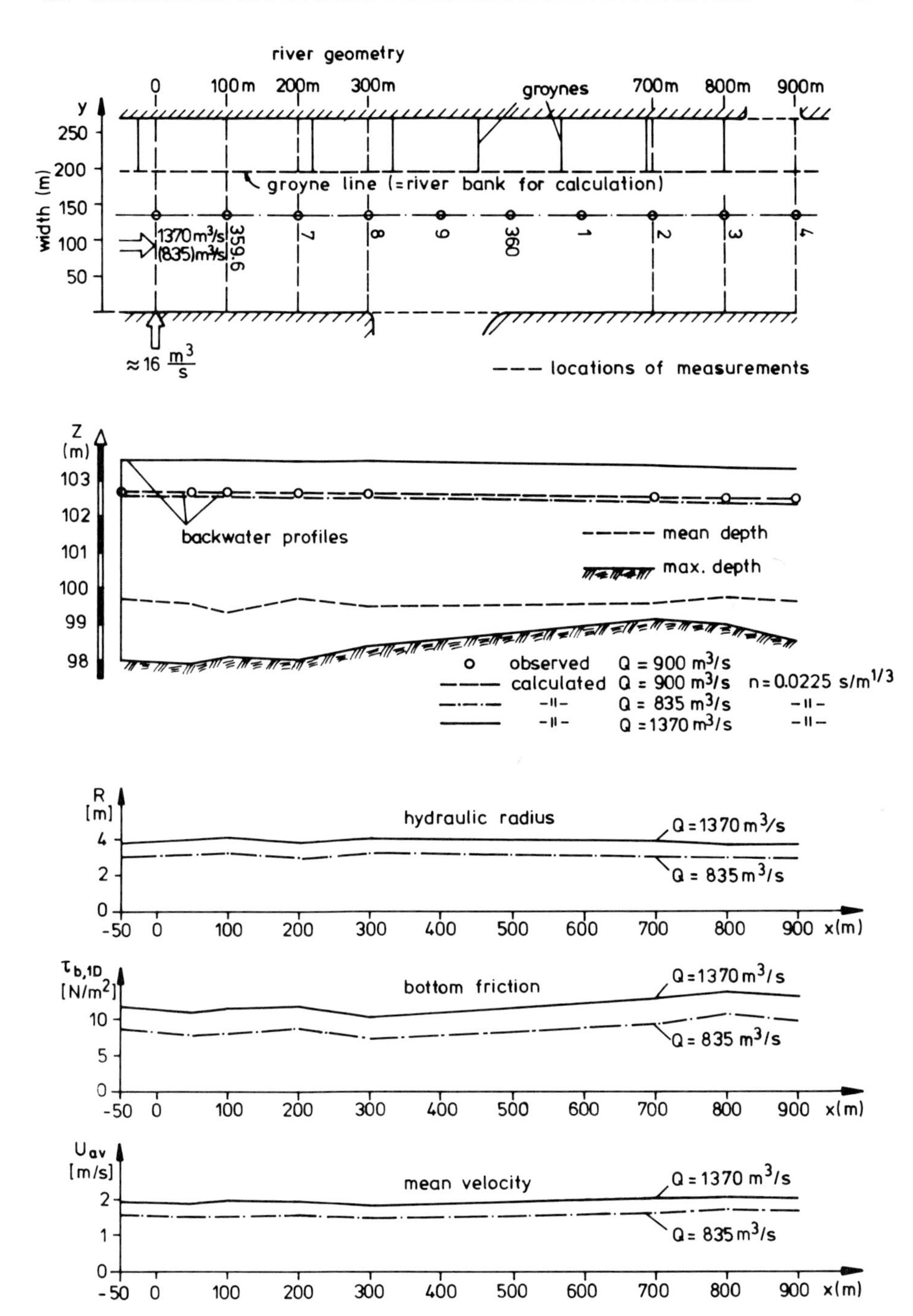

Figure 18. One-dimensional backwater profile calculations for the River Rhine.

3.2.4 Computational details of the 2D-calculation. The elliptic region extended from 50 m upstream of the discharge (the first measurement station) to 180 m downstream. From there to $x = 900$ m (the last measurement station) the parabolic model was applied. The cross section was assumed constant for the relatively short elliptic region and was taken as an average of that given at $x = -50$ m, $x = 50$ m and $x = 100$ m. The numerical grid used is shown in Fig. 22 where the grid points can be seen to be concentrated in the discharge and near-bank region. The actual bed profiles were used in the parabolic calculations, with linear interpolation between the measurement stations. The forward steps were the same as in the Neckar case. The entry to the Karlsruhe harbor 300 m downstream of the discharge was ignored, as measurements had indicated that there is little exchange between the river flow and the stagnant water in the harbor. The heads of the groynes were treated as the river bank on the opposite side. In the elliptic model, the computation domain could have been extended to the dead-water zone between the groynes (see Srivatsa and Rodi, 1980b), but since the groynes are on the opposite bank and are therefore quite far away from the thermally polluted zone, such a detailed treatment appeared not warranted. In the present calculations, the groynes have only an indirect influence in that their increased flow resistance will affect the value for the Manning roughness factor n and thus the bed shear and the effective diffusivity.

The velocity profile measured at $x = -50$ cm for $Q = 900 \mathrm{m}^3/\mathrm{s}$ was adjusted and taken as initial profile for the elliptic calculations. The excess temperature is of course zero at the inflow boundary. The submerged discharge configuration leads to a strongly three-dimensional flow field very near the discharge, and it is difficult to do justice to this configuration in a 2D-depth-average model. The discharge geometry has to be idealized strongly. For Cases 1 and 2 the same treatment of the discharge was used as described for the Neckar case in Section 3.1.4. This means that the discharge was assumed to be at the surface (see Fig. 19), having the same cross section as the real discharge. This shift to the surface also means that the outlet is moved landward several meters and is located at the edge of the river. Case 3 was first also calculated with this discharge configuration, but the resulting flow pattern in the near-field, which is very similar to that shown in Fig. 14, did not resemble the flow observed during the field measurements. As was mentioned already, reverse flow was not observed behind the discharge while the calculations with the discharge at the river edge did, of course, produce a recirculation zone. Hence, an attempt was made to idealize the discharge in such a way that a more realistic flow pattern be generated. The discharge location was shifted from the edge ($y = 0$) to a position y midway between the points where the real discharge channel intersects with the sloping bank (see Fig. 19). At this y-location, the discharge was represented by a 9.7 m long line source that emits mass, momentum and heat only in the positive y-direction. Of course, the emission occurs now over the local water depth (2.68 m) and not, as in reality, from a channel of 1 m depth, but care was taken that the correct amount of mass, momentum and heat was added.

The elliptic calculation took 174 iterations to reach convergence, and the computing time was 14 minutes. The parabolic calculations required a computing time of only 1 minute.

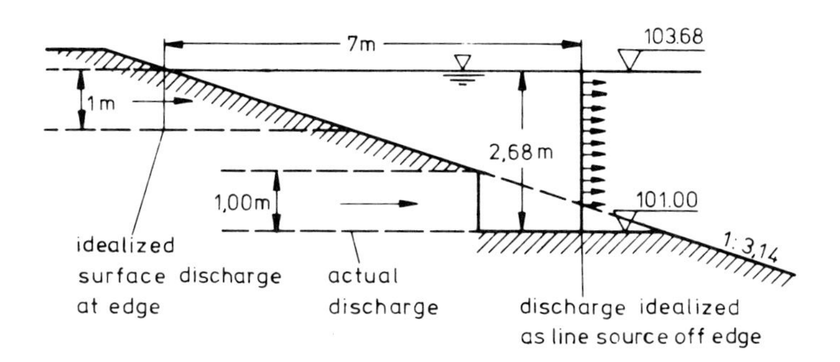

Figure 19. Idealization of discharge.

3.2.5 Comparison of 2D-calculations with field-measurements. For the Cases 1 and 2, predicted depth-average velocity and excess-temperature profiles are shown respectively in Figs. 20a and 20b, and are compared with Grimm-Strele's (1976) field-measurements whenever possible. The difference between Cases 1 and 2 is the river discharge. For Case 1 with the lower discharge the concentration profiles agree on the whole reasonably well with the measurements, although there appears to be somewhat more spreading and dilution in the measurements in the region between 100 and 300 m downstream of the discharge. This may be due to the fact that the immediate vicinity of the discharge is not calculated entirely satisfactorily by the 2D-model, a remnant influence from the groynes located at the right bank upstream of the discharge which may cause increased lateral mixing. A value of $e^* = 0.6$ was used as empirical input into the turbulence model. As in the Neckar case, a value of 0.15 gave too little spreading in the far-field, but there was little influence of e^* up to $x = 300$ m.

Velocity measurements for this river discharge were taken only at $x = 100$ m and $x = 800$ m. Figure 20a shows clearly that the model simulates well the velocity profile and in particular the shift of the main flow from the right to the left bank, which is caused by the change in the river-bed geometry.

For Case 2 with high river discharge there is also reasonable agreement when the predictions are compared with the corrected experimental profiles. The model predicts correctly that there is faster dilution. There are no velocity measurements available for this case. The predicted velocity profiles have the same trend as described above.

In Case 3 the river discharge was even higher. In this case only the near-field is considered, so that only elliptic calculations were carried out. Figure 21 compares predicted temperature profiles and isotherms with the measurements of Grimm-Strele. Results are included for both types of discharge idealizations investigated. When the discharge is placed at the river edge, there are, of course,

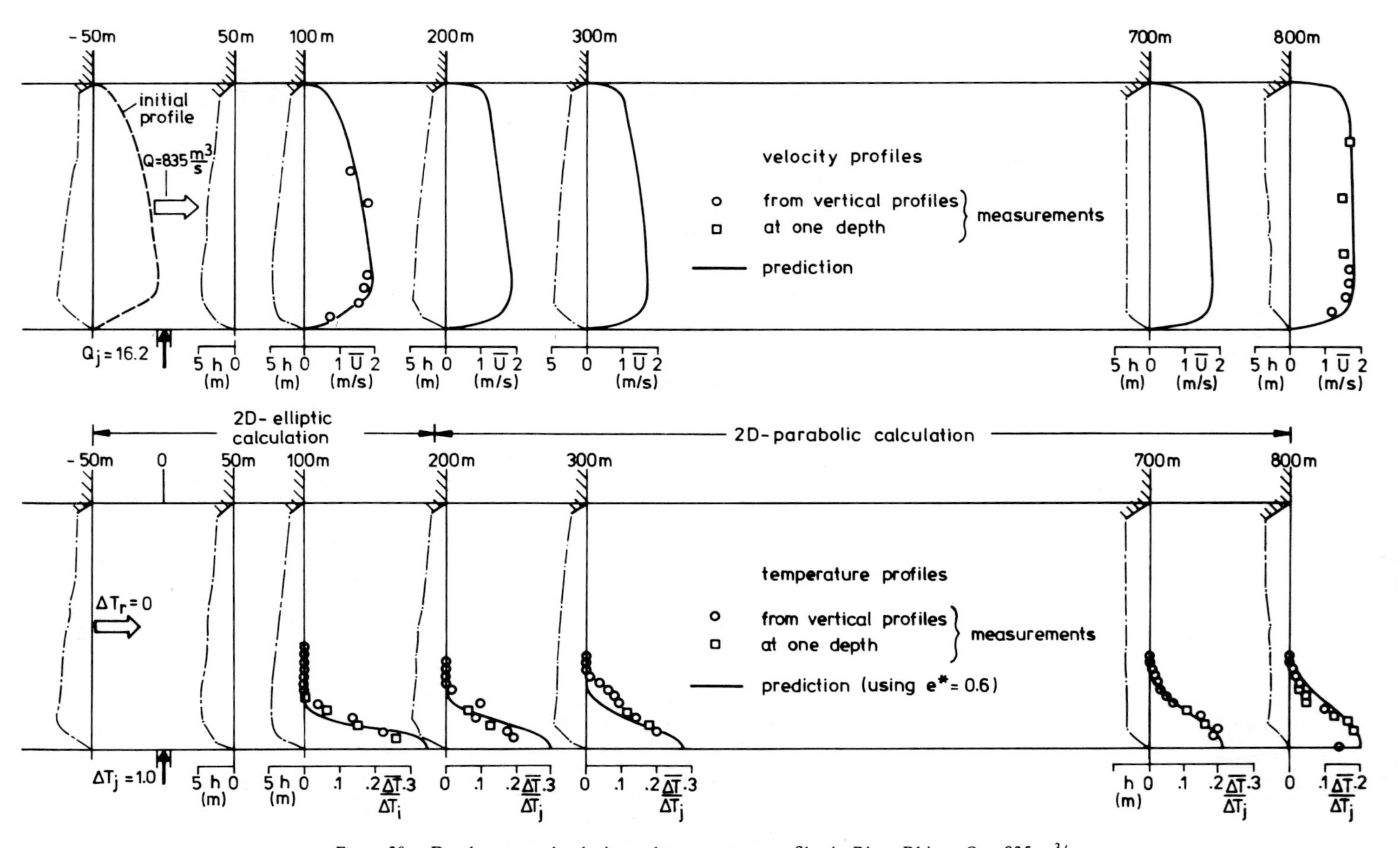

Figure 20a. Depth-averaged velocity and temperature profiles in River Rhine, $Q = 835\ m^3/s$.

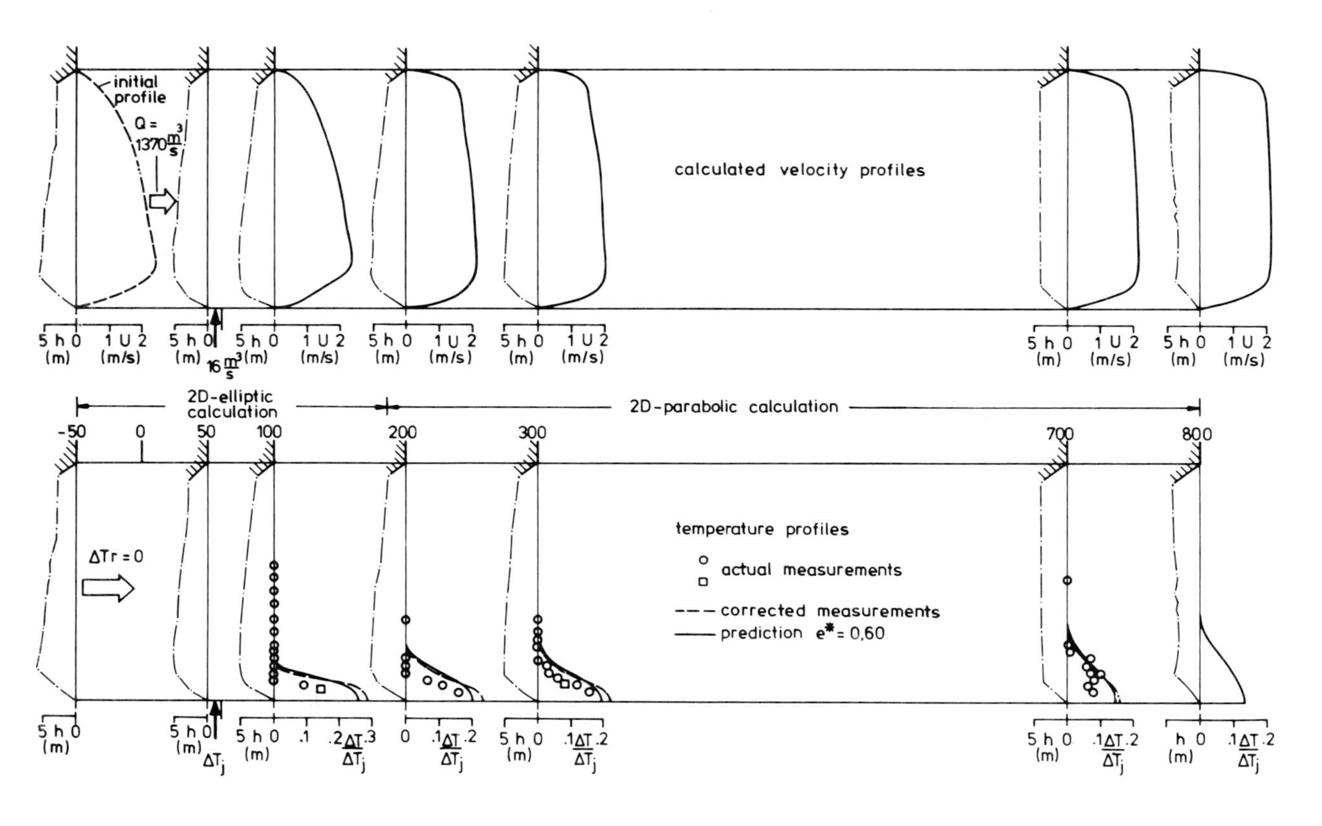

Figure 20b. Depth-averaged velocity and temperature profiles in River Rhine, $Q = 1370\ m^3/s$.

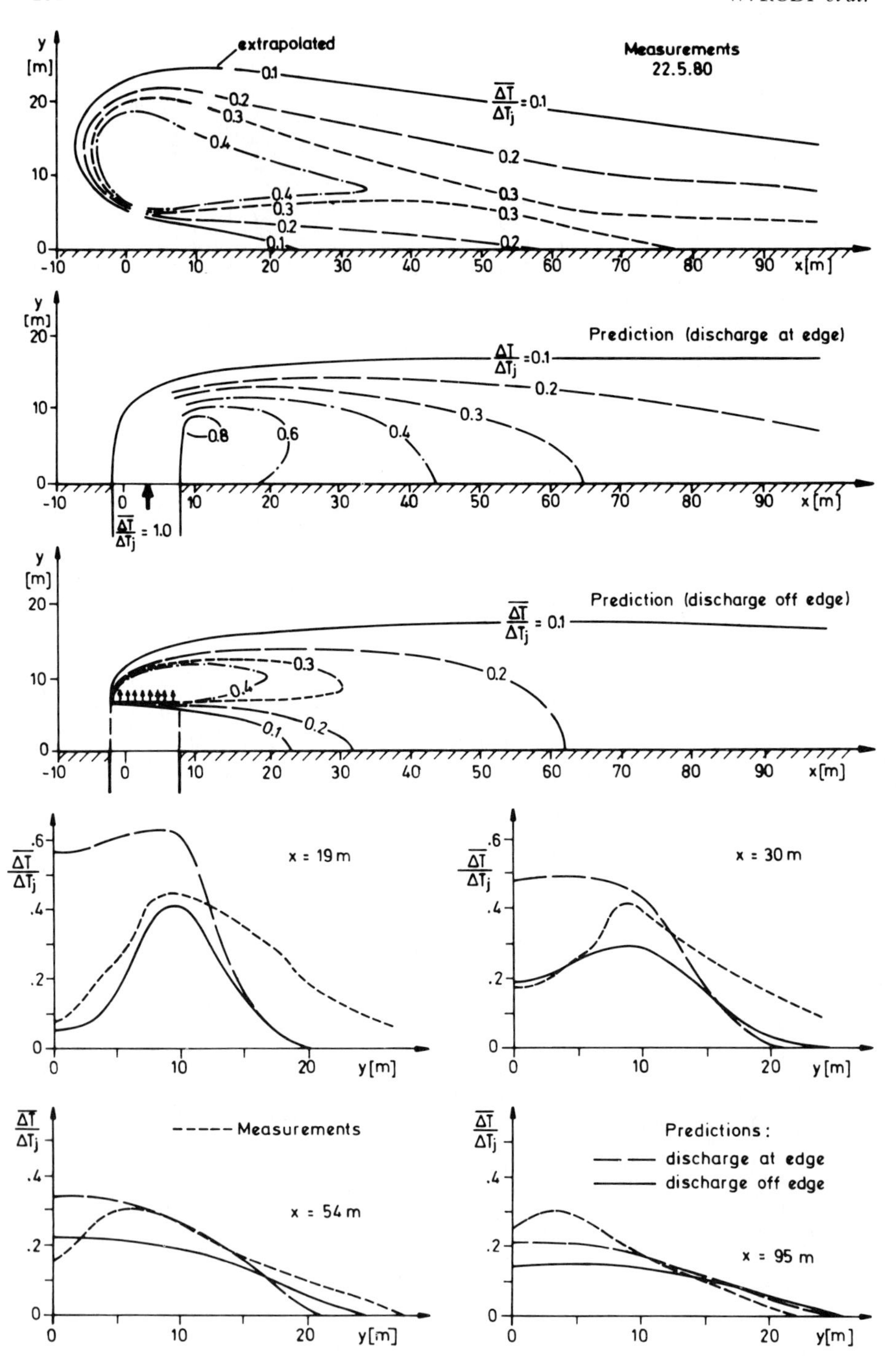

Figure 21. Temperature distribution in the near field of the cooling-water discharge into the Rhine, Case 3.

significant differences between the predicted and the measured temperature field in the immediate vicinity of the discharge. The area where the submerged jet surfaces is clearly discernible from the isotherms, and the center of this area is about 13 m away from the edge of the river. Hence, the apparent heat source is significantly further away from the river edge in the real situation than in either of the calculations, and therefore the isotherms lie further away from the edge of the river. When the discharge is placed off the edge, the near-field behavior is predicted more realistically. In particular, the existence of an initial low-temperature region near the bank is reproduced. In the intermediate region, the prediction with the discharge located at the river edge produces closer agreement with the experiments since it yields generally a higher temperature level (owing in this case to the reverse flow). Heat-flux evaluations indicate, however, that the fairly high experimental values are not depth-average values but pertain to a warm-water layer near the surface (the measurements were taken only at 40 cm depth). Initially, the heat flux determined from the measured temperature profiles is significantly larger than the heat input and approaches this only at $x = 100$. From there on the differences between the two calculations using different discharge idealizations also become small.

The conclusion of the near-field study of Case 3 is that strongly three-dimensional phenomena, which were present in the case, cannot be simulated realistically in a two-dimensional model. A three-dimensional model as described in the introduction is required in such situations.

Figure 22 shows the predicted streamline pattern for Case 3 with the discharge located off the river edge. As can be seen, water flows downstream between the discharge and the river edge, and there is no reverse flow. This flow picture is qualitatively in agreement with the observed one, and it is markedly different from the flow pattern obtained with the discharge located at the edge, for which Fig. 14 is representative.

4. CONCLUDING REMARKS

The novel feature of the mathematical model described in this paper is that it can resolve the details of the near field of a waste-water discharge but can also be applied to longer river stretches without excessive computational cost. The economy is achieved by the use of the model-chain concept in which the river is divided into characteristic regions and the simplest possible model component is employed for each region. A further important feature is the use of a refined turbulence model which is suitable for resolving the details of the near field where a complex interaction takes place between discharge-jet-generated and bed-generated turbulence. In the far-field (except near the banks) this model leads to the same diffusivity relation as used in conventional diffusion models, that is, the nondimensional diffusivity e^* assumes the value introduced into the model as empirical input. None of the empirical coefficients of the turbulence model have been tuned to match the experimental results but were simply adopted from the literature. The value of $e^* = 0.6$ adopted from measurements in rivers yields

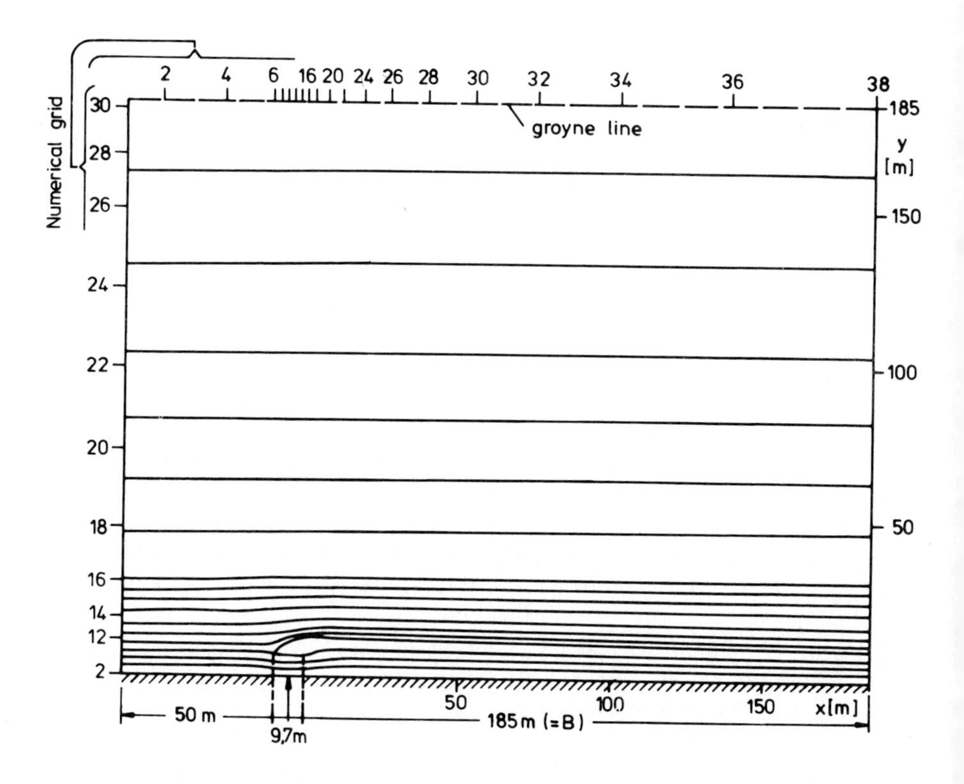

Figure 22. Predicted streamlines in the vicinity of the cooling-water discharge into the River Rhine for Case 3, $Q = 1420\ \mathrm{m}^3/\mathrm{s}$.

reasonable results for the far-field spreading of pollutants; the near-field behavior was found to be influenced little by this empirical input. In this context it should also be remembered that e^* may not represent just the turbulent diffusion but also the dispersion due to vertical non-uniformities and may thus account for the effect of secondary motions that are due, for example, to upstream bends.

Unfortunately, few detailed field measurements are available in the immediate vicinity of waste-water discharges so that the comparison between predictions and measurements for this region can only be sketchy. However, when the discharge is located at the surface, the behavior appears to be predicted well. For submerged discharges, on the other hand, three-dimensional effects were found to be strong so that the field near the jet exit is difficult to simulate with a two-dimensional model. The difficulties experienced point to the limits of the two-dimensional approach and call for the use of three-dimensional models.

Acknowledgements. The authors should like to thank Dr. M. A. Leschziner for his valuable advice on the calculations, Miss Gaby Bartman for typing the manuscript efficiently and Mrs. Susi Issel for producing the drawings. The calculations were carried out on the UNIVAC 1108 computer of the University of Karlsruhe and the work reported here was funded jointly by the Deutsche Forschungsgemeinschaft via the Sonderforschungsbereich 80 and by the Umweltbundesamt through contract 102 04 307.

REFERENCES

Carter, H. H. (1969). A Preliminary Report on the Characteristics of a Heated Jet Discharged Horizontally into a Transverse Current. Part I: Constant Depth. Chesapeake Bay Institute, Johns Hopkins University, Tech. Rep. No. 61.

Fischer, H. B., List, E. J., Koh, R.C.Y., Imberger, J., and Brooks, N. H. (1979). "Mixing in Inland and Coastal Waters." Academic Press, New York.

Flokstra, C. (1976). Generation of Two-dimensional Horizontal Secondary Currents. Delft Hydraulics Laboratory Rep. S163, Part II.

Glover, R. E. (1964). Dispersion of Dissolved or Suspended Materials in Flowing Streams. U. S. Geological Survey Prof. Paper 433-B.

Gosman, A. D., and Pun, W. M. (1974). Calculation of Recirculating Flows. Imperial College, London, Rep. No. HTS/74/2.

Grimm-Strele, J. (1976). Naturmessdaten zur Quervermischung in Flüssen. Universität Karlsruhe, Sonderforschungsbereich 80, Rep. SFB 80/E/75.

Holly, F. M., Jr. (1975). Two-dimensional Mass Dispersion in Rivers. Colorado State University, Fort Collins, Hydrology Papers No. 78.

Käser, F., Tödten, H., Hahn, H.H., and Küppers, L. (1980). Naturmessungen über die Konzentrationsverteilung von konservativen und suspendierten Wasserinhaltsstoffen unteralb einer Einleitung in einen Fluss. Universität Kalsruhe, Sonderforschungsbereich 80, Report SFB 80/ME/157.

Kuipers, J., and Vreugdenhil, C. B. (1973). Calculations of Two-dimensional Horizontal Flow. Delft Hydraulics Laboratory Rep. S163, Part I.

Lau, Y. L., and Kishnappan, B. G. (1978). Transverse dispersion in trapezoidal channels. *Transport Processes and River Modelling Workshop, Burlington, Ontario, 1978.*

Launder, B. E., and Spalding, D. B. (1974). The numerical computation of turbulent flow. Computer Methods in Applied Mechanics and Engineering, 3, pp. 269-289.

Leschziner, M. A., and Rodi, W. (1979). Calculation of a strongly curved open channel flow. *J. Hydraul. Div. Proc. Am Soc. Civ. Eng.* **105,** No. HY10, 1297-1314.

McGuirk, J. J., and Rodi, W. (1978). A depth-averaged mathematical model for the near field of side discharges into open channel flow. *J. Fluid Mech.* **86,** 761-781.

McGuirk, J. J., and Rodi, W. (1979). Calculation of unsteady mass exchanges between a main stream and a deadwater zone. *Proc. Cong. Int. Assoc. Hydraul. Res. 18th,* B.A.4.

McGuirk, J. J., and Spalding, D. B. (1975). Mathematical modelling of thermal pollution in rivers. *Proc. Int. Conf. Math. Models for Environmental Problems, Southampton, 1975.*

Patankar, S. V., and Spalding, D. B. (1970). "Heat and Mass Transfer in Boundary Layers," 2nd Ed., Intertext Books, London.

Patankar, S. V., and Spalding, D. B. (1972). A calculation procedure for heat, mass and momentum transfer in three-dimensional parabolic flows. *Int. J. Heat and Mass Transfer.* **15,** 1787-1806.

Pavlović, R. N., and Rodi, W. (1979). Berechnung der Wärmausbreitung in Flüssen mit einem tiefengemitelten parabolischen Modell, Universität Karlsruhe, Institut für Hydromechanik, Report No. 569.

Powell, R. W., and Posey, C. J. (1959). Resistance experiments in a triangular channel. *J. Hyd. Div. Proc. Am. Soc. Civ. Eng.*, HY5, 31-66.

Pratap, V. S., and Spalding, D. B. (1976). Fluid flow and heat transfer in three-dimensional duct flows. *Int. J. Heat and Mass Transfer* **19,** 1183-1188.

Rastogi, A. K., and Rodi, W. (1978). Predictions of heat and mass transfer in open channels. *J. Hydraul. Div. Proc. Am. Soc. Civ. Eng.* **104,** HY3, 397-420.

Rastogi, A.K., Rodi, W., and Yen, B. C. (1975). Backwater Surface Profile Computer Program. Universität Karlsruhe. Report SFB 80/T/48.

Rodi, W. (1980a). Berechnung der Abwärme- und Abwasserausbreitung in Flüssen. *Wasserwirtschaft*, 185-190.

Rodi, W. (1980b). "Turbulence Models and their Application in Hydraulics: A State-of-the-Art Review." International Association for Hydraulic Research Book Publication, Delft.

Rodi, W., and Srivatsa, S. K. (1980b). A locally elliptic calculation procedure for three-dimensional flow and its application to a jet in a cross flow. *Computer Methods in Applied Mechanics and Engineering,* **23,** 67-83.

Rodi, W., and Srivatsa, S. K. (1980b). A Mathematical Model for the Flow in Channels Containing Groynes. Universität Karlsruhe, Report 80/T/160.

Yotsukura, N., and Sayre, W. W. (1976). Transverse mixing in natural channels. *Water Resources Res.* **12,** No. 4, 695-704.

Yotsukura, N., Fischer, H. B., and Sayre, W. W. (1970). Measurement of mixing characteristics of the Missouri River between Sioux City, Iowa, and Plattmuth, Nebraska. U. S. Geological Survey Prof. Paper 1899-G.

DISCUSSION

Vladan Milisic

Institute Jaroslav Cerni, Belgrade, Yugoslavia

In connection with Dr. Rodi's lecture, I would like to mention the possibilities of application of the same basic mathematical model for definition and prediction of transport and settling processes of sediments.

The main differences between passive pollutant transport and sediment transport are the following: (1) a negative buoyancy term appearing in the dynamic transport equation, and (2) in the closing turbulent equations, it is necessary to add the influence of particles on the redistribution of turbulent flow energy.

The same types of numerical models are used (GENMIX and TEACH) and the steady 2-D parabolic and elliptic flow has been considered, but in the vertical plan (case of settling basin).

Two different levels of closure turbulence models were used: (1) differential flux model which entails the solution of transport equations for turbulent stresses and turbulent flux of particle concentration (κ-ϵ-$\overline{u'v'} - \overline{s'^2} - \overline{v's'}$) for parabolic flow, and (2) "κ-ϵ" model for recirculating flow. Interesting outcome: profiles of hydrodynamic field (mean velocity, turbulent quantities); concentration profiles.

It is also interesting to note that the models predict the interaction between the hydrodynamic field and particles, i.e., the turbulence damping and consequent continuous development of the flow due to the particles settling.

The results obtained have been compared in the first phase of study with some measurements done in steady flow conditions in the Llaui Dujailah Irrigation Canal (Iraq). Some of these results were presented at the 130 EUROMECH Colloquium, June 1980, Belgrade, Yugoslavia.

REPLY

We were pleased to hear that Dr. Milisic uses a similar mathematical model for sediment transport calculations, and we should like to add that work of this kind is in progress also at the University of Karlsruhe. Attention needs to be drawn to the differences between the models: Dr. Milisic's 2-D model considers processes in the vertical plane while the model presented in the paper is for 2-D depth-average calculations.

Y. L. Lau

National Water Research Institute, Canadian Department of the Environment.

I would like to comment that although none of the empirical constants used in the model had to be tuned to experimental data, the non-dimensional diffusivity e^* is a quantity which is site-specific as it represents not only turbulent diffusive transport but also the effect of non-uniformity of velocity profiles. This quantity can vary significantly for different river geometries and the value of 0.6 used in this paper may not apply in many cases. This may not be a problem for the three-dimensional model.

REPLY

In response to Dr. Lau's comment we should like to say that the non-dimensional diffusivity e^* is certainly site-specific. We have found, however, that the value of 0.6 recommended by Fischer, *et al.*, (1979) works surprisingly well for a number of different sites. In addition to the simulation of discharges into the rivers Neckar and Rhine described in the paper, Pavlović (1981) performed calculations of cooling-water discharges into the Danube near Ingolstadt and into the Po near Piacenza. In these cases, the uses of $e^* = 0.6$ led also to predictions in fairly good agreement with field measurements. Hence, we believe that $e^* = 0.6$ is a reasonable value for many rivers, except perhaps when strong irregularities and bends are present. It would, of course, be desirable to determine e^* for each site from dye-spreading experiments.

Peter Mangarella

Woodward Clyde Consultants, San Francisco, California, USA

One of your initial slides showed a rather dramatic enhancement of lateral transport of heat around a river bend, yet your model does not seem to address this problem. Could you explain the physical mechanism(s) responsible for this phenomena, and the system of equations (with reference to your existing model) required to model this feature? Are you aware of (and if so, please describe) other models which attempt to take into account the effect of river bends in lateral transport.

REPLY

We should like to reply to Dr. Mangarella that the increase in lateral spreading of warm water in a river bend apparent from an infrared photograph of a section of the Rhine is probably due to the secondary motion in the bend. This motion, which is induced by centrifugal forces, moves fluid near the surface from the inner to the outer bank and is particularly effective in increasing the spreading of discharged warm water when this has risen to the surface due to the action of buoyancy. These processes can be described properly only by a three-dimensional model, where the secondary motion is determined by solving the momentum equations for the transverse and vertical directions. Leschziner and Rodi (1979; reference cited in the paper) have developed such a model which predicts realistically the secondary motion. The model has not been applied so far to calculate the warm water spreading. In the 2-D depth-average model the effects of secondary motion on the spreading of heat are represented by the dispersion term in Eq. (5), but as this model is incapable of predicting the secondary motion (nor the vertical variation of Φ), the dispersion term cannot be evaluated. Hence, in this model, the only way to account for the influence of bends on heat spreading is by use of empirical dispersion terms in Eq. (5)., e.g., by an increase of the value of e^*.

Timothy J. Smith

Institute of Oceanographic Sciences,
Crossway, Taunton, Somerset, England

In view of the author's statements regarding the use of uncalibrated coefficients, I would like to question the author regarding the use of a turbulent Schmidt number (Sc) of 0.5 and the sensitivity of the results to the parameter e^*. I understand that the value of $Sc = 0.5$ was chosen as having been shown to be representative of free shear turbulence. However, in view of the small depth to width ratio in the channel, the turbulence structure must be dominated by wall turbulence generated at the bottom boundary. Some indication for this is the sensitivity of the model results to the parameter e^* which scales the rate of production of dissipation rate by wall generated shear. Consequently, does the author think that a higher value of Sc representative of boundary turbulence might be more appropriate? I would also like to suggest that an examination of the relative magnitudes of the terms in the governing equations at various points within the flow field might help to evaluate the importance of the bed shear production terms and identify those regions of the flow where sophisticated turbulence models are essential to accurate prediction.

REPLY

Dr. Smith comments that the turbulent Schmidt number (σ_t in the paper) of 0.5 used in the model is suitable for free shear turbulence and that a higher value representative of near-wall turbulence would be more appropriate. The value of 0.5 was chosen to simulate correctly the initial discharge-jet region where the lateral mixing is influenced strongly by the free shear turbulence in the jet. Further downstream, where the mixing is governed by bottom-generated turbulence, the value of the turbulent Schmidt number has no influence as the diffusivity $\tilde{\Gamma}_t$ is determined solely by the non-dimensional diffusivity e^* specified as input via the relation $\tilde{\Gamma}_t = e^* U_* h$.

REFERENCE

Pavlović, R.N. (1981). Mathematische Simulierung von Wärme- und Stoffausbreitung in Flüssen. Ph.D. thesis, Universität Karlsruhe.

MODELING OF LAGOONS: THE EXPERIENCE OF VENICE

G. Di Silvio and G. Fiorillo

Institute of Hydraulics,
University of Padua, Italy

1. INTRODUCTION

In November 1966, the attention of the public in Italy and all over the world was attracted by a large flood suffered by the city of Venice, at the same time that the city of Florence was also struck by a far more serious flood of the Arno River. Although the city had often experienced the phenomenon of "Acqua Alta" (High Water, that is, a more or less severe flood due to the combination of storms, seiches and large astronomical spring tides), the maximum water level on that occasion (1.94 m above sea level), as well as the duration of the flood (many hours), were exceptionally large. Moreover, the frequency and gravity of minor inundations has become increasingly alarming in the last decades.

Two main causes for the worsening of the "Acqua Alta" have been recognized: (a) the progressive rising of the sea-level, produced by the melting of the polar pack (eustatism); and (b) soil subsidence, produced by excessive withdrawing of groundwater, especially in the industrial area of Marghera. As an order of magnitude we may say that, in the last fifty years, the mean sea level in Venice has increased 25 cm, one-third of which should be attributed to the eustatism. Although 25 cm difference may appear negligible, it must be kept in mind that most of the city's pavement is placed at elevation hardly higher than 0.75 m above the present mean sea level and that many squares and streets of Venice become flooded even with ordinary spring tides. Due to the raising of the mean sea level, then, not only is the extent of the city liable to be flooded increased, but a tidal event which fifty years ago had a frequency of only once in ten years, takes place nowadays practically every year.

The "Acqua Alta", in conclusion, has become a very common show on the Venetian scene, with obvious annoyance for pedestrians, troubles in boat navigation under the canal bridges, and increased discomfort due to the water invasion of shops, courts and ground floors.

Different measures have been taken or proposed to eliminate the floods of Venice or to reduce their frequency. A first step has been the interruption of

ISBN: 0-12-25852-0

Figure 1. St. Mark Square flooded in 1966.

groundwater withdrawal, in order to slow down or stop soil subsidence. The reduction of withdrawal from about 500 l/s to 200 l/s from 1972 to 1975, has produced a remarkable recovery of piezometric levels in the area. As a consequence soil subsidence in Venice has come to a stop, and even a small rebound of about 2 cm has been observed. Nevertheless, the city is still liable to frequent flooding, so means of shielding the city and the lagoon of Venice from the sea by adequate dams or barrages must be considered. A number of proposals have been made in the last fifteen years, ranging from a simple embankment around the city to a system of gates on the three inlets of the lagoon.

In 1976 the Italian Ministry of Public Works conducted an international bid for the design and construction of control works in the three inlets of the lagoon, fixing some constraints for the solution. The works are in principle aimed at reducing the tidal oscillation by means of a permanent narrowing of the three inlets; at the same time, however, provisions were to be made for eventual movable barrages to be inserted later in the inlets in order to protect the city during the exceptionally large tides. Five large groups of contractors have submitted their proposals for the examination, but none of them has been considered completely satisfactory.

Inevitably, since the memorial event of 1966, interminable discussions have arisen both on the diagnosis and the treatment of the "diseases" of Venice. On the one hand, emphasis was put on the universal reputation of Venice as a unique product of art and culture, on its peculiar character of an insular city, and on its incomparable environment. On the other hand, concerns were shown about the everyday necessities of the Venetians, the exigencies of the industrial port of Marghera, and the need for a social and economic development of the entire region. Opposition against the planned works has been expressed by various groups for different reasons: increased pollution, interruptions of maritime traffic, undesirable ecological modifications, excessive costs, etc. Many objections, moreover, came forward not only on the measures to be taken to protect the city from the "Acqua Alta", but also on the construction in the lagoon during the last fifty years. Factors that have been claimed to have possibly increased the water levels in Venice include the dredging of a new navigation canal from the inlet of Malamocco to the industrial area of Marghera, the embanking of some large peripheral areas of the lagoon for fish production, and the reclamation of tidal flats for industrial settlements (Fig. 2). Extensive studies are required to evaluate the real effects of these projects, as well as to predict the consequences of the planned protection against flood, both by physical and mathematical models.

2. PHYSICAL AND MATHEMATICAL MODELS

Whenever a detailed description of a restricted portion of the flow field is required, a non-distorted model with an appropriate reduction scale is still the most reliable tool of investigation. For instance, the local consequences of a gated barrier in the inlets of the lagoon (energy losses, bottom scouring, navigation problems, etc.) have been studied on a 1:60 scale, movable bed model (Adami, A., *et*

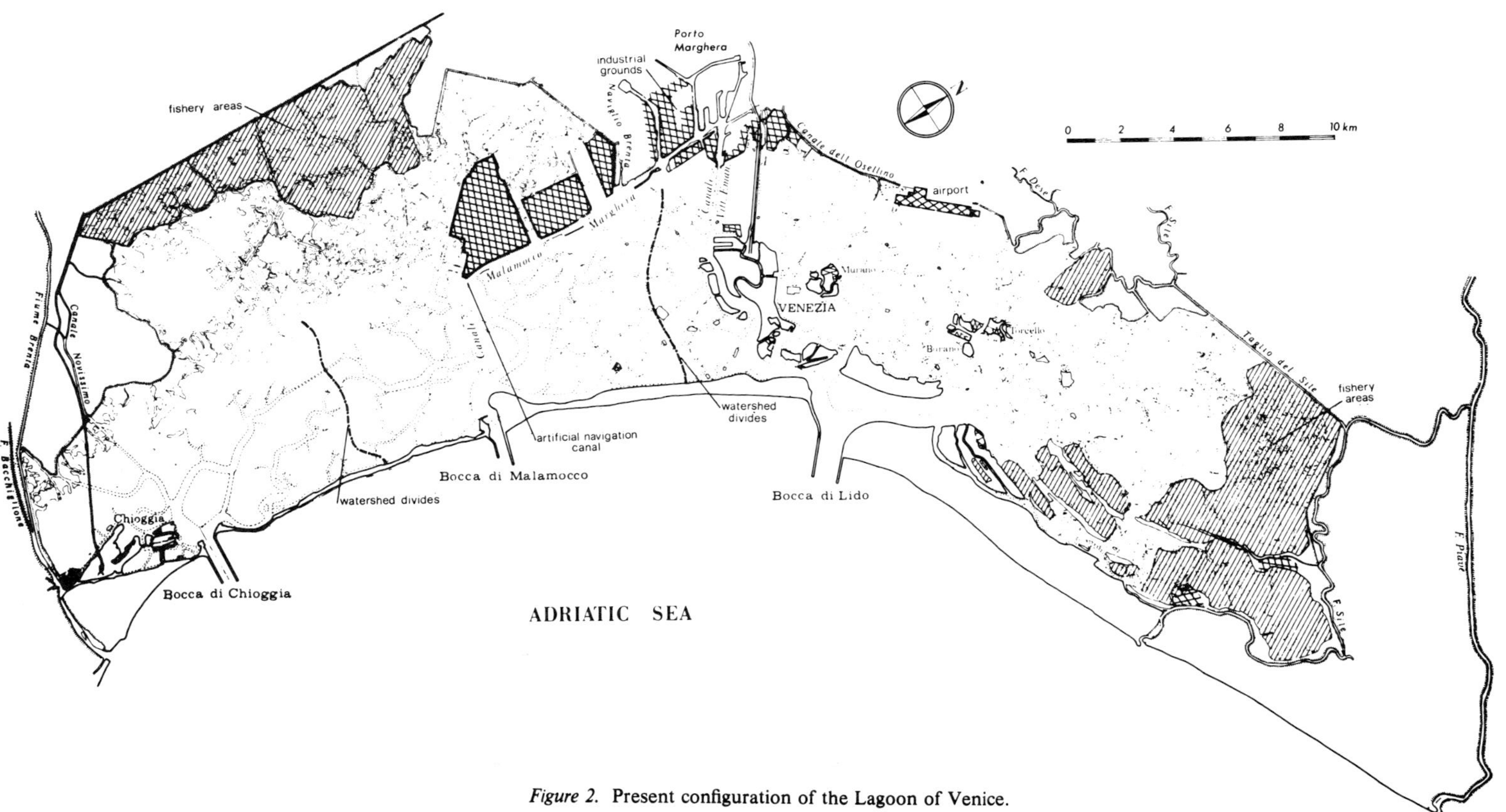

Figure 2. Present configuration of the Lagoon of Venice.

al. 1977). On the other hand, answering many of the questions posed in the previous paragraph requires the reproduction of the entire lagoon of Venice, that is an area of about $500km^2$. This requires a very large distorted model, like the one recently completed by the Ministry of Public Works (vertical scale 1:20; horizontal scale 1:250, model size $16,000m^2$).

In comparison with large distorted physical models, mathematical models often represent a much more manageable and less expensive alternative, especially if a large number of simulations are required. A mathematical model, moreover, can be set up with different degrees of detail depending upon the particular feature to be investigated. In this way, the maximum amount of information can be economically obtained by resorting each time to the appropriate configuration of the model.

In estuary hydrodynamics, the most general equation of fluid motion (Navier-Stokes equations) and waste transport (mass conservation) are always reduced to more or less simplified forms: the description of the flow field and concentration distribution is correspondingly more or less accurate and detailed, but in many cases even extremely simplified equations are sufficient to answer particular questions.

Successive simplifications of the general equations proceed from successive operations of averaging time and space. The first unavoidable step of these simplifications is the time-averaging aimed at eliminating the turbulent fluctuations of pressure, velocity and concentration. Following this operation, the original equations are transformed in the so-called Reynolds equations, where the resulting turbulent shear-stresses and diffusion coefficients are in their turn to be expressed in terms of time-averaged velocity components. Reynolds equations are able to accurately describe three-dimensional fluid motion (atmosphere, ocean, lakes), and they can, in principle, be applied to predict the velocity field of an estuary. Although some models of this type are mentioned in the literature (Leendertse, *et al.*, 1973-1977), the numerical integration of three-dimensional equations becomes intolerably heavy as soon as the estuary is large and its morphology complex.

Contrary to the atmosphere, deep bays and lakes, however, estuaries are invariably characterized by very small depths in comparison with horizontal dimensions, so that the vertical acceleration is negligible, while the concentration is in general uniformly distributed along the depth (no stratification). This circumstance allows a further simplification, consisting of averaging these quantities over the depth, leading to the so-called *two-dimensional models.* In estuaries with no stratification or small stratification (such as the Lagoon of Venice), a two-dimensional model gives practically as much information as a three-dimensional model, with a substantial reduction of computational work.

Even so, a two-dimensional schematization of an estuary still requires a very large number of grid-points for a computational grid fine enough to reproduce the complicated morphology of gullies, shoals and tidal flats peculiar to natural lagoons (Fig. 3). On the other hand, the very same complex morphology of a lagoon permits another important simplification of the equations, based on the fact that natural lagoons may be schematized as a network of channels, flanked by broad and shallow basins becoming more or less dried during the tidal cycle. In such a case the flow direction is mainly controlled by the channels, so that a one-dimensional dynamic equation (along the axis of the channel) is sufficient to describe the water motion. The one-dimensional equation proceeds from appropriate averaging operations over the cross section composed by the channel and the lateral basins. If the tidal flow is predominately up and down channels, then *one-dimensional models* are preferable to more expensive two-dimensional models. The last, though, are to be applied whenever the velocity field continuously changes its orientation, as it often occurs in an ample basin of almost constant depth.

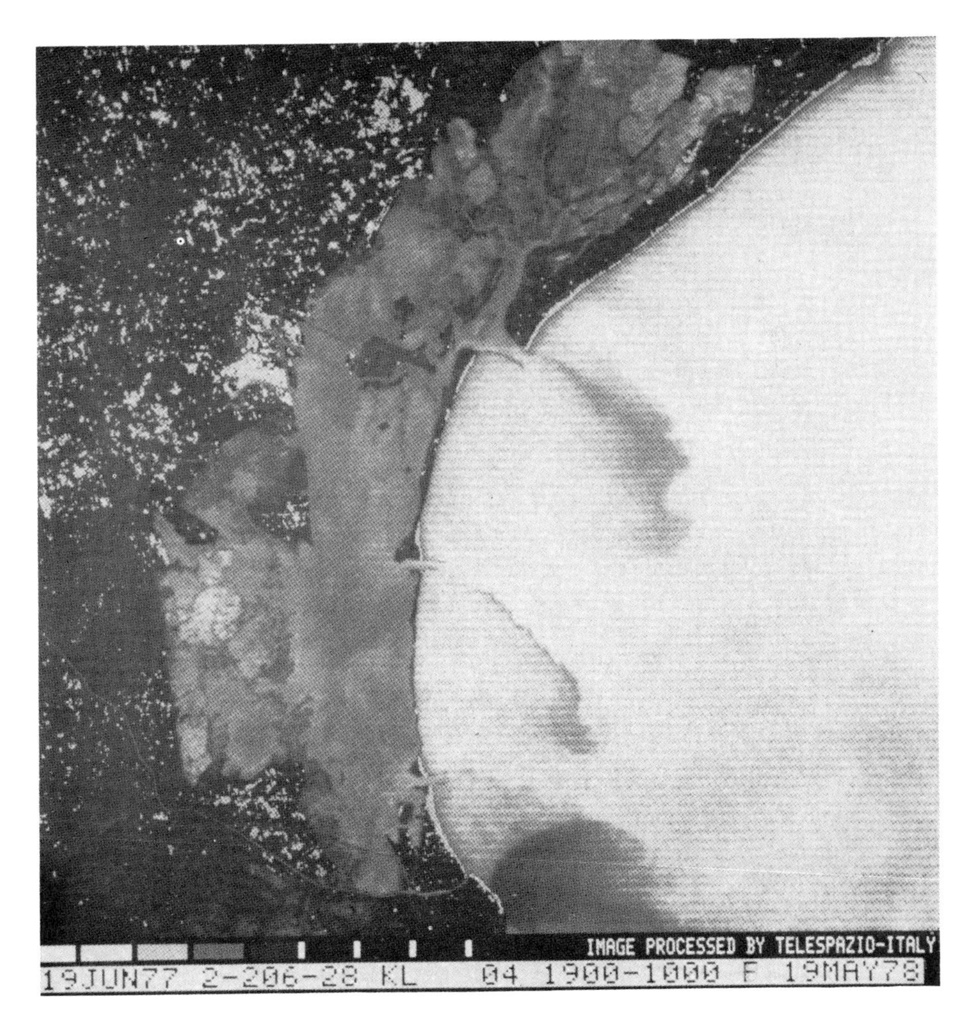

Figure 3. The Lagoon of Venice from satellite.

A last radical simplification of the equations is obtained by averaging over the entire surface of the estuary. By suppressing the spatial variables, the equations become ordinary differential equations and the spatial hydrodynamic characteristics of the basin (assumed to be concentrated in one point) are only varying with the time. *Zero-dimensional models* give, of course, less detailed information but they can often be profitably employed to predict with reasonable accuracy time-variable quantities in a certain location of the basin, especially if results are checked against the results of more sophisticated models.

With the exception of the three-dimensional approach, which is definitely impractical for a general description of the Lagoon of Venice, all the above mentioned mathematical models have been carried out, compared and jointly employed in different applications.

2.1 Zero-Dimensional Model

The zero-dimensional model has been used to investigate some basic hydrodynamic features related to the planned restriction of the three inlets (Ghetti, *et al.,* 1972; Ghetti, 1974). If energy losses are supposed to be concentrated in the inlets, the dynamic equation of the zero-dimensional model is reduced to

$$Q = \pm\ C_q A \sqrt{2g\,|h_s - h_b|} \tag{1}$$

where Q is the instantaneous flood or ebb-discharge through the inlet; A is the inlet cross section; h_s and h_b are the instantaneous water levels in the sea and within the basin; C_q is the discharge coefficient of the inlet structure, obtained by theoretical considerations or, better, by experiments on the physical model. In addition, the zero-dimensional continuity equation is

$$Q = S\,\frac{dh_b}{dt} \tag{2}$$

where S is the surface of the lagoonal basin relevant to the inlet. For a given tidal oscillation in the open sea, the numerical integration by standard methods of Eqs. (1) and (2) provides the tidal oscillation within the lagoon, as well as the discharge and water velocity through the inlet.

A zero-dimensional model has been extensively applied, in behalf of the Ministry of Public Works, for statistical predictions of water level lowering, flushing-volume reduction, interruption of navigation, as consequences of different types of intervention. The accuracy of this model has been tested by a comparison with the following more detailed one-dimensional model.

2.2 One-Dimensional Model

A one-dimensional model of the Lagoon of Venice has been set up since 1970, not only for checking the results of the zero-dimensional model, but also to get some information on the *spatial distribution,* over the lagoonal basin, of the water level, velocity and pollutant concentration.

The hydrodynamic model is based on the numerical integration of the momentum and continuity equations for a channel having a composed cross section

$$\frac{\partial h}{\partial x} + \alpha \frac{V}{g}\frac{\partial V}{\partial x} + \frac{\beta}{g}\frac{\partial V}{\partial t} + j = 0 \tag{3}$$

$$\frac{\partial Q}{\partial x} + \frac{\partial A}{\partial t} = 0 \tag{4}$$

where h is the water level above the datum-line; Q is the total discharge flowing through the entire cross section; $V = Q/A$ is the average velocity on the composed cross section; j is the energy gradient; α and β are the correction factors taking into account the non-uniform velocity distribution in the composed cross section. In Eq. (3), the assumption has been made that the instantaneous velocity distribution over the composed cross section is essentially controlled by friction forces, that is by resistance and depth of the main channel and the lateral shoals.

It has been assumed, in other words, that in each subsection i ($i=l$, left shoal; $i=c$, main channel; $i=r$, right shoal), longitudinal discharge represents a certain portion R_i of the total discharge Q

$$R_i = \frac{Q_i}{Q} = \frac{\chi_i A_i \sqrt{a_i}}{\sum\limits_i \chi_i A_i \sqrt{a_i}} \tag{5}$$

where χ_i is the Chezy coefficient, A_i the area and a_i the mean depth of each subsection. By the same token, the energy gradient and velocity-distribution coefficients are

$$j = \frac{Q^2}{\left[\sum\limits_i A_i \chi_i^2 \sqrt{a_i}\right]}; \quad \alpha = A^2 \sum_i (R_i^3/A_i^2); \quad \beta = A \sum_i (R_i^2/A_i) \tag{6}$$

Solution of Eqs. (3) and (4) provide the longitudinal discharge in each subsection, as well as the transversal flow between main channel and lateral shoals ($i=l,r$)

$$q_i = \frac{\partial Q_i}{\partial x} + b_i \frac{\partial h}{\partial t} = Q \frac{\partial R_i}{\partial x} + \frac{\partial h}{\partial t}(b_i - R_i b) \tag{7}$$

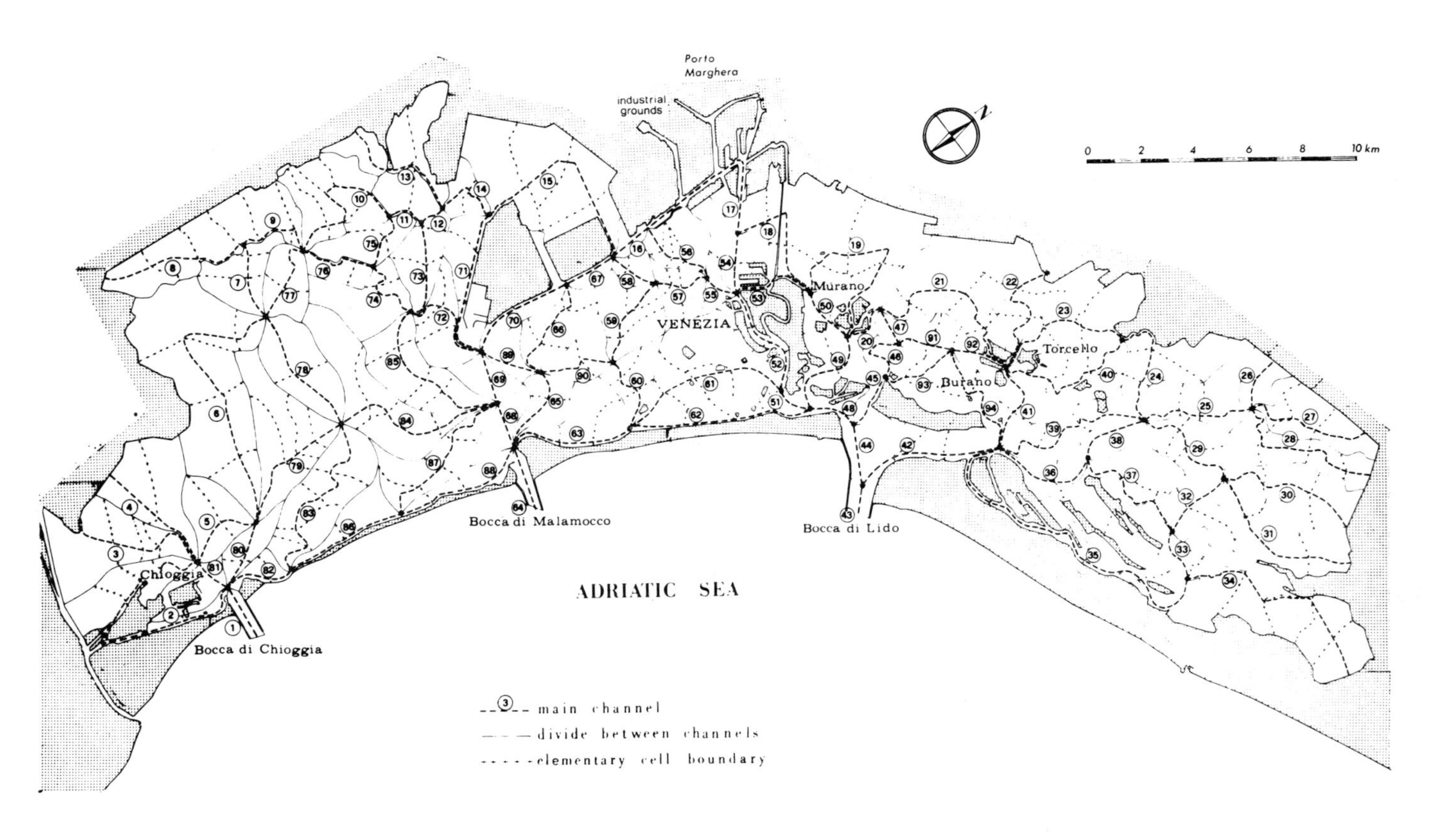

Figure 4. One-dimensional schematization of the Lagoon of Venice.

b_i and b being respectively the subsection's and the total width. This equation shows that transversal flow can be produced in two ways: (a) by longitudinal variations of the channel geometry (even in steady conditions), and (b) by time variation of water level (even in prismatic channels).

Equations (3) and (4), after some manipulation, have been written in finite difference form and numerically integrated by an implicit method. The Lagoon of Venice has been schematized as a network of about 100 connected channels (Fig. 4), each one divided in a number of reaches (flanked by the pertinent shoals both on the left and the right side), according to the local morphology. Boundary conditions are given by the tidal oscillation in the Adriatic Sea (DiSilvio, *et al.*, 1972a; 1972b).

The schematization of water-flow between adjacent subsections as given above permits large economies in the network size; should each subsection be considered an independently connected element of the network, computational cost would be dramatically increased.

The one-dimensional model is then linked with the equation of pollutant transport (DiSilvio, *et al.*, 1977). Longitudinal transport along the main channel and the lateral shoals is considered together with the transversal exchange between channels and shoals. Transversal exchange appears to be a very effective cause of mixing in lagoons, as well as the phase-lags of the tidal currents in confluent channels. In Fig. 5 the mechanism of dispersion due to the transversal exchange between shoals and main channel is represented.

The one-dimensional model has been especially used to evaluate the hydrodynamic effects of the constructions made in the Lagoon of Venice in the last 50 years and to predict the movement of waste and pollutants, either in the actual situation or following the planned control works in the inlets.

A "Lagrangian" version of the model has been also set-up to follow the trajectories of floating particles during successive tidal cycles, up to their eventual expulsion from the lagoon into the open sea (D'Alpaos, *et al.*, 1977).

2.3 Two-Dimensional Model

Due to different causes, the central part of the Lagoon of Venice has been subjected in recent decades to a remarkable deterioration of the original network of channels. While the gullies tend to be silted, the lateral shallows tend to be deepened, so that a general flattening has taken place and is very likely still in progress. In this area a two-dimensional approach is definitely preferable for a good reproduction of the velocity field and the dispersion of pollutants.

A two-dimensional model of the Central Lagoon has been set up in collaboration with the Delta Service of the Dutch Ministry of Waters. The model is essentially the one proposed by Leendertse (1970-1975) for Jamaica Bay (U.S.A.), and is based on the numerical integration of the following equations of momentum (respectively in x and y directions) and continuity

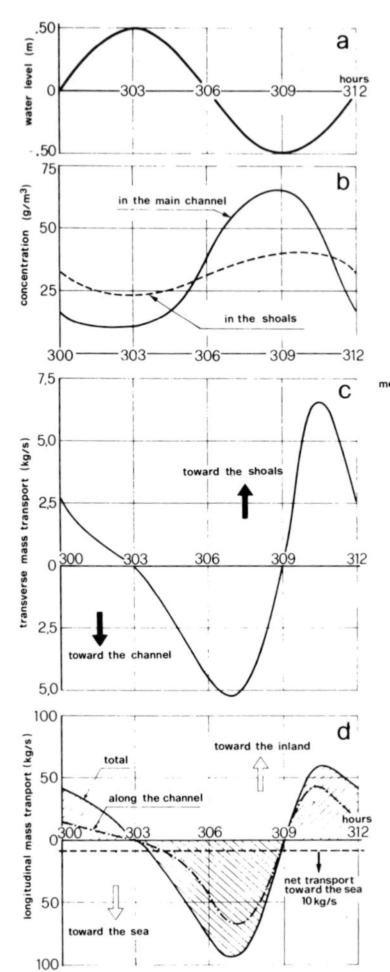

Figure 5. Dispersion mechanism in a rectangular lagoon with a continuous injection of mass (quasi-permanent conditions). Graphs on the left show (a) sinusoidal vertical tide; (b) periodical variation of concentration within the cell in the middle of the channel and within the adjacent shoals; (c) alternate transversal transport of mass between channels and shoals; (d) alternate longitudinal transport of mass along the channel and shoals. Note that the net transport of mass toward the sea during a tidal cycle, equal to the mass input, is produced by longitudinal and lateral advection only (dispersion coefficient equal to zero).

$$\frac{\partial h}{\partial x} + \frac{u}{g}\frac{\partial u}{\partial x} + \frac{v}{g}\frac{\partial u}{\partial y} + \frac{\partial u}{\partial t} + j_x = 0 \tag{8}$$

$$\frac{\partial h}{\partial y} + \frac{u}{g}\frac{\partial v}{\partial x} + \frac{v}{g}\frac{\partial v}{\partial y} + \frac{\partial v}{\partial t} + j_y = 0 \tag{9}$$

$$\frac{\partial h}{\partial t} + \frac{\partial}{\partial x}(au) + \frac{\partial}{\partial y}(av) = 0 \tag{10}$$

where h is the water level with respect to a datum; u and v are the velocity components in x and y directions; a is the water depth; and, j_x and j_y are the components of the energy gradient.

The two-dimensional model covers only the central part of the lagoon, which has been schematized with a 225 m size mesh, rather smaller than the breadth of the major channels of this area (Fig. 6). In this way the total number of grid-points has been reasonably limited, still allowing an accurate representation of the local morphology. Note that an equally accurate representation of the peripheral areas of the lagoon, cut by a maze of narrow channels, would have required a much finer mesh size.

The level oscillations at the open boundaries of the two-dimensional model (boundary conditions) are provided by the one-dimensional model, which is both less expensive and better fitted to describe the complicated morphology of the entire lagoon.

Due to the high computational cost of the two-dimensional model, however, only a moderate use has been made of it for studying the current pattern and the dispersion processes in the central area (where the application of the one-dimensional model is undoubtedly more questionable), while it has been primarily employed to compare its results with those of the one-dimensional model.

3. CHECKING THE MODELS

Due to the approximations inherent to any mathematical model, no matter how detailed and realistic it may be, a certain amount of unfaithfulness in reproducing the real-life phenomena is to be expected. The same is true, on the other hand, for the physical models as well, where a perfect reproduction of all phenomena is prevented by effects of scale reduction and distortion.

A peculiar abiding feature of mathematical models, though, are the inaccuracies proceeding from the numerical techniques of integration. In many cases, computational errors can be reduced by increasing the number of computational steps, either in space or time, but obviously at the expense of increased computational work, and very often with the risk of giving place to numerical instabilities. For the transport-diffusion equation a substantial improvement of the solution can be achieved by introducing an appropriate negative coefficient to counteract the effects of the numerical diffusion (Boris, *et al.*, 1976; Ghetti, *et al.*, 1978). In any case a somewhat inaccurate numerical solution of the equations is to be accepted as soon as the physical processes are satisfactorily reproduced by the model and dominate the solution itself. An example of the relative importance of numerical and physical dispersion and of the possibility of reducing the first by an appropriate correction is given in Fig. 7.

The problem, indeed, is not the complete elimination of numerical errors, but rather how to make an estimate of them. A first idea of the numerical accuracy can usually be achieved by a *sensitivity analysis,* that is, by testing how large the difference is among results obtained by the same model by modifying the

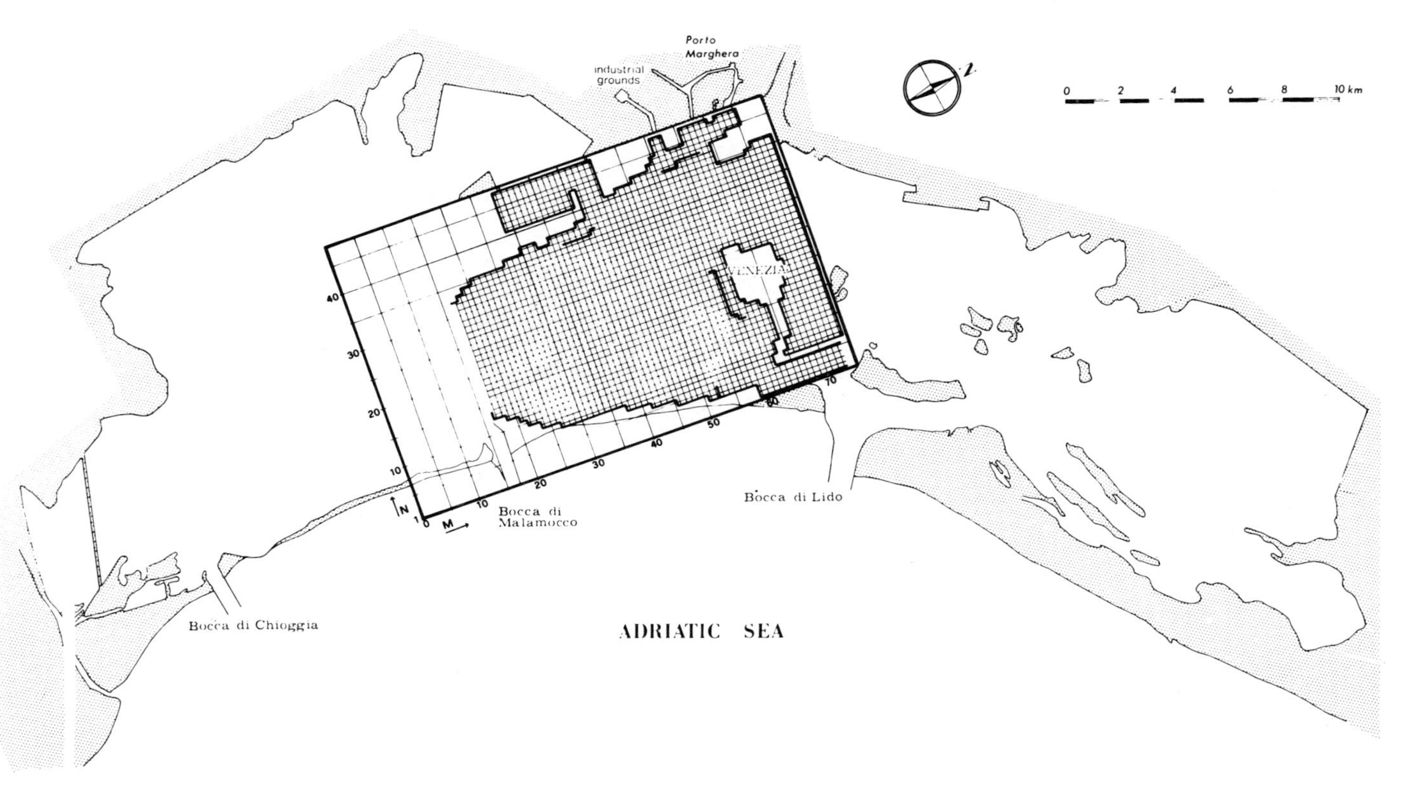

Figure 6. Two-dimensional schematization of the central part of the Lagoon.

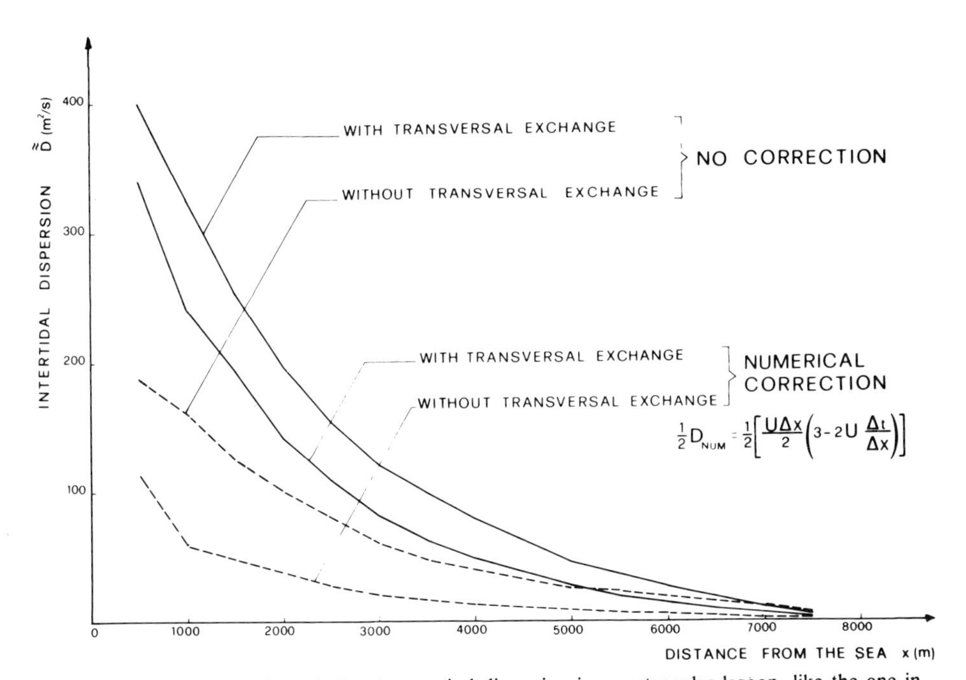

Figure 7. Effects of physical and numerical dispersion in a rectangular lagoon, like the one in Fig. 5. The intertidal dispersion coefficient $\tilde{\tilde{D}}$, at different distances from the sea, is defined as the ratio between tide-averaged mass-transport and tide-averaged concentration-gradient. With no correction (graphs above), the effects of numerical dispersion are even larger than the physical effects due to the alternate transversal transport between channel and shoals. By introducing a suitable negative dispersion-coefficient (graphs below) the effects of numerical dispersion become much smaller than the physical effects. Unfortunately, the solution becomes unstable if the full theoretical correction is introduced in the computation. The numerical integration of the transport equation is based on an implicit "donor-cell" (backward) scheme, with central time-difference.

computational time and space steps. Consistent results should also be obtained by introducing "plausible modifications" of the geometric schematization and of the resistance and dispersion coefficients.

Another useful way to evaluate the accuracy of the computational technique is to compare the finite-difference solution with an analytical solution available for a more schematic, still similar to the actual, configuration. By this procedure, the hydrodynamic one-dimensional model has been checked against the analytical solution for the case of a rectangular lagoon crossed by a deep rectangular channel (Datei, 1972; Datei, *et al.,* 1972; Dronkers, 1972).

The most convincing way of checking a model (either physical or mathematical), however, is to compare the model results with measurements carried out on the prototype, under different tidal conditions. Usually some of the experimental data are initially used to adjust the calibration parameters (like resistance and dispersion coefficients); subsequently, a verification of the model is made by resorting to an independent set of observations.

As for the hydrodynamic quantities the numerical models of the Lagoon of Venice have been extensively checked by reproducing tidal oscillation recorded at different gauges, discharge measurements made through the inlets and the major channels, surface velocity distribution in several areas of the basin detected by aerophotogrammetric methods (Ministero dei Lavori Pubblici, 1979).

The capability of the models to simulate dispersion processes has subsequently been tested by tracer experiments. The results of this comparison will be described in the following paragraph. Both models have been found insensitive to the value of the longitudinal dispersion coefficient. It has been set equal to zero in the one-dimensional model and equal to 20 m^2/s (only for numerical reasons) in the two-dimensional model.

4. AN EXAMPLE OF VERIFICATION

4.1 Tracer and Velocity Measurements

The one-dimensional and two-dimensional models are both intended to simulate the large-scale dispersion processes in the Lagoon of Venice, essentially controlled by the horizontal tidal currents. For the verification of the models it is necessary to carry out tracer experiments over a broad area of the basin and a long period of time. There is no significance in a test consisting in following for a few hours the evolution of a small cloud of tracer, since the tracer dispersion is controlled in this case by sub-grid size mechanisms. On the other hand, a measurement campaign covering an area of hundreds of square kilometers and lasting several weeks, raises enormous financial and organizational problems.

After a number of tests, performed on the one-dimensional model, a compromise choice has been made of an area of about 20km^2 in the proximity of the Marghera industrial port. Since this area stretches near the divide between the watersheds of Lido and Malamocco, the currents are relatively small and the transport phenomena can be followed for several days with a moderate input of tracer. Moreover, the test area is a part of the Central Lagoon, where the performances of the one-dimensional and two-dimensional models could be compared. The test area with the twelve sampling sections is shown in Fig. 8. The models schematization is given in Fig. 10.

Concentration measurements went on for three days, from 24 to 26 October 1977. 10 kg of Rhodamine B were injected on October 24 from 9 a.m. up to 12 o'clock; from the model tests this amount has appeared to be sufficient for detecting, all over the area and for the entire duration of the campaign, concentrations larger than 5.10^{-12} kg/l (background level of fluorescence).

One of the test purposes was the verification of the concentration distribution within the channel cross sections, which in the models is supposed to be uniform. In order to simulate a similar initial condition in the prototype, the tracer injection has been designed to obtain the maximum pre-mixing. The injection device was a 25 l/s boat-mounted pump. A 30% solution of Rhodamine B was discharged at a constant rate from astern of the boat at a concentration of about 4.10^{-5} kg/l. The ballasted pipe was hauled up and down at different depths while the boat followed a zigzag route within the injection cell, constituted by a reach of the navigation

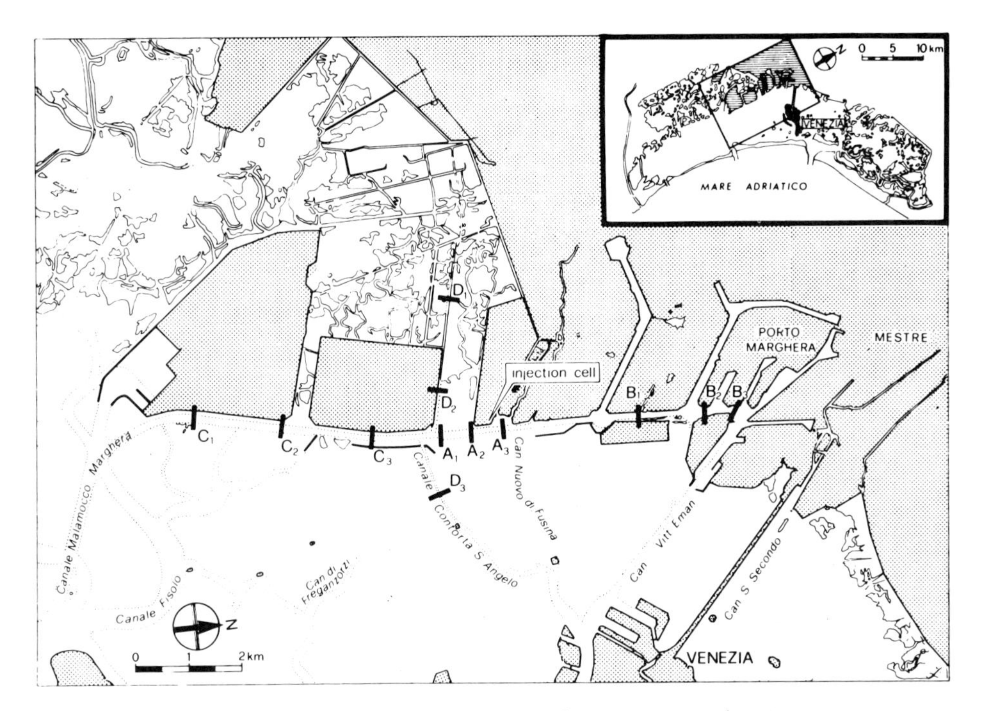

Figure 8. Test area and sampling sections for the tracer experiments.

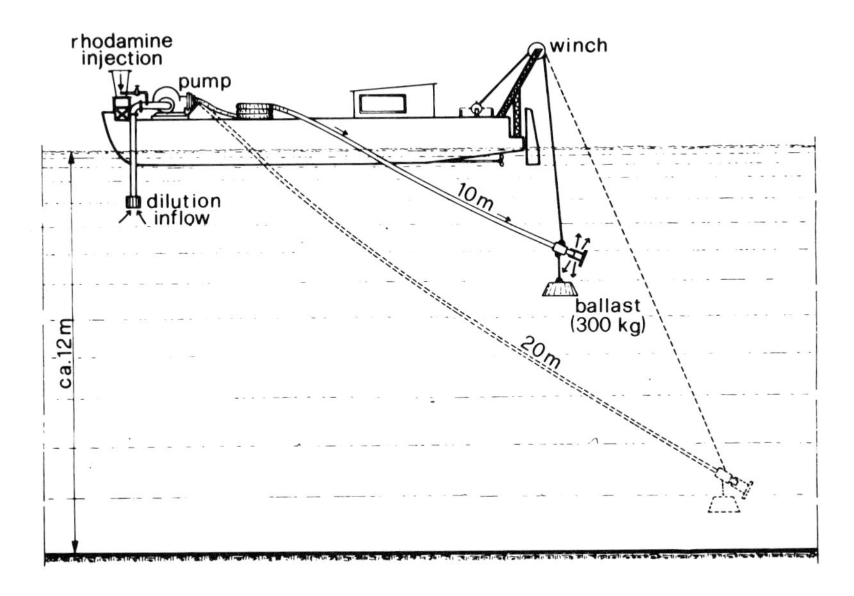

Figure 9. Pre-mixing technique for the tracer injection.

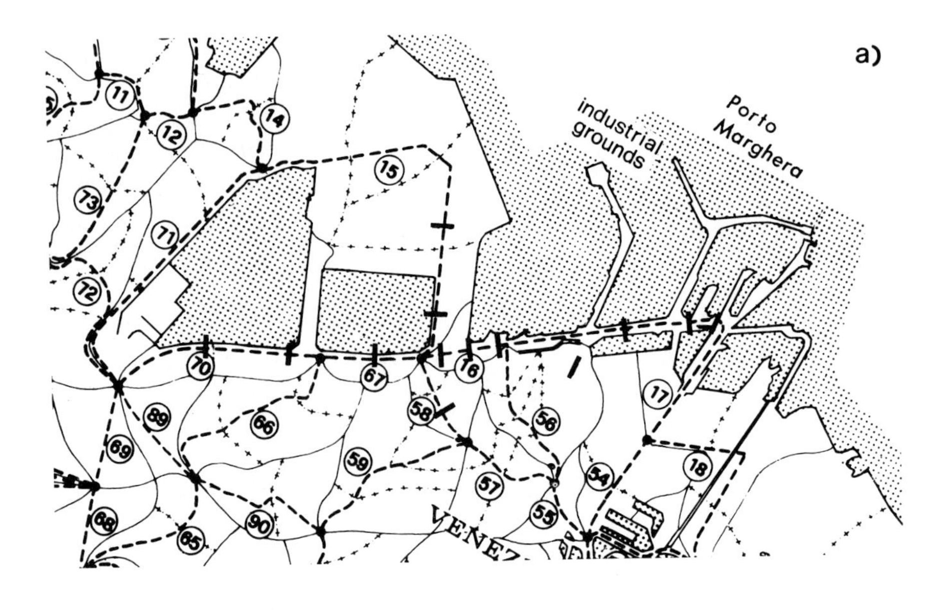

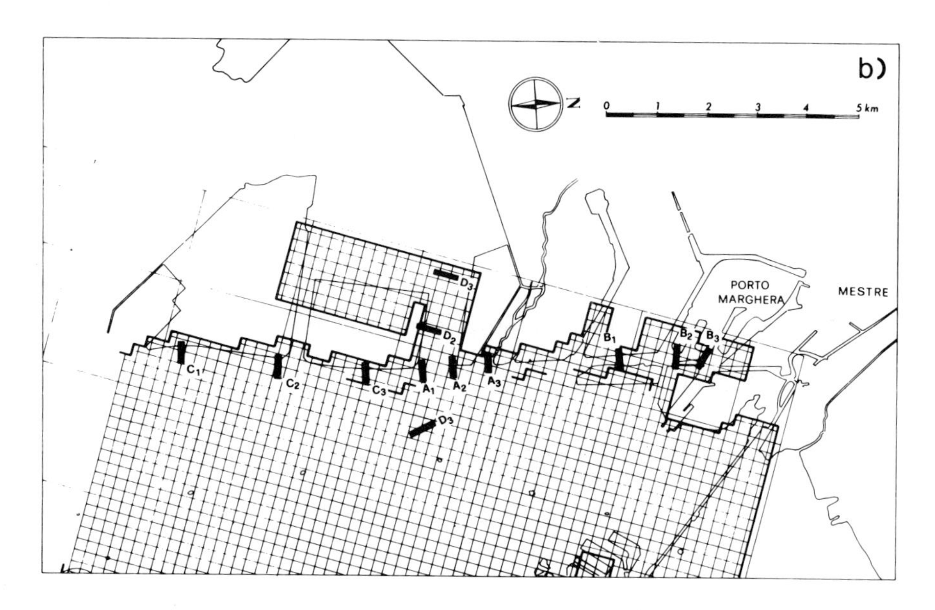

Figure 10. Test area schematization: (a) One-dimensional and (b) Two-dimensional.

canal (1200 m long, 130 m wide and 12 m deep). The tracer injection started at slack tide and finished at full flow conditions (Fig. 9).

In the first day, samples were taken at one-hour intervals from 11 a.m. up to dusk, and in the following two days from 9 a.m. to 4 p.m. Samples were taken by three boats, each one controlling a group of sections; because of the scarcity of boats, sampling in some sections was limited to one or two days. Along the large navigation canal, samples were collected at five points of each cross section in order to check the concentration distribution. Samples of water were taken by means of a winch-suspended sampling bottle and kept in labeled plastic vessels sheltered from the light. At the end of the measurements campaign, the samples were analyzed by a fluorimeter.

A week after the measuring campaign, check samples in all the sections showed that the fluorescence of the water was back to the same values as before the injection.

During the October 1977 tracer experiments, no waterflow measurements were carried out. However, surface velocity data in the Central Lagoon are available (Ministero dei Lavori Pubblici, 1979) for very similar tidal conditions (September 1973). Tidal conditions in both occurrences are shown in Fig. 11.

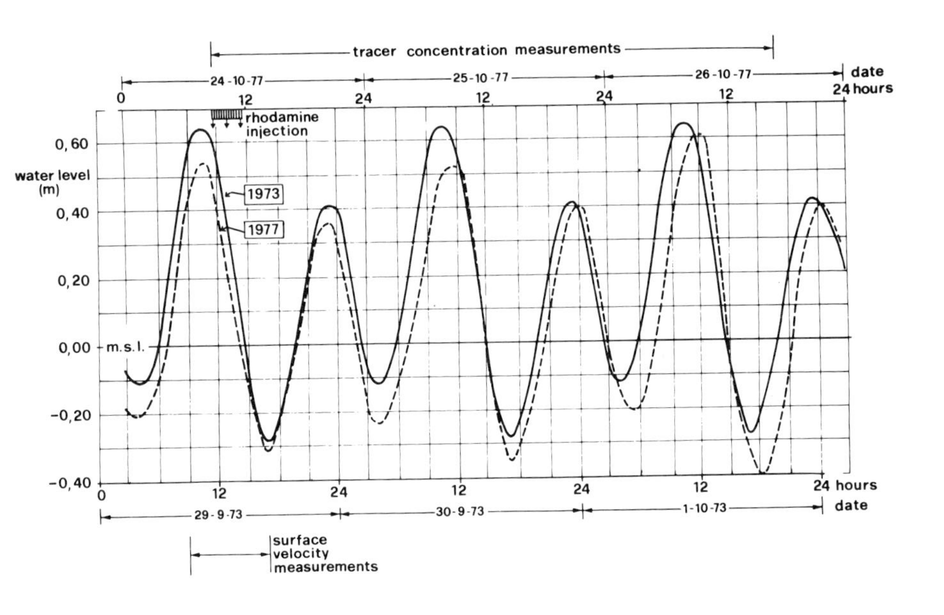

Figure 11. Vertical tide in the Adriatic Sea during tracer experiments and velocity measurements.

The pattern of tidal currents during the tracer experiments can be inferred by the measurements of September 1973. In Figs. 12, 13 and 14 measured and computed velocities are given for the slack period (starting of the injection), as well as for the flow and ebb-tide.

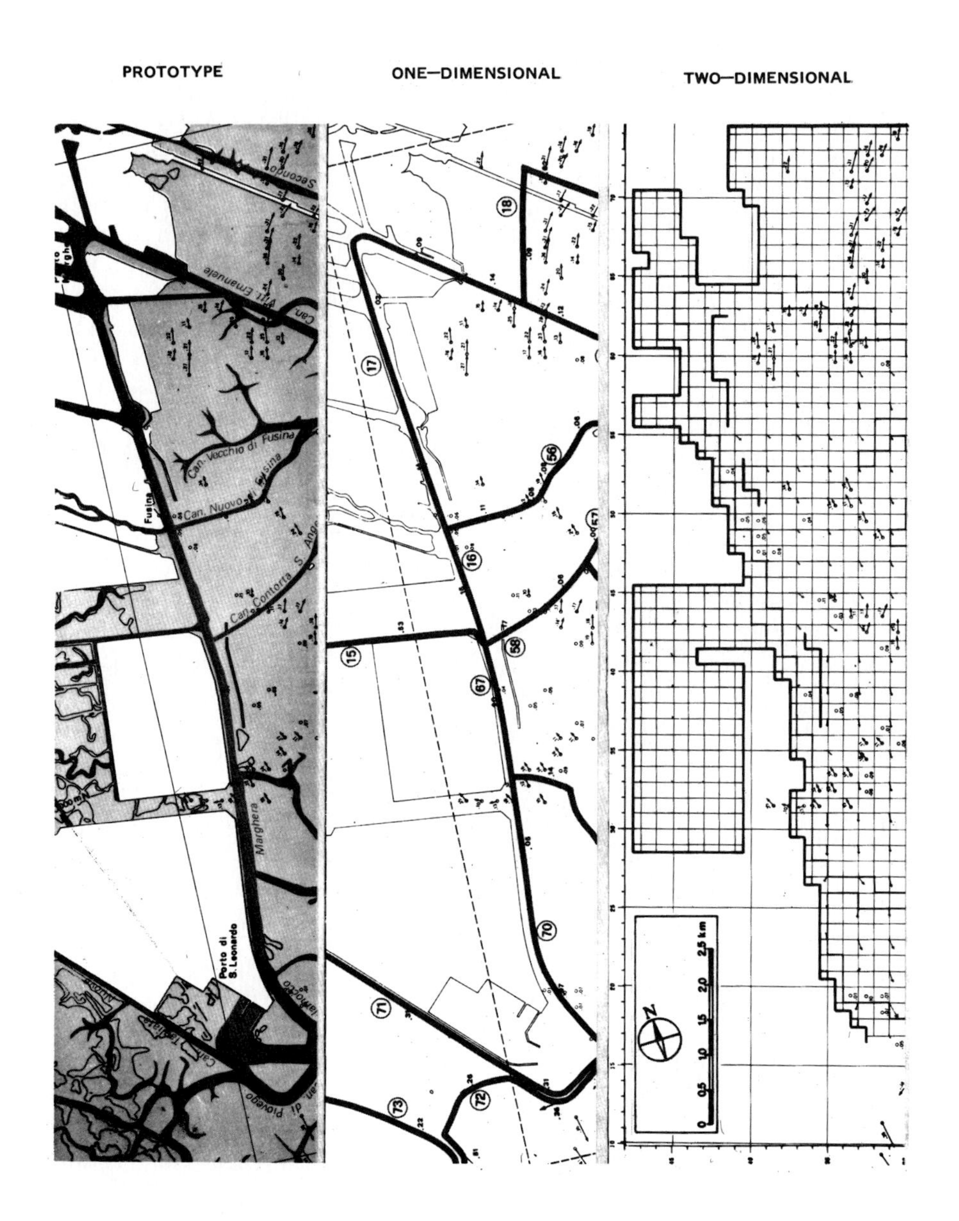

Figure 12. Tidal currents during the tracer experiments. Slack tide.

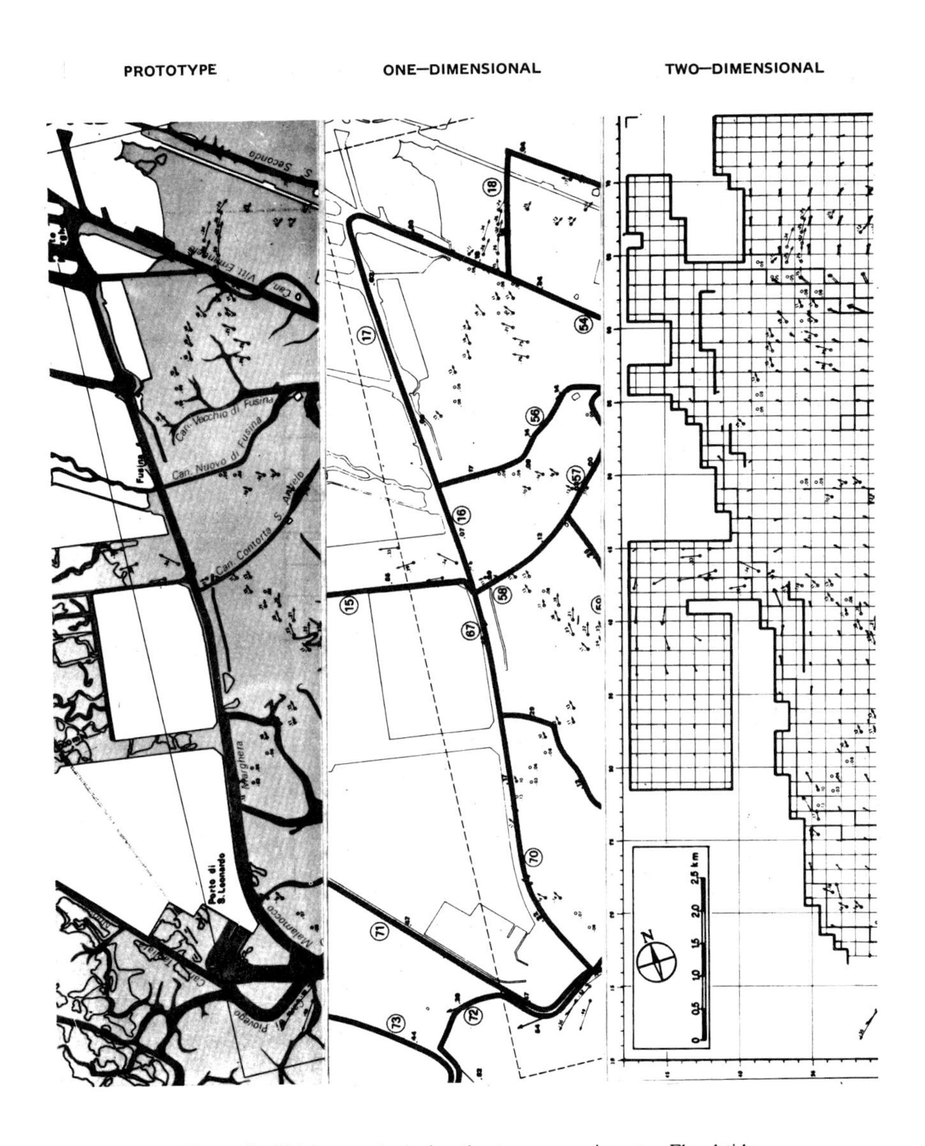

Figure 13. Tidal currents during the tracer experiments. Flood tide.

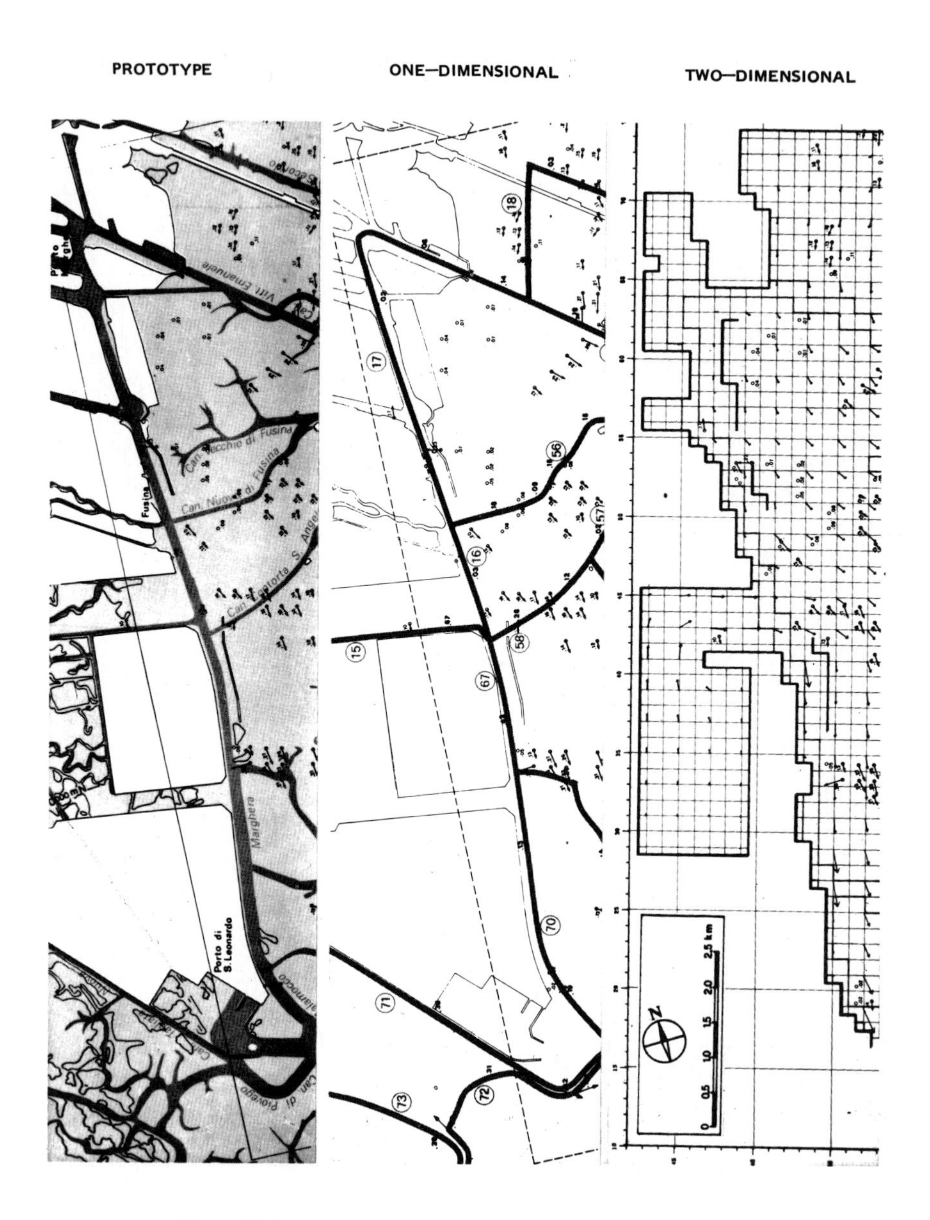

Figure 14. Tidal currents during the tracer experiments. Ebb tide.

4.2 Results of the Experiments and Comparison with the Models

The results of the tracer experiments are given for each section in Figs. 15 and 16, together with the corresponding outputs of the models. The experimental values plotted in the figures represent the average concentration of the samples collected in each section at a given time. No matter how great the effort was to get a strong initial mixing, concentration distribution in the cross sections is not always uniformly distributed, especially on the first day of measurements in the vicinity of the injection cell. Experimental values corresponding to a non-uniform distribution (i.e. with a variation coefficient larger than 0.5) have been marked by a white dot on the figures.

A combined analysis of velocity field and concentration distribution, in the models and prototype, allows the following observations.

In the neighborhood of the injection cell (Sections A_1, A_2, A_3), the two-dimensional model generally gives smaller concentrations than the one-dimensional model, especially at the beginning of the injection. This is probably due to the fact that the velocity components in the transversal direction to the navigation canal (which tend to disperse the tracer toward the middle of the lagoon) are rather pronounced in the two-dimensional model, while the longitudinal components are relatively weaker.

This observation is still more apparent in the northern end of the navigation canal (Sections B_1, B_2, B_3) where the concentrations given by the two-dimensional model are even much smaller and delayed, because the longitudinal currents are too feeble to push the tracer up to here in a short time.

On the contrary, moving southward along the navigation canal (Sections C_2 and C_3) where direction and intensity of the currents are comparable in the two models, the tracer concentrations are also similar. In the southernmost section (Section C_1), transversal currents of the two-dimensional model are again stronger and concentration correspondingly lower.

As a matter of fact, while the transversal currents in the two-dimensional model are more or less evenly distributed along the navigation canal, in the one-dimensional model they are concentrated in correspondence to the few branching lateral channels. As a consequence the tracer dispersion toward the middle of the lagoon can only take place through these branches.

The transport of the tracer out of the navigation canal has been detected in Section D_1, D_2 and D_3, located along two transversal channels. In these sections, the results of the one-dimensional and two-dimensional models are rather similar, although in the latter concentrations seem to be a little larger.

It is interesting to observe that in Sections D_1 and D_2, the one-dimensional model shows a peak of concentration, just after the starting of the tracer injection, which is not duplicated by the two-dimensional model. This peak is because at the beginning of the injection the one-dimensional model's tidal currents are still

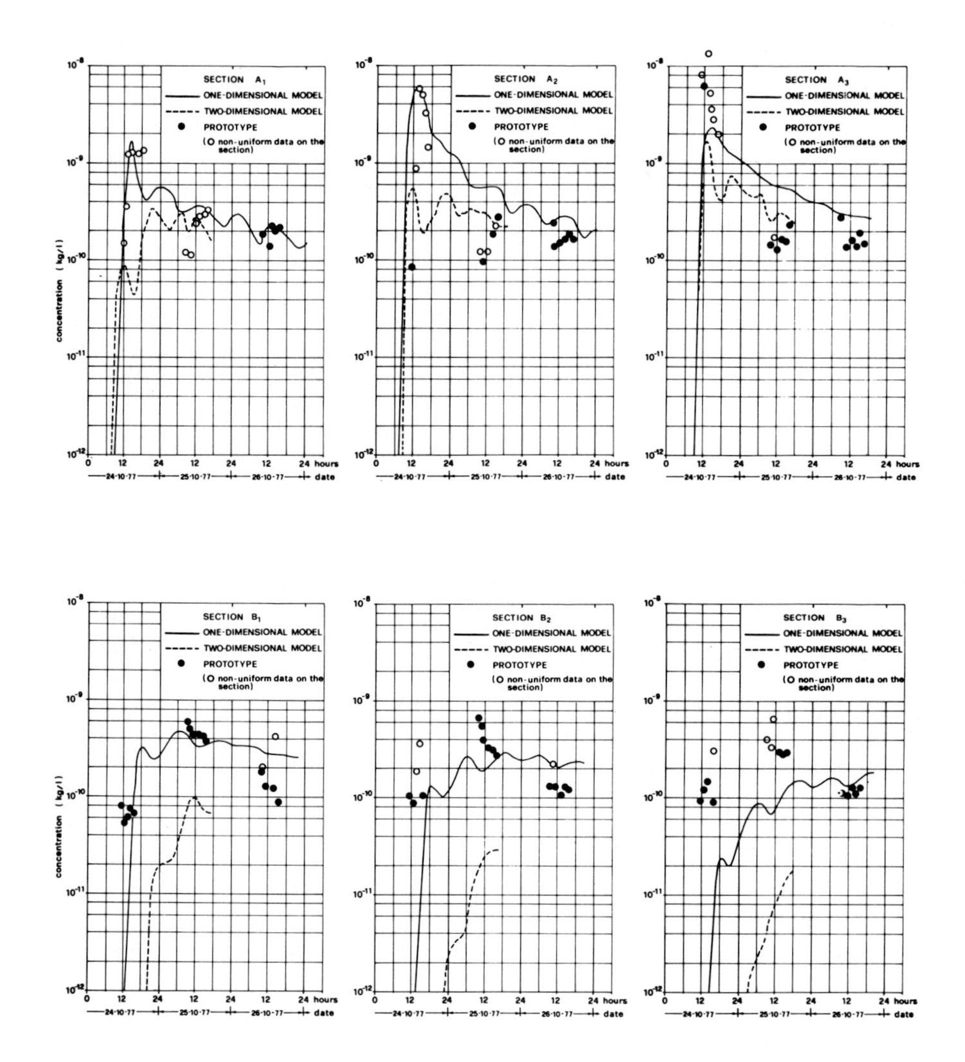

Figure 15. Tracer concentration in the sampling sections.

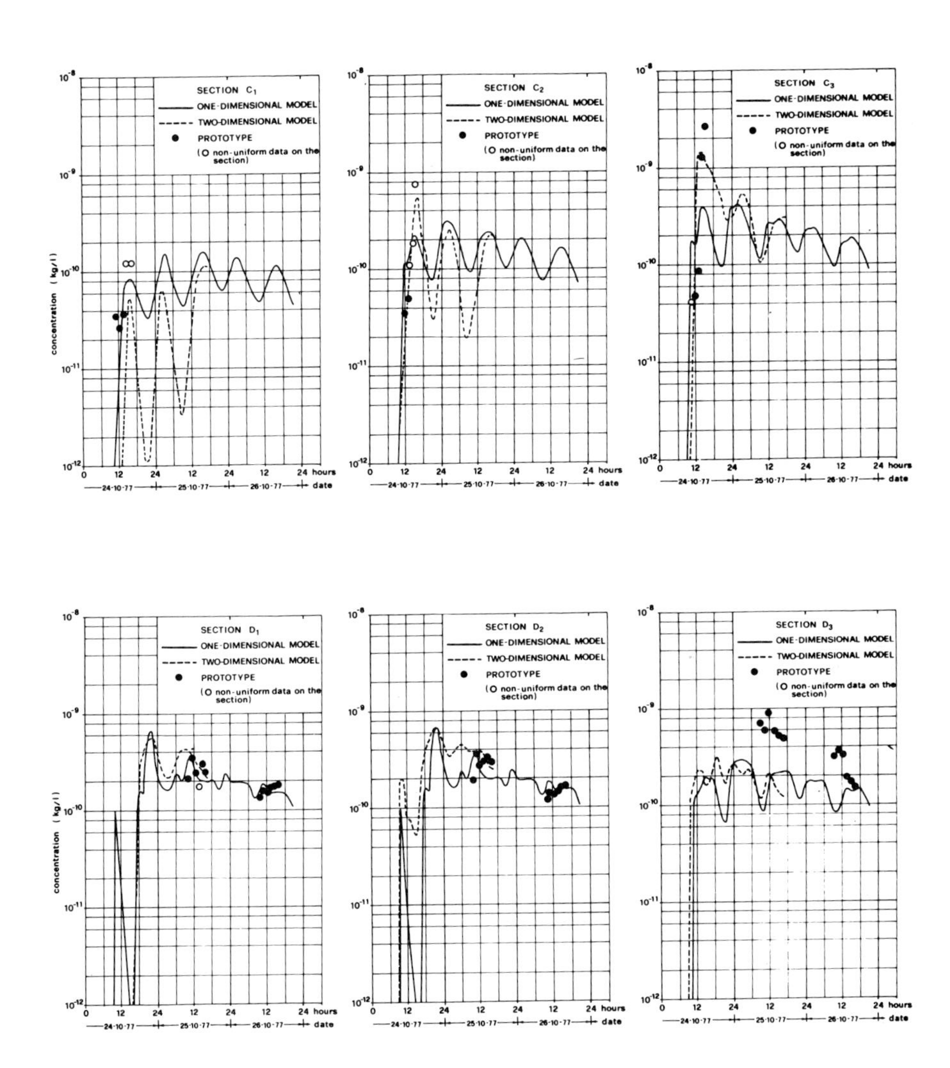

Figure 16. Tracer concentration in the sampling sections.

flowing from the injection cell towards Sections D_1 and D_2, while the two-dimensional model is already in ebb conditions (Fig. 12). As already pointed out in a previous paper (Ministero dei Lavori Pubblici, 1979), the two-dimensional model has a tendency to anticipate the prototype, while the one-dimensional model is slightly delayed.

As for the comparison models versus prototype, the following observations can be made. In the proximity of the injection cell (Sections A_1, A_2 and A_3), the one-dimensional model seems a bit more able to simulate the transport-dispersion process at the beginning of the injection, while the performances of the two-dimensional model offer a better comparison in the following days. At the northern end of the navigation canal (Sections B_1, B_2 and B_3) the one-dimensional model is definitely more accurate, although in the long-run the behavior of both models may possibly converge. In the southern reach of the canal (Sections C_1, C_2 and C_3) prototype results, available only for the first day, are reasonably reproduced by both models. A good duplication has also been obtained by both models in the area east of the canal (Sections D_1 and D_2). Concentrations provided by the models are too low in the area west of the canal (Section D_3), especially on the second day.

4.3 Conclusions

Neither model is able to duplicate the *detailed* prototype's distribution of tracer concentration because of their inability to duplicate the detailed pattern of the instantaneous velocity field. On the other hand, either model can give useful information on the *overall dispersion* process, since both predict the concentration in different parts of the lagoon with an accuracy of 200-400% (not much worse than the scattering of the experimental data), while reproducing temporal and spatial variations of concentration of a factor of 100. The overall dispersion processes, controlling the tide-averaged spatial distribution of concentration produced by a long-lasting injection of a pollutant is probably the only important mechanism which should be reproduced by a model aimed at evaluating the environmental effects of large modifications in the lagoon (barrages for flood control, permanent restrictions of the inlet, dikes, navigation canals, etc.). The overall dispersion process does not so much depend on the instantaneous velocity field as on the lumped amount of water exchanged in ebb and flow conditions among adjacent areas of the lagoon, and above all on the phase-lag of those exchanges, according to a mechanism similar to the one described in Fig. 5. Apparently both models are reasonably able to simulate those exchanges although the instantaneous velocity field reproduced by the two-dimensional model is undoubtedly closer to the real one, the tracer experiment results are perhaps better reproduced by the one-dimensional model. The difference between the model results is probably irrelevant for many practical problems, as can be seen in Fig. 17. In this figure the tendency of the one-dimensional model to exaggerate the transport along the deepest channels is once more confirmed, contrary to the tendency of the two-dimensional model to underestimate it. But as a whole the corresponding pictures are fairly similar.

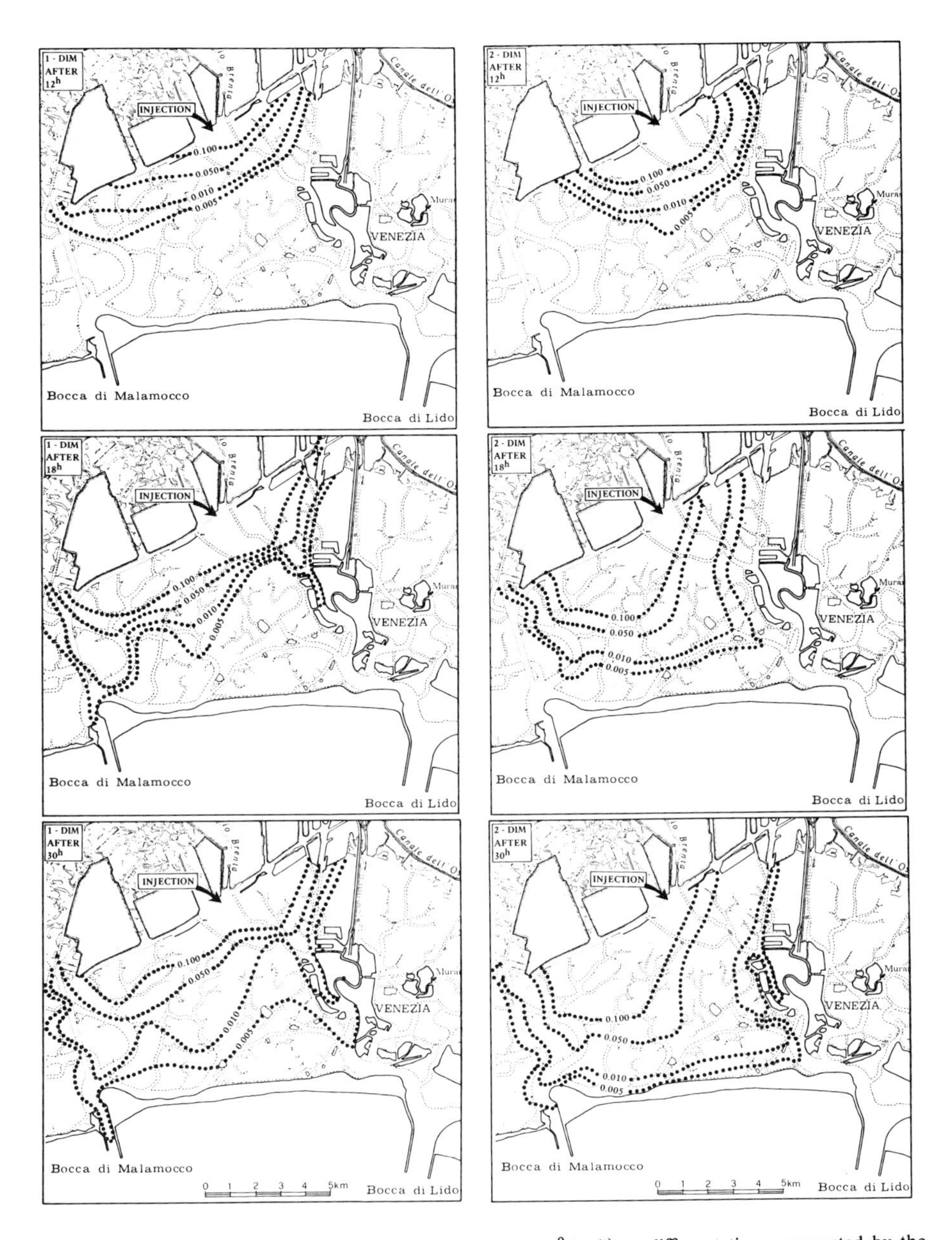

Figure 17. Concentration distribution (p.p.b. or 10^{-9} kg/l) at different times computed by the one-dimensional (left) and two-dimensional (right) models.

Whenever the predictive abilities are comparable, economy becomes the crucial argument. As a consequence, a two-dimensional model is definitely more appropriate for simulating the current pattern in a *specified* tidal or wind condition, as only a relatively short simulation pattern is required. The opposite is true for simulating the build-up of concentration all over the lagoon, produced by a *continuous* output of different wastes, as in this case a week-long simulation is required to reach quasi-steady conditions. For the sake of example, the computational cost (commercial fee) of a two-weeks simulation of a typical estuary with a sufficiently detailed description (5-10,000 meshes for a two-dimensional model and 1,000-2,000 cells for a one-dimensional model) jumps from tens of thousands of dollars (two-dimensional) to less than one thousand dollars (one-dimensional).

REFERENCES

Adami, A., Avanzi, C., Baroncini, E., Da Deppo, L., and Lippe, E. (1977). Esame su modello idraulico degli effetti di un eventuale sbarramento della bocca di Lido. Commissione di studio dei provvedimenti per la conservazione e difesa della laguna e della città di Venezia. Istituto Veneto di Scienze, Lettere ed Arti. No. 7.

Avanzi, C., and Fiorillo, G. (1978). Esperimenti di dispersione nella laguna di Venezia: confronto con un modello matematico unidimensionale. XVI Convegno di Idraulica e Costruzioni Idrauliche, Torino.

Boris, J.P., and Book, D.L. (1976). Flux corrected transport: minimal error FCT algorithms. *J. Comp. Phys.* **20,** 397.

D'Alpaos, L., and Degan, F. (1977). Un modello lagrangiano della Laguna di Venezia per lo studio del movimento delle particelle. Commissione de studio dei provvedimenti per la conservazione e difesa della laguna e della città di Venezia. Istituto Veneto di Scienze, Lettere ed Arti. No. 7.

Datei, C. (1972). Sulla propagazione della marea in una laguna schematica secondo l'impostazione bidimensionale. Commissione di studio dei provvedimenti per la conservazione e difesa della laguna e della città di Venezia. Istituto Veneto di Scienze, Lettere ed Arti. No. 5.

Datei, C. and Dronkers, J.J. (1972). Considerazioni sulla nota del prof. G. Supino: la marea nella Laguna di Venezia. Commissione di studio dei provvedimenti per la conservazione e difesa della laguna e della città di Venezia. Istituto Veneto di Scienze, Lettere ed Arti. No. 5.

Degan, F. (1972). Propagazione della marea in una rete complessa di canali: una programma FORTRAN per gli studi della Laguna di Venezia. Commissione di studio dei provvedimenti per la conservazione e difesa della laguna e della città di Venezia. Istituto Veneto di Scienze, Lettere ed Arti. No. 5.

Di Silvio, G., and D'Alpaos, L. (1972a). Il comportamento della Laguna di Venezia esaminato col metodo propagatorio unidimensionale. Commissione di studio dei provvedimenti per la conservazione e difesa della laguna e della città di Venezia. Istituto Veneto di Scienze, Lettere ed Arti. No. 5.

Di Silvio, G., and D'Alpaos, L. (1972b). Validità e limiti di un modello unidimensionale nella propagazione di onde lunghe in campi di moto con caratteri parzialmente bidimensionali. XIII Convegno di Idraulica e Costruzioni Idrauliche, Milano.

Di Silvio, G., Fiorillo, G., and Degan, F. (1977). A dispersion model for the Lagoon of Venice. *Proc. Congr. Int. Assoc. Hydraul. Res., 17th, Baden-Baden, Germany*

Dronkers, J.J. (1972). Des considérations sur la marée de la Lagune de Venise. Commissione di studio dei provvedimenti per la conservazione e difesa della laguna e della città di Venezia. Istituto Veneto di Scienze, Lettere ed Arti. No. 5.

Fiorillo, G. (1978). Modello unidimensionale per lo studio della dispersione di inquinanti nella Laguna di Venezia (considerazioni su alcuni schemi alle differenze finite). *Atti del Convegno: Metodologie numeriche per la soluzione di equazioni dell'Idrologia e dell'Idraulica, Bressanone.* Patron Editore, Bologna.

Ghetti, A, D'Alpaos, L., and Dazzi, R. (1972). La regolziaone della bocche della Laguna di Venezia per l'attenuazione delle acque alte indagata col metodo statico. Commissione di studio dei provvedimenti per la conservazione e difesa della laguna e della città di Venezia. Istituto Veneto di Scienze, Lettere ed Arti. No. 5.

Ghetti, A. (1976). I problemi idraulici della Laguna di Venezia. (Hydraulic problems of the Lagoon of Venice, with English translation). *Giornale Economico,* Venezia, Aprile 1974. Also in, *CNR-- Quaderni della ricerca scientifica,* **94** Roma, 1976.

Ghetti, A. (1979). Etudes concernant les problèmes hydrodynamiques de la Lagune de Venise. *Proc. Congr. Int. Assoc. Hydraul. Res. 18th, Cagliari* **4.**

Leendertse, J.J., *et al.* (1977). A Three-Dimensional Model for Estuaries and Coastal Areas. Vols. 1-7. The Rand Corporation. Santa Monica, California.

Ministero dei Lavori Pubblici (1979). Le correnti di marea nella Laguna di Venezia. (Tidal currents of the Lagoon of Venice, with English translation). A cura dell'Istituto d'Idraulica dell'Università di Padova, Padova.

TECHNIQUES FOR FIELD VERIFICATION OF MODELS

Donald W. Pritchard

Marine Sciences Research Center,
State University of New York,
Stony Brook, New York

1. INTRODUCTION

1.1 Some General Comments

This paper describes techniques used for the measurement and analysis of field data to be employed in the verification of either hydraulic or numerical models of the flow and of the physical transport of dissolved and finely divided material in estuaries and other tidal waterways. It is therefore less general in scope than might be implied by the title. However, much of what will be discussed here can be applied, at least in part, to the field verification of models of other natural water bodies. Problems encountered in verification of these models are also described.

Much of what is presented here is based on the personal experience of the author and of his colleagues at the Chesapeake Bay Institute, The Johns Hopkins University, and at the Marine Sciences Research Center, State University of New York at Stony Brook. Detailed experimental evidence in support of some of the conclusions given here has not yet appeared in the refereed literature. Because of the general, broad scope of this presentation no attempt is made here to present such evidence.

This presentation is not intended to be a review of the literature in this field. Different conclusions than those presented here may be applicable to estuaries or tidal waterways other than those which have been studied by this author. For an alternate approach to the question of field verification of models, the reader is directed to Wang (1980).

ISBN: 0-12-25852-0

1.2 Types of Models to be Considered

The character of the field program and the techniques for data reduction and analysis required for model verification depend upon the characteristics of the model under consideration. In this presentation the following types of models will be considered:

(a) Hydraulic models of estuaries and other coastal tidal water bodies. These models are fully three-dimensional in space, and may be exercised in a pseudo steady-state (i.e., repeated hydraulic boundary conditions from tidal cycle to tidal cycle) and in true transient state. Only models which cover the full estuarine reach of a coastal waterway, that is, which extend from the ocean to the upper limit of sea salt intrusion, which cover a major portion of the tidal reach of such waterways, will be considered. Such models are used to determine the possible effects upon the tide, the tidal and non-tidal flow pattern, the distribution of salinity, the location and amount of shoaling, and the distribution of pollutants in the waterway, of such actions as: (i) alteration of the time history of fresh water inflow, through control and or diversion; and (ii) deepening of existing navigation channels, the construction of new channels, or the physical alteration of the waterway in other ways such as the construction of dikes.

(b) Numerical models of the hydrodynamic and kinematic processes which control the rise and fall of the tide, the oscillatory tidal flows and the density-driven and wind-driven non-tidal flows, in estuaries and other tidal waterways. The equations which are numerically treated in these models are considered to be ensemble-averaged, such that the short term turbulent fluctuations enter the equations only statistically, via the stress terms. Otherwise, the models are considered to be "real time" models, in that time variations in the smoothed (ensemble-averaged) field of flow are treated for time steps which are small compared to a tidal cycle. The model may be a coupled dynamic/kinematic model, whereby the computed velocity field continuously serves as input to the advective/diffusive salt balance equation, which in turn is employed via an equation of state to compute the density-induced portion of the pressure force in the equations of motion. On the other hand, the dynamic equations are sometimes solved for surface elevation and flow distribution with a given, constant density distribution. Consideration will be given here to sectionally-averaged one-dimensional models; laterally-averaged two-dimensional models; and fully three-dimensional models.

(c) Water quality models of the distribution of conservative, water borne constituents of estuaries and tidal waterways. The modeled parameters would include such truly conservative constituents as salinity, as well as some nominal non-conservative constituents for which the decay or exchange constants involve time intervals long compared to the effective "age" of the constituent within the modeled area. An example might be the excess heat or excess temperature in the near field of a thermal plume, where the distribution is usually controlled by mixing rather than by heat loss to the atmosphere. The problems of verification of models of the distribution of non-conservative properties are outside the scope of this paper. Input to the water quality models considered here is the velocity distribution in space and time; produced either by an interpolation of an observed data set or as output of a hydrodynamic model of the waterway. As with the hydrodynamic model, spatially averaged forms of the water quality model will be considered.

1.3 Parameters to be Measured for Model Verification

Techniques of measurement, and of data reduction and analysis of the distributions, in space and time, of the following parameters will be discussed in this presentation: water surface elevation, velocity, salinity, and a locally introduced inert, waterborne tracer.

2. TECHNIQUES FOR MEASUREMENT

2.1 Water Surface Elevation

The National Ocean Survey (formerly the USC & SGS) maintains tide gauges in most estuaries and other coastal tidal waterways. Many of these are well maintained and records are reasonably promptly processed and made available. However, often the number of such gauges and the location of these gauges are not adequate for use in adjustment and verification of either hydraulic or numerical models. Also, the legislated mission of NOS dictates that navigation needs control the priorities with respect to which gauges are to be maintained in good working order, and also with respect to which records are to be promptly processed. Recently legal matters dealing with the location of the boundary between public and private wetlands have also received some priority with regard to the selection of tide gauge records to be processed in a timely manner. Scientists and engineers studying these waterways are not considered high priority clients of NOS, and obtaining records from the well-placed and functioning gauges in a reasonable time period is uncertain at best.

It is therefore often necessary to install temporary tide gauges to augment the water surface elevation data provided by the NOS tide gauges. The most suitable instrument to use is either of the two paper tape digital recording gauges used by NOS. However, there are several analog recording gauges which would provide adequate data, and might be easier to install on a temporary basis. One such device is a bottom mounted bubbler type pressure gauge, which eliminates the need for a piling and pier location for the gauge. A line of levels should be run from the temporary gauge to the nearest first-order bench mark.

2.2 Velocity

My intent in this presentation is to give some pertinent criteria which should be satisfied by the instruments and methods used for obtaining the measurements, and not with the detailed description of a particular field procedure. For instance, the number of positions at which simultaneous measurements of current speed and direction should be made, and the period over which such observations should be obtained, generally require that a number of *in situ* recording current meters be deployed in an array of vertical moorings. The mooring systems must be such as to provide reasonable security against having the array be moved by the combined action of currents and wind waves, even in severe weather, or be subject to casual vandalism. The moorings must be suitably marked to avoid having them run down by shipping. Such considerations also place restrictions on the locations where the moorings may be placed. There are several different acceptable mooring schemes used by different groups engaged in field measurements of currents. Some of these are described by Beardsley, *et al.*, (1977) in a report on an intercomparison of current meters conducted off Long Island in February-March, 1976. Some of the references given in that report contained even more detailed discussions of these mooring systems.

The above cited report by Beardsley, *et al.*, (1977) presents data which can be used to indicate desirable features of current meters for use in estuaries and tidal waterways, even though the current meter intercomparisons were carried out on the continental shelf. My own experience with various types of current meters added to the information in that report leads me to conclusions extending beyond those stated in the subject report. The conclusions I have reached regarding the type of current meter best suited for estuarine studies are strengthened by the results of a yet unpublished current meter intercomparison study conducted by William Boicourt of the Chesapeake Bay Institute in the lower estuary of the Potomac River.

In narrow, well protected estuaries or segments of estuaries, or of other coastal tidal waterways, there are a number of *in situ* recording current meters which would provide suitable data for model verifications. However, in wider, more open reaches, where wind waves can reach a foot or more in amplitude, severe contamination of the current meter record can occur for current meters having Savonius rotors, Savonius type rotors or other direction independent speed sensors, such as horizontal rotating cups. Such contamination results from the short period oscillatory motion induced by wind waves acting both directly on the current meter and indirectly via induced mooring motion. As a result, rotor speed sensors are "spun up" by the short term wave induced motions. In the presence of wind waves, rotor type current meters read too high, and the effect is most pronounced for low current speeds. This effect can readily be demonstrated in an elongated estuary in which the tidal currents are primarily oscillatory (as distinct from rotary) in nature, and pass through zero when the current changes from ebb flow to flood flow and again when the current changes from flood flow to ebb flow. In the presence of significant wind waves, current speed records from meters having rotors as speed sensors never dip to zero, and the shape of the speed record is more square than half-cosine in appearance. This effect shows even more clearly when the speed and direction data are used to resolve the current along the principle flood-ebb direction. The resulting record looks like a noisy square wave, rather than having the more characteristic cosine appearance.

Vector averaging current meters (VACM's) have a small direction sensing vane adjacent to the Savonius rotor speed sensor, and sample speed and direction in short bursts of closely spaced observations. These meters resolve each pair of speed and direction observations into the north and east vector components, and average these components over the sampling interval, which is of the same order as the wave period. The records from such meters show considerably less "spin up" problems than other meters which use rotors as speed sensors.

The experience of several of my associates, such as Bill Boicourt and Richard Whaley at the Chesapeake Bay Institute, and Harry Carter at the Marine Sciences Research Center, who have been involved in the collection, reduction and analysis of a large number of current meter records in estuaries and other tidal waterways, together with my own experience, has led us to conclude that there is a better alternative than the use of VACM's in estuaries and tidal waterways. This alternate is the use of a modified ENDECO Model 105 or Model 174 ducted propeller current meter. This modification involves replacement of the tethered connection between mooring cable and meter with a fixed, broad, U-shaped bridle which connects to the taut wire mooring above and below the centerline of the meter. This modification is described by Beardsley, *et al.*, (1977).

Although at the present time we consider this meter to be the best available meter for current measurements in estuaries and other tidal waterways, the meter certainly could be improved, particularly with respect to reliability. There are some recently developed current meters which use non-mechanical speed sensors, which may result in improved measurement systems, but at significantly higher costs than the presently used meters. Another meter which shows promise for the use described here is the Vector Measuring Current Meter, which employs two cosine-response propellers set at right angles to each other. A meter of this type was first described by Cannon and Pritchard (1971).

2.3 Salinity

The most cost-effective method of obtaining salinity measurements is to use devices which simultaneously measure temperature and electrical conductivity. Ideally such measurements should be made at a number of locations in the estuary at closely spaced time intervals over several tidal cycles. Such observations should be repeated a sufficient number of times so as to encompass the full normal range of those factors, such as fresh water inflow, which control the distribution of salinity.

Some current meters which record on magnetic tape can be purchased with temperature and conductivity sensors. Hence, time series of salinity data can be obtained at these locations where such meters are deployed. However, it has been difficult to obtain an adequate set of such time series data for a large estuary, and other types of salinity data has proved more useful in model verification. For instance, in the case of the hydraulic model of the Chesapeake Bay the most useful salinity data from the standpoint of model verification has been a set of same-slack runs made along the axis of the Bay, from the mouth at Cape Henry to the head at Turkey Point, at approximately monthly intervals, over a seven-year period. This period included the unique, record-setting fresh water inflow incident associated with the hurricane AGNES.

2.4 A Waterborne Tracer of Advective and Diffusive Processes

The most effective method of obtaining data for use in the verification of pollutant plume models is the Rhodamine fluorescent dye tracer technique first described by Pritchard and Carpenter (1960). Details of the current practice in the use of this technique are well described by Carter (1974).

3. THE REDUCTION, ANALYSIS AND INTERPRETATION OF DATA

3.1 Time Series Data

Included in this category are the records of surface water elevations obtained from tide gauges; records of current speed and direction obtained from *in situ* recording current meters, and time series salinity data obtained by processing the temperature and conductivity records obtained from some of the *in situ* recording current meters, and/or by other means.

For the most part, these records exhibit certain common features, which can be demonstrated by the record of water surface elevation given in the upper part of Fig. 1. This record was obtained from a tide gauge located on the eastern shore of the Chesapeake Bay approximately opposite the Calvert Cliffs Nuclear Power Plant. This location is about 133 km from the head of the Bay and 163 km from the mouth of the Bay. The float for the tide gauge that provided this record, as is common with all such gauges, was located in a stilling well which provided for an hydraulic filtering of high frequency variations in water surface elevation, such as would be produced by wind waves. Because of this filtering this type of record is here referred to as a low pass filtered (LPF) record.

The semidiurnal tidal variation in water surface elevation, of approximately sinusoidal shape, can be clearly seen in the upper part of Fig. 1. There is also evident an aperiodic variation in the water level with time intervals between maxima or between minima much longer than a tidal cycle. When this record is subjected to a sharp, low pass filter with a half power point at about 34 hours, the record shown in the lower part of Fig. 1 is obtained. This record is here designated by the term low-low pass filtered (LLPF) record. This record does not show

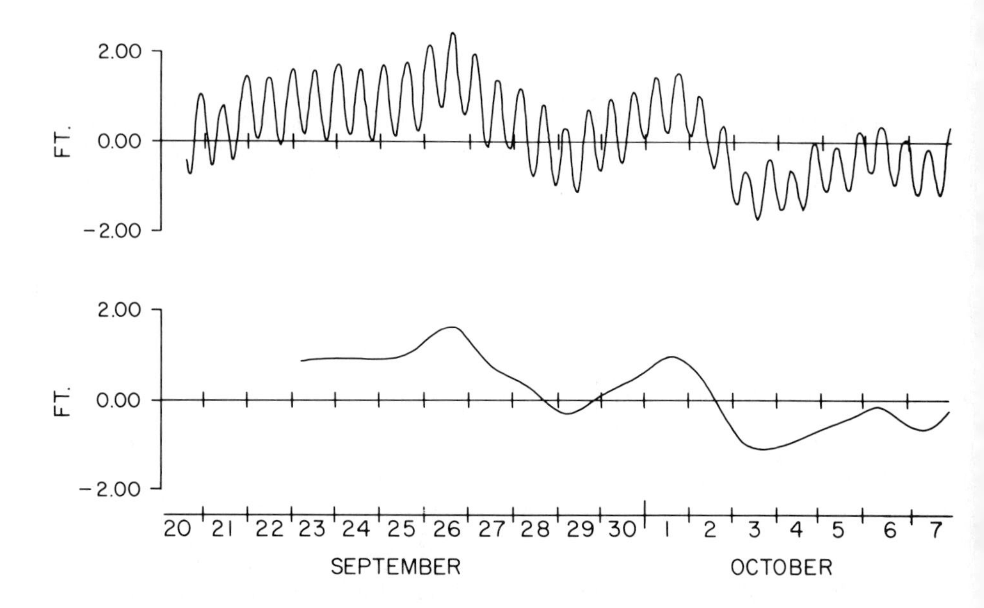

Figure 1. Record of the water surface elevation obtained for the period 20 September through 7 October from the tide gauge installed on the Honga River, plotted about the mean elevation for the period of record of this gauge (upper), and the low-low pass filtered data of this record (lower).

fluctuations which can be identified with any semi-diurnal or diurnal tidal variation.

The aperiodic variations evident in the lower diagram of Fig. 1 have characteristic time intervals between major events, that is, between maxima or between minima, which, based on longer records from this and other gauges, are 2 ½ days, 4 days, and between 6 and 10 days. Power spectra of tidal records from a number of gauges in the Chesapeake Bay and also from gauges located all along the Atlantic seaboard show peaks at these inverse frequencies.

Figure 2 shows a LPF water surface elevation record from a gauge located on the western shore about 20 km further north than the gauge from which the record shown in Fig. 1 was obtained. Note the close similarity in the LLPF records shown in the lower parts of the two figures. Almost perfect coherence exists for the LLPF water level data over this distance.

Note that the amplitude of these aperiodic variations shown in the lower part of Fig. 1 and Fig. 2 are of the same order as the amplitude of the tidal variations. This can be perhaps more easily seen by comparing the lower part of Fig. 1 and Fig. 2 with the two records shown in Fig. 3. The curve shown in the upper part of Fig. 3 was obtained by subtracting the curve given in the lower part of Fig. 1 from the curve given in the upper part of Fig 2. These curves in Fig. 3 are called the band pass filtered (BPF) records, and represent essentially the astronomical tide.

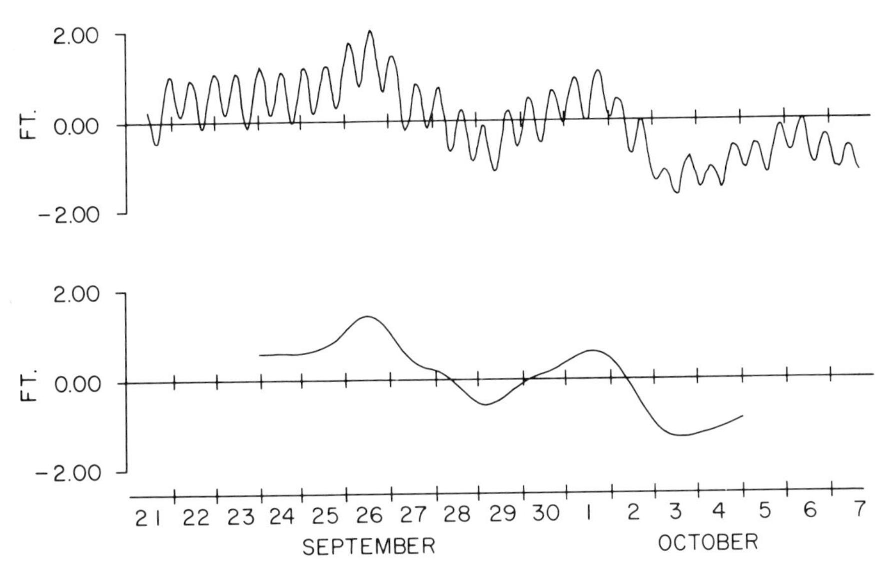

Figure 2. Record of the water surface elevation obtained for the period 21 September through 7 October from the tide gauge installed at Kenwood Beach, plotted about the mean elevation for the period of record of this gauge (upper), and the low-low pass filtered data of this record (lower).

Quite similar features are exhibited by records from current meters deployed on vertical moorings in the Chesapeake Bay near the location of the tide gauges which produced the records discussed above. Records of speed and directions from these current meters were resolved into vector components along and perpendicular to the principal ebb and flood directions. In the case of the current meters in this reach of the Bay and, in fact, over most of the Bay, the principal ebb and flood axis is oriented very nearly along the central axis of the Bay. The component directed along the central axis is termed the longitudinal component. Figure 4 gives a plot of the longitudinal component of the velocity versus time for two meters located on a vertical mooring in the deep channel east of Cove Point. The upper part of this figure shows the data for a depth of 3 meters and the lower part is for a depth of 12.2 meters. In addition to fluctuations of clearly semidiurnal tidal period, these records show both high frequency noise, as well as lower frequency non-tidal variations. The current meter does not function as a low pass filter as is the case with the tide gauges. Therefore, to obtain a record which is comparable to the record shown in the upper part of Fig. 1, the data plotted in Fig. 4 must be subjected to a low pass filter with a half power point of about 2 hours. This LPF record is then subjected to the low-low pass filter with a half power point of 34 hours. The resulting LLPF record for the curve given in the upper part of Fig. 4 is shown in the lower part of Fig. 5, and the LLPF record for the curve in the lower part of Fig. 4 is shown in the upper part of Fig. 6. Clearly, these LLPF current meter records have features very similar to the LLPF tide record.

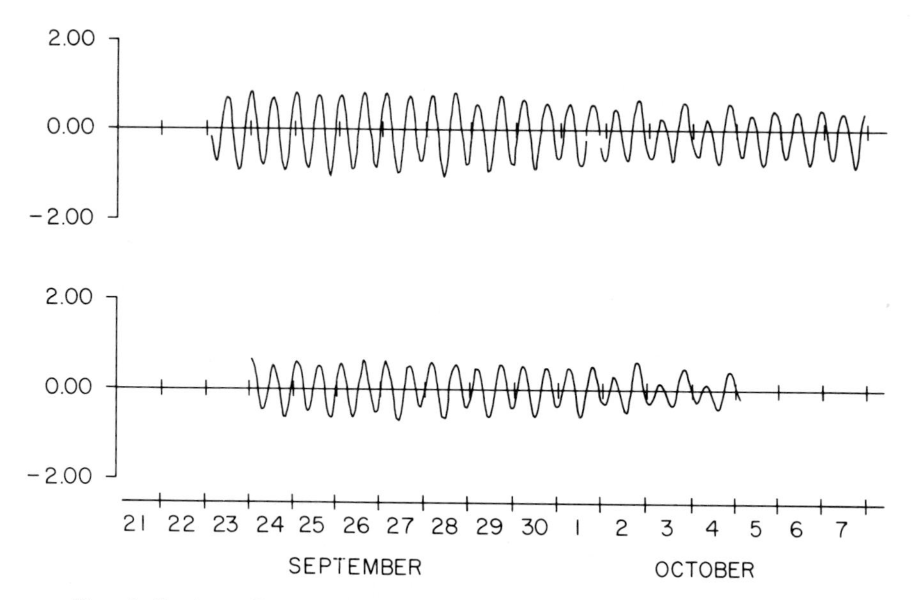

Figure 3. Band pass filtered tide gauge data from the Honga River (upper) and from Kenwood Beach (lower) for a period in late September and early October, plotted about the mean water level for each gauge for this period of record.

Not shown in this paper are examples the BPF current meter records obtained when the LLPF record is subtracted from the LPF record. Such records are, however, composed primarily of astronomical tidal currents of semi-diurnal and diurnal periods.

A number of recent studies have shown that the LLPF records of the tide gauge date, that is, the non-tidal variations in the low pass filtered water surface elevation, are related to variations in local and far field winds. A summary prepared by Carter, *et al.*, (1979a) contains most of the pertinent references, but particular attention should be given to the studies by Elliott and Wang (1978), by Elliott, *et al.*, (1978), by Wang (1979), and by Wang and Elliott (1978). These studies show that the power spectrum of the LLPF record of the longitudinal component of the wind has peaks at the same 4 day to 6 to 10 day inverse frequencies as is observed for water surface elevation. The same is also the case for many current meter records.

An inspection of the curve in the lower part of Fig. 5 shows that averaged over the full length of this record the flow is positive, or by the convention used here, directed seaward, at this depth of 3.0 meters. Likewise, the record shown in the lower part of Fig. 6 shows that the average longitudinal current over the full record at a depth of 18.3 meters is directed up the Bay. These long term mean flows represent the classical two-layered density driven estuarine circulation pattern. However, at each of these depths there is superimposed meteorologically

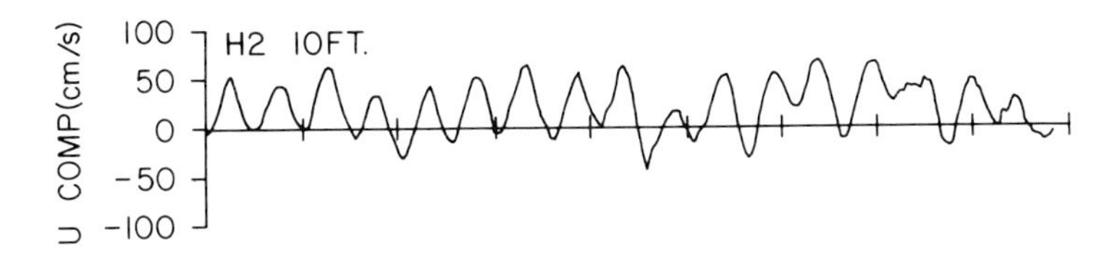

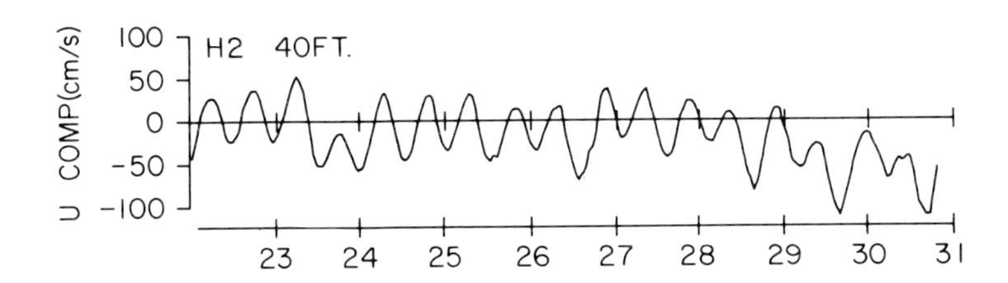

Figure 4. Low pass filtered longitudinal component of the current velocity as measured at depths of 10 feet and 40 feet, at Station H2, located 4.4 km east of Cove Point, at mid-channel in the Chesapeake Bay.

driven fluctuations which are of sufficient amplitude to sometimes cause a complete reversal of the classical flow pattern.

Continuous time series records of salinity at a given location have certain features in common with the water surface elevation records and the records of the longitudinal component of the velocity, particularly in an elongated estuary. In such a waterway the salinity decreases steadily along the axis of the estuary, from the junction of the estuary with the ocean to the furthest extension of measurable sea derived salt up the estuary. The advection of this pattern back and forth by the tidal currents results in salinity variation of tidal period, with maximum salinities normally occurring at the end of the flood interval and minimum salinities at the end of the ebb interval. This tidal fluctuation is superimposed on a non-tidal variation which exhibits changes primarily in response to changes in fresh water inflow. There are no clear relationships between the time variations in the LLPF record of salinity and of wind, as there often appear to be with surface water elevations and currents. Winds do contribute to changes in the vertical stratification in estuaries, and there must then be an indirect relationship since the wind clearly affects the current, which must then lead to a shift in the advective/diffusion salinity balance.

3.2 Synoptic and Pseudo Synoptic Measurements

With a given total resource in manpower, equipment and operating funds, collection of data for model verification could proceed primarily along the line of

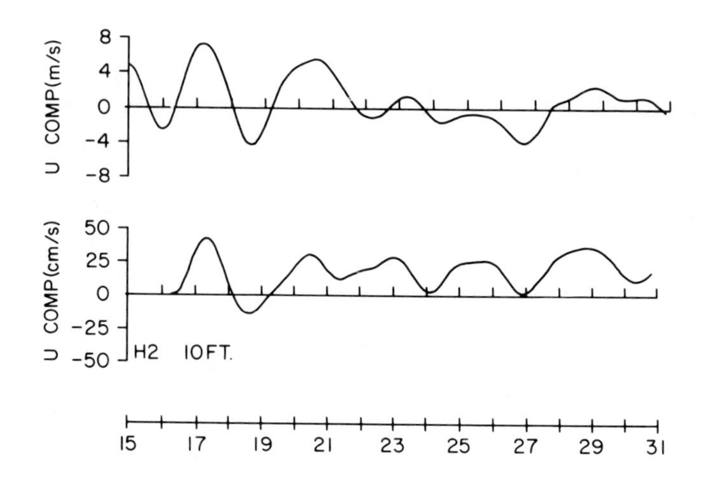

Figure 5. LLPF longitudinal component of the wind velocity (upper), and of the current velocity at Station H2 at a depth of 10 feet (lower).

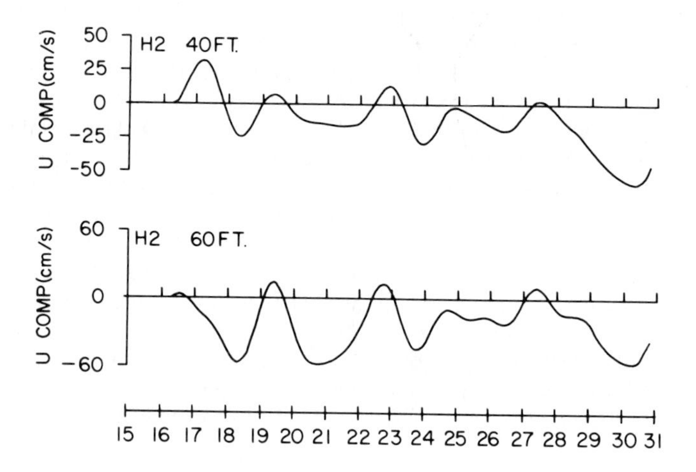

Figure 6. LLPF longitudinal component of the current velocity for Station H2 at depths of 40 feet and 60 feet.

establishment of a limited number of key stations or sampling locations, at each of which measurements of all pertinent parameters would be made continuously over a long period of time. Alternately, a large number of stations could be established, and a single set of observations of all parameters might be made at all stations, over as short a time period as is possible. Thus, in this latter mode, an effort would be made to obtain as nearly a synoptic set of measurements as is possible. This synoptic or nearly synoptic data set would then be repeated from time to time, depending on the resources available, in order to cover a range of controlling conditions.

For some parameters and some types of models, synoptic observations are not required. Thus, a transient state, coupled hydrodynamic/kinematic model being run to verify the computations of salinity, could be exercised over a simulated period corresponding to the prototype time interval during which the set of observations which will be used for verification was obtained. Then the comparison of observed and model produced salinities, for purposes of verification, would be made using model produced salinity for the model times and locations corresponding to the prototype times and locations at which the observed data was obtained.

These same comments apply to hydraulic models as well as numerical models.

Field dye tracer studies designed to obtain data necessary for verification of a transient state plume model in a tidal waterway involve measurements which are a combination of synoptic and time series observations. The field study may involve the injection of the fluorescent tracer dye into an existing plant discharge or possibly into an unused outfall. Alternately, the dye can simply be continuously released over the period of the study as a local continuous source. In any case, the dye plume will oscillate, extending in seaward direction during the ebb interval, and toward the head of the estuary during the flood interval. Our experience indicates that attempts to obtain a synoptic snapshot of the concentration distribution are generally of little use. Instead, sections which cross the plume at right angles should be located and marked with buoys. Vessels equipped to continuously sample, measure and record fluorescent intensity along the track of the moving vessel should then occupy one section located up the estuary from the source during each flood interval and some section located down the estuary from the source during each ebb interval. During the period that a particular cross section is occupied, the dye measuring vessel moves back and forth across the plume, obtaining as many realizations as possible of the shape of the plume and the peak concentration in the cross section. The manner in which such data can be used for model verification will be described later in this presentation.

4. EXAMPLES OF FIELD VERIFICATION OF MODELS

4.1 Transient State, One-Dimensional Dynamic Model of the Chesapeake and Delaware Canal

This section describes work reported by Rives and Pritchard (1978), plus continuing as yet unpublished work done by these investigators.

The C&D Canal system extends from Turkey Point in the upper Chesapeake Bay to Reedy Point in the Delaware River, a distance of 42 km. As seen from the sketch map given in Fig. 7 there is a reach of some 15 km at the Chesapeake end of the Canal which traverses the Elk River. From a location between Old Town Point and Chesapeake City, the Canal extends eastward for about 27 km to its junction with the Delaware Bay at Reedy Point. Over most of this reach the Canal is nearly straight with a more or less uniform cross section. The authorized minimum channel dimensions are 35 ft (10.7 m) deep, 450 ft (137 m) wide, with a 2:1 slope. Most sections are somewhat wider than the 590 ft (180 m) indicated by these minimum project dimensions.

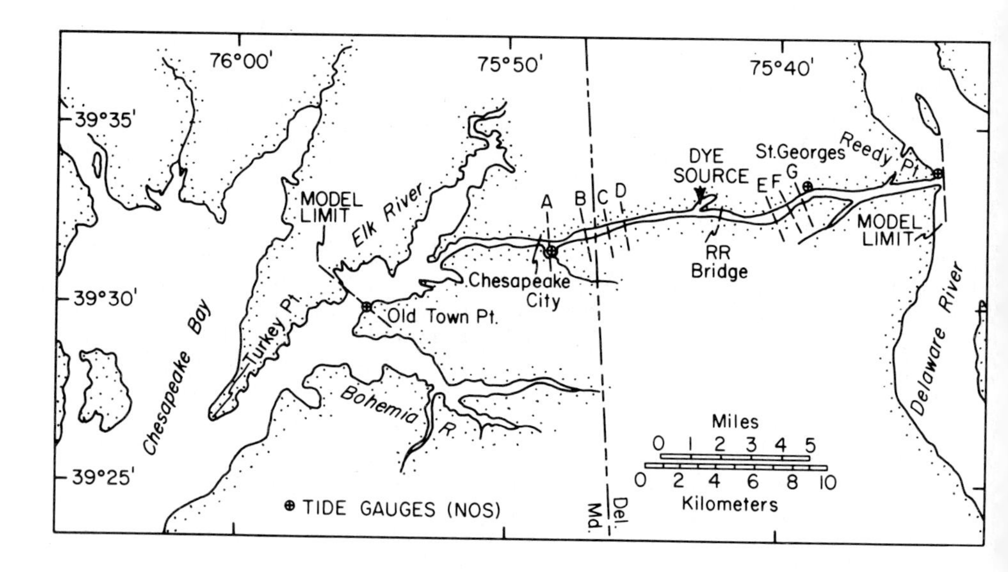

Figure 7. Chesapeake and Delaware Canal system showing model limits and referenced locations.

Over much of the year (December through July) salinities of less than one part per thousand occur at the Chesapeake end of the Canal, with salinities at the Delaware end during this 8 month period ranging from slightly less than one to about three parts per thousand. During the 4 month period, August through

November, salinities at the Chesapeake end reach values as high as 3.5 parts per thousand, with the values at the Delaware end running about 2 to 3 parts per thousand higher. With maximum tidal velocities at control sections averaging 1.2 m s^{-1} (2.3 knots), vertical and lateral homogeneity occurs at all cross sections over most of the year.

The mean tidal range at the Chesapeake end of the canal is about 2 ft (0.61 m) and that at the Delaware end is about 5 ft (1.52 m). The tidal phase at the Delaware end precedes that at the Chesapeake end by about 11 hours, although, since the semi-diurnal tidal cycle is just over 12 hours, casual inspection would suggest that the tidal phase at the Chesapeake end precedes the tide at the Delaware end by just over one hour. Since there is a considerable diurnal inequality in the tide, at least during part of the 28-day tidal month, the tide with the lower high water and higher low water at the Chesapeake end occurs close to the time interval of the tide at the Delaware end which has the high high water and the lower low water. Consequently there is a time-varying head difference over the length of the Canal and an associated hydraulically driven flow through the Canal. Superimposed on this semi-diurnal tidal head difference are meteorologically driven, aperiodic variations in head difference having characteristic time intervals between maxima and between minima considerably longer than the tidal period. The distance from Old Town Point on the Elk River, which marks the Chesapeake end of the Canal as modeled by Rives and Pritchard (1978), to the mouth of the Chesapeake Bay is 300 km, while the distance from Reedy Point at the eastern end of the Canal to the mouth of the Delaware Bay is 108 km. The distance between the two bays over much of their lengths seaward of the Canal is sufficiently large so that the direction of the wind prevailing over each bay at any time can be very different. Consequently, the simultaneous wind set-up in the two bays result in a head difference between the two ends of the Canal, which at times is of the same order as that produced by the tide. The local wind blowing over the surface of the Canal seldom contributed any significant head difference.

A one-dimensional numerical hydrodynamic model, similar to that described by Harlemann and Lee (1969), should adequately model this waterway. Pritchard and Rives (1978) made use of such a model, as modified by Hunter (1975), and initially applied to the C&D Canal by Gardner and Pritchard (1974).

Figure 8 presents pertinent water surface elevation data for the C&D Canal for a 34-day period in April and May 1975. The upper diagram shows the records of water surface elevation for the Old Town Point at the Chesapeake end of the Canal and for Reedy Point at the Delaware end of the Canal. The difference in amplitude and phase of the tides at the two ends is clearly shown by this diagram. The middle diagram shows the time varying slope of the water surface along the length of the Canal. This curve was obtained by simply dividing the difference between the water surface elevations at Old Town Point and Reedy Point by the distance along the axis of the Canal between these two gauges. It is this slope which drives the instantaneous (tidal plus non-tidal) flow in the Canal. The lower diagram shows the time variations, over this 34-day period, of the low-low pass filtered record of the slope. It is this LLPF slope which drives the non-tidal, or meteorologically forced, flow in the Canal.

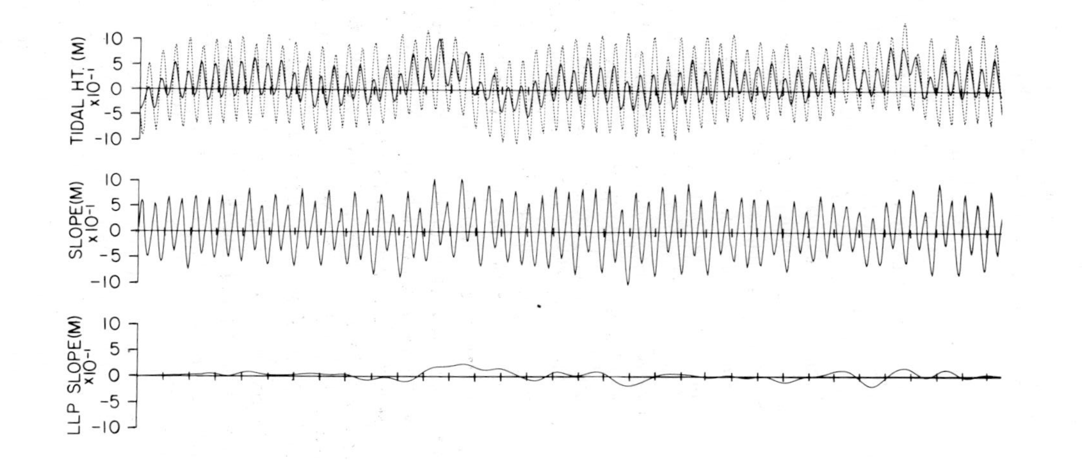

Figure 8. Upper: Observed record of water surface elevation from tide gauges at Old Towne Point (solid line) at the Chesapeake end of the C&D Canal, and at Reedy Point (dashed line) at the Delaware end, for a 65-tidal cycle interval in April and May 1975. Middle: Variation in instantaneous values of water surface slope over length of Canal over same time period. Bottom: LLPF record of slope.

Current meter measurements were made at two cross sections in the Canal over a 55 tidal cycle interval within the same 34-day period in April and May, 1975, represented by the water surface elevation data described above. A total of twelve current meters were deployed in the two cross sections, with three meters mounted on each of two vertical moorings in each cross section. The two moorings in each section were located just inside the outer edges of the dredged channel, since the ships that pass through the canal essentially sweep over the full width of the channel. During an earlier study involving a similar deployment of current meters in cross sections located between the two discussed above, the Canal was closed to ship traffic for one 24-hour period to allow the deployment of an additional six current meters mounted on two moorings located in the channel at one of the two cross sections. The data collected during this interval was used to establish the character of the lateral velocity distribution (see Pritchard and Gardner, 1974).

The current meters employed in this study recorded readings of speed and direction every 20 minutes. The speed value represented an average over the sampling interval and the direction information includes the predominant direction of the current during the sampling interval together with an indication of the range of direction which occurred during this interval. The velocity data for each 20 minutes was used to compute the flux, in cubic meters per second, through each section, as a function of time.

A subset of this data encompassing eight tidal cycles was first used to adjust the model. Figure 9 shows the observed and computed water surface elevations at two tide gauges within the canal, for a quadratic frictional coefficient based on a Manning's n of 0.020. This value of Manning's n gave the best fit to the data. Because of the short length of the Canal, and the fact that the model is driven by the observed water level variations at the two ends, the comparison of water level data in the interior of the Canal does not provide a sufficiently critical test. Figures 10, 11 and 12 show the observed and computed values of volume flux through each section, for Manning's n values of 0.018, 0.020, and 0.022. For a variety of reasons the authors placed greater reliance on the data for Section F, and consequently chose a value of 0.020 for Manning's n, to be used for further exercising of the model, since this value gives the best match between the observed and computed flux value for that section.

During these runs it was found that, in order for the computed net non-tidal flux through the Canal over the full eight tidal cycle interval to match the average of the observed values through the two cross sections, the elevation of the zero level of the tide gauge at Reedy Point, relative to the National Sea Level Datum for this area, had to be adjusted from the value provided to us by the National Ocean Survey by 0.04 ft (1.2 cm). In an earlier study, Gardner and Pritchard (1974) had encountered a similar problem with this gauge, which had been relocated just prior to that study because of the widening and deepening of the Canal which was then in progress. There have been several changes in the "official" elevation of this gauge as provided by NOS subsequent to its relocation, and on the basis of other evidence concerning the accuracy of the first order level net in this region, an uncertainty in the relative elevations of the two gauges at the ends of the Canal of 0.04 ft (1.2 cm) is not surprising.

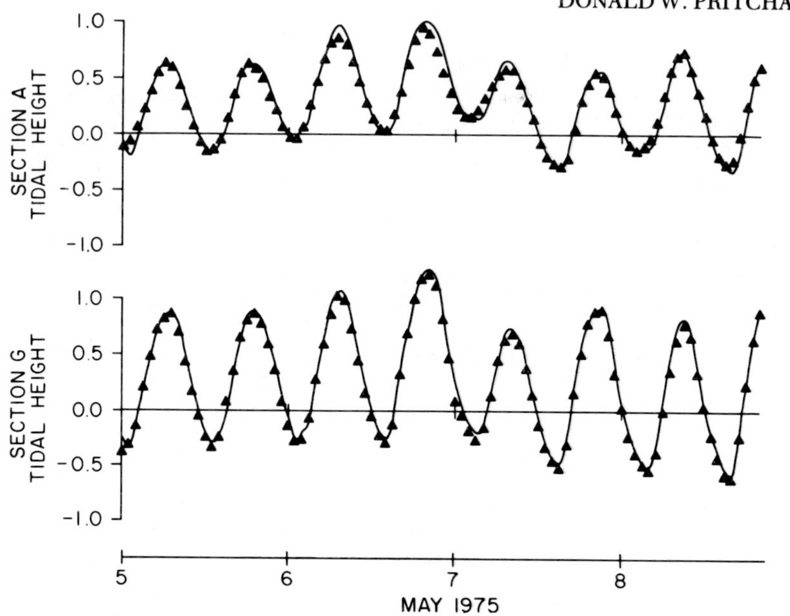

Figure 9. Tidal elevations (meters) at sections A and G in the C&D Canal for Manning's n = 0.020. Solid line denotes model prediction; and triangles denote observations from NOS tide gauges. See Figure 7 for location of section.

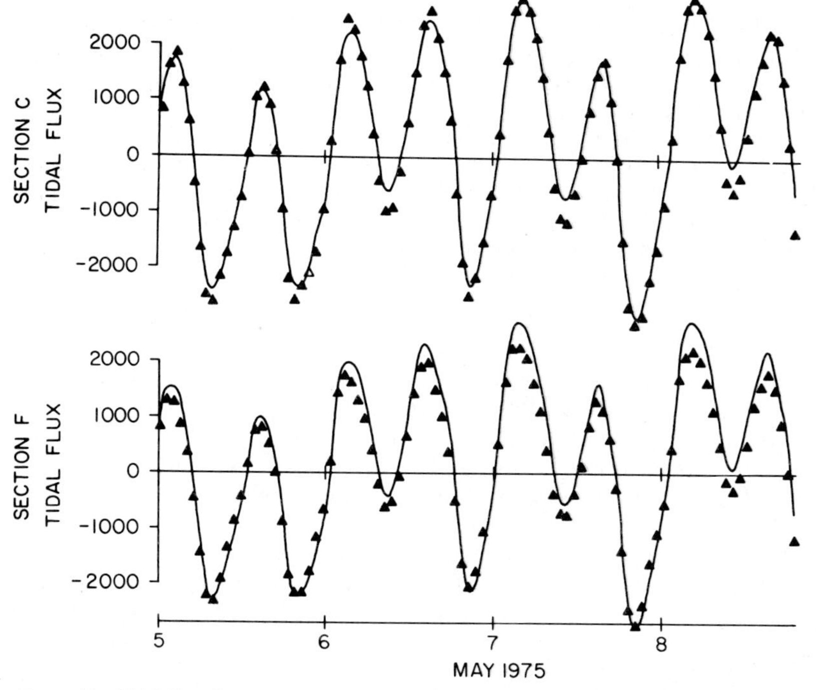

Figure 10. Tidal flux (cubic meters per second) at sections C and F in the C&D Canal for Manning's n = 0.018. Solid line denotes model predictions; and triangles denote observed values (integrated current meter records). Positive fluxes are eastward. See Figure 7 for location of sections.

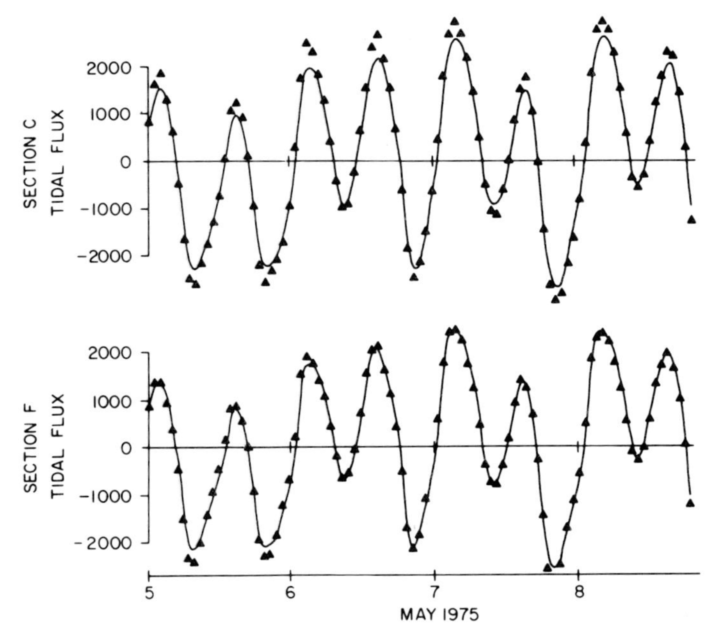

Figure 11. Tidal flux (cubic meters per second) at sections C and F in C&D Canal for Manning's $n = 0.020$. Solid line denotes model predictions; and triangles denote observed values (integrated current meter records). Positive fluxes are eastward. See Figure 7 for location of sections.

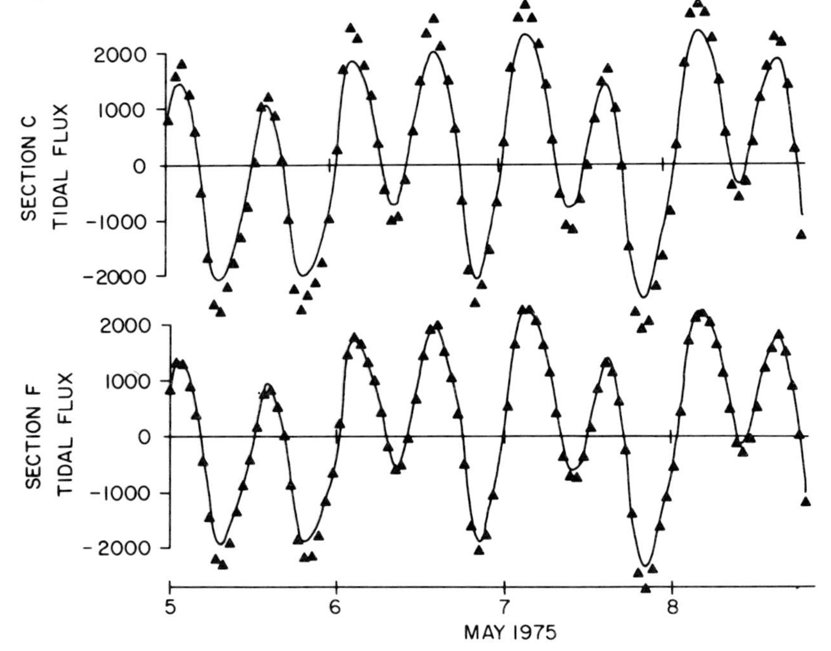

Figure 12. Tidal flux (cubic meters per second) at sections C and F in C&D Canal for Manning's $n = 0.022$. Solid line denotes model predictions; and triangles denote observed values integrated current meter records). Positive fluxes are eastward. See Figure 7 for location of sections.

The model was then exercised for a two-month period, which included the 55-tidal cycle interval over which current meter data had been obtained, using the 0.020 value for Manning's *n* found during the 8-tidal cycle long calibration runs, and also using the adjustment of ± 0.04 ft (1.2 cm) to the elevation of the Reedy Point gauge. Figures 13 and 14 show, for each of the two cross sections, the computed flux values for a 34 day segment of the 2 month period. The upper part of these figures give the computed instantaneous flux, while the lower portions show the time variations in the LLPF flux values. Also entered on these figures, as small filled triangles, are hourly values of the observed fluxes. The departure of the observed values of the instantaneous flux from the computed values, at peak ebb and flood flows, particularly for Section C, gives a distorted impression of the goodness of fit of the computed values to the observed values. Close inspection reveals that the majority of observed values fall on or immediately adjacent to the computed curve. Also, the fit of the peak values at Section C could be made as good as that for Section F, if a value for Manning's *n* of 0.018 were employed for the segment of the Canal containing Section C. For both sections, the fit of the computed LLPF flux values to the observed values is very good.

In order to test the hypothesis that the non-tidal flux through the Canal is driven primarily by differences in the meteorologically forced water level variations in the two bays, a simple linear correlative model was tested. In this model the LLPF water surface elevation at Old Town Point was computed from tide gauge data at the mouth of Chesapeake Bay and the longitudinal component of the square of the wind velocity as observed at a point about half way down the Bay. Similarly, the LLPF water surface elevations at Reedy Point was computed from tide gauge data at the mouth of Delaware Bay and the longitudinal component of the square of the wind velocity as observed at a point about half way down that waterway. The results of these computations, together with the LLPF water surface elevation record obtained from the observed data at Old Town Point and at Reedy Point, as shown in Fig. 15, clearly supports the hypothesis of meteorological forcing.

4.2 A Transient State, One-Dimensional Water Quality Model of the C&D Canal

This model is an extension of the hydrodynamic model of the C&D Canal described in the previous section, and the reference (Rives and Pritchard, 1978) serves as the basis for the material presented here.

A 20 percent solution of Rhodamine WT dye was continuously injected into the Canal starting at 11:20 1 May 1975, and continued through 8 May 1975, from a location along the northern side of the C&D Canal at about midway between Chesapeake City and Reedy Point. The position of the dye release is marked on the sketch map given in Fig. 7. From 5 May through 8 May, dye measurements were made at four cross sections (A through D) to the west of the dye source and at three cross sections to the east of the source, at the sections (E, F, and G), as shown in Fig. 7. In the following discussion the data for the closest section to the

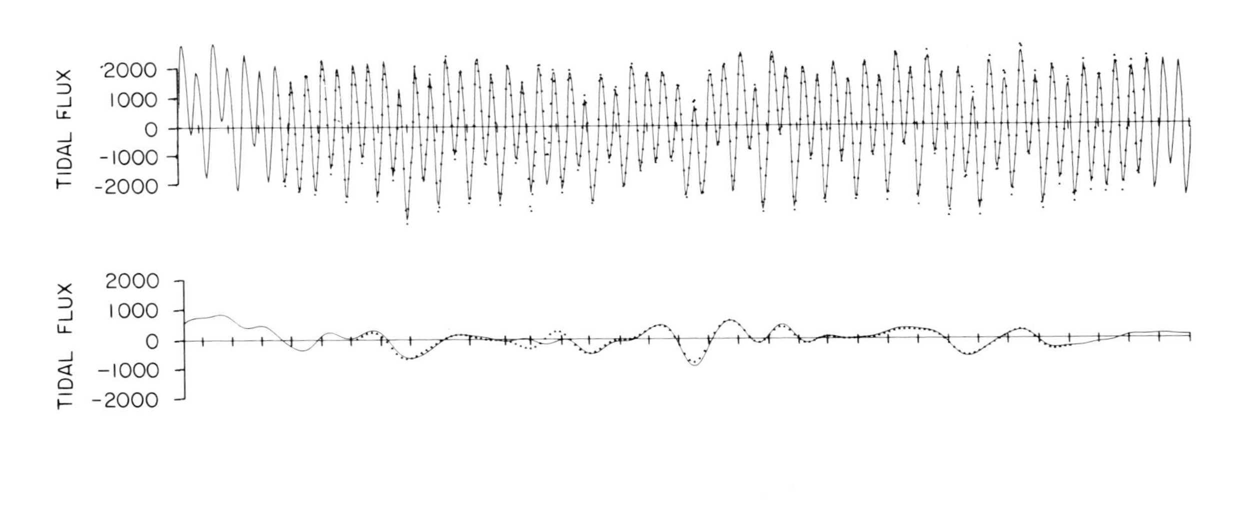

Figure 13. Upper: Variation in the instantaneous flux through Section F in the C&D Canal over a 65-tidal cycle interval in April, May 1975. Solid line, computed; triangles, observed values. Lower: LLPF record of flux. Solid line, computed; triangles, observed. See Figure 7 for location of section.

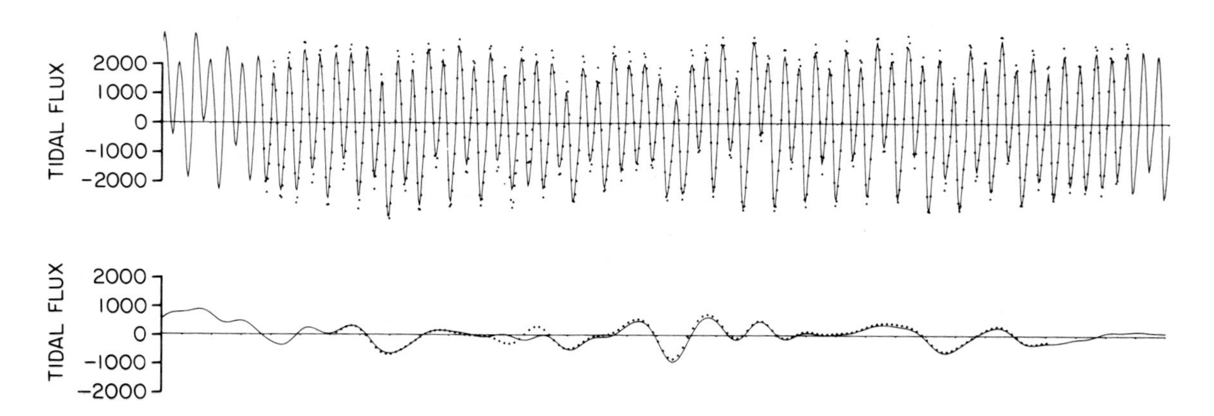

Figure 14. Upper: Variation in the instantaneous flux through Section C in the C&D Canal over a 65-tidal cycle interval in April, May 1975. Solid line, computed; triangles, observed values. Lower: LLPF record of flux. Solid line, computed; triangles, observed.

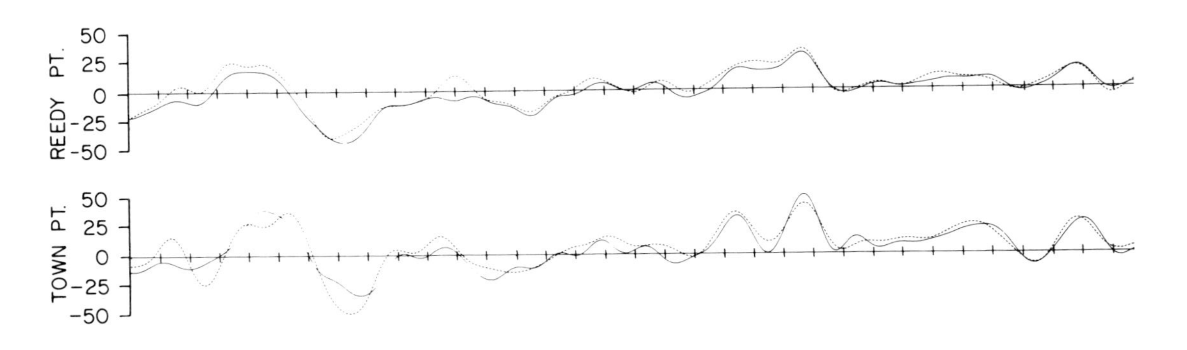

Figure 15. LLPF records of water surface elevation at Reedy Point (upper) and at Old Towne Point (lower). Solid lines, from observed tide gauge data. Dashed line, computed from simple correlative model.

west of the dye source is not considered, since there is evidence that lack of complete mixing of the dye over this cross section invalidates a one-dimensional treatment.

The model uses as input the volume flux data computed by the hydrodynamic model discussed in the previous section. The diffusion factor was selected to give the best fit to the data, and hence this comparison might better be described as a calibration rather than a verification. However, to my knowledge, this is the first comparison of this type of a model in a time varying field of motion.

The model was run for the eight-day period over which the dye was discharged. The results of the computations for the period 5 May through 8 May, during which time observations of dye concentration were made, are shown in Fig. 16. Again note that the observed data for Section D is not included because of lack of sectional homogeneity. The computed curves are in reasonable agreement with the plotted observed values.

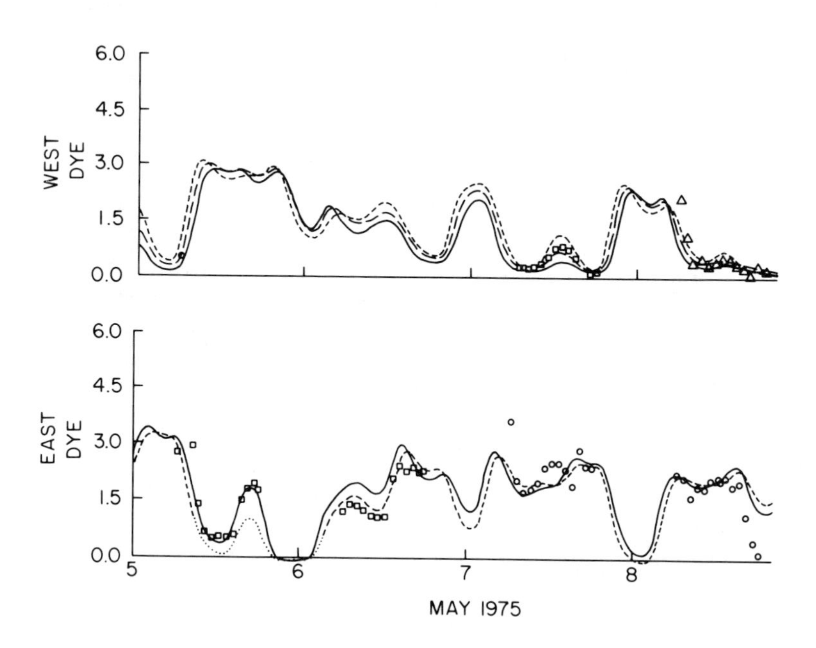

Figure 16. Dye concentrations normalized by the injection rate of dry dye (parts per million). West plot: solid line and plus signs denote predicted and observed values at section A; long dashed line and triangles denote predicted and observed values at section B; short dashed line denotes predicted values at section D. East plot: solid line and plus signs denote predicted and observed values at section E; short dashed line and hexagons denote predicted and observed values at section G. See Figure 7 for location of sections.

4.3 A Transient State, Two-Dimensional Thermal Plume Model

This section is based on the study reported by Pritchard and Wilson (1980). The model used in the study is described in greated detail in Carter, *et al.*, (1979b) than is given in the above reference.

The model was applied to a heated discharge from a power plant located at Vienna, Maryland. The discharge is made into the Nanticoke River, a tidal tributary to the Chesapeake Bay. In the vicinity of this discharge, the salinities are usually less than one part per thousand, and are very nearly uniform with depth.

The model couples a near field computation, which takes into account the momentum entrainment, as well as the centerline trajectory of the plume, which are the result of the transverse momentum introduced into the waterway by the discharge, with a far-field computation, which makes use of the Okubo-Pritchard Diffusion Equations (Pritchard, 1960).

The model was run to simulate the conditions associated with a study conducted by Carter and Regier (1975). In that study, Rhodamine WT dye was introduced continuously into the heated discharge of the Vienna Electric Generating Station on the Nanticoke River, for a 17-day period beginning 9 April 1973. Measurements were made at a series of cross sections in both the flood and ebb directions from the point of discharge. At each of these sections, continuous measurements were made of the dye concentration on repeated crossings of the plume over nearly one full tidal cycle.

The model can be used to produce the horizontal distribution of concentration of a contaminant contained in the discharge, or of excess temperature, at a given time. Example of this use are shown in Figures 17, 18 and 19. These examples are for the discharge from a cooling tower for a planned new unit at the Vienna Plant, and do not apply to the verification study.

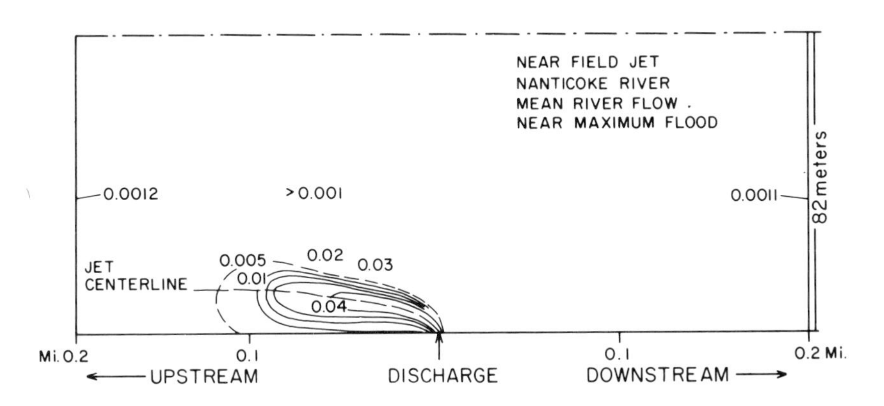

Figure 17. Isolines of relative concentration in the near field plume for mean river flow, for near maximum flood tidal flow. Example of model computation for Nanticoke River estuary.

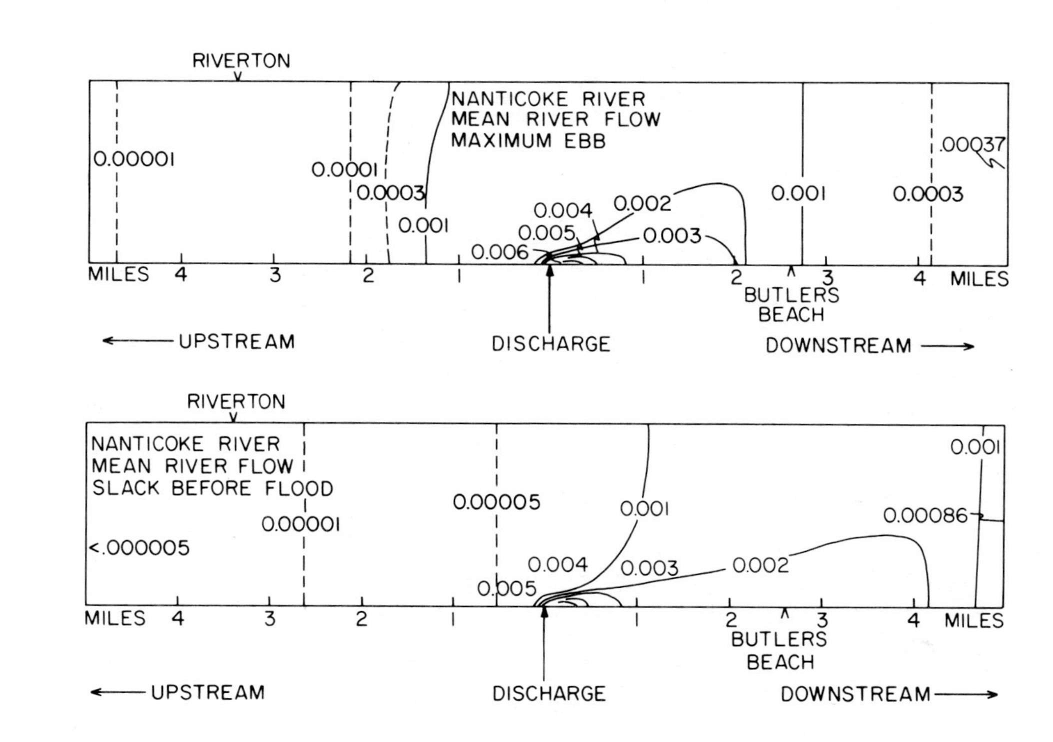

Figure 18. Isolines of relative concentration in far field for mean river flow (810 cfs), for maximum ebb tidal flow and for near slack before flood. Computations are for a discharge flow of 2570 gal min^{-1}, representative of a 600 MW plant. Example of computations for Nanticoke River estuary.

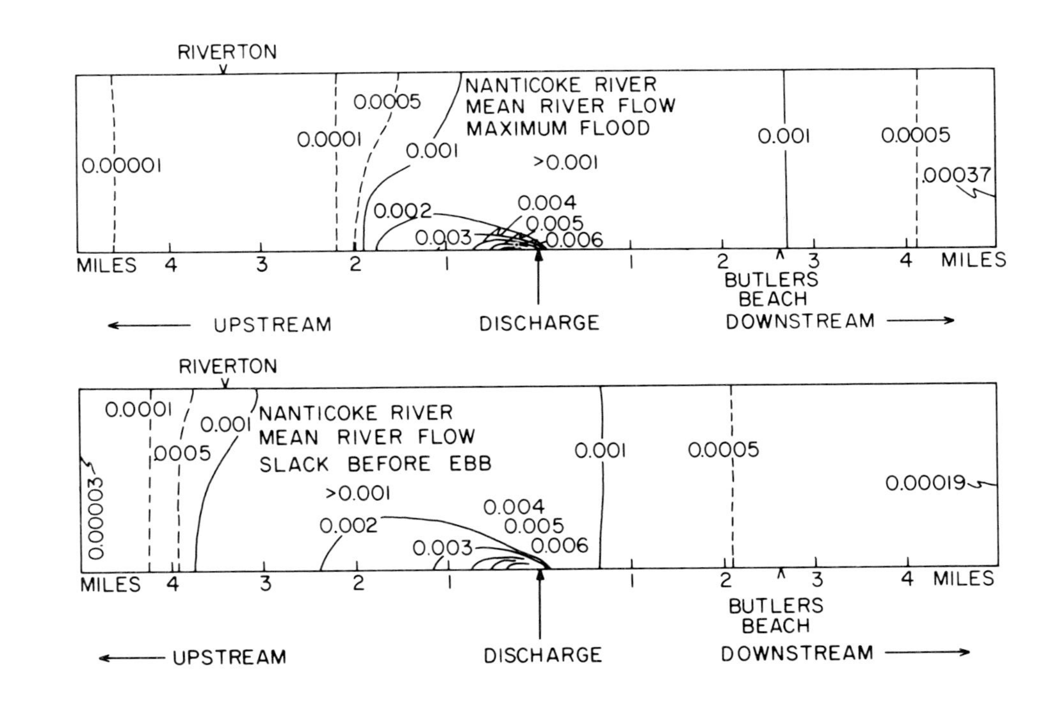

Figure 19. Isolines of relative concentration in far field for mean river flow (810 cfs), for maximum flood flow for near slack before ebb. Computations are for a discharge flow of 2570 gal min^{-1} representative of a 600 MW plant. Example of computations for the Nanticoke River estuary.

The model may also be used to produce the time history of concentration of a contaminant contained in the discharge for a given point. It is this mode of output which was used for model verification. Figures 20, 21 and 22 give the results of these verification runs.

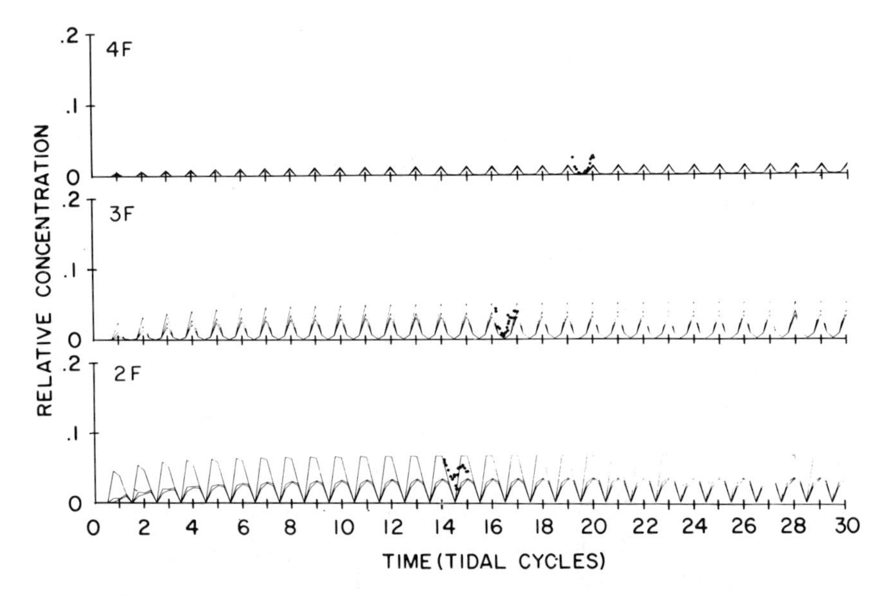

Figure 20. A comparison of observed time histories of relative dye concentration (solid circles) with simulated time histories at sections 4F, 3F and 2F, Nanticoke River estuary near Vienna, Maryland.

The model was run for the 17-day period corresponding to the dye release. The model requires velocity data as input. For these runs in the Nanticoke the velocity was assumed to be directed along the axis of the waterway, to be uniform across a given section, and to be in pseudo steady state; that is, to be a repeated function of time from tidal cycle to tidal cycle. At any given time within a tidal cycle, the velocity was assumed to be the sum of a constant, non-tidal velocity and zero-centered, sinusoidal tidal velocity. The non-tidal term was obtained by dividing the fresh water inflow applicable to the study period by the cross sectional area. The phase of the tidal component was taken from data given in "Tidal Current Tables, 1975" issued by the National Ocean Survey. Tidal amplitude was based on current meter measurements made at several moorings in the area by the Chesapeake Bay Institute in 1979, and on the current meter data reported by Carter and Regier (1975) obtained at a single mooring located just off-shore from the location of the discharge from the Vienna Electric Generating Station.

Four of the cross sections at which observations of dye concentration were made are located up the estuary from the source, and are labeled F1, F2, F3 and F4. The distance from the source to each of these sections is 450 m, 2060 m,

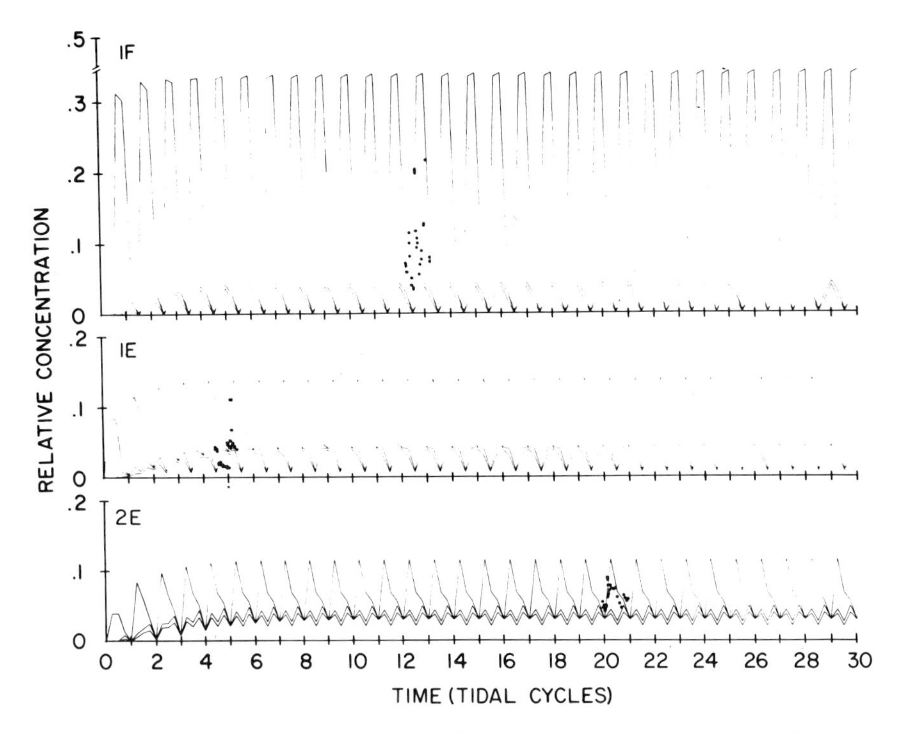

Figure 21. A comparison of observed time histories of relative dye concentration (solid circles) with simulated time histories at sections 1F, 1E and 2E, Nanticoke River estuary near Vienna, Maryland.

3790 m, and 7320 m, respectively. There were also five sections down the estuary from the source at which dye measurements were made. These are labeled E1, E2, E3, E4 and E5. The distance from the source to each of these sections is 1110 m, 2835 m, 6080 m, 10710 m and 15000 m, respectively.

The observed dye concentrations were divided by the concentration in the undiluted discharge. This ratio is called the relative concentration. The model output is also expressed in terms of relative concentration.

Computed values of relative concentration as a function of time, for each of three positions in each of the nine cross sections, are shown in Figures 20, 21 and 22. The three positions in each cross section for which the computed values of relative concentration are given in these figures included the computational grid point at which the highest values of the concentration ratio occurred, and two adjacent grid points.

These figures show the computed time history of relative concentration for the full 35 tidal-cycle time period over which the model was exercised. This is the same time period over which the dye discharge was made. Also plotted on the

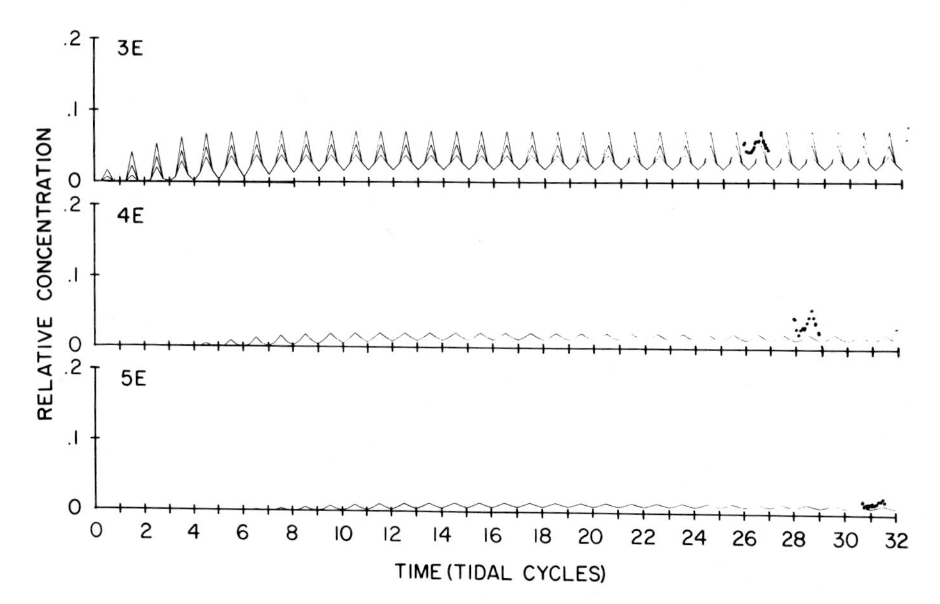

Figure 22. A comparison of observed time histories of relative dye concentration (solid circles) with simulated time histories at sections 3E, 4E, and 5E, Nanticoke River estuary near Vienna, Maryland.

diagrams for each section are the maximum concentration ratios observed on each crossing of the dye plume. These data, plotted as solid circles, are entered on each diagram at the appropriate simulated time corresponding to the actual time of each traverse of the section by the research vessel.

An inspection of Figures 20, 21 and 22 reveals that the observed relative dye concentrations at section 4F, which is 7320 meters up the estuary from the source, was considerably higher than the computed values, though the observed time sequence of maximum to minimum to maximum is reflected in the computed curves. There is a reasonably good fit of the observed and computed values at Section 2F (3790 m from the source) and at Section 3F (2060 m from the source). Section 1F is only 450 m up the estuary from the source. The plume at this distance is very narrow, and the turbulent fluctuations are quite large, resulting in a very "noisy" time history. The computational grid may not be sufficiently dense to assure that the maximum concentration is computed. In view of these facts, the observed data is not inconsistent with the model results.

Section 1E is the nearest section in the ebb direction from the source, at a distance of 1110 meters. Comparison of observed and computed relative concentrations at this section is made difficult by the fact that the observations were not obtained over a time period during which maximum concentrations occurred. The two highest values are, however, in good agreement with the computed curve. The fit of the observed and computed relative concentrations are reasonably good at Section 2E (2835 m from the source) and 3E (6080 m from the source). The observed data considerably exceed the computed values at Section 4E (10470 m from the source) though the observed phasing of the minimum values and the maximum value is well reproduced by the model results. At Section 5E (15000 meters from the source) the difference between the observed and computed values is of the same order as the uncertainty in the measured values.

There are some severe problems associated with any effort to verify a transient state water quality model of the type discussed here. This model, and others having similar features, give as output what is essentially an ensemble-averaged distribution, in space and time, of the relative concentration. Assume for the moment that there existed a large number of worlds, each with a Nanticoke River estuary, each with a Vienna Electric Generating Station. Let this large number be designated by k. Each of these k Nanticoke River estuaries is assumed to have the same time history of river inflow, of tidal rise and fall of the water surface and hence of tidal flow, and each subject to the same time history of meteorologically driven flows. In each of these Nanticoke River estuaries it is assumed that a tracer dye experiment is conducted, in exactly the same manner in each waterway. If all the resulting observations of relative dye concentrations from all k such experiments were superimposed in space and time, and the average taken over all realizations, the result would be an ensemble average. The difference between the ensemble average and one of the individual realizations, is non-deterministic in character. It is a stochastic quantity which cannot be individually computed. However, certain statistical statements can sometimes be made about the population of such differences.

In essence the values of the various parameters computed by numerical models are, in all cases, ensemble-averaged values of these parameters. There is no way that a true ensemble average can be obtained from measured data. However, there are ways to approximate the ensemble average from measurements made in a single system. For example, if in the Nanticoke such external driving processes as fresh water inflow and the wind were constant or only slowly varying over times long compared to a tidal cycle, and the tidal processes were repetitive, or nearly so, from tidal cycle to tidal cycle, then observed data taken over a number of tidal cycles (say, n) could be used to approximate the ensemble average. The data collected on each of the n tidal cycles is then considered as one realization of data taken over the same time epoch on each of n different Nanticoke estuaries. The observed time history at each point of measurement would then be divided into n records, each one tidal period long. These n records for each position at which observations were made would be superimposed with respect to tidal time, and averaged over n.

The larger the n the more stable the estimate of the ensemble average, provided, however, that external driving process do not change significantly over the n tidal cycles. I believe n should be at least 5 and probably should be 10, in order to obtain a reasonable approximation of the ensemble average.

Note that verification of models which represent averages over space or over time, using observed data averaged in the same way, is facilitated by the fact that, at least according to some theories, averages over space or time tend towards the ensemble average.

Also note that data obtained from runs of an hydraulic model do not represent an ensemble average, but rather a single realization in one world. The prototype represents another world. The modeling laws can, at best, only provide that the ensemble average applicable to the model is, after use of appropriate scaling factors, the same as the ensemble average applicable to the prototype. No single realization or sequence of realizations in the model can be expected to be exactly the same as in the prototype. Thus the problem of verification is doubly complicated for hydraulic models. It follows from this argument, however, that predictions of the consequence of man's manipulation of the environment using hydraulic models should be based on several repeats of the model experiments.

REFERENCES

Beardsley, R. C., Boicourt, L. C., and Scott, J. (1977). CMICE 76: A current meter intercomparison experiment conducted off Long Island in February-March 1976. Tech. Rep. WHOI 77-62, Woods Hole Oceanographic Institution.

Cannon, G. A., and Pritchard, D. W. (1971). A biaxial propeller current meter system for fixed-mount applications. *J. Mar. Res.,* **29** No. 2, 181-190.

Carter, H. H. (1974). The measurement of rhodamine tracers in natural systems by fluorescence. *Rapp. R.-v. Réun. Cons. Int. Explor. Mer.* **167,** 193-200.

Carter, H. H., and Regier, R. (1974). The three-dimensional heated surface jet in a cross flow. Tech. Rep. 88, Ref. 74-8, Chesapeake Bay Institute, The Johns Hopkins University.

Carter, H. H., Najarian, T. O., Pritchard, D. W., and Wilson, R. E. (1979a). The dynamics of motion in estuaries and other coastal water bodies. *Rev. Geophy. Space Phys..* **17,** No. 7.

Carter, H. H., Wilson, R. E., and Carroll, G. E. (1979b). An assessment of the thermal effects on striped bass larvae entrained in the heated discharge of the Indian Point generating facilities Units 2 and 3. Spec. Rep. 24, Ref. 79-7. Marine Sciences Research Center, State University of New York at Stony Brook.

Elliott, A. J., and Wang, D.-P. (1978). The effect of meteorological forcing on the Chesapeake Bay: the coupling between an estuarine system and its adjacent coastal waters. *In* "Hydrodynamics of Estuaries and Fjords" (J. C. J. Nihoul, ed.), Elsevier, Amsterdam. 127-145.

Elliott, A. J., Wang, D.-P., and Pritchard, D. W. (1978). The circulation near the head of Chesapeake Bay. *J. Mar. Res.,* **36,** 643-655.

Gardner, G. B., and Pritchard, D. W. (1974). Verification and use of a numerical model of the Chesapeake and Delaware Canal. Tech. Rep. 87, Ref. 74-7. Chesapeake Bay Institute, The Johns Hopkins University.

Harlemann, D.R.F., and Lee, C. H. (1969). The computation of tides and currents in estuaries and canals. Tech. Bull. 16, Committee on Tidal Hydraulics, Corps of Engineers, U. S. Army, Vicksburg, Mississippi.

Hunter, J. R. (1975). A one-dimensional dynamic and kinematic numerical model suitable for canals and estuaries. Spec. Rep. 47, Ref. 75-10. Chesapeake Bay Institute, The Johns Hopkins University.

Pritchard, D. W. (1960). The application of existing oceanographic knowledge to the problem of radioactive waste disposal into the sea. *In* "Disposal of radioactive wastes. Vol. 2". International Atomic Energy Agency, Vienna, 229-248.

Pritchard, D. W., and Gardner, G. B. (1974). Hydrography of the C&D Canal. Tech. Rep. 85, Ref. 74-1, Chesapeake Bay Institute, The Johns Hopkins University.

Rives, S. R., and Pritchard, D. W. (1978). Adaptation of J. R. Hunter's one-dimensional model to the Chesapeake and Delaware Canal System. Spec. Rep. 66, Ref. 78-6. Chesapeake Bay Institute, The Johns Hopkins University.

Wang, D.-P. (1979). Wind-driven circulation in the Chesapeake Bay, Winter 1975. *J. of Phys. Ocean.*, **9**, 564-572.

Wang, D.-P., and Elliott, A. J. (1978). Non-tidal variability in the Chesapeake Bay and Potomac River: evidence for non-local forcing. *J. of Phys. Ocean.*, **8**, 225-232.

Wang, John (1980). Analysis of tide and current meter data for model verification. *In* "Mathematical Modeling of Estuarine Physics" (J. Sündermann and K.-P. Holz, eds.) Lecture Notes on Coastal and Estuarine Studies. Springer-Verlag, New York.

SPECTRA PRESERVATION CAPABILITIES OF GREAT LAKES TRANSPORT MODELS

Keith W. Bedford

Department of Civil Engineering
The Ohio State University
Columbus, Ohio

1. INTRODUCTION

The Great Lakes are one of the most heavily used systems of lakes in the world; the large and often conflicting demands by industrial, commercial, recreational, and municipal interests require rational methods for managing this scarce resource. To insure proper resource use, a complicated web of state, federal, and international agreements are in place, whose satisfaction demands the use of the best available analytical methods for computing optimal solutions. A most widely used analysis tool is the numerical transport model; a count by the author of reports and papers indicates that over one hundred different aquatic transport models have been used in the Great Lakes. The verification of these models is a prime concern, but the success rate is to date quite low. Early studies by Allender (1975, 1976) indicated that a prediction of zero was a least squares better answer than three-dimensional model predictions for a Lake Michigan study, while Thomann and Winfield (1976) had to accept two standard deviations before they could verify a three-dimensional Lake Ontario phytoplankton model.

Two verification procedures are commonly used in such studies; the first is the point by point comparison of computed state variables and observed data while the second is the comparison of predicted surface transport patterns with remotely acquired imagery. The first method is used most frequently for quantitative verification as to date it yields the best capability of performing comparisons of horizontal and vertical transport structures. There are two problems with this type of verification. First, data for this type of whole lake verification is usually collected from moored chains of current meters and thermistors, and due to economic limits, data storage limitations, and required long-term monitoring, they are deployed with large separation distances and relatively coarse sampling time increments. At no time in the foreseeable future will there ever be as intense a data gathering effort as on Lake Ontario during the 1972 International Field Year

ISBN: 0-12-25852-0

on the Great Lakes (IFYGL) where many millions of dollars were spent collecting such whole lake data. Though masterfully taken and prepared, this data has already been shown as insufficient to meet the needs of certain three-dimensional model verifications (Simons 1974, 1975; Bennett, 1977). Improvements in the whole lake data gathering density will probably not be economically possible and this will place a limit on the data and phenomena resolution possible from such procedures. The second problem is that models verified to the scale of these data are only valid to these scales and, therefore, very little confidence can be placed in model predictions of activity whose scales are this size or smaller. The ability even to parameterize the effect of small-scale activity on the mean flow is prohibited without necessary and often unavailable small scale sampling.

Model complexity is far outstripping the collection density of data necessary for point by point verification. Modeling improvements include schemes such as the finite element method, coefficient estimators for eddy viscosities and growth rates, and feedback and control forecasting strategies. The most troublesome model improvement has been the heavy use of three-dimensional models, motivated by the rationale that mesh refinement yields more exact calculations. These models are expensive to operate and the decisions based upon the model output also incur vast expense; therefore, the modeler must be sure his calculations are "correct." However the whole lake data required for such 3-D verifications are never going to be available at the scales resolved in these models.

The problem is apparent: other methods of verifying 3-D models must also be used. Imagery comparisons are fine but to date lack the vertical resolutions necessary for 3-D models.

Modelers in the turbulence community have long recognized that verification of turbulent computations consists of not only time and space pattern comparisons but also the necessity of reproducing the theoretically known and observed statistics of the flow. With this criteria the adequacy of structural features in the model can be verified and predictions placed on a less empirical basis. From theory and experimental observation a great deal is known about the statistical and particularly the spectral structure of geophysical flows, and it is these spectra that Great Lakes and for that matter all lake models should be able to reproduce.

2. OBJECTIVES

It is the assertion of this paper that significant progress in the structuring and verification of lake turbulent transport models can be made by requiring model output be verified not only by the point comparison method but by the spectral statistics of turbulent flow. This paper presents a study designed to determine whether models, structured with the existing assumptions used in available Great Lakes models, reproduce such spectral relationships.

Attention is directed only to three-dimensional models in this work and both rigid lid and free surface transport models are used. A large eddy simulation approach is used to develop these analogs; by comparison with theory, field data, and traditionally prepared lake models the conditions under which spectra are properly calculated will be demonstrated and structural improvements required in existing models will be delineated.

This paper is organized into seven subsequent sections. The third consists of a review of current Great Lakes transport models and verifications; the fourth describes the conceptual background of turbulence including theoretically known and observed spectral shapes; the fifth derives the governing equations for both rigid lid and free surface lake analogs as prepared by large eddy simulation methods; the sixth presents the numerical formulation; while the seventh summarizes the data used to implement the numerical comparisons; the eighth presents a summary of the results; while the final sections identify and discuss the major conclusions and implications.

3. VERIFIED GREAT LAKES TRANSPORT MODELS

Several of the first three-dimensional transport models were developed for the Great Lakes. This section presents a review of such models and their verifications, followed by a general summary of their success.

3.1 Review of Transport Models and Verification Research

The first three-dimensional Great Lakes model was by Gedney and Lick (1970, 1972) who, using a steady state linearized rigid lid formulation, simulated Lake Erie currents and compared the results to a few select current readings from FWPCA data. Qualitative agreement was reached. This model structure was also used by Bonham-Carter and Thomas (1973) in a study of the Rochester, New York Embayment. Again, qualitative agreement was reached. These models are, however, only of historical interest as within three years they were superseded by radically more complicated dynamic models, all of which included nonlinear, Coriolis, and dispersion effects. These models are unique not only in their being amongst the first 3-D models to be published, but more importantly, they were the first such models to be subject to a published verification analysis. Among the many papers published by these researchers are the exemplary works by Simons (1975, 1975), Bennett (1977), Allender (1977), and Thomann, Winfield and Segna (1979) and the continuing work of Lick's group (for example, Sheng *et al.* 1978, Paul and Lick, 1976, and Lick 1976). Since most of Lick's work has been in Lake Erie where no whole basin data are available, verifications have been limited to pattern comparisons for whole basin work. Therefore, this review will concentrate on the other four authors' work.

Simon's model uses four thermally homogeneous layers, with permeable interfaces. A free surface formulation for currents, elevations and temperature is derived and decomposed into barotropic and baroclinic modes. Horizontal diffusion is included but only to smooth the solution. The finite difference solution uses leap frog time differencing; forward differencing for advection, centered space differencing for other terms and two Richardson lattices for computational mode suppression.

The model was applied to the Lake Ontario IFYGL data with particular attention drawn to Hurricane Agnes. Stress coefficients were determined from the decrease in kinetic energy after the storm and the subsequent shear was calculated from buoy observations as interpolated over the lake by an inverse square law. Model output time series were then compared to available data time series at selected points. A feature of Simon's work is the considerable attention paid to the proper filtering of data in order to compare the model to observations prepared with length and time scales consistent with the model assumptions.

Because the major conclusions from this work are to be shared with all the Great Lakes models, they will be explained in the next subsection. However (and of particular significance to this work), it was found that inertial frequency oscillations were important and were governed by the vertical shear. Also, for time scales greater than the inertial period, water levels and currents agree with observation in both thermally homogeneous or stratified conditions. Internal horizontal momentum fluxes could vary by an order of magnitude before affecting the result. Stratification effects prohibited the separation of modes, while the layer approach did not permit the adequate resolution of summertime stratification. No quantitative measures for statistically testing a verification of a time history were reported.

Bennett presents an eight permeable layer, rigid lid Lake Ontario model. The rigid lid approach permitted a much larger step. A unique bulk viscosity form of smoothing was applied to compensate for the underestimation of rotation effects in limiting vertical motion. This term effectively filters out small-scale motion containing large-scale divergence. Temperature fields were computed but only advective transport was considered. The velocity field employed DuFort Frankel marching while Adams Bashforth was used for advection and forward time for vertical diffusion. One lattice with mid-side velocity definitions was employed. Data for comparisons was from the IFYGL coastal chain stations, unlike Simons, who used whole lake data. Wind stress was interpolated by an inverse square relation.

Conclusions specific to this study are that uniform grids did not work at all well in that they would not permit prediction of internal Kelvin wave propagation and the proper shore normal temperature gradient. Variable resolution near the coastline provided a substantial improvement. Qualitative comparisons showed that gyres were properly positioned and that the response to storms was appropriately larger and direct with well resolved coast boundary layers.

Unlike both these works which used qualitative methods for comparison, both Allender and Thomann attempted quantitative statistical verifications of three-dimensional models. Allender's paper reports the intercomparison of four Lake Michigan hydrodynamic models against field data collected by the FWPCA in 1963. Simons and Bennett's model was used along with a two impermeable layer model by Kizlauskas and Katz (1974) and a simple alternating direction implicit one layer formulation. The philosophy of the research was to compare the models to the data, not to each other.

Input shear data was prepared for neutral conditions and interpolated by a modified Cressman weighting technique. Comparisons were quantitatively compared by two norms, a Fourier rms norm for velocity magnitudes and a direction norm for trajectory comparisons. The specific conclusions are that the norm of computed and observed data always exceeded the norm of the observed data and zero. The largest decrease in norms occurred when near inertial motions were removed. Vertical viscosity does not affect the norms for comparisons at time scales greater than the inertial period while subinertial motions are so affected. Allender also states that space scale activity at less than 20 km exerts a significant effect on the model performance.

Finally Thomann, Winfield and Segna's exemplary quantitative verification of 3-D phytoplankton models for Lake Ontario represents the latest and perhaps most unique quantified verification study in the lake transport literature. Very simple 1-D and complex 3-D models of Lake Ontario were used as well as a 3-D small-scale model of Rochester Embayment. IFYGL data was used for both model initialization calibration, and verification. Three measures of verification were used, standard statistical comparisons of means from data and models, regression analyses of observed and computed values, and relative error of observed and computed output. The results share general implications from all these models; therefore, a summary of these features follows.

3.2 Summary of Major Verification Results

In general these research works are excellent verification studies. Each study was thoroughly presented and prepared. Great care was taken to filter, average, or prepare the observed data commensurate with the time and length scales implied in the basic structure of their models. This point is almost routinely neglected by many model practitioners and in many cases is at the heart of the necessity for perpetual model calibration. All these works demonstrate excellent qualitative agreement with their respective data. Whole basin patterns are reproduced within theoretical expectations as evidenced by Bennett's work. Time series comparisons by Simons also exhibit good comparison. Allender reported qualitative agreement with remote imagery.

The quantitative comparisons are, with one exception, at best mediocre. Allender's norm studies do not demonstrate good results. His results indicate that a predicted answer of zero would be a better Fourier norm result than any model prediction. Thomann reports essentially the same experience; within the Rochester Embayment relative median error is greater than 50%. Thomann also reports, however, that for whole Lake Ontario simulations with coarse grids compared to properly aggregated data his errors could be calibrated to 10%, quite a precise calculation.

A review of the remaining conclusions indicates that these models shared four problems. The first problem shared by Allender and Simons is the parameterization of the wind forcing function. Both authors mention that this was a difficult problem and the results were sensitive to the wind shear parameters. Simons had to use quite large values to make his results compare favorably. Storm surge modeling in Lake Erie (Schwab, 1978) has shown that drag coefficients must account for large unstable air/water temperature differences and neither of these two authors included this effect. As opposed to the shorter term calculations of Simons and Allender, Bennett's long-term simulations found very little sensitivity to shear fields or, for that matter, viscosity.

The second problem shared by these authors is bad or incomplete data. Three types of bad data are reported. The first is data taken by poor or faulty instrumentation. Allender, in Lake Michigan, had missing data and wide scatter. A second category of bad data is that taken by several collection groups with poor quality control and intercomparison checks. This was a problem with the data Thomann had to use.. The EPA and CCIW collected the Lake Ontario data during IFYGL and these data only agreed within very coarse limits. The third type of bad data is that not resolved properly in either spatial or temporal density. This was also a particular problem with Simon's, Allender's, and Thomann's work and resulted in the inability to assess vertical momentum transfer structures in the models. The first two types of bad data introduce scatter in the measured results; data must be overly aggregated or low pass filtered, or highly averaged in order to see any deterministic structure with which to compare model output. The effective length scales of these averaged data are usually very coarse, as would be the verified length scale of any model verified with such data.

A third major finding of these works is the curious inability of the models to make any kind of reliable calculation at all under weak no stress conditions. Allender reports this problem for velocity and temperature fields as well. This seems a curious result, but a systematic relationship will be hypothesized at the end of this paper.

The fourth and most important shared result is the inability to either predict or parameterize the effect of smaller scale activity. The second paragraph in Simon's first paper states that a major problem requiring much more detailed work is the inability to predict the smaller subgrid scale activity on the resolved flow. Allender and Simons also report that all such smaller scale activity ($\leqslant$ 14 hour) was poorly simulated. The most striking result is Thomann's work whereby using his Lake Ontario model in the Rochester Embayment he increased his error by fivefold. Indeed, the results in the first three major results from these works are implied comments of the same type, i.e., the more model complexity, as represented by denser space and time nodes, the worse the results become. Thomann rightly attributes this problem to the model resolution of small-scale random hydrodynamic activity which to date has not been parameterized or investigated.

Numerical or computational mathematics indicate that the more nodes the closer the answer will converge to the correct one. However, these reports indicate the contrary, and state or imply that the traditional state variable data is not even available to either formulate or verify proposed parameterizations. Indeed, the experience with bad data indicates that the contradiction exists even when very dense state variable data are collected for very dense models. It will have enough scatter introduced such that it will have to be low pass filtered to the extent where all relevant desirable small-scale activity is suppressed. The resolution of smaller scale physical phenomena by the small grids used in 3-D models requires very detailed descriptions of the fluid mechanical turbulence. Simple parameterizations for shear and dispersion such as eddy viscosities and diffusivities which are derived for, and are consistent with, the coarser grids, are now inefficient in resolving the smaller-scale fluctuating activity.

It is natural to ask these models to reproduce the fluctuating activity, but the expense of collecting dense data will particularly prohibit the traditional kind of point verification. Spectral or statistical properties of this activity are both theoretically known and observed; therefore one way to verify 3-D models is to ask that they reproduce spectral structures.

4. A REVIEW OF SPECTRAL TURBULENCE THEORY AND OBSERVATIONS

This section, abstracted from Bedford and Babajimopoulos (1980) and Babajimopoulos and Bedford (1980), summarizes the known theoretical spectral distributions for velocity, temperature, and contaminant fluctuations, and the aquatic field data results supporting or rejecting this theoretical spectra.

4.1 Theoretical Spectral Distributions

Velocity field energy spectra contain a non-homogeneous range at low wavenumber (Fig. 1) at which energy enters the flow field and an equilibrium range through which this energy cascades until dissipated in the high frequency or short wavelength range. Within the equilibrium range there is an extensive region called the inertial subrange. As derived by Kolmogorov (1941) and Obukhov (1949a, b), the power spectral density $E(k)$ of the turbulence kinetic energy is proportional to wavenumber raised to the -5/3 power. For stationary homogeneous and, therefore, three-dimensional turbulence, then

$$E(k) = C\epsilon^{2/3}k^{-5/3} \tag{1}$$

where C is a constant, ϵ is the viscous dissipation, and k is the wavenumber. Fjortoft (1953) and Kraichnan (1967) formulated another theoretical derivation for energy spectra in non-equilibrium (accelerated) or two-dimensional flows. According to this formulation energy cascades to the higher frequencies as

$$E(k) = C_2\xi^{2/3}k^{-3} \tag{2}$$

where C_2 is a coefficient, ξ is the enstrophy (vorticity squared) dissipation, and k

is the wavenumber. This was later modified by Kraichnan (1971) to include saturation effects. If energy enters the system at wavenumber k, then

$$E(k) = C_2 \xi^{2/3} k^{-3} \left[\ln \frac{k}{k_i}^{-1/3} \right] \tag{3}$$

Fig. 2a contains a qualitative summary of these shapes.

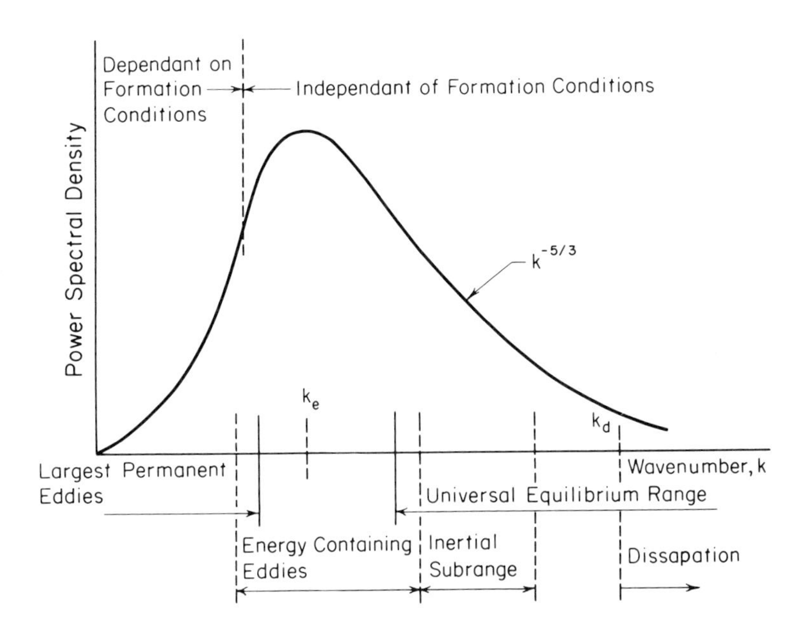

Figure 1. Generalized spectrum.

The spectra of temperature fluctuations was theoretically derived by Corrsin (1951) and Batchelor (1959a, b). As conceived both spectra follow a -5/3 shape in the equilibrium range, i.e.

$$\Gamma(k) = C_T \chi \epsilon^{-1/3} k^{-5/3} \tag{4}$$

where $\Gamma(k)$ is the wavenumber spectral density function for thermal fluctuations, χ is the rate of destruction of thermal variance, ϵ is the rate of kinetic energy dissipation, and k is the wavenumber. At the dissipation range, however, two different shapes could occur; for the case where the thermal conductivity, α, is much greater than the viscosity, ν, the shape is proportional to $k^{-17/3}$; and for the case where $\nu >> \alpha$, the shape is proportional to k^{-1}. Lumley (1946) extended

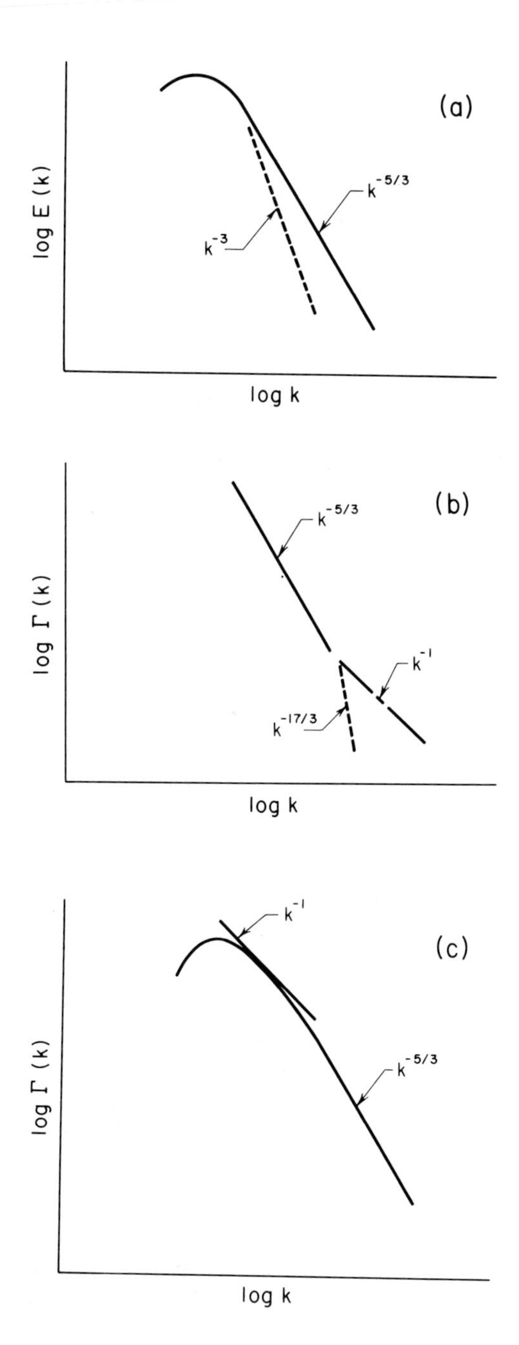

Figure 2. Theoretical spectral shapes for (a) velocity, (b) passive scalar, and (c) reactive substance fields.

this work to predict temperature fluctuation spectra for stably stratified flows. Fig. 2b contains qualitative descriptions of the two temperature spectra.

Finally, for mass transport fluctuations without source/sink terms, i.e., the case of sediment transport, Batchelor and Corrsin's work cited above applies. If, however, there is a biochemical reaction, such as would occur with each species equation in an ecosystem model, the work of Corrsin (1961) and Denman *et al.* (1977) applies. For first order reaction kinetics controlled by a rate coefficient r, there are again two theoretically derivable regions: the first is the familiar "5/3" region with formulae similar to Eq. (4); and the second occurs at smaller wavenumbers as a function of the size of 4 and is proportional to k_{-1}. Figure 2c contains a typical distribution.

4.2 Field Observations of Spectral Distributions

The observation of these spectral shapes in field data has a long and valued literature. Monin and Yaglom (1975) provide a good summary of atmospheric observations. In aquatic situations, velocity or kinetic energy spectra have been observed since 1962. Grant (1962, 1968), in a benchmark series of papers, used hot film anemometry to confirm the existence of the -5/3 region in both tidal estuary and open sea situations. Cross stream and down stream components were also found to obey this law. Webster (1969) found the five-thirds law behavior in the ocean at wavelengths ranging as high as 5000 meters. Palmer (1973) used both current meter and hot film anemometry to measure Lake Ontario energy spectra. Measurements located at a depth of 5.8 meters were taken at two locations: the first was located one kilometer offshore in unstratified nine-meter deep water; while the second station was well offshore in water 22 meters deep. The 5/3 slope was well developed in the deeper water, while a -3.0 slope characterized the spectral shape in the nearshore region. Lemmin *et al.* (1974) collected energy spectra during and after a brief wind event in Lake Ontario. During the acceleration period a -3.0 slope prevailed which relaxed back to a -5/3 shape after the wind eased. The work of the Lake Tahoe Group (Dillon and Powell, 1976; Leigh-Abbot *et al.* 1978, and Powell and Richerson *et al.* 1975) documents current spectra with the full five-thirds spectra prevailing at wavelengths between 10 and 100 meters.

With the exception of the previously cited papers by Lemmin, Dillon and Powell, and Powell and Richerson, temperature spectra were also reported, and all reported spectra behave in the inertial range as the kinetic energy spectra. Additional observations of temperature spectra are reported by Gregg (1976, 1977), Nasmyth (1970), Elliot and Okey, (1976), and Mamarino and Caldwell (1978). These measurements also support the Batchelor viscous-convective spectra by observation of the linear range. All the high frequency range was classified as the convective transport subrange with $\nu >>> \alpha$ with a slope of -1. This is in direct agreement with Batchelor's and Corrsin's work. Additional confirmation is reported by Dillon and Caldwell (1980) who report data obtained from vertical microstructure profiles. These profiles were taken by a controlled rate of descent device developed by Caldwell *et al.* (1975). This paper summarizes and synthesizes spectra behavior in the very high frequency dissipation range. Batchelor's spectrum is confirmed for high Cox numbers.

Suspended sediment transport is classified, for clays, silts, and fines, as a passive scalar substance and, therefore, fluctuation spectra also should behave as the above temperature spectra. Willis and Kennedy (1977) and Willis and Bolton (1979) report sediment concentration spectra in laboratory river channels. Work by Lavelle, *et al.* (1978) in Long Island Sound support the parallelism between the velocity spectra slopes and the sediment spectra slopes.

Perhaps the most intensely studied reacting transport fluctuations are those found in phytoplankton or chlorophyll studies. Using continuously measured chlorophyll, temperature, and/or current readings, fluctuation spectra are calculated from these data and compared to Denman's or Corrsin's theoretical spectra. The works of Fasham and Pugh (1976), Denman (1976), Platt (1972), Leigh-Abbott *et al.* (1978), and Powell and Richerson *et al.* (1975) provide the most substantial confirmation of the theory. In the 1975 paper currents and chlorophyll were measured in Lake Tahoe; at wavelengths of 100 meters and less phytoplankton variability followed the -5/3 laws completely. At greater wavelengths, a radical departure of chlorophyll and current spectra occurs, attributed to chlorophyll variability resulting from biological sources. This was found in the later paper reported by Leigh-Abbot *et al.* (1978), but the chlorophyll spectra did not support the theoretical behavior of Denman's (i.e., the -1 source/sink slope at long wavelengths). A NATO Conference edited by Steele (1978) summarizes much of this work.

4.3 Summary of Spectral Characteristics

From this background work the general spectral characteristics that any lake transport model should reproduce are as follows: (1) The maximum spectral content should occur at wavelengths proportional to the basin length; (2) For stationary, homogeneous conditions energy should cascade through to the dissipatory range with a -5/3 slope with indications that this range should spread over several decades of wavenumbers (3) During acceleration a -3 slope should be observed; (4) During die away a -5/3 slope should return; and (5) Passive contaminant spectra should follow the velocity spectra for all ranges except the very high frequency dissipatory region.

With these known spectral features in mind a sequence of numerical experiments was designed for the wind driven lake circulation and transport problem. The goals of this experiment are to determine if and when the traditional lake model formulation will reproduce these known spectral shapes and the behavior of these spectra under a variety of loading and basin conditions.

5. PREPARATION OF THE TRANSPORT MODEL EQUATIONS

The philosophy motivating the models to be presented here is quite simple. By using the technique of large eddy simulation, the traditional lake transport equations for rigid lid and free surface models can be derived to preserve spectral statistics. A number of test cases are then performed, spectra are compared with known theoretical and field results to determine the adequacy of the basic physical assumptions used to formulate the equations and to see under what conditions the spectral patterns from Section 4 are reproduced. Then by comparison to these computed spectra, output from traditional models will be analyzed and the conditions under which traditional models can be used will be determined.

5.1 Filtration and Large Eddy Simulation

5.1.1 Review of large eddy simulation models. The problem of resolving all the length scales of turbulent flow is much too overwhelming for even the largest computer. The analyst is forced to numerically resolve by proper averaging all the mean flow fluctuating activity of importance. Combinations of mean flow field variables are then used to parameterize the effect of all momentum and transport flux activity at or below the grid scale level. This model development pattern is in some sense followed by all modelers with the basic method for equation preparation consisting of formulations based upon the consistency of time and length scales between mathematical terms and the phenomena to be computed. In data analysis the rational elimination of high frequency or wavenumber information is routinely done by averaging or in generalized form by low pass filtration. Recognition that a formal connection between the preparation of the governing equations by proper uniform grid volume averaging and the creation of subgrid scale effects was first achieved for channel flow by Deardorff (1970). He used a grid size and strain rate subgrid scale (SGS) energy dissipation formulation first developed by Smagorinsky (1965). Similar treatments with uniform grid volume averaging were used in geophysical models by Spraggs and Street (1975) and Bedford and Shah (1977) in a Lake Erie sediment transport calculation.

In Bedford and Shah's work considerable fluctuating activity was found for waves of wavelength of grid nodal spacings of eight or less and, although upwind differencing helped, the calculations were very sensitive to small changes in forcing functions and stability was difficult to control. Deardorff (1973) was not able to extend the 1970 methodology to thermally affected three-dimensional atmospheric models and had to resort to additional equations for the four sets of subgrid scale coefficients. It remained, however, for Leonard (1974) to clearly define the grid volume average as a special case of the general low pass filter concept. His advanced smooth low pass filter was used to define a new sequence of corrective nonlinear terms, a technique successfully applied by Reynolds and his co-workers at Stanford (Ferziger *et al.* 1977, and Kwak, *et al.* 1975, for example). Using very coarse grids theoretical spectra were repeatedly and inexpensively reproduced in computations.

Using these methods additional work has been done on Burger's equation by Love and Leslie (1977) and on filtered energy transfers by Leslie and Quarini (1979). Schumann (1975) and Grotzbach and Schumann (1977) have been particularly successful in LES-SGS modeling of annuli and channel flows. The application of this technique to natural aquatic flows was first done by Bedford and Babajimopoulos (1978, 1980) and Babajimopoulos and Bedford (1980) for 3-D lake circulation, and for two-dimensional stratified flows by Findikakis and Street (1979).

5.1.2 The basic large eddy filter operation. The higher order filter or averaging operation is defined by a convolution integral, i.e., for a vector function $f(\underline{x},t)$ the spatially filtered or averaged value is

$$\bar{f}(\underline{x},t) = \int_{-\infty}^{\infty} G(\underline{x} - \underline{x}',t) f(\underline{x}',t)\, dx' \tag{5}$$

in which G is defined as a filter or weighing function such that $\int_{-\infty}^{\infty} G(x)\, dx = 1$.

The first and second moments must exist. Figure 3a contains the constant weighting function used in Deardorff's, Spraggs', Paul's and Bedford's (1977) and Findikakis' work, while Fig. 3b contains a plot of the symmetric Gaussian filter function used by Leonard. In this filter

$$G(x - x') = \sqrt{\frac{\gamma}{\pi}}\, \frac{1}{\Delta} \exp\left\{\frac{-(x - x')^2}{\Delta^2}\right\} \tag{6}$$

where $\gamma = 6.0$ and Δ is called the filter length and is set equal to twice the grid length. Clark (1977) reports that this filter produces excellent results as it suppresses high frequency oscillations and provides resolution of waves at almost the Nyquist frequency.

Since lake circulation is marked by highly distorted vertical scales a Gaussian low pass filter for distorted cells must simply incorporate a different length scale Δ_i in each of the three directions. If γ is constant and Δ_i is twice the grid length in the x_i direction, then

$$G_i(x_i - x_i') = \left\{\sqrt{\frac{\gamma}{\pi}}\right\} \frac{1}{\Delta_i} \exp\left[\frac{-\gamma (x_i - x_i')^2}{\Delta_i^2}\right] \tag{7}$$

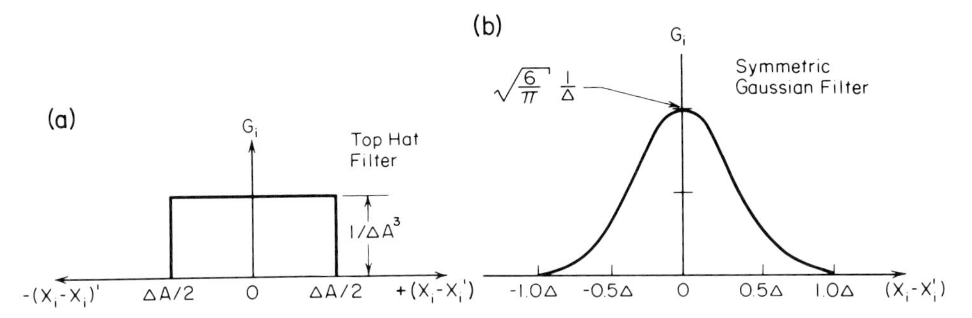

Figure 3. Filter functions for (a) top hat and (b) Gaussian filters.

In the next section the basic physical assumptions used in the lake equations are noted and the derivation of the filtered equations summarized.

5.2 *Filtered Lake Transport Equation*

5.2.1 Equation filtration for general equation. The general method for filtraton is as follows. The Gaussian filter for distorted cells is passed through the governing equations and after making the general decomposition $\alpha = \overline{\alpha} + \alpha'$, the general 3-D hydrodynamic equations are of the form

$$\frac{\partial u_i}{\partial t} + \frac{\partial}{\partial xj} \overline{\overline{u}_i \overline{u}_j} = -\frac{1}{\rho} \frac{\partial P}{\partial x_i} + \nu \nabla^2 \ \overline{u}_i - \frac{\partial}{\partial xj} N_{ij} \tag{8}$$

where

$$N_{ij} = \overline{\overline{u}_i u_j'} + \overline{u_i' \overline{u}_j} + \overline{u_i' u_j'} \tag{9}$$

As pointed out in Clark (1977) and Kwak *et al.* (1975), it is customary for modelers to assume that $\overline{\overline{u}_i \overline{u}_j} = \overline{u}_i \overline{u}_j$; $\overline{u}_i u_j' + u_i' \overline{u}_j = 0$, and $\overline{u_i' \overline{u}_j'} = f(\overline{u}_i, \overline{u}_j)$. In the LES filtration scheme, $\overline{\overline{u}_i \overline{u}_j}$ is decomposed by a Taylor series expansion of the filter convolved with the product. This decomposition leads to a second order correct differential expression of the form

$$\overline{\overline{u}_i \overline{u}_j} = \overline{u}_i \overline{u}_j + \frac{\Delta^2}{4\gamma} \frac{\partial}{\partial x_k} (\overline{u}_i \overline{u}_j) \tag{10}$$

In work up to Clark's report the cross product terms have either been ignored or subsumed into the subgrid scale correlation. Clark found that for "smooth" u_j, i.e., $\bar{u}_j$ being close to u_j, an expression for the cross products could be found and parameterized by grid scale parameters. This form will not be used here because it is not possible to assume smoothness as this will ultimately imply the presence of well behaved non divergent "waves." Secondly, Clark estimated these terms to be at most half the importance of the Leonard term. Thirdly, as Clark points out there are no physics in the term. Finally, many of the comparisons to be presented here were underway before Clark's report was issued and for the sake of economy and uniform comparison we have not used this term.

Subgrid scale parameterizations have been investigated by many including Rose (1977), and Leslie and Quarini (1979) and for thermal flows by Findikakis and Street (1979). However, Clark's detailed analysis (Clark, *et al.* 1979), concluded that for thermally homogeneous flows no improvement in Smagorinsky's formulation could be made. This general large eddy simulation methodology is now used to derive the lake transport equations.

5.2.2 Lake assumptions. Before filtration the following assumptions are imposed on the governing equations. A traditional coordinate system (Fig. 4) is placed in the still water level with z positive downward. The wind stress is specified at the surface and an arbitrary bottom and planform topography is, of course, assumed. The Coriolis force is not included (its neglect will not affect the goals of this study) and hydrostatic pressure assumed. Molecular and viscous effects are also ignored. Thermal homogeneity is assumed and growth/death kinetics are ignored in the transport equation for the time being, but could be represented by a simple cell volume averaged sources/sink term.

5.2.3 Coordinate stretching. Since lakes can have radically varying depths a vertical coordinate stretching procedure is used. This is defined by the following map: $t=t$; $x=x$; $y=y$; and $\sigma = (h-z)/(h+\xi)$. In this system the x, y, and t variables remain unchanged, but the coordinate z is stretched into σ which, regardless of the x,y location, will now vary from 0 to 1. This transformation permits automatic vertical mesh refinement in shallow water, but will cause difficulty in restricted time marching steps. The governing filtered equations are derived in this new coordinate system and used to formulated both free surface and rigid lid models.

5.2.4 Nondimensionalization. The momentum and passive contaminant equations are nondimensionalized with reference quantities: velocity, U; horizontal reference length, L; vertical reference length, D; characteristic viscosity A_H; reference concentrations, C_m; and time L^2/A_H; These result in a group of familiar nondimensional terms, i.e., Re, a Reynolds Number = UL/A_H; Fr, a Froude Number $U\sqrt{gD}$; and Pr, a Prandtl Number.

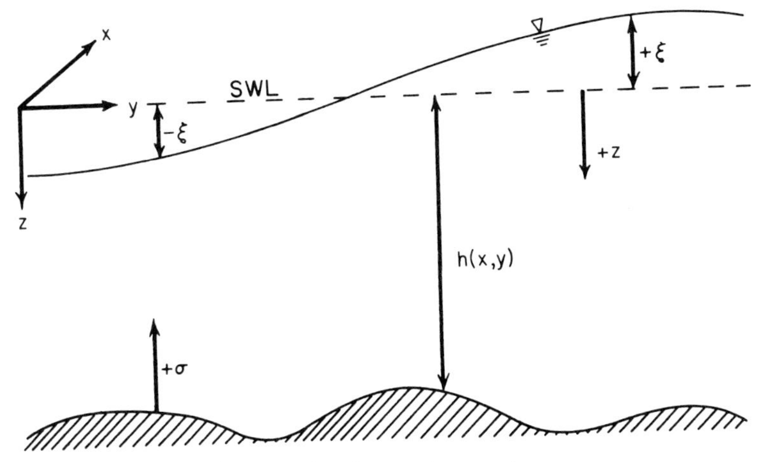

Figure 4. Coordinate system notation.

5.3 Governing Lake Transport Equations

For u, v, and w being the averaged velocities in the x, y, and z direction, for p being pressure, and c being passive contaminant concentration the final filtered equations are: *Continuity*

$$\frac{\partial u}{\partial x} + \frac{\partial v}{\partial y} + \frac{\partial w}{\partial z} = 0 \tag{11}$$

x-Momentum

$$\begin{aligned} \frac{\partial u}{\partial t} + \mathrm{Re} \Bigg[\Bigg(& \frac{1}{H} \frac{\partial}{\partial x} (Huu) + \frac{\Delta_1^2}{4\gamma} \frac{1}{H} \frac{\partial^3}{\partial x^3} (Huu) \\ & + \frac{\Delta_2^2}{4\gamma} \frac{1}{H} \frac{\partial^3}{\partial y^2 \partial x} (Huu) + \frac{\Delta_3^2}{4\gamma} \frac{1}{H^3} \frac{\partial^3}{\partial \sigma^2 \partial x} (Huu) \Bigg) \\ & + \Bigg(\frac{1}{H} \frac{\partial}{\partial y} (Huv) + \frac{\Delta_1^2}{(4} \gamma \frac{1}{H} \frac{\partial^3}{\partial x^2 \partial y} (Huv) \\ & + \frac{\Delta_2^2}{4\gamma} \frac{1}{H} \frac{\partial^3}{\partial y^3} (Huv) + \frac{\Delta_3^2}{4\gamma} \frac{1}{H^3} \frac{\partial^3}{\partial \sigma^2 \partial y} (Huv) \Bigg) \\ & + \Bigg(\frac{-1}{H^2} \frac{\partial}{\partial \sigma} (Huw) - \frac{\Delta_1^2}{4\gamma} \frac{1}{H^2} \frac{\partial^3}{\partial x^2 \partial \sigma} (Huw) \\ & - \frac{\Delta_2^2}{4\gamma} \frac{1}{H^2} \frac{\partial^3}{\partial y^2 \partial \sigma} (Huw) - \frac{\Delta_3^2}{4\gamma} \frac{1}{H^4} \frac{\partial^3}{\partial \sigma^3} (Huw) \Bigg) \Bigg] \end{aligned}$$

$$= \frac{-\partial p}{\partial x} + \text{Re}\left(\frac{1}{H^3}\frac{\partial}{\partial x}\left[KH^2\left(\frac{\partial}{\partial x}(Hu) + \frac{\partial}{\partial x}(Hu)\right)\right]\right.$$
$$+ \frac{1}{H^3}\frac{\partial}{\partial y}\left[KH^2\left(\frac{\partial}{\partial y}(Hu) + \frac{\partial}{\partial x}(Hv)\right)\right]$$
$$\left. + \left(\frac{L}{D}\right)^2 \frac{1}{H^2}\frac{\partial}{\partial \sigma}\left(K\frac{\partial u}{\partial \sigma}\right)\right) \tag{12}$$

y-Momentum

$$\frac{\partial v}{\partial t} + \text{Re}\left[\left(\frac{1}{H}\frac{\partial}{\partial x}(Hvu) + \frac{\Delta_1^2}{4\gamma}\frac{1}{H}\frac{\partial^3}{\partial x^3}(Hvu)\right.\right.$$
$$\left. + \frac{\Delta_2^2}{4\gamma}\frac{1}{H}\frac{\partial^3}{\partial y^2 \partial x}(Hvu) + \frac{\Delta_3^2}{4\gamma}\frac{1}{H^3}\frac{\partial^3}{\partial \sigma^2 x}(Hvu)\right)$$
$$+ \left(\frac{1}{H}\frac{\partial}{\partial y}(Hvv) + \frac{\Delta_1^2}{4\gamma}\frac{1}{H}\frac{\partial^3}{\partial x^2 \partial y}(Hvv)\right.$$
$$\left. + \frac{\Delta_2^2}{4\gamma}\frac{1}{H}\frac{\partial^3}{\partial y^3}(Hvv) + \frac{\Delta_3^2}{4\gamma}\frac{l}{H^3}\frac{\partial^3}{\partial \sigma^2 \partial y}(Hvv)\right)$$
$$+ \left(\frac{-1}{H^2}\frac{\partial}{\partial \sigma}(Huw) - \frac{\Delta_1^2}{4\gamma}\frac{1}{H^2}\frac{\partial^3}{\partial x^2 \partial \sigma}(Huw)\right.$$
$$\left.\left. - \frac{\Delta_2^2}{4\gamma}\frac{1}{H^2}\frac{\partial^3}{\partial y^2 \partial \sigma}(Hvw) - \frac{\Delta_3^2}{4\gamma}\frac{1}{H^4}\frac{\partial^3}{\partial \sigma^3}(Hvw)\right)\right]$$
$$= \frac{-\partial p}{\partial y} + \text{Re}\left(\frac{1}{H^3}\frac{\partial}{\partial x}\left[KH^2\left(\frac{\partial}{\partial x}(Hv) + \frac{\partial}{\partial y}(Hu)\right)\right]\right.$$
$$+ \frac{1}{H^3}\frac{\partial}{\partial y}\left[KH^2\left(\frac{\partial}{\partial y}(Hv) + \frac{\partial}{\partial y}(Hv)\right)\right]$$
$$\left. + \left(\frac{L}{D}\right)^2 \frac{1}{H^2}\frac{\partial}{\partial \sigma}\left(K\frac{\partial v}{\partial \sigma}\right)\right) \tag{13}$$

Hydrostatic Pressure

$$\frac{\partial P}{\partial \sigma} = \frac{\text{Re}}{Fr^2} \tag{14}$$

Transport

$$
\begin{aligned}
&Pr\left\{\frac{\partial c}{\partial t} + \mathrm{Re}\left[\left[\left(\frac{1}{H}\frac{\partial}{\partial x}(Hcu) + \frac{\Delta_1^2}{4\gamma}\frac{1}{H}\frac{\partial^3}{\partial x^3}(Hcu)\right.\right.\right.\right. \\
&\quad + \left.\frac{\Delta_2^2}{4\gamma}\frac{1}{H}\frac{\partial^3}{\partial y^2 \partial x}(Hcu) + \frac{\Delta_3^2}{4\gamma}\frac{1}{H}\frac{\partial^3}{\partial \delta^2 \partial x}(Hcu)\right) \\
&\quad + \left(\frac{1}{H}\frac{\partial}{\partial y}(Hcv) + \frac{\Delta_1^2}{4\gamma}\frac{1}{H}\frac{\partial^3}{\partial x^2 \partial y}(Hcv)\right. \\
&\quad + \left.\frac{\Delta_2^2}{4\gamma}\frac{1}{H}\frac{\partial^3}{\partial y^3}(Hcv) + \frac{\Delta_3^2}{4\gamma}\frac{1}{H}\frac{\partial^3}{\partial \delta^2 \partial y}(Hcv)\right) \\
&\quad + \left(\frac{1}{H}\frac{\partial}{\partial z}(Hcw) + \frac{\Delta_1^2}{4\gamma}\frac{1}{H}\frac{\partial^3}{\partial x^2 \partial z}(Hcw)\right. \\
&\quad + \left.\left.\left.\left.\frac{\Delta_2^2}{4\gamma}\frac{1}{H}\frac{\partial^3}{\partial y^2 \partial z}(Hcw) + \frac{\Delta_3^2}{4\gamma}\frac{1}{H}\frac{\partial^3}{\partial \delta^3}(Hcw)\right)\right]\right]\right\} \\
&\quad = Pr\ \mathrm{Re}\left\{\frac{1}{H^3}\frac{\partial}{\partial x}\left[KH^2\frac{\partial}{\partial x}(Hc)\right]\right. \\
&\quad + \left.\frac{1}{H^3}\frac{\partial}{\partial y}\left[KH^2\frac{\partial}{\partial y}(Hc)\right] + \left(\frac{L}{D}\right)^2\frac{1}{H^2}\frac{\partial}{\partial \delta}\left(K\frac{\partial}{\partial \delta}\right)\right\}
\end{aligned}
\tag{15}
$$

where $H = h + \xi$. K is a sub grid scale eddy viscosity. Following Smagorinsky (1965), if the quantity Δ is set equal to $(\Delta x \Delta y \Delta z)^{1/3}$ and $\bar{c}$ is a basin specific coefficient, then K is given by

$$
\begin{aligned}
K = \frac{(\bar{c}\Delta)^2}{H}&\left\{2\left[\left(\frac{\partial}{\partial x}(Hu)\right)^2\right.\right. \\
&+ \left.\left(\frac{\partial}{\partial y}(Hv)\right)^2 + \frac{1}{H^2}\left(\frac{\partial}{\partial \sigma}(Hw)\right)^2\right] \\
&+ \left[\frac{\partial}{\partial x}(Hv) + \frac{\partial}{\partial y}(Hu)\right]^2 \\
&+ \left[\frac{L}{D}\frac{\partial}{\partial y}(Hw) - \frac{L}{D}\frac{1}{H}\frac{\partial}{\partial \sigma}(Hv)\right]^2 \\
&+ \left.\left[\frac{L}{D}\frac{1}{H}\frac{\partial}{\partial \sigma}(Hu) + \frac{L}{D}\frac{\partial}{\partial x}(Hw)\right]^2\right\}^{1/2}
\end{aligned}
\tag{16}
$$

The boundary conditions are a wind shear specified on the surface and, as in the previously cited Great Lakes models, the no slip condition on solid boundaries.

6. NUMERICAL FORMULATION

6.1 Grid Selection

The staggered mesh grid developed at the Los Alamos Laboratory (Welch *et al.*, 1966) is used for generating discretizations. Also used is the modified staggered grid developed by Paul and Lick (1976) that defines horizontal velocities at grid nodal points that the vertical velocities defined at half nodal points in the vertical and horizontal direction. Surface elevations are defined at half nodal points in the horizontal. Both systems were used and results for the presented comparisons were indistinguishable.

6.2 Derivative Approximations

Since the physics of these equations places heavy reliance upon the nonlinear terms, a careful discretization of them is required. The overall accuracy of all nonlinear terms is to be fourth order. Therefore, the first derivative terms being first order differentially correct requires a fourth order accurate energy and momentum conservative expression. The method developed by Kwak *et al.* (1975) as used by Babajimopoulos and Bedford (1980) is used here.

The basic fourth order operator is written as $\partial u/\partial x = 4/3 \partial u/\partial x|^{\Delta x} - 1/2 \partial u/\partial x|^{2\Delta x}$. A typical expansion of the first order nonlinear term would be

$$\frac{\partial}{\partial x}(Huv) = \frac{4}{3}\frac{\partial}{\partial x}(Huv)\,|^{\Delta x} - \frac{1}{3}\frac{\partial}{\partial x}(Huv)\,|^{2\Delta x} \tag{17}$$

Asymmetric fourth order correct expressions from Collatz (1960) are used at the boundaries. The Leonard terms are second order differentially correct and therefore overall fourth order accuracy is ensured if a centered second order approximation is used. All the dissipation terms are discretized by second order centered approximations.

6.3 Time Marching and Splitting

Two time marching schemes were employed, a simple first order accurate explicit scheme and a second order Adams Bashforth method. The small vertical scale of the problem puts a severe restriction on the time step. Therefore, a vertical implicit diffusion time splitting is performed. In this procedure the governing equations are rewritten in the following form

$$\frac{\partial u}{\partial t} - \mathrm{Re}\left(\frac{L}{D}\right)^2 \frac{1}{H^2}\frac{\partial}{\partial \sigma}\left(K\frac{\partial u}{\partial \sigma}\right) = G(u) - \frac{\partial p_s}{\partial y} \tag{18}$$

$$\frac{\partial v}{\partial t} - \mathrm{Re}\left(\frac{L}{D}\right)^2 \frac{1}{H^2}\frac{\partial}{\partial \sigma}\left(K\frac{\partial u}{\partial \sigma}\right) = F(v) - \frac{\partial p_s}{\partial y} \tag{19}$$

$$Pr\,\frac{\partial c}{\partial t} - Pr\,\mathrm{Re}\left(\frac{L}{D}\right)^2 \frac{1}{H^2}\frac{\partial}{\partial \sigma}\left(K\frac{\partial c}{\partial \sigma}\right) = E(c) \tag{20}$$

The left-hand sides are solved implicitly while the right-hand sides, $G(u)$, $F(v)$, $E(c)$, containing the nonlinear terms and the horizontal diffusion terms, are solved explicitly.

6.4 *Pressure Field Formulation*

6.4.1 *Free surface model.* Surface pressure specification for the free surface model takes advantage of the hydrostatic pressure equation to use that $\partial p_s/\partial x = \rho g \partial \xi/\partial x$ and $\partial p_s/\partial y = \rho g \partial \xi/\partial y$. From the continuity equation ξ is found from

$$\frac{\partial \xi}{\partial t} = -\mathrm{Re}\left(\frac{\partial}{\partial x}\left[H\int_0^1 u\,d\sigma\right] + \frac{\partial}{\partial y}\left[H\int_0^1 v\,d\sigma\right]\right) \tag{21}$$

6.4.2 *Rigid lid model.* The traditional rigid lid formulation for pressure proceeds by differentiating the x and y momentum equations by x and y, respectively, and then requiring that the sum of these equations satisfy the time derivative of the continuity equation. This procedure yields an elliptic differential equation for surface pressure. In the model herein the numerical divergence is taken and used to satisfy the time discretized continuity equation. Truncation errors do not accumulate as quickly in this procedure. The method, first implemented in lake models by Paul and Lick (1976), is tedious and is detailed in Bedford and Rai (1977).

7. MODEL IMPLEMENTATION

7.1 Basin Data and Configurations

The implementation of this model initially requires the specification of a test basin which retains the salient features of the Great Lakes, yet does not cloud the overall computational objective with minutiae specific to each Great Lake. The primary criteria for selecting the test basin is that it be of sufficient size and shape to permit the assumptions in Section 5.2.2 to be valid. Coriolis effects are excluded as they neither add nor detract from the dynamics being tested here. The second and equally important criteria is that the basins and forcing functions be of sufficient size and strength to permit the fluid dynamics to develop the spectral shapes summarized in Section 4.3. Of their spectral behaviors the existence of the inertial subrange becomes the key criteria for basin selection. Tennekes and Lumley (1972) present (as redrawn in Fig. 5) a simple criteria for the existence of the inertial subrange; the study basins were selected with the goal of ensuring that this criteria is satisfied. To be sure the Great Lakes *rms* Reynold Numbers are far greater than the 10^5 necessary for the inertial subrange. However, the nodal density required for prototype simulations becomes far too costly. The basins used herein require fewer nodes and have smaller size, but they retain the physics, save money and still permit the study objectives to be met. The relevant basin parameters are summarized in Table 1.

Three different basins were employed in this study. The motivation for these three rectangular flat bottom basins is that for turbulence/spectral calculations a uniform mesh with constant spacing is necessary. This minimizes phase and amplitude shifts caused by unequal grid lengths. Rectangular basin shapes with uniform grids also permit fourth order accuracy to be preserved in the discretizations. Additionally, since the basic physics of lake models is still manifest in these basins, the original objective has certainly not been avoided. As will be seen, the rectangular basins are complex enough without the difficulties of unequal grid lengths.

7.2 Grid Spacings

Four grid spacings and four time steps were used. Tables 1 and 2 summarize these grids. Time steps of 1, 5, 30, and 120 seconds were used for time marching. The 1 and 5 second steps are commensurate with the sampling frequencies of the cited data and, therefore, comparisons with predictions must be made at these time intervals. Comparisons with a more "typical" time step used in geophysical modeling are permitted with the two-minute and thirty-second time step.

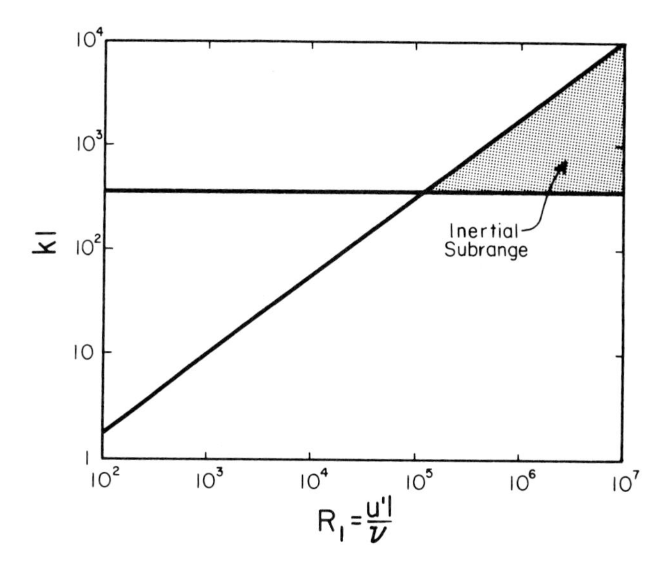

Figure 5. Inertial subrange region.

Table 1.
Basin Specifications

BASIN SIZE				
No.	Cross Shape	Length (m)	Width (m)	Depth (m)
1	rect.	1700	900	16
2	rect.	1360	900	16
3	rect.	2040	1080	16
4	rect.	1700	900	16
5	rect.	1700	900	16
6	rect.	1700	900	16

GRID			
No.	Δx (m)	Δy (m)	Δz (m)
1	100	100	2
2	80	80	2
3	120	120	2
4	75	75	2
5	100	100	1.33
6	50	50	2

SCALE FACTORS					
No.	L	D	A_H (m^2/sec)	U (m/sec)	C
1	1700	16	0.1	0.1	0.1
2	1360	16	0.1	0.1	0.1
3	2040	16	0.1	0.1	0.1
4	1700	16	0.1	0.1	0.0185
5	1700	16	0.1	0.1	0.0175
6	1700	16	0.1	0.1	0.0185

Table 2.
List of Numerical Experiments

Experiment No.	Basin No.	Time Step (sec)	Pressure Form	Filtered	Wind Speed (MPH)	Noise	Contam. Transp.	Pr	Elevation Traces
1	1	5	Free Surface	Yes	5	No	No	-	Yes
2	1	5	Free Surface	Yes	5	1%	No	-	Yes
3	1	5	Free Surface	Yes	5	3%	No	-	Yes
4	1	5	Free Surface	Yes	10	No	No	-	Yes
5	1	5	Free Surface	Yes	10	1%	No	-	Yes
6	1	5	Free Surface	Yes	10	3%	No	-	Yes
7	1	5	Free Surface	Yes	15	No	Yes	1.0	Yes
8	1	5	Free Surface	Yes	15	1%	No	-	Yes
9	1	5	Free Surface	Yes	15	3%	No	-	Yes
10	1	5	Free Surface	Yes	25	No	No	-	Yes
11	4	5	Free Surface	Yes	15	No	No	-	Yes
12	5	5	Free Surface	Yes	15	No	No	-	Yes
13	1	5	Free Surface	Yes	0	1%	No	-	Yes
14	1	5	Free Surface	No	15	No	No	-	Yes
15	1	5	Rigid Lid	Yes	5	No	No	-	No
16	1	5	Rigid Lid	Yes	5	1%	No	-	No
17	1	1	Rigid Lid	Yes	5	1%	No	-	No
18	1	5	Rigid Lid	Yes	10	No	No	-	No
19	1	5	Rigid Lid	Yes	10	1%	No	-	No
20	1	5	Rigid Lid	Yes	10	3%	No	-	No
21	1	5	Rigid Lid	Yes	15	No	Yes	1.0	No
22	1	5	Rigid Lid	Yes	15	1%	Yes	1.0	No
23	1	5	Rigid Lid	Yes	15	3%	No	-	No
24	1	1	Rigid Lid	Yes	15	1%	No	-	No
25	6	5	Rigid Lid	Yes	15	No	No	-	No
26	6	5	Rigid Lid	Yes	15	1%	No	-	No
27	1	30	Rigid Lid	Yes	5	1%	No	-	No
28	1	120	Rigid Lid	Yes	5	1%	Yes	1.0	No
29	1	120	Rigid Lid	Yes	5	1%	Yes	1.5	No
30	1	120	Rigid Lid	Yes	5	1%	Yes	2.0	No
31	2	120	Rigid Lid	Yes	5	1%	No	-	No
32	3	120	Rigid Lid	Yes	5	1%	No	-	No
33	1	120	Rigid Lid	No	5	1%	No	-	No

7.3 Surface Wind Shear

Both noisy and noiseless wind shears are imposed at the surface. Noiseless shears are imposed as constant values representative of 5, 10, 15, 20 and 25 miles per hour winds. Methods for relating wind speed to shear are found in Lick (1976).

Noisy winds are created by requiring that the fluctuations have a specified character. The method of Shinozuka and Jan (1972) is used to generate spectrally correct shear fields. The time series generator for wind shear is

$$\tau(t) = 2\sum_{k=1}^{N} S_o\ (\omega_k)\Delta\omega^{1/2} \cos(\omega_k{}'t + \phi_k) \tag{22}$$

where $S_o(\omega)$ is the target spectrum which is to be simulated. S_o is insignificant outside the region $0 \geqslant \omega \geqslant \omega_u$ where ω_u is an upper limit frequency. The "discretization" interval $\Delta\omega$ is defined as $\Delta\omega = (\omega_u - 0)/N$ where N = the intervals in the digitized target spectrum. ϕ_k is an independent random phase uniformly distributed on $[0,2\pi]$. $\omega_k = (k - 1/2)\Delta\omega$, for $k = 1,2, \cdots N$ and $\omega'_k = \omega_k + \delta\omega$ for $k = 1,2,...N$. $\delta\omega$ is a small random frequency introduced to avoid the periodicity of the simulated process. It is uniformly distributed on $-\Delta\omega'/2,\Delta\omega'/2$ with $\Delta\omega'<<<\Delta\omega$. Values generated for this study were done with $N = 50$ and $\Delta\omega' = \omega/20$. The *rms* value of τ is adjusted through the change in $S_o(\omega)$. One percent and three percent noise levels were used in this study. Figure 6 presents a target and simulated spectrum for a -5/3 shear and Table 2 summarizes the wind fields.

7.4 Extraction and Statistical Preparation of Model Results

Wind is permitted to blow at equal strength in both the positive x and y directions. Steady or stationary conditions were established and then a minimum of 500 c, u, uv, w, and elevation readings were stored at each vertical node in the basin center (Fig. 7). Additional time series of these readings were collected for all vertical nodes and two horizontal nodes on either side of the basin center. After a minimum of 500 time steps the wind was released and during die away the same variables were stored for a minimum of 300 points in the free surface model and only 100 in the rigid lid. This data collection is identical to the process used to collect the majority of time series data reported in the literature.

The velocity traces were detrended and a power spectra computed for each of u, v, w, c. Using Taylor's frozen turbulence hypothesis two- and three-dimensional wavenumber spectra were computed by adding each trace (Tennekes and Lumley, 1972). Since it is difficult to assume that the fluctuations are homogeneous in the vertical direction spectra were restricted to two-dimensional horizontal spectra. A complete index to all the numerical experiments is found in Table 2.

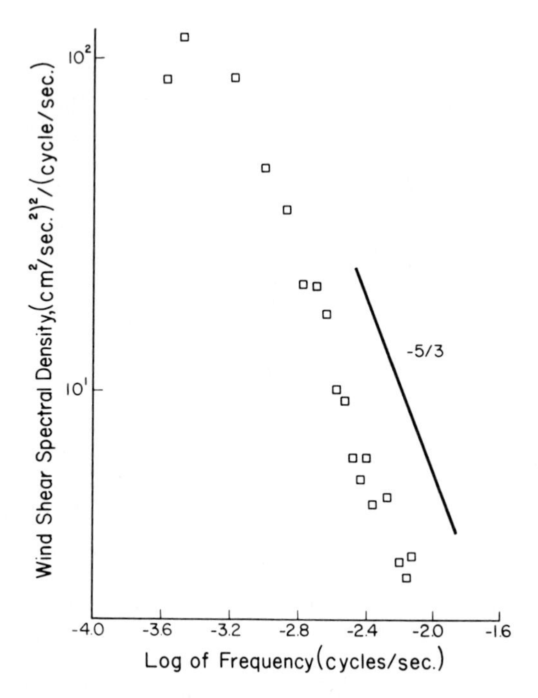

Figure 6. Wind shear spectra.

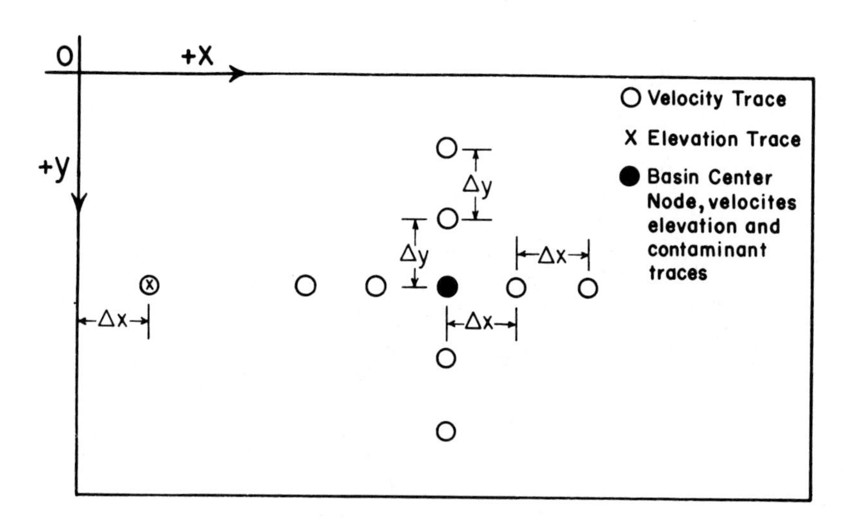

Figure 7. Basin data collection points.

Table 3.
Index of Results

A. MEAN FLOW FIELD (Sec. 8.2)	Type of Plot	Experiment Used (Table 2)	Figure Number	Planform Location of Plot	Non-Dimensionalizing Variables (cgs units)
1. Velocity					
u	vert. profile	21	8(a)	center	u_m = 88.67
v	vert. profile	21	8(a)	center	v_m = 80.51
2. Vertical Grad					
du/dz	vert. profile	21	8(a)	center	(du/dz) m = .143
dv/dz	vert. profile	21	8(a)	center	(dv/dz)m = .143
3. Strain Rates					
Sxx	vert. profile	21	8(b)	center	Sxx_m = .00007
Sxy	vert. profile	21	8(b)	center	Sxy_m = .00013
Sxz	vert. profile	21	8(b)	center	Sxz_m = .143
Syz	vert. profile	21	8(b)	center	Syz_m = .143
4. Inertia					
$\overline{u}\,\overline{u}$	vert. profile	21	8(c)	center	$\overline{u}\,\overline{u}_m$ = 7863.
$\overline{u}\,\overline{v}$	vert. profile	21	8(d)	center	$\overline{u}\,\overline{v}_m$ = 7139.
5. Leonard Stress					
L_{11}^x	vert. profile	21	8(c)	center	L_{11m}^x = 1.0
L_{11}^y	vert. profile	21	8(c)	center	L_{11m}^y = 202.
L_{11}^z	vert. profile	21	8(c)	center	L_{11m}^z = 305.
L_{12}^x	vert. profile	21	8(d)	center	L_{12m}^x = 0.48
L_{12}^y	vert. profile	21	8(d)	center	L_{12m}^y = 90.4
L_{12}^z	vert. profile	21	8(d)	center	L_{12m}^z = 302.

B. SPECTRAL RESULTS (Sec. 8.3)	Effect on Spectra Due To Variation Of	Experiments Used	Figure Number	
1. Velocity				
	Horizontal grid	7, 11	10(a)	
	Horizontal grid	21, 25	10(c)	
	Vertical grid	7, 12	10(b)	
	Time Step	27, 28	9(a)	
	Time Step	22, 24	9(b)	
	Depth	26	11	
	Basin Size	30,31,32	12(a)(b)	
	Planform Location	7	13(a)(b)	
	Wind Speed (Free Surface vs. Rigid Lid)	1,4,7/15,18,21	14(a)(b)(c)(d)	
	Wind Speed (Noise Level)	21,23/8,9	15(a)(b)	
	Rigid Lid/Free Sur.	7,21/9,23	16(a)(b)	
	Die Away (a)	28/7/7	17(a)(b)(c)	
	(b)	7,8/7,8/7,8	18(a)(b)(c)	
	Filtration	33/14/14	19(a)(b)(c)	
2. Passive Contaminant	Depth	28,29,30/28,29,30	20(a)	
	Prandtl Number	28,29,30/28,29,30	20(b)(c)	
3. Production and Dissipation	D.N.A.	21	21	

C. ANCILLARY RESULTS (Sec. 8.4)	Type of Plot	Experiment Used (Table 2)	Figure Number	Planform Location	Non-Dimensionalizing Variables
1. Dissipation	vert. profile	21	22	center	$\varepsilon_m = .192\ cm^2/sec^3$
2. Smagorinsky viscosity	vert. profile	21	22	center	$K_m = 4.62\ cm^2/sec$
3. Regular Eddy Viscosity K_{13}	vert. profile	21	22	center	$K_{13m} = 442.0\ cm^2/sec$
4. Dissipation vs. Wind Speed	D.N.A.	15,18,21	23	D.N.A.	D.N.A.

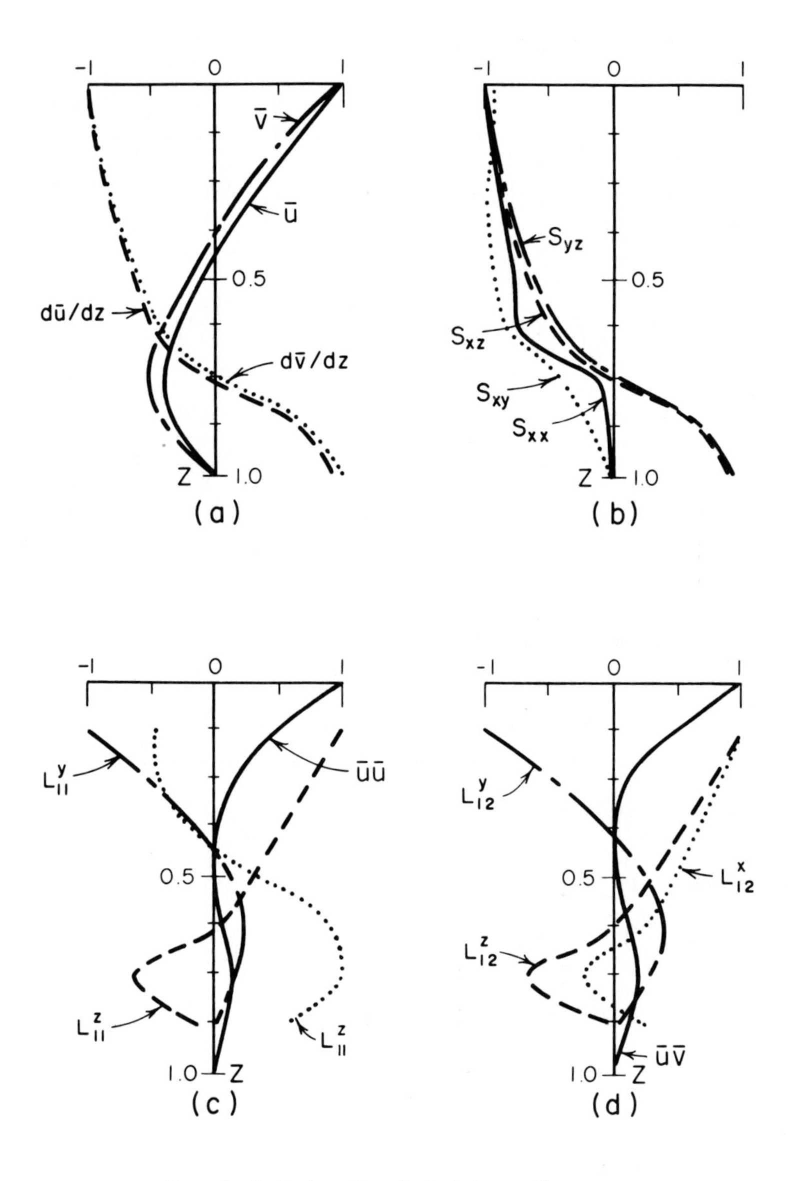

Figure 8. Vertical profiles of selected mean flow paraemeters.

8. RESULTS

8.1 Overview and Results Index

As can be seen from Table 2 a large selection of information is available for review. This section will present results in three general categories: first, simple vertical profiles of typical resolved mean flow field parameters; second, computed spectra from a variety of flow field conditions; and third, calculations for ancillary variables such as dissipation and eddy viscosities. In accordance with the objectives of this paper, the second portion is emphasized. Table 3, an index of types, formats, and experiments used in the construction of figures, has been compiled.

8.2 Mean Flow Field Behavior

Two different methods of looking at the resolved or mean flow field exist. The first is to check the whole basin circulation features. Since graphic displays of these features take considerable space and since methods of checking these field have been presented in Babajimopoulos and Bedford (1980) we will not review them here. Suffice it to say that with the exception of slightly higher curvatures on vertical gradients, introduced by higher order approximations whole basins patterns are similar to those produced by more traditional models.

The second viewpoint concentrates on the vertical profiles of selected mean flow field variables at the basin center node, the site of most active data collection. Figure 8 portrays a variety of such profiles. Each curve has been nondimensionalized by the maximum value of that variable. These maximum variables are tabulated in the index, and although this portrayal complicates intercomparison, it does permit compact qualitative comparisons. Under the assumption that $i = 1, 2, 3$ for x, y, and z, coordinate directions, the strain rates, $S_{ij} = 1/2(\partial u_i/\partial x_j + \partial u_j/\partial x_i)$, and the Leonard stresses, N_{ij}, are calculated for example purposes from the x momentum equation, i. e.,

$$L_{11}^{x} = \frac{\Delta_1^2}{4\gamma}\frac{\partial^2(\overline{uu})}{\partial x^2}; \quad L_{11}^{y} = \frac{\Delta_2^2}{4\gamma}\frac{\partial^2(\overline{uu})}{\partial y^2}; \quad L_{11}^{z} = \frac{\Delta_3^2}{4\gamma}\frac{\partial^2(\overline{uu})}{\partial z^2} \tag{23}$$

and

$$L_{12}^{x} = \frac{\Delta_1^2}{4\gamma}\frac{\partial^2(\overline{uv})}{\partial x^2}; \quad L_{12}^{y} = \frac{\Delta_2^2}{4\gamma}\frac{\partial^2(\overline{uv})}{\partial y^2}; \quad L_{13}^{z} = \frac{\Delta_3^2}{4\gamma}\frac{\partial^2(\overline{uv})}{\partial z^2} \tag{24}$$

It should be noted that although these data were prepared from Experiment 21, the vertical distributions reported share the same shapes and relative magnitudes for all runs except the 30 sec and 120 sec time step runs.

Briefly these profiles indicate the following physical features. First the $\bar{u}$ and $\bar{v}$ profiles are almost identical, and as is expected the crossover point is a little deeper than traditional crossovers. The vertical gradients are almost identical with slowly decreasing behavior followed by increased shear below the crossover. A comparison of the shears with *Sxz* and *Syz* indicates that they are identical, which is a result of the presumption of small vertical activity, i.e., $\partial w/\partial x$ and $\partial w/\partial y$ are negligible. The profiles of *Sxx* and *Sxy* are again almost identical, but several orders less in magnitude than *Sxz* and *Syz*. Above crossover, the correct clockwise distortion is noted while practically zero distortion occurs below crossover. The inertia ($\overline{uu}$, and $\overline{uv}$) and Leonard (L_{11}, L_{12}) profiles reveal that in general the Leonard stresses are very active in zones of high inertia profile curvature, particularly near nodes 3 and 7. The Leonard stress maxima is roughly 5 percent of the inertia maxima, but this is misleading as the maxima could be drawn from different vertical locations. For this numerical experiment the maximum ratio is 32 percent occurring at node 5. In general, L_{11}^{z}, L_{12}^{z}, L_{21}^{z} and L_{22}^{z} are the largest Leonard terms, again reflective of the weakened vertical activity assumption.

8.3 Spectra Results

8.3.1 Grid selection. Figures 9 and 10 present a sequence of plots showing how and why the grids and times were selected. Figure 10a demonstrates the effect of 100 and 75 m horizontal grids on the free surface model while 10c demonstrates 100 and 50 m grids on the rigid lid model. Figure 10b shows the impact of 2 m and 1.33 m vertical spacing for the free surface model. It can be seen particularly in Fig. 10c, that regardless of a coarse (100 m) or fine grid (50 m) the spectral plots are quite similar and contain a two slope region. Vertical mesh refinement adds more energy to the high frequency components, but the overall two slope profile is maintained. Since the relatively coarse grid (Number 1 of Table 1) maintains the overall spectral placements and shapes of more refined meshes and at an economical cost, most of the subsequent calculations were done on this basis.

The effect of time step is seen in Figs. 9a and 9b. The coarse time step completely eliminates all the -5/3 high frequency information, while the smaller time steps permit the high wavenumber range to be much more fully resolved. A small increase in the total variance is noted. The smaller time step spectra preserve the two-slope configuration and since the five second time step calculation yields a sufficient length of record to resolve both high and low wave number regions, it was used in a large majority of these results.

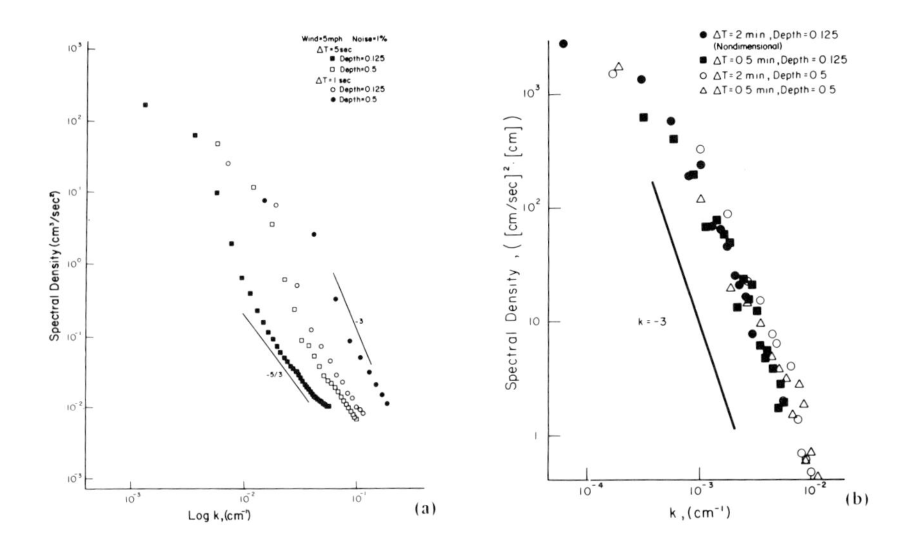

Figure 9. Effect on velocity spectra of fine (a) and coarse (b) time steps.

8.3.2 Spectral results. The spectra from Figs. 9a through 18c and 20a and b display the following features. The overall spectral shape is as depicted by field results and theory. The maximum energy content occurs at the large basin size wavelengths and is resolved to the limit of the time step or the Nyquist frequency. For this paper this high wave number limit is somewhat less than the Kolmogorov scales. Of most importance all simulations contain a well defined region where the slope is proportional to k^{-3} while in the five and one second time step calculations, there is a second region at a wave length of half the total basin depth and less where a $k^{-5/3}$ region exists.

From the figures the following specific patterns are noted as influenced by the variation of individual basin input and physics. (1) Variations with depth are marked by noisy spectra at the crossover node (Figure 11, Depth = 0.5) with the very top depths containing more vigorous long wavelength activity. Spectra at 0.25

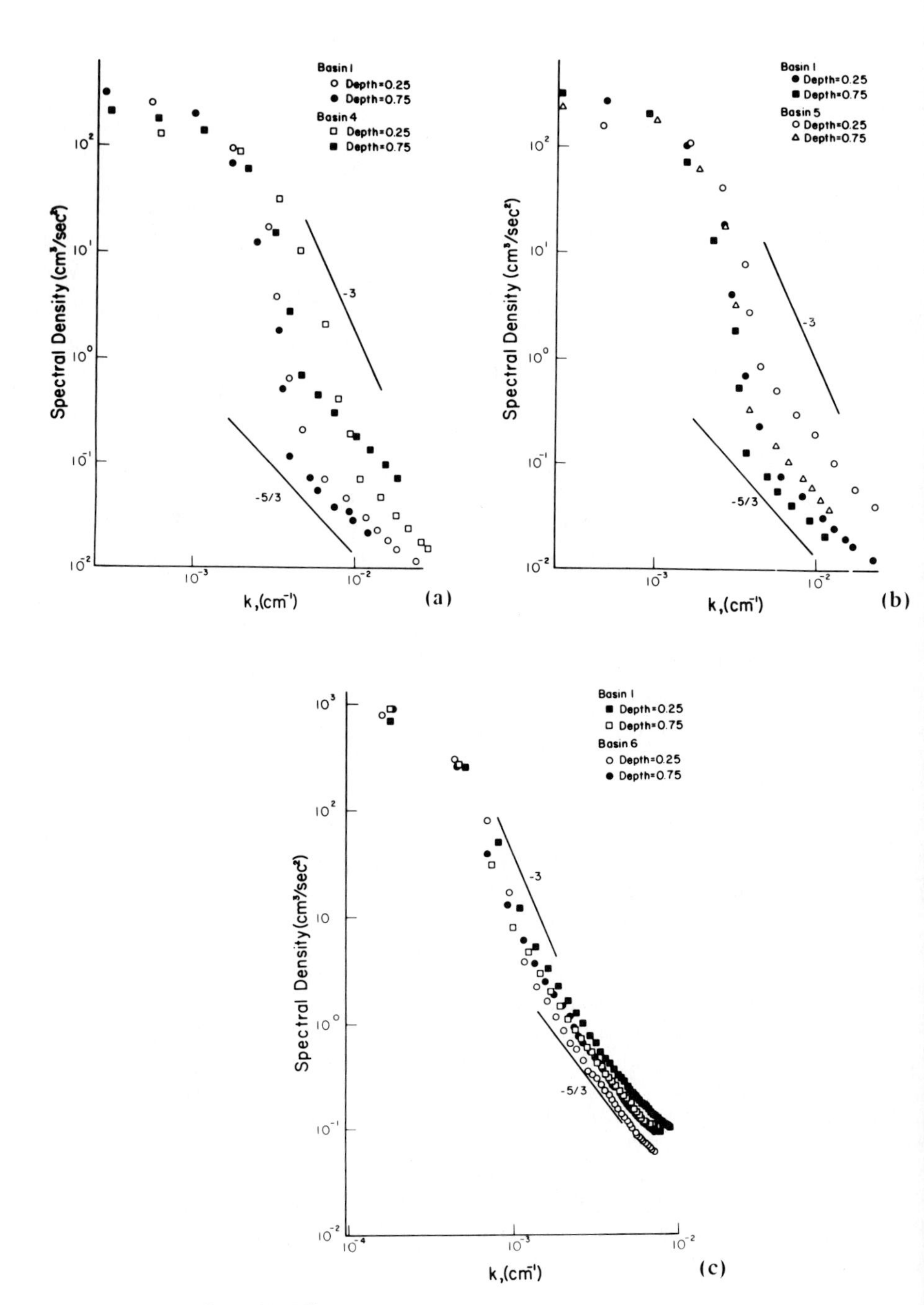

Figure 10. Effect on velocity spectra of coarse and fine spatial mesh.

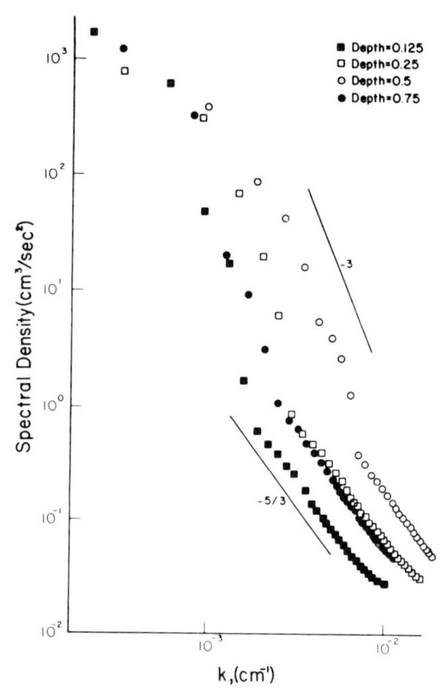

Figure 11. Velocity spectra variation with depth.

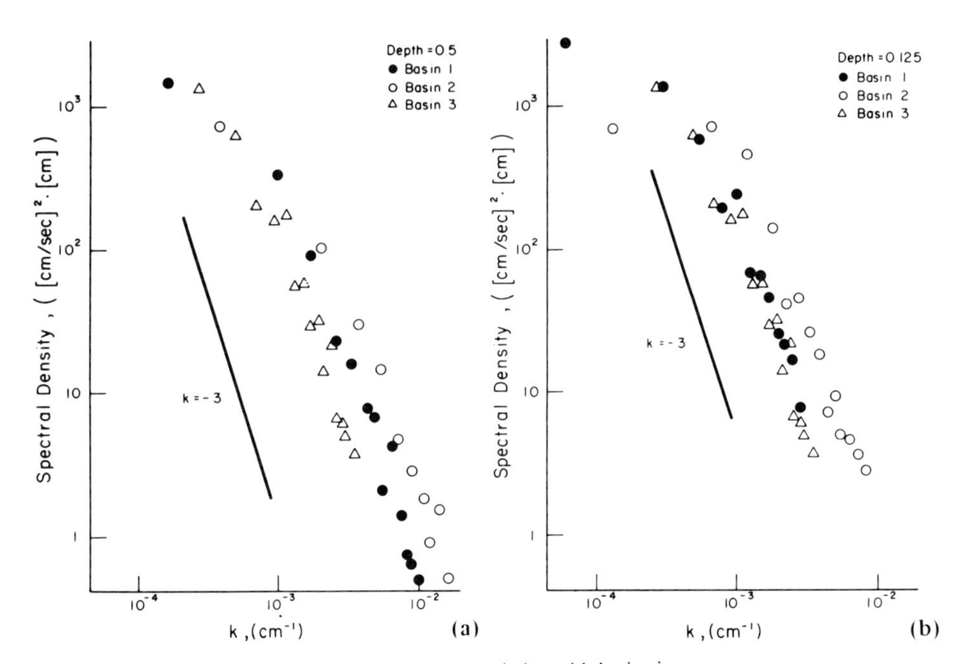

Figure 12. Velocity spectra variation with basin size.

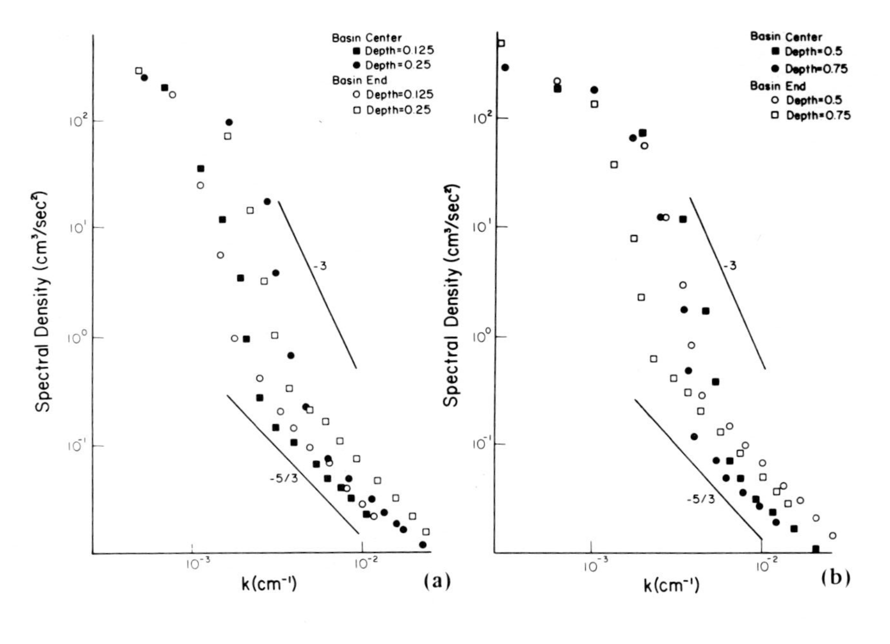

Figure 13. Velocity spectra variation with planform location.

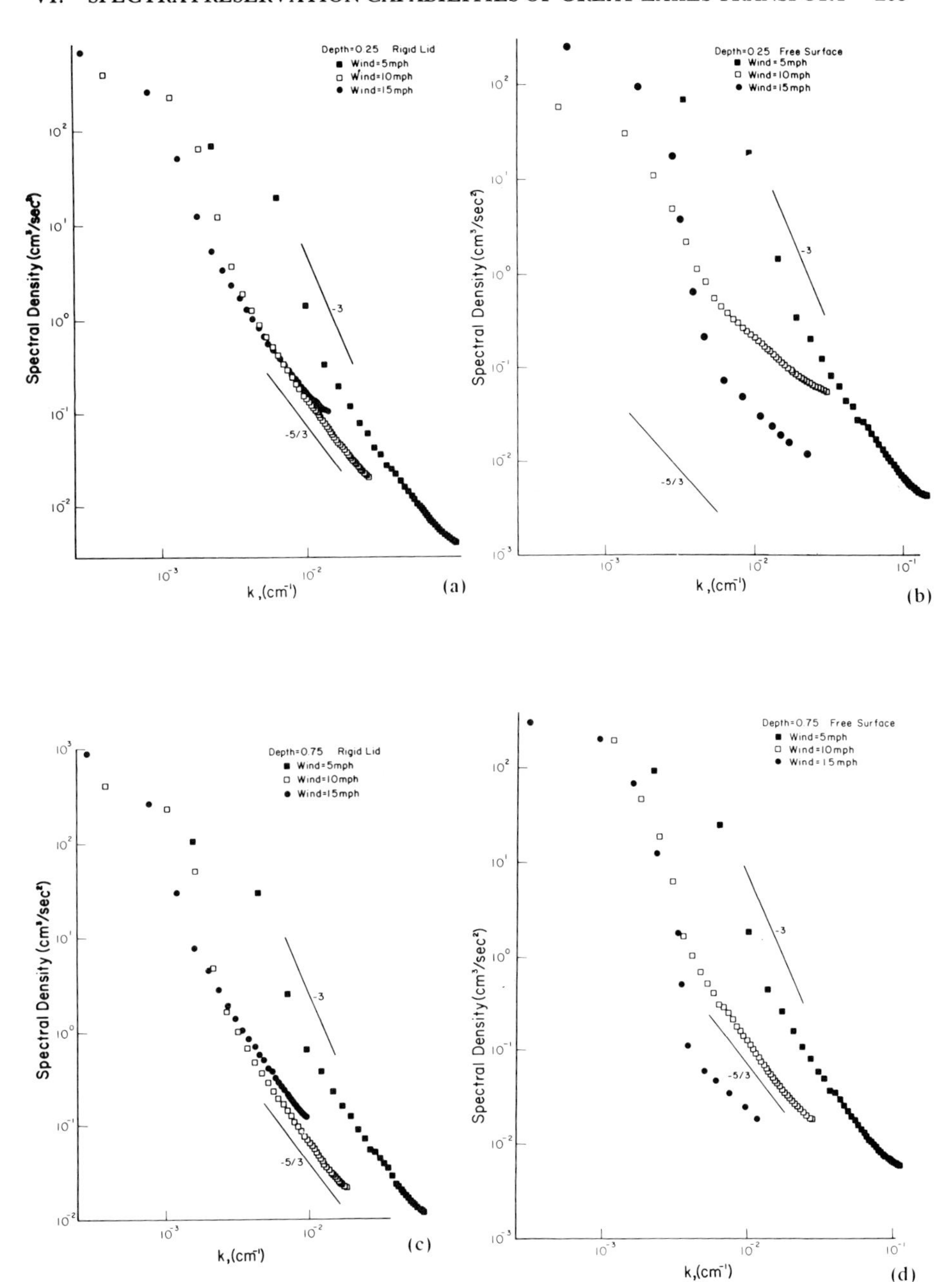

Figure 14. Velocity spectra variation with wind speed for free surface and rigid lid formations

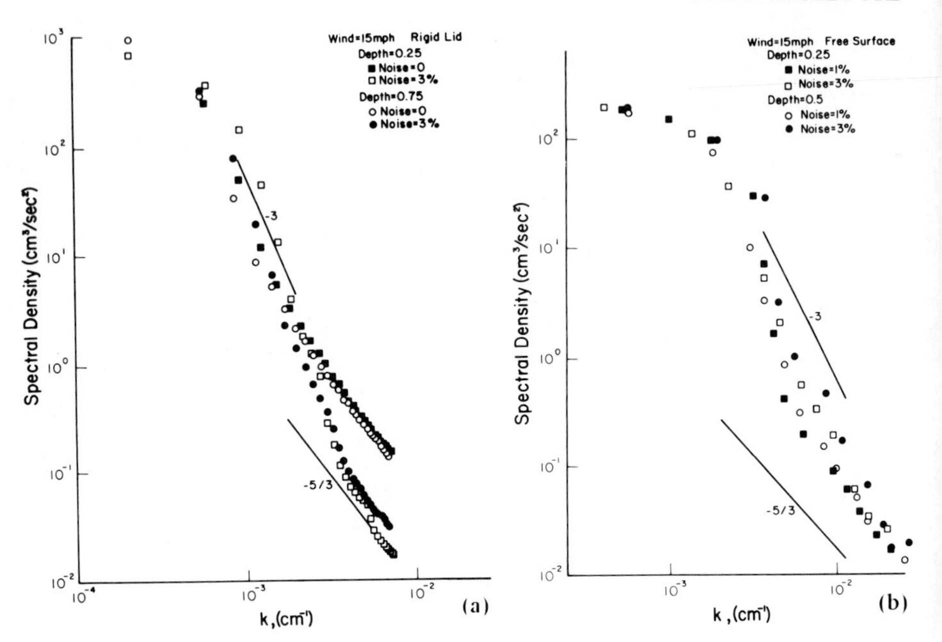

Figure 15. Velocity spectra variation with wind shear noise level.

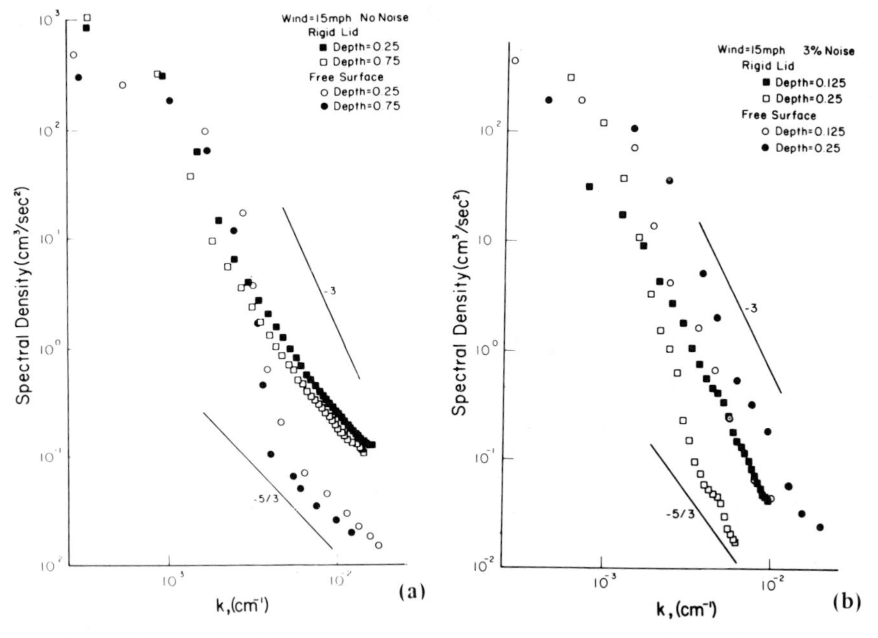

Figure 16. Effect of rigid lid and free surface pressure formulations on the velocity spectra.

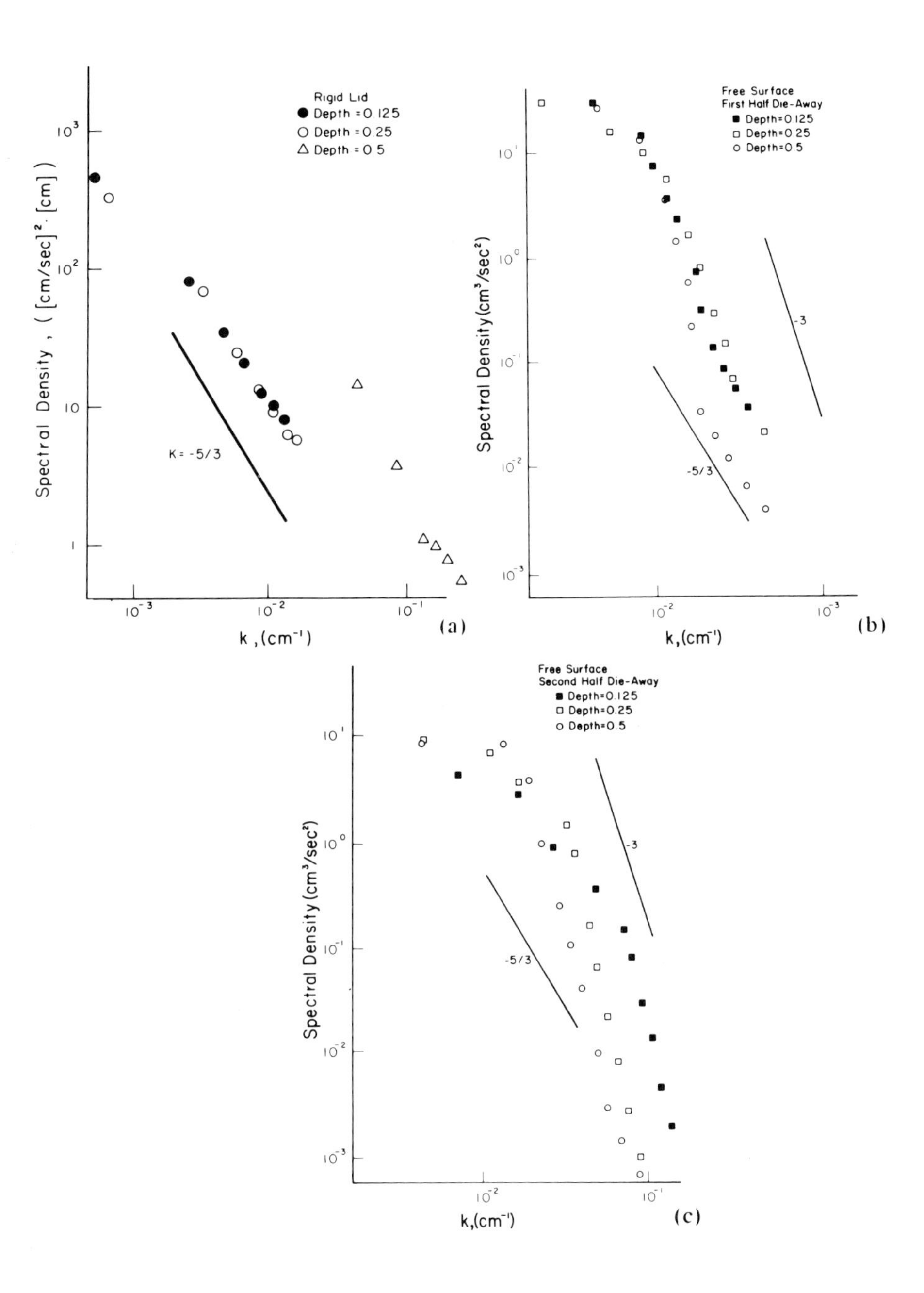

Figure 17. Velocity spectra during die-away.

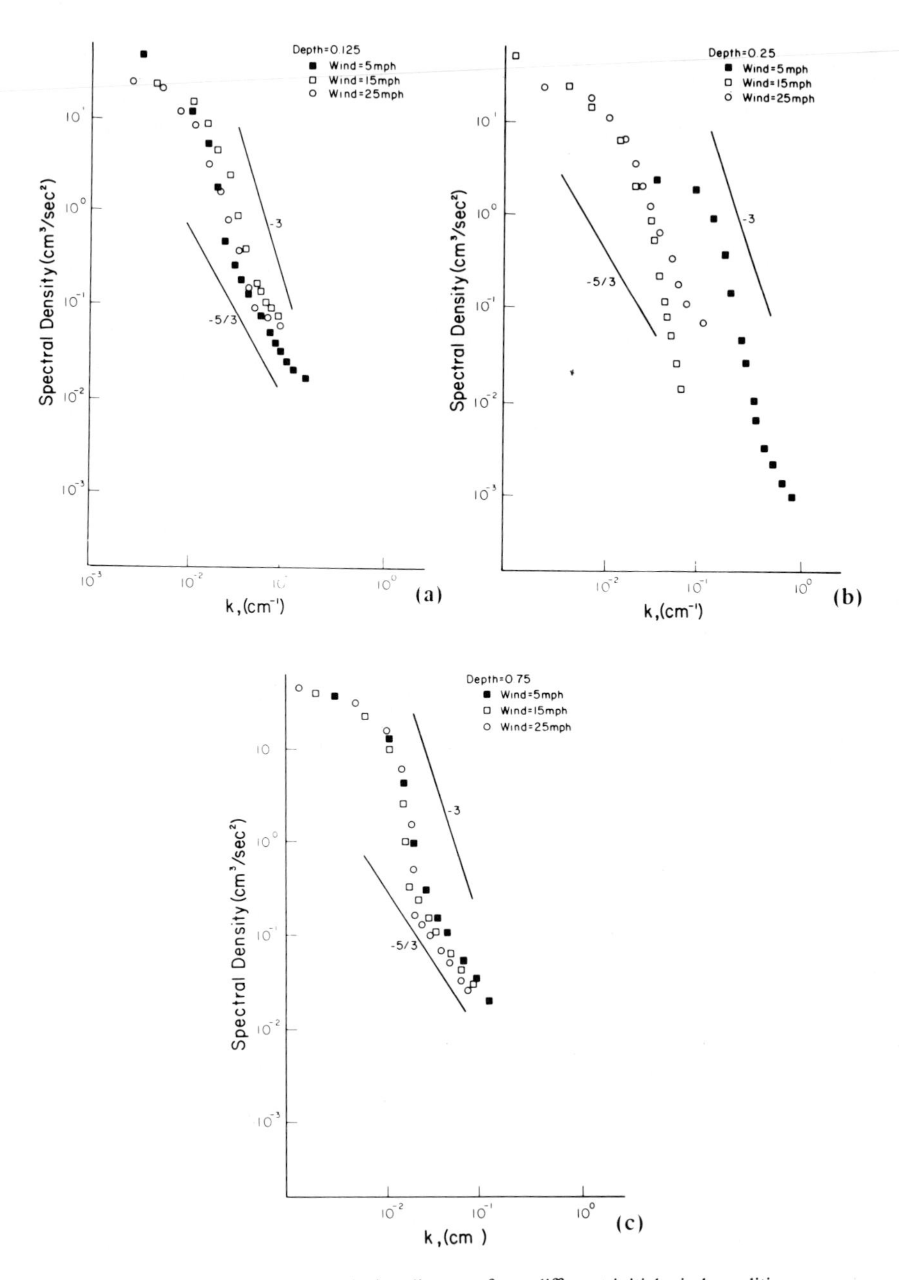

Figure 18. Velocity spectra during die-away from different initial wind conditions.

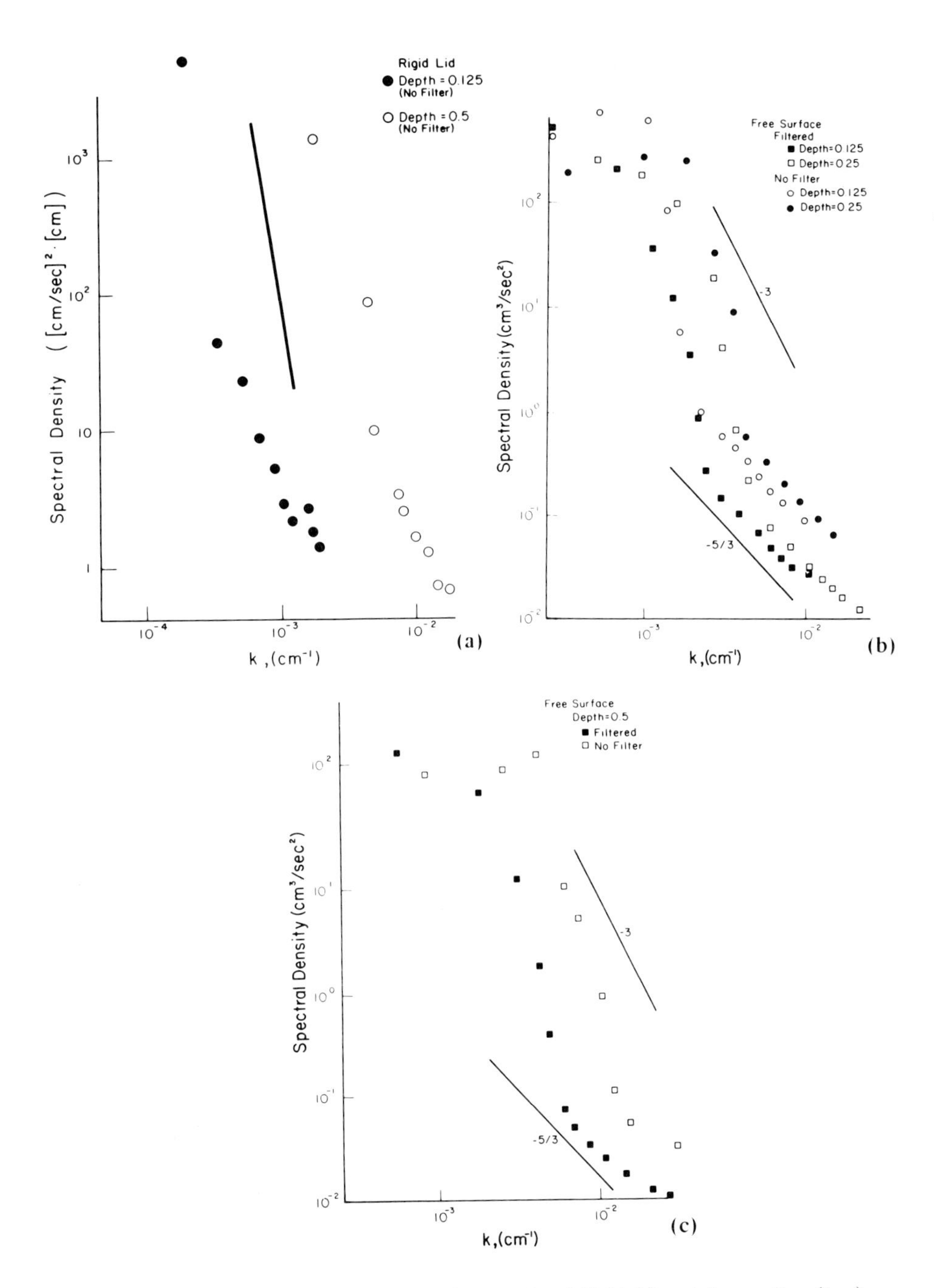

Figure 19. Effect on filtration vs. non-filtration for rigid-lid (a) and free surface (b, c) models.

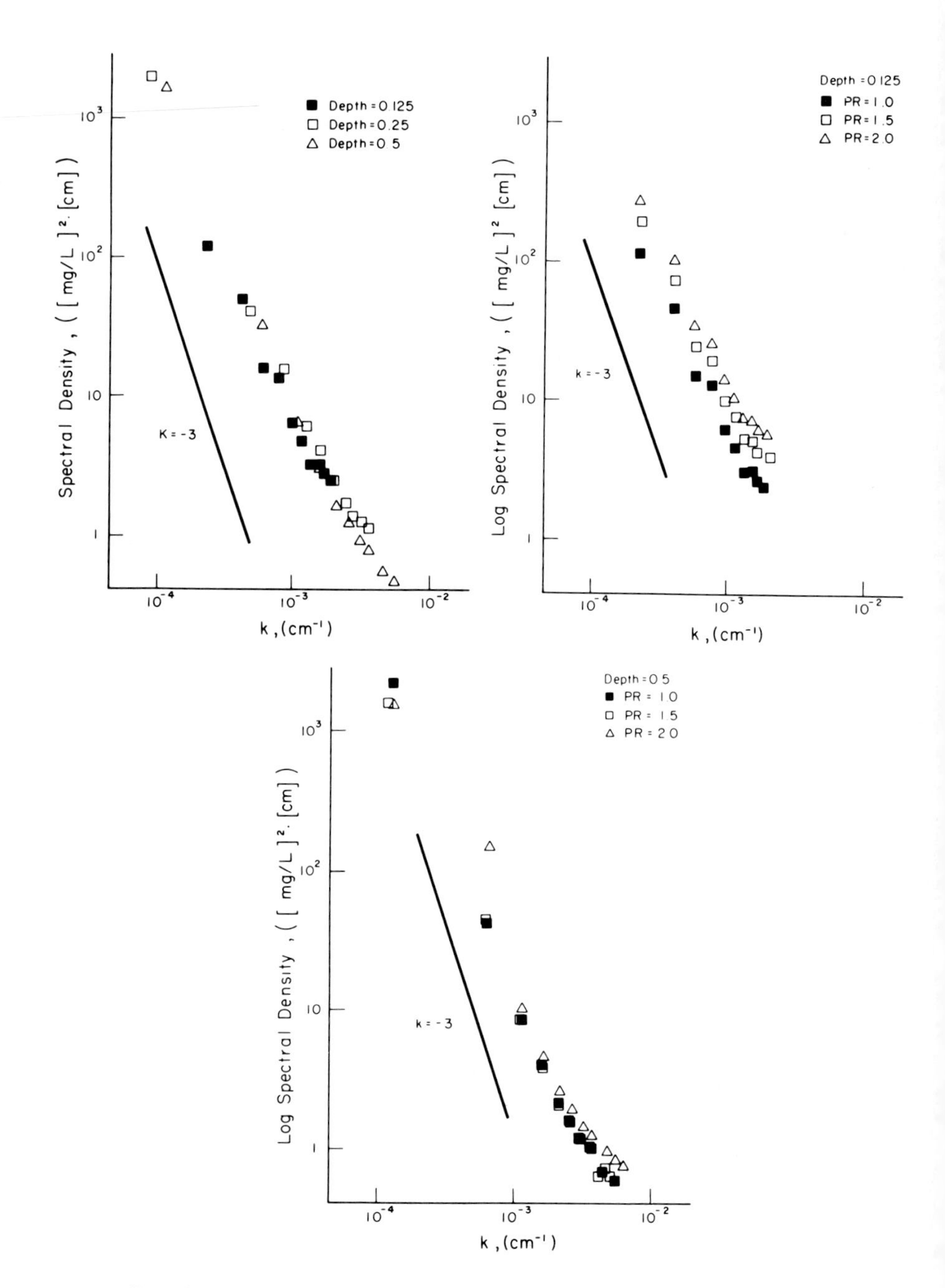

Figure 20. Contaminant spectra variation with depth (a) and Prandtl number (b, c).

and 0.75 are similar. (2) Smaller basins give higher noise and lower long wave length activity, *vice versa* for larger basins. (3) Basin center or end spectra are similar in shape, location and variation with depth. (4) Wind speed variations (Figure 14) cause similar responses in free surface and rigid lid models, both have the same noisy highly shifted response to the 5 mph wind while for stronger winds both model spectra respond with even higher long wave variance. The rigid lid high frequency response is much reduced in comparison to the free surface response. (5) The presence of noise in the wind causes no change in the two-slope spectral shape, but in the rigid lid there is more variance in the higher wavenumber range with no-noise simulations than with noise simulations. (6) The rigid lid model absorbs boundary noise into the longer wave or mean flow field wave number range yielding a depressed high wave number spectrum; non-noisy winds give much noisier high wave number responses. The free surface model has a less noisy high k region and a sightly reduced low k energy region in comparison to the rigid lid. (7) With regard to Figs. 20a and 20b and as calculated by the 120 sec rigid lid model, the passive contaminant spectra do have a -3 slope at longer wavelengths, but the high frequency end of the graph contains a smaller sloped region at much lower k than expected. Therefore, the contaminant spectra departure behaves as a $\gamma >>> \alpha$ case but at the wrong wavenumber range. (8) Increasing the Prandtl Number decreases or flattens the slope and exacerbates the problem in (7).

Again, as reported in Babajimopoulos and Bedford (1980), and Bedford and Babajimopoulos (1980), the die-away and filtration experiments proved most interesting. In general after release of the mean wind field the following behavior is observed. (1) The rigid lid model almost instantaneously yielded a -5/3 slope over the wavenumber range. (2) The rigid lid model had virtually no organized motion at long wave lengths in the deeper depths during die-away (Figure 17a). (3) Figures 17b and 17c show that free surface die-away is dominated by a degeneration into a -3 two-dimensional flow and the -3 region region develops in time from the water surface to the bottom. (4) Die-away behavior from different wind speeds is, for free surface models (Figures 17a, b, and c), dominated by a degeneration to -3 in the upper layers, particularly from higher winds. Again the progression develops from the top down. Notice that the spectra, particularly at the top and bottom, are quite similar in that low wave number activity is gone and the high frequency information is almost identical.

In order to check the adequacy of traditional models in preserving theoretically known spectra, two tests were performed. First the filter coefficients are set to zero, which yields an almost traditional model with fourth order inertia discretization and a Smagorinsky viscosity. The rigid lid spectra in Fig. 19a reveals an almost discontinuously sloped spectra. This region is contained at the wavelength of the grid scale and indicates a failure to propagate energy properly. Figures 19b and 19c are the same tests repeated for the free surface model and the same steep plummeting spectra is seen in both figures. Finally, if second order discretizations are used for the inertia term along with no filters we return to a very traditional approach mentioned earlier and these models blow up before the required time series data can be collected. Obviously traditional models have difficulty propagating energy and preserving spectra.

8.3.3 Production and dissipation spectra. To ensure that an inertial subrange is developed the total production and total dissipation, $kE(k)$ and $k^3E(k)$ (Grant, *et al.*, 1962), are plotted. If the inertial subrange exists a wavenumber separation of a decade or more should be noted as it is in Figure 21. These plots are made after each run and used to determine whether the energy transformations are correct. A run is accepted if the production and dissipation spectra are as theoretically expected while a run rejected for non-compliance with theory is checked for errors. The dominant graphical indication of such errors is considerable energy dissipation at the grid scale wavenumber.

8.4 Ancillary Results

Three pieces of ancillary information have been calculated from experiment 21 and are presented to demonstrate how turbulence models might be used to validate hypotheses presented in the geophysical literature. First the dissipation, ϵ, is calculated. Under certain high Reynolds Numbers, turbulence ϵ is estimated from (Deardorff, 1970)

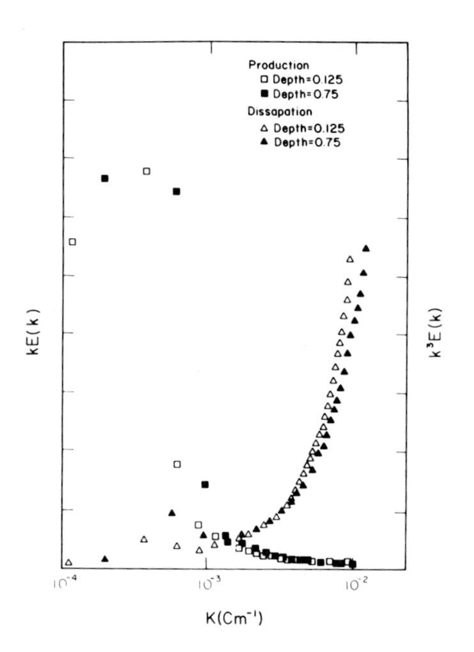

Figure 21. Total production and dissipation spectra.

$$\epsilon = K \frac{\partial \bar{u}_i}{\partial x_j} \left(\frac{\partial \bar{u}_i}{\partial x_j} + \frac{\partial \bar{u}_j}{\partial x_i} \right) \tag{26}$$

For the center node the vertical profile of ϵ is plotted as in Figure 22. It smoothly decreases to zero at the no-shear point and then increases near the bottom. Secondly, Figure 22 also contains a plot of the vertical profile of the Smagorinsky viscosity K. Notice that it, too, steadily decreases to the no-shear point and then increases briefly near the bottom.

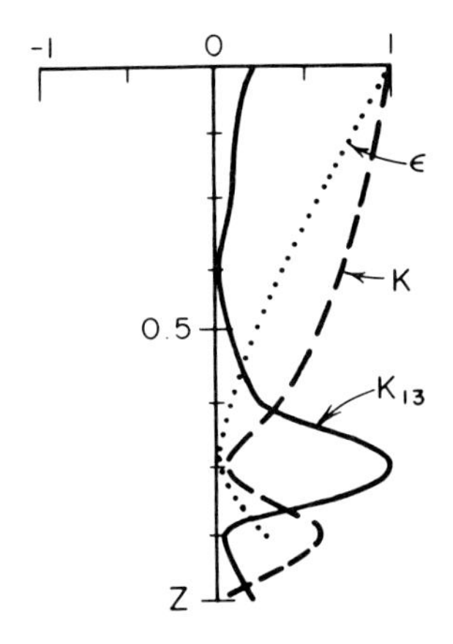

Figure 22. Vertical profile of the dissipation, ϵ, Smagorinsky viscosity, K, and traditional viscosity, K_{13}.

The third result is an attempt to calculate the traditional vertical eddy viscosity K_{13} that would be necessary to reproduce the observed output. This means invoking the traditional Boussinesq closure for Reynolds stresses with K_{13} accounting for the three Leonard stresses, and the subgrid scale flux. Therefore, from a flux balance as defined from Eqs. (8), (9), and (10)

$$L_{13}^{x} + L_{13}^{y} + L_{13}^{z} + N_{13} = K_{13} \frac{\partial \bar{u}}{\partial z}. \tag{27}$$

The stored computer output permitted evaluation of these terms by the discretization used in the program. K_{13} then represents an equivalent viscosity necessary to yield the fluctuating activity without resorting to the Leonard approach. The profile is plotted in Fig. 22 and the results are interesting.

9. DISCUSSION

Obviously a great deal of information can be obtained from these experiments. All the velocity spectra contain a -3 region indicative of 2-D turbulence while for smaller Δt's a -5/3 or 3-D turbulent region is observed. The contaminant spectra do follow the -3 slope behavior in the low wavenumber range but fail badly in the calculation of high wavenumber shapes. The wavenumber range at which this occurs is much too low. Therefore, rather than attribute this behavior to inertial dominant transfer it is more plausible to assume it is wrong and to look to future improvements in the SGS preparation for passive scalars. In passing it should be noted that some simple first order growth rate structures were also tried with no success. The simulation times required to resolve heterogeneity were much too long (and expensive!). There are three remaining items of discussion to be dealt with; first is the dissipation (ϵ) behavior; second is the rigid lid and free surface model behavior during changes in input energy levels; and third is how results derived from complicated models such as these are beneficial to more traditionally structured models.

9.1 Dissipation

As one more check on the validity of these calculations a comparison between the calculated dissipation and field data on dissipation rates gained from other basins and experiments can be made. The dissipation was calculated from experiment 21 and is similar in magnitude and vertical distribution to those generated in the other experiments. Dillon and Powell (1976) have plotted all field reported values of ϵ on a log plot of depth versus ϵ. Figure 24 is this plot wherein the dark lines envelope all values reported and plotted by Dillon and Powell. Selected values from experiment 21 fall within this envelope. The top layer dissipations are high, of the same order as Palmer's (1973) Lake Erie values, but deeper water values are smaller. Our simulations also support the conclusions of Dillon and Powell that the $z^{-0.7}$ dependence suggested by Webster (1969) is not correct. ϵ does decrease in nearly a log constant fashion near the surface, but in a $z^{-2.4}$ fashion. Finally, it has recently been suggested (Oakey and Elliott, 1980) that surface layer dissipation is a function of the wind speed cubed. Calculated values of ϵ obtained at the one-quarter depth indicate that there is a smooth relation but not a precise cubic fit. These dissipation checks do add alternate sources of credence to the results presented in Section 8.

9.2 Modification of the Spectral Shapes

It is apparent that under steady wind conditions and with a small enough time step both 2-D and 3-D fluctuating fields exist. Modifications to this broken spectra shape are clues as to both the adequacy of the physics as well as the mathematical structure of the model. Changes in spectral shape are introduced by

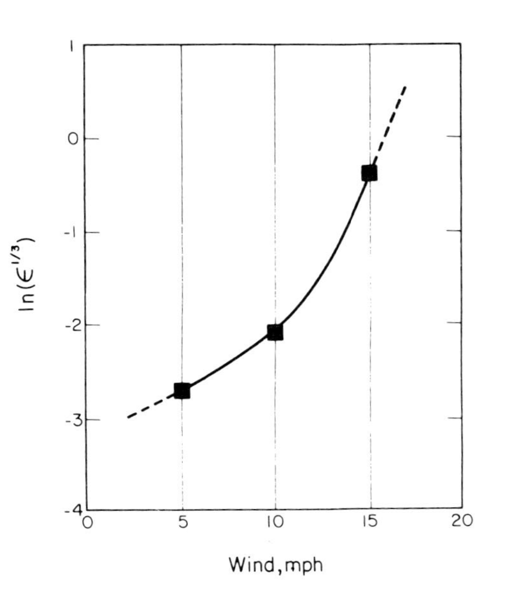

Figure 23. Dissipation vs. wind speed.

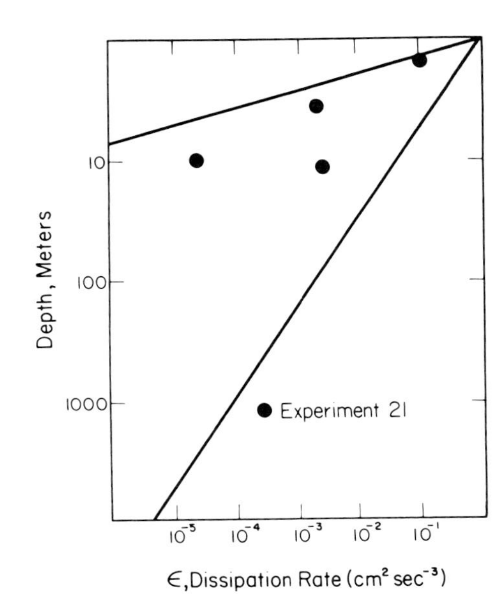

Figure 24. Envelope of dissipation vs. depth (redrawn from Dillon and Powell, 1976).

changes in the physical boundary loadings and by the action of modelers themselves. Following is a discussion of each type of modification.

9.2.1 Spectral response to changes in energy input. Significant spectral changes occurred in response to changes in basin input parameters and comparisons between theory, field data, and model calculations reveal interesting behavior patterns about these models. Changes in boundary loadings consist of changes in either the mean wind speed or changes in the noise level. Die-away is also a change in the mean wind field, and it, too, is a measure of how the model behaves when going from one mean wind speed to another.

Turbulence theory states that in 3-D turbulence, energy entering a system at wavenumber k_i flows toward higher wavenumbers. For two-dimensional turbulence energy flows from wavenumber k_i towards lower wavenumbers (longer wavelength) and enstrophy flows from k_i towards higher wavenumbers. In 2-D flow enstrophy flow is zero in the energy cascade range and energy flow is zero in the enstrophy cascade range. It has already been established by inspection of the spectral slopes that for any steady wind the flow in both model types is 2-D and if the time period is small enough, 3-D. Therefore, inspection of high and low wavenumber ranges during changes in boundary loadings will permit categorizing the type of model response during non-equilibrium events.

In general increasing the wind speed in both free surface and rigid lid models causes an variance increase in the low wavenumber end of the spectrum. This is a 2-D response. For each particular wind the free surface long wave energy is less vigorous than the rigid lid long wave energy, but the 2-D response still exists. Since with no noise the effective k_i (energy input wavenumber) is at the basin length, the 2-D hypothesis is appropriate in terms of known spectral theory. As additional confirmation of the 2-D response note, particularly in Figs. 14a and 14c, that no high frequency changes have taken place as they would in 3-D turbulence.

The response of the high frequency portion of the spectra adds additional support to the 2-D spectral behavior but in so doing demonstrates significant differences between the free surface and rigid lid models. For noiseless winds (Fig. 16a) the high frequency 3-D portion of the free surface spectrum has lower variance than the rigid lid spectra. The free surface spectrum contains more energy than the rigid lid spectrum in the mid-k range. Noisy winds create free surface spectra (Figs. 14b, 15a, 15b, and 16a) that contain small variance increases in the -5/3 range and the mid-k range. For noisy winds the rigid lid high frequency variance drops dramatically to the level of the free surface high frequency variance accompanied by a small increase in mid-k range variance, and apparent 2-D energy transfer. Noise is entering the wind shear time series at 5-second intervals, and this translates into an additional energy input wavenumber, k^*_i, at the high wavenumber end of the spectrum. It is apparent that the free surface model propagates all the high high frequency wind energy back up the cascade which is a 2-D response. Rigid lid spectra created with a noisy wind shear behave as the free surface spectra and therefore a 2-D response is indicated. However, the high frequency spectral "drop" observed when going from non-noisy to noisy winds is certainly suspect.

Die-away in the rigid lid is a curious event. The spectra indicate a strictly three-dimensional energy cascade. This is identical to what is observed in the field, but the dynamics of 3-D turbulence are not within the model at the wavenumber scales where the 5/3 range exist and it is doubtful this 3-D response is a physically realistic circumstance. More work is required to explain this result. The free surface die-away spectra shows a progressive deepening of a transformation from a combined -5/3, -3 spectra shape to a complete -3 spectra which is a 2-D nonequilibrium response. Both model responses during die-away are suspect, particularly in light of the complete disappearance of the -5/3 range in the free surface model.

From the results and discussion it is hypothesized that these models respond in every respect as 2-D models; to call them 3-D models is a misnomer, and to numerically resolve 3-D activity is incorrect.

9.2.2 Spectral modifications by modelers. Simple physical reasoning says that the maximum possible wavelength, γ_B, at which the -3 to -5/3 slope break occurs is roughly equal to the depth. The time scale, t_B, or time step equivalent to γ_B, is found from Taylor's hypothesis and the average current, i.e., $t_B = \gamma_B/\bar{U}(z)$ where γ_B equals the wavelength (which is equal to the basin depth) and $\bar{U}(z)$ is the total average velocity at the depth where the spectra is being calculated. Therefore, it is possible for the time scale at which the -5/3 region appears to fluctuate according to the average velocity. The smaller the average velocity the longer the time scale at the -5/3 region occurs. Modelers most often select grid scales on the basis of the "largest permissible for stability" criteria. However, with 3-D free surface models, some very small time steps are being used. Since for time series spectra the 2-D/3-D break occurs at a time step or scale inversely proportional to velocity, two problems are posed which must be addressed by modelers.

The first is the problem of establishing an initial Δt. If, as has been argued in this paper, these models respond to energy changes in a 2-D fashion, then the resolution of the 3-D portion of the spectrum is at best superfluous and at worst violates implicit assumptions made in the formulation of the governing equations. Therefore, Δt's should be selected to resolve that portion of the spectrum for which the model's assumptions and uses are valid, i.e., $\Delta t > t_B$. Since t_B fluctuates, selecting Δt becomes a problem of anticipating the smallest average velocity, and selecting a Δt which keeps the computation stable and resolves the correct portion of the spectra.

The second problem is inadvertent resolution of 3-D physics during the model calculations because of simple reduction of average velocity caused by changes in boundary loadings. Even if a Δt is established commensurate with the 2-D flow assumption, a simulation which contain a rapid velocity fall off during the simulation (for example, a post-storm lull), could inadvertently resolve a 3-D flow because the 3-D scale, t_B, exceeded Δt. The model might not be prepared for such a calculation in that the basic model numerics, such as eddy viscosities, might be incorrect, or the wrong assumptions might have been used, i.e., hydrostatic pressure, in the preparation of the governing equations. A simple review of the field spectra presented in the review section indicates that t_B during weak flow may be as large as 5-10 minutes easily within the range of quasi-2-D model time steps. This becomes a source of future research.

9.3 Use of Turbulence Models

In addition to the ability to make spectral comparisons, the use of turbulence models is encouraged because ultimately they can identify, by proper analysis and averaging of the output, more exact expressions for the many coefficients that abound in traditional models. As an example, consider the calculation for the traditional eddy viscosity K_{13}. A multitude of empirical forms for K_{13} exist and many debates ensue over the adequacy of one form versus another. Depth variations form simple linear and exponential decay to Von Karmen Prandtl mixing length forms have been proposed. Yet, it appears that for all but the top portion of the water column a form as simple as

$$K = \frac{\tau_0/\rho}{\partial \bar{U}/\partial z}$$

suffices. Simple tests such as these demonstrate the value of turbulence modeling.

10. CONCLUSIONS

The difficulties of precise spectra curve fitting and the lack of any but isolated specific field data make quantitative conclusions difficult, but three general conclusions are drawn.

The first conclusion is that special steps, aside from mesh refinement, are necessary in order to prepare a model for economical, high Reynolds Number, spectrally correct turbulence predictions. Coarse grid calculations can be made, but, at minimum, a higher order spatial averaging treatment of the nonlinear terms is necessary along with a grid size dependent method for proper subgrid energy dissipation. Filtration is necessary, otherwise energy is not cascaded properly. For the models used here the Δt's must fall in the inertial subrange. It is particularly important for the time step to also be placed in the frequency spectra inertial subrange and thereby prevent the inadvertent resolution of hydrodynamic activity that model assumptions prohibit. Finally, noisy winds are not necessary to introduce turbulence, but much better wall boundary conditions must be developed and used.

The second conclusion is that the spectral behavior of lake models indicate they propagate energy as 2-D models and therefore are not, as so often referred to, 3-D. The general spectral shapes are properly portrayed and contained between the proper maximum and minimum scales. Dissipation rates are also of proper magnitude as is their variation with depth and basin size, and their correlation with speed. The shapes contain a two slope range dominated by a -3 or 2-D range at wavelengths greated than the depth and a 3-D or -5/3 region at smaller scales. Since vertical fluctuation dynamics are overly reduced by use of the hydrostatic pressure assumption, the 3-D spectra portion should be interpreted with caution and avoided if possible. An increase or die-away of the wind causes the free surface model to cascade energy as a two-dimensional flow while the rigid lid gives a much too rapid 3-D response during die-away.

Finally, as an extension of the second conclusion the research reported here points out two serious flaws in traditional models which must be remedied by structural improvements in the model physics, not by improvements in numerical methods or empirical coefficients.

Lake models, although 3-D in coordinate system structure, transmit energy as a two-dimensional flow field. They simply do not have the proper physics to adequately predict full 3-D fluctuations. Yet modelers, by improperly selecting time steps, can indeed resolve a 3-D fluctuation field which might be justifiable relative to the imposed model assumptions, but is not correct in terms of known theoretical and field observed behavior. The hydrostatic pressure approximation limits not only the physics that can be resolved but sets bounds on the permissible grid sizes used in lake models. Therefore, if model time steps resolve physics with length scales the size of the depth then the hydrostatic assumption must be abandoned as model results are erroneous.

Traditional models, because they do not reproduce theoretically expected spectral shapes, improperly transmit energy and are therefore inherently unstable. To control this instability the modeler must either refine the grid scales to the Kolmogorov scales, which is economically impossible, or use a variety of ad-hoc schemes to impose stability on the computation. Ad-hoc schemes, such as eddy viscosities and up-wind or down-wind differencing, are often used to remedy such problems but as is well known there are numerous and intense struggles waged in the literature over which method or coefficient form is superior. Further, arguments are also waged over the adequacy of finite elements or finite differences in solving such problems. It is the author's opinion that these arguments are all superfluous in the face of the evidence presented here. Not one of the above remedies corrects the inability of traditional models to adequately propagate energy. Therefore, it is time to begin reassessing the very basic structure of the equations that are used to represent the desired physics.

REFERENCES

Allender, J., Berger, M., and Saunders, K. (1975). Symposium of Modeling of Transport Mechanisms in Oceans and Lakes. National Water Research Institute, Burlington, Ontario, Canada.

Allender, J., and Saunders, K. (1976). Nineteenth Conference on Great Lakes Research Programme. Guelph, Ontario.
Allender, J. (1977). *J. Phys. Oceanogr.* **7**, 711-718.
Babajimopoulos, C. and Bedford, K. (1980). *J. Hydraul. Div. Proc. Am. Soc. Civ. Eng.* **106**, 1-19.
Batchelor, G. (1959). *J. Fluid Mech.* **5**, 113-133.
Bedford, K. and Babajimopoulos, C. (1978). Filtered aquatic transport. *IAGLR Conf. on Coastal Processes.* Sundridge, Ontario, Canada.
Bedford, K. and Babajimopoulos, C. (1980). *J. Hydraul. Div. Proc. Am. Soc. Civ. Eng.* **106**, 21-38.
Bedford, K. and Rai, I. (1978). *J. Hydraul. Div. Proc. Am. Soc. Civ. Eng.* **104.**
Bedford, K. and Shah, B. (1977). Maumee Bay Sediment Transport Mechanisms During Spring Flood. EPA/IJC Project Completion Report, PLUARG.
Bonham-Carter, G., Thomas, J. and Lockner, D. (1973). Rochester Embayment Project. International Field Year for the Great Lakes, Rep. No. 1. University of Rochester, New York.
Bennett, J. (1977). *J. Phys. Oceanogr.* **7**, 591-601.
Caldwell, D., Wilcox, S., and Matsler, M. (1975). *Limnol. Oceanogr.* **20**, 1035-1047.
Clark, R., Ferziger, J., and Reynolds, W. (1977). Evaluation of Subgrid Scale Turbulence Models Using a Fully Simulated Turbulent Flow. NASA Res. Rep. NGR-05-020-622, Thermosciences Division, Stanford University.
Clark, R., Ferziger, J., and Reynolds, W. (1979). *J. Fluid Mech.* **91**, 1-16.
Collatz, L. (1960). The Numerical Treatment of Differential Equations, Springer-Verlag.
Corrsin, S. (1951). *J. Appl. Phys.* **22**, 469-473.
Corrsin, S. (1961). *J. Fluid Mech.* **11**, 407-416.
Deardorff, J. (1970). *J. Comp. Phys.* **7**, 120-133.
Deardorff, J. (1973). *J. Fluids Eng. Trans. Am. Soc. Mech. Eng.*, 429-438.
Denman, K. (1976). *Deep Sea Res.*, **23**, 539-550.
Denman, K., Okubo, A., and Platt, T. (1977). *Limn. Oceanogr.* **22**, 1033-1038.
Dillon, T. and Powell, T. (1976). *J. Geophys. Res.* **81**, 6421-2427.
Dillon, T. and Caldwell, D. (1980). *J. Geophys. Res.* **85**, 1910-1916.
Elliot, J. and Oakey, N. (1976). *J. Fish. Res. Bed. Can.*, **33**, 2296-2306.
Fasham, M. and Pugh, P. (1976). *Deep Sea Res.* **23**, 527-538.
Ferziger, J., Mehta, U. and Reynolds, W. (1977). *Symp. Turb. Shear Flow, University Park, Pennsylvania,* Sec. 14, 31-40.
Findkakis, A. and Street, R. (1979). *J. Atmos. Sci.* **36**, 1934-1949.
Fjortoft, R. (1953). *Tellus* **5**, 225-230.
Frost, W. (1978). "Handbook of Turbulence," (W. Frost and T. Mouldon, eds.), Vol. 1, Plenum Press.
Gedney, R. and Lick, W. (1970). *J. Geophys. Res.* **77**, 2714-2723.
Grant, H., Stewart, R., and Moilliet, A. (1962). *J. Fluid Mech.* **12**, 241-268.
Grant, H., Moilleit, A., and Vogel, W. (1968). *J. Fluid Mech.* **34**, 443-448.
Gregg, M. C. (1976). *J. Phys. Oceanogr.* **6**, 528-555.
Gregg, M. C. (1976). *J. Phys. Oceanogr.* **7**, 436-454.
Grotzbach, G. and Schumann, U. (1977). *Symp. Turb. Shear Flow, University Park, Pennsylvania,* Sec. 14, 11-20.
Haq, A. and Lick, W. (1975). *J. Geophys. Res.* **80**, 431-437.
Hinze, J. (1975). "Turbulence." McGraw-Hill, New York.
Kizlauskas, A. and Katz, P. (1974). *Arch. Meteor. Geophys. Bioklim.* **23**, 181-197.
Kolmogorov, a. (1941). *Dokl. Akad. Nau, SSR* **32**, 19-21.
Kraichnan, R. (1967). *The Physics of Fluid* **10**, 1417-1423.
Kraichnan, R. (1967). *J. Fluid Mech.* **47**, 525-535.
Kwak, D., Reynolds, W., and Ferziger, J. (1975). Three-Dimensional time Dependent Computation of Turbulent Flow. Department of Mechanical Engineering, Stanford University, Tech. Rep. No. TF-5.
Lavelle, J., Young, R., Swift, R., and Clark, T. (1978). *J. Geophys. Res.* **83**, 6052-6062.
Leigh-Abbott, M., and Coil, J. (1978). *IAGLR Conf. on Coast. Proc.*, Sundridge, Ontario, Canada.
Leigh-Abbott, M., Coil, J., Powell, T., and Richerson, P. (1978), *J. Geophys. Res.* **83**, 4668-4672.
Lemmin, U., Scott, J., and Czapski, U. (1974). *J. Geophys. Res.* **79**, 3442-3448.

Leonard, A. (1974). *Adv. Geophys.* **18A,** 237-248.
Leslie, D. and Quarini, G. (1979). *J. Fluid Mech.* **91,** 65-91.
Lick, W. (1976). Numerical Models of Lake Currents. U.S. EPA Ecological Research Report Series, Rept. EPA-600/3-76-020.
Love, M. and Leslie, D. (1977). *Symp. Turb. Shear Flow, University Park, Pennsylvania,* Sec. 14, 1-10.
Lumley, J. (1964). *J. Atmos. Sci.* **21,** 99-102.
Mamorino, G. and Caldwell, D. (1978). *Deep Sea Res.* **25,** 1073-1106.
Monin, A. and Yaglom, A. (1971). "Statistical Fluid Mechanics: Mechanics of Turbulence." Vol. 1, The MIT Press, Cambridge, Massachusetts.
Monin, A. and Yaglom, A. (1975). "Statistical Fluid Mechanics: Mechanics of Turbulence." Vol. 2, The MIT Press, Cambridge, Massachusetts.
Nasmyth, P. W. (1970). Oceanic Turbulence. Ph.D. Thesis, Institute of Oceanography, University of British Columbia, Vancouver, Canada.
Obukhov, A. (1941). *Izvestiya An SSSR. Geogr. Geofiz.* No. 4-5, 453-466.
Obukhov, A. (1949a). *Izv. Akad. Nank. SSSR, Ser. Geogr. Geofiz.* **13,** 58-69.
Obukhov, A. (1949b). *Dokl. Akad. Nauk. SSSR* **66,** 17-20.
Okey, N. and Elliot, J. (1980). Dissipation in the mixed layer near Emerald Basin. *In* "Marine Turbulence" (J. Nihoul, ed.), pp. 123-234. Elsevier Publishing Company.
Orszag, S. (1978). *In* "Handbook of Turbulence," (W. Frost and T. Mouldon, eds.), Volume 1, Plenum Press.
Palmer, M. (1973). *J. Geophys. Res.* **28,** 3585-3595.
Paul, J. and Lick, W. (1976). Lake Erie International Jetport Model FeasibilityInvestigation: Application of Three-Dimensional Hydrodynamic Model to Study Effects of Proposed Jetport Island on Thermocline Structure in Lake Erie. Lake Erie Regional Transportation Authority, Contract Rep. H-75-1. U. S. Army Corps of Engineers, Waterways Experiment Station.
Phillips, O. (1978). "The Dynamics of the Upper Ocean," 2nd Edition, Cambridge University Press.
Platt, T. (1972). *Deep Sea Res.* **19,** 183-187.
Powell, T., Richerson, P., Dillon, T., Agee, B., Dozier, B., Godden, D., and Myrup, L. (1975). *Science* **189,** 1088-1090.
Rose, H. (1977). *Symp. Turb. Shear Flow, University Park, Pennsylvania,* Sec. 14, 41-48.
Schumann, U. (1975). *J. Comp. Phys.* **18,** 376-404.
Schwab, D. (1978). *Monthly Weather Rev.* **96,** 1476-1487.
Sheng, U., Lick. W., Gedney, R., and Molls, F. (1978). *J. Phys. Oceanogr.* **8,** 713-727.
Shinozuka, M. and Jan C. (1972). *J. Sound Vibration* **25,** 111-128.
Simons, T. (1974). *J. Phys. Oceanogr.* **4,** 507-523.
Simons, T. (1975). *J. Phys. Oceanogr.* **5,** 98-110.
Smagorinsky, J., Manabe, S., and Holloway, J. Jr., (1965). *Monthly Weather Rev.* **93.**
Spraggs, L. and Street, R. (1975). Three-Dimensional Simulation of Thermally Influenced Hydrodnamic Flows. Technical Report No. 190, Stanford University.
Steele, J. H. (Editor). "Spatial Pattern in Plankton Communities," NATO Conference Series, Plenum Press.
Taylor, G. (1915). *Phil. Trans. Roy. Soc.* **A215,** 1-26.
Taylor, G. (1932). *Proc. Roy. Soc.* **A135,** 685-706.
Tennekes, H. and Lumley, J. (1972). "A First Course in Turbulence." MIT Press, Cambridge, Massachusetts.
Thomann, R. and Winfield, R. (1976). *Proc. EPA Conf. Envir. Model. Simul.* (W. R. Ott, ed.), Cincinnati, Ohio.
Webster, F. (1969). *Deep Sea Research* **16,** (Suppl.), 357-368.
Welch, J., Harlow, F., Shannon, J., and Daly, B. (1966). "The MAC Method." Los Alamos Scientific Laboratory, Los Alamos, New Mexico.
Willis, J. and Bolton, G. (1979). *J. Hydraul. Div. Proc. Am. Soc. Civ. Eng.,* **105,** 1-15.
Willis, J. and Kennedy, J. (1977). "Sediment Discharge of Alluvial Streams Calculated from Bed-Form Statistics." Project Department, Iowa Institute of Hydraulic Research, No. 202.

NUMERICAL MODELING OF FREE-SURFACE FLOWS THAT ARE TWO-DIMENSIONAL IN PLAN

M. B. Abbott

International Institute for Hydraulic
and Environmental Engineering
Delft, The Netherlands

A. McCowan and I. R. Warren

Danish Hydraulic Institute
Horsholm, Denmark

1. INTRODUCTION

The modeling of flows that are two-dimensional in plan is now widely practiced and its results are accepted for many predictive purposes. The principles of the modeling work have been set out in some dozens of papers (e.g., Miller and Yevjevich, 1975; Abbott, 1979) while one simple model has been described in considerable detail, in several very extensive publications (Leendertse, 1967, *et seq.*). As the explicit objective of this symposium was to provide a level of detail that would permit an independent observer to reproduce the results, it has been necessary to restrict the present discussion to only a limited class of models and to concentrate on just a few of their more important features. However, the class of models considered (the class of models built upon time-centered implicit schemes) is that most widely used for predictive purposes, and includes the Leendertse model, while the features considered are those most commonly discussed in the literature. These features are those of amplification error, phase error, numerical vorticity generation, numerical sensitivity to zig-zagging, truncation error correction and the related introduction of vertical accelerations. Since a certain amount of transport is represented in these models at sub-grid scales, a brief discussion of sub-grid scale modeling has been included. Some of these features are then illustrated by examples of recent applications of free-surface flow models in engineering practice.

ISBN: 0-12-25852-0

NOTATION

D	local grid-resolved deformation rate
f	generalized level (mass, momentum, energy, etc.)
g	acceleration due to gravity (ms^{-2})
h	water depth (m)
h_0	mean value of h
H	elevation of bed measured above a reference horizontal plane
k	address in y-coordinate
p	volume flux density in x-direction
q	volume flux density in y-direction
s	generalized distance (m)
t	time (δ)
u, v	velocity components
x, y	orthogonal distance coordinates (m)
ν	eddy-viscosity
ω	vorticity
Ω	depth-integrated vorticity

2. A CLASS OF DIFFERENCE SCHEMES FOR TWO-DIMENSIONAL FLOWS

The difference schemes considered are those which: (a) are consistent with the equations of free-surface flow

$$\frac{\partial f}{\partial t} + A_x \frac{\partial f}{\partial x} + A_y \frac{\partial f}{\partial y} = F(f)$$

(b) when locally linearized, have one-dimensional descents

$$\frac{(f_j^{n+1} - f_j^n)}{\Delta t} + A_x \left[\frac{(f_{j+1} - f_{j-1})^n}{4\Delta x} + \frac{(f_{j+1} - f_{j-1})^{n+1}}{4\Delta x} \right] = F(f)$$

in the x-direction and

$$\frac{(f_k^{n+1} + f_k^n)}{\Delta t} + A_y \left[\frac{(f_{k+1} - f_{k-1})^n}{4\Delta y} + \frac{(f_{k+1} - f_{k-1})^{n+1}}{4\Delta y} \right] = F(f)$$

in the y-direction. In the event that $F(f) = 0$ we have the schemes consistent with the primitive equations of nearly-horizontal flow.

The following are representatives of this class of schemes for the case of nearly-horizontal flow:

(1) the linearized ideal scheme, with three component equations

$$\frac{p_{j,k}^{n+1} - p_{j,k}^{n}}{\Delta t} + gh\left(\frac{h_{j+1,k}^{n+1} - h_{j-1,k}^{n+1}}{4\Delta x} + \frac{h_{j+1,k}^{n} + h_{j-1,k}^{n}}{4\Delta x}\right) = 0$$

$$\frac{h_{j,k}^{n+1} - h_{j,k}^{n}}{\Delta t} + \left(\frac{p_{j+1,k}^{n+1} - p_{j-1,k}^{n+1}}{4\Delta x} + \frac{p_{j+1,k}^{n} - p_{j-1,k}^{n}}{4\Delta x}\right)$$

$$+ \left(\frac{q_{j,k+1}^{n+1} - q_{j,k-1}^{n+1}}{4\Delta y} + \frac{q_{j,k+1}^{n} - q_{j,k-1}^{n}}{4\Delta y}\right) = 0$$

$$\frac{q_{j,k}^{n+1} - q_{j,k}^{n}}{\Delta t} + gh\left(\frac{h_{j,k+1}^{n+1} - h_{j,k-1}^{n+1}}{4\Delta y} + \frac{h_{j,k+1}^{n} - h_{j,k-1}^{n}}{4\Delta y}\right) = 0$$

This scheme is called "ideal" because it is the simplest conceivable member of the class of schemes considered. However, it admits of no algorithmic structure, necessitating an iterative solution, so that it is not an implicit scheme in the sense of Abbott (1979, p. 170; see also Leonard, 1979, p. 7).

(2) The linearized Leendertse (1967) scheme, with six component equations

$$\frac{u_{j,k}^{n+1/2} - u_{j,k}^{n}}{1/2.\Delta t} + g\,\frac{h_{j+1,k}^{n+1/2} - h_{j-1,k}^{n+1/2}}{2.\Delta x} = 0$$

$$\frac{v_{j,k}^{n+1/2} - v_{j,k}^{n}}{1/2.\Delta t} + g\,\frac{h_{j,k+1}^{n} - h_{j,k-1}^{n}}{2\Delta y} = 0$$

$$\frac{h_{j,k}^{n+1/2} - h_{j,k}^{n}}{1/2.\Delta t} + h_0\left(\frac{u_{j+1,k}^{n+1/2} - u_{j-1,k}^{n+1/2}}{2.\Delta x} + \frac{v_{j,k+1}^{n} - v_{j,k-1}^{n}}{2.\Delta y}\right) = 0$$

$$\frac{u_{j,k}^{n+1} - u_{j,k}^{n+1/2}}{1/2.\Delta t} + g\,\frac{h_{j+1,k}^{n+1/2} - h_{j-1,k}^{n+1/2}}{2.\Delta x} = 0$$

$$\frac{v_{j,k}^{n+1} - v_{j,k}^{n+1/2}}{1/2.\Delta t} + g\,\frac{h_{j,k+1}^{n+1} - h_{j,k-1}^{n+1}}{2.\Delta y} = 0$$

$$\frac{h_{j,k}^{n+1} - h_{j,k}^{n+½}}{½.\Delta t} + h_0\left(\frac{u_{j+1,k}^{n+½} - u_{j-1,k}^{n+½}}{2.\Delta x} + \frac{v_{j,k+1}^{n+1} - v_{j,k-1}^{n+1}}{2.\Delta y}\right) = 0$$

(3) The linearized Abbott (1968) scheme (Sobey, 1970), with four component equations

$$\frac{u_{j,k}^{n+1} - u_{j,k}^{n}}{\Delta t} + g + \frac{h_{j+1,k}^{n+½} - h_{j-1,k}^{n+½}}{2.\Delta x} = 0$$

$$\frac{h_{j,k}^{n+½} - h_{j,k}^{n}}{½.\Delta t} + h_0\left(\frac{u_{j+1,k}^{n+1} - u_{j-1,k}^{n+1}}{4.\Delta x} + \frac{u_{j+1,k}^{n} - u_{j-1,k}^{n}}{4.\Delta x}\right.$$

$$\left. + \frac{v_{j,k+1}^{n} - v_{j,k-1}^{n}}{2.\Delta y}\right) = 0$$

$$\frac{v_{j,k}^{n+1} - v_{j,k}^{n}}{\Delta t} + g\left(\frac{h_{j,k+1}^{n+1} - h_{j,k-1}^{n+1}}{4.\Delta y} + \frac{h_{j,k+1}^{n} - h_{j,k-1}^{n}}{4.\Delta y}\right) = 0$$

$$\frac{h_{j,k}^{n+1} - h_{j,k}^{n+½}}{\Delta t} + h_0\left(\frac{u_{j+1,k}^{n+1} - u_{j-1,k}^{n+1}}{4\Delta x} + \frac{u_{j+1,k}^{n} - u_{j-1,k}^{n}}{4.\Delta x}\right.$$

$$\left. + \frac{v_{j,k+1}^{n+1} - v_{j,k-1}^{n+1}}{2.\Delta y}\right) = 0$$

and (4) the linearized S21 scheme (Abbott, Damsgaard and Rodenhuis, 1973), which, through linearization, reduces from eight component equations to four component equations

$$\frac{(h_{j,k}^{n+½} - h_{j,k}^{n})}{½\Delta t} + \left(\frac{p_{j+1,k}^{n+1} - p_{j-1,k}^{n+1}}{4\Delta x} + \frac{p_{j+1,k}^{n} - p_{j-1,k}^{n}}{4\Delta x}\right)$$

$$+ \left(\frac{q_{j,k+1}^{n+1/2} - q_{j,k-1}^{n+1/2}}{4\Delta y} + \frac{q_{j,k+1}^{n-½} - q_{j,k-1}^{n-½}}{4\Delta y}\right) = 0$$

$$\frac{p_{j+1,k}^{n+1} - p_{j+1,k}^{n}}{\Delta t} + gh_0\left(\frac{h_{j+2,k}^{n+1/2} - h_{j,k}^{n+1/2}}{2\Delta x}\right) = 0$$

$$\frac{h_{j,k}^{n+1} - h_{j,k}^{n+1/2}}{1/2\Delta t} + \left(\frac{p_{j+1,k}^{n+1} - p_{j-1,k}^{n+1}}{4\Delta x} + \frac{p_{j+1,k}^{n} - p_{j-1,k}^{n}}{4\Delta x}\right) = 0$$

$$+ \left(\frac{q_{j,k+1}^{n+3/2} - q_{j,k-1}^{n+3/2}}{4\Delta y} + \frac{q_{j,k+1}^{n+1/2} - q_{j,k-1}^{n+1/2}}{4\Delta y}\right) = 0$$

$$\frac{q_{j,k+1}^{n+3/2} - q_{j,k+1}^{n+1/2}}{\Delta t} + gh_0\left(\frac{h_{j,k+2}^{n+1} - h_{j,k}^{n+1}}{2\Delta y}\right) = 0$$

For the purposes of analysis, it is convenient to reduce such schemes to their 2-level forms. The ideal scheme does not allow of any alternating direction algorithmic structure so we have written it directly in 2-level form. The algorithmically tractable forms have the following equivalent 2-level forms:

(2) Leendertse (1967)

$$\frac{(u^{n+1} - u^{n})}{\Delta t} + g\left[\frac{(h_{j+1} - h_{j-1})_{k}^{n+1}}{4\Delta x} + \frac{(h_{j+1} - h_{j-1})_{k}^{n}}{4\Delta x}\right]$$

$$+ \frac{gh_0\Delta t}{16\Delta x\Delta y}\left[\{(v_{k+1} - v_{k-1})_{j+1}^{n+1} - (v_{k+1} - v_{k-1})_{j-1}^{n+1}\}\right.$$

$$\left. - \{(v_{k+1} - v_{k-1})_{j+1}^{n} - (v_{k+1} - v_{k-1})_{j-1}^{n}\}\right] = 0$$

$$\frac{(h^{n+1} - h^{n})_{j,k}}{\Delta t} + h_o\left[\frac{(u_{j+1} - u_{j-1})_{k}^{n+1}}{4\Delta x} + \frac{(u_{j+1} - u_{j-1})_{k}^{n}}{4\Delta x}\right.$$

$$\left. + \frac{(v_{k+1} - v_{k-1})_{j}^{n+1}}{4\Delta y} + \frac{(v_{k+1} - v_{k-1})_{j}^{n}}{4\Delta y}\right] = 0$$

$$\frac{(v^{n+1} - v^{n})_{j,k}}{\Delta t} + g\left[\frac{(h_{k+1} - h_{k-1})_{j}^{n+1}}{4\Delta y} + \frac{(h_{k+1} - h_{k-1})_{j}^{n}}{4\Delta y}\right] = 0$$

The Leendertse scheme is seen to differ from the ideal scheme only through the term

$$\frac{gh\Delta t}{16\Delta x\Delta y}\Big[\{(v_{k+1}-v_{k-1})_{j+1}^{n+1}-(v_{k+1}-v_{k-1})_{j-1}^{n+1}\}$$

$$-\{(v_{k+1}-v_{k-1})_{j+1}^{n}-(v_{k+1}-v_{k-1})_{j-1}^{n}\}\Big]$$

which arises during the elimination of terms with time address $n+½$. Writing ¢ for "which is consistent with", the Leendertse scheme differs from the ideal scheme only through a term

$$¢\ \frac{gh\Delta t^2}{4}\ \frac{\partial^3 v}{\partial x\partial y\partial t}$$

in the x-Euler equation.

(3) Abbott (1968/1970)

$$\frac{(u^{n+1}-u^n)_{j,k}}{\Delta t}+g\left[\frac{(h_{j+1}-h_{j-1})_k^{n+1}}{4\Delta x}+\frac{(h_{j+1}-h_{j-1})_k^{n}}{4\Delta x}\right]$$

$$+\frac{gh_0\Delta t}{16\Delta x\Delta y}\left[\left\{(v_{k+1}-v_{k-1})_{j+1}^{n+1}-(v_{k+1}-v_{k-1})_{j-1}^{n+1}\right\}\right.$$

$$\left.-\left\{(v_{k+1}-v_{k-1})_{j+1}^{n}-(v_{k+1}-v_{k-1})_{j-1}^{n}\right\}\right]=0$$

$$\frac{(h^{n+1}-h^n)_{j,k}}{\Delta t}+h_0\left[\frac{(u_{j+1}-u_{j-1})_k^{n+1}}{4\Delta x}+\frac{(u_{j+1}-u_{j-1})_k^{n}}{4\Delta x}\right.$$

$$\left.+\frac{(v_{k+1}-v_{k-1})_j^{n+1}}{4\Delta y}+\frac{(v_{k+1}-v_{k-1})_j^{n}}{4\Delta y}\right]=0$$

$$\frac{(v^{n+1}-v^n)_{j,k}}{\Delta t}+g\left[\frac{(h_{k+1}-h_{k-1})_j^{n+1}}{4\Delta y}+\frac{(h_{k+1}-h_{k-1})_j^{n}}{4\Delta y}\right]=0$$

This scheme is then identical to the Leendertse scheme, again differing from the ideal scheme only through a term

$$¢ \; \frac{gh_0\Delta t^2}{4}\left(\frac{\partial^3 v}{\partial x \partial y \partial t}\right)$$

in the x-Euler equation. This term is then of second order in the truncation error.

4) Linearized S21*

$$(p_{j,k}^{n+1} - p_{j,k}^{n}) + \frac{gh\Delta t}{4\Delta x}\left[(h_{j+1} - h_{j-1})^{n+1} + (h_{j+1} - h_{j-1})^{n}\right]_k$$

$$- \frac{(gh)^2\Delta t^3}{64\Delta y^2\Delta x}\left[\left\{(h_{k+2} - 2h_k + h_{k-2})_{j+1}\right.\right.$$

$$\left. - (h_{k+2} - 2h_k + h_{k-2})_{j-1}\right\}^{n+1}$$

$$\left. + \left\{(h_{k+2} - 2h_k + h_{k-2})_{j+1} - (h_{k+2} - 2h_k + h_{k-2})_{j-1}\right\}^{n}\right] = 0$$

$$(h_{j,k}^{n+1} - h_{j,k}^{n}) + \frac{\Delta t}{4\Delta x}\left\{(p_{j+1} - p_{j-1})^{n+1} + (p_{j+1} - p_{j-1})^{n}\right\}_k$$

$$+ \frac{\Delta t}{4\Delta y}\left\{(q_{k+1} - q_{k-1})^{n+1} + (q_{k+1} - q_{k-1})^{n}\right\}_j$$

$$+ \frac{gh\Delta t^2}{16\Delta y^2}\left\{(h_{k+2} - 2h_k + h_{k-2})^{n+1}\right.$$

$$\left. + (h_{k+2} + 2h_k + h_{k-2})^{n}\right\}_j = 0$$

*This form follows when the q-values are readdressed $\Delta t/2$ back in time. Such a change in notation alters neither the algebraic nor the algorithmic properties of the scheme.

$$(q_{j,k}^{n+1} - q_{j,k}^{n}) + \frac{(gh)\Delta t}{2\Delta y}\,(h_{k+1} - h_{k-1})_j^{n+1} = 0$$

As set up here, the two-level scheme has extra terms on x-momentum and mass, as compared with the ideal scheme. This extra truncation error can, of course, be redistributed in order to restore $x-y$ symmetry.

3. CHARACTERIZATION OF SCHEMES THROUGH THEIR PROPAGATION PROPERTIES

1) The ideal scheme has the following Fourier transform, for $\Delta x = \Delta y = \Delta s$

$$\begin{bmatrix} (\phi-1) & \frac{i\cdot gh\Delta t}{2\Delta s}(\phi+1)\sin\sigma_1\Delta s & 0 \\ \frac{i\Delta t}{2\Delta s}(\phi+1)\sin\sigma_1\Delta s & (\phi-1) & \frac{i\Delta t}{2\Delta s}(\phi+1)\sin\sigma_2\Delta s \\ 0 & \frac{igh\Delta t}{2\Delta s}(\phi+1)\sin\sigma_2\Delta s & (\phi-1) \end{bmatrix} \begin{bmatrix} p^* \\ h^* \\ q^* \end{bmatrix} = 0$$

where ϕ is a complex amplification factor. When ϕ is 1, the determinant of the above matrix is 0, so that ϕ_1 is 1. Introduce

$$\frac{gh\cdot\Delta t^2}{\Delta s^2}\sin^2\sigma_1 = \alpha^2\,, \quad \frac{gh\cdot\Delta t^2}{\Delta s^2}\sin^2\sigma_2 = \beta^2$$

Then the transform determinant reads

$$(\phi-1)^3 + (\phi-1) - (\phi+1)^2\beta^2 + (\phi+1)^2\alpha^2(\phi-1) = 0$$

or

$$(\phi-1)^2 + (\phi+1)^2\,\frac{(\alpha^2 + \beta^2)}{4} = 0$$

giving

$$\left(\frac{\phi-1}{\phi+1}\right)^2 = -\,\frac{(\alpha^2 + \beta^2)}{4}$$

or

$$\left(\frac{\phi-1}{\phi+1}\right) = \pm\, i\,\frac{(\alpha^2 + \beta^2)^{1/2}}{2}$$

Writing

$$\frac{(\alpha^2 + \beta^2)^{1/2}}{2} = A$$

we find

$$\phi_2 = \frac{1 - iA}{1 + iA}, \quad \phi_3 = \frac{1 + iA}{1 - iA}$$

i.e.,

$$\phi_2 = \frac{(1 - iA)^2}{1 + A^2} = \frac{(1 - A^2) - i2A}{1 + A^2}$$

$$\phi_3 = \frac{(1 + iA)^2}{1 + A^2} = \frac{(1 - A^2) + i2A}{1 + A^2}$$

Then the amplification factor is 1 for all wave numbers and the celerity ratio Q is given by Abbott (1979, p. 164) as

$$Q = \frac{\text{Arc tan}(\text{Im}(\phi)/\text{Re}(\phi)}{2\pi C_r/N_s}$$

Thence

$$Q(\phi_2) = \frac{\text{Arc tan}(-2A/1-A^2)}{2C_r/N_s}$$

with (reverting for a moment to the general case of independent Δx and Δy)

$$A = \left\{(C_r/2)^2 \sin^2 \frac{2\pi k_x \Delta x}{2l} - (C_r/2)^2 \sin^2 \frac{2\pi k_y \Delta y}{2l}\right\}^{1/2}$$

$$= (C_r/2)\left\{\sin^2 \frac{2\pi}{N_x} + \sin^2 \frac{2\pi}{N_y}\right\}^{1/2}$$

Thus, when propagation is along a grid line (say, an x-line)

$$A = C_r/2 \sin \frac{2\pi}{N_x}$$

while, when propagation is at 45° to grid lines

$$A = C_r/2 \ (2 \sin^2 \frac{2\pi}{N_s})^{1/2} = \sqrt{2} \cdot C_r/2 \sin \frac{2\pi}{N_s}$$

where C_r is corrected to $\sqrt{2 \cdot C_r}$, corresponding to the real celerity if propagation is along a line at 45° to the grid. The celerity ratios for other angles to the grid lines are bounded by these two values, for each and every wave number. The phase portraits are graphed in Figs. 1 and 2.

2, 3) For the Leendertse (1967) and Abbott (1968) schemes the Fourier transform determinant now becomes

$$\begin{vmatrix} (\phi-1) & i(\frac{g\Delta t}{2\Delta s}\sin\sigma_1\Delta s(\phi+1)) & -(\frac{gh_0\Delta t^2}{4\Delta s}\sin\sigma_1\Delta s\sin\sigma_2\Delta s(\phi-1)) \\ i(\frac{h_0\Delta t}{2\Delta s}\sin\sigma_1\Delta s(\phi+1)) & (\phi-1) & i(\frac{h_0\Delta t}{2\Delta s}\sin\sigma_2(\phi+1)) \\ 0 & i(\frac{g\Delta t}{2\Delta s}\sin\sigma_2\Delta s(\phi+1)) & (\phi-1) \end{vmatrix} = 0$$

We again introduce

$$\frac{gh_0\Delta t^2}{\Delta s}\sin^2\sigma_1 = \alpha^2, \quad \frac{gh_0\Delta t^2}{\Delta s}\sin^2\sigma_2 = \beta^2$$

Then the transform determinant becomes

$$(\phi-1)^3 + (\phi-1)(\phi+1)^2\frac{\beta^2}{4} + (\phi-1)(\phi+1)^2\frac{\alpha^2}{4} + (\phi-1)(\phi+1)^2\frac{\alpha^2\beta^2}{16} = 0$$

or

$$(\phi-1)\left[(\phi-1)^2 + (\phi+1)^2(\frac{\alpha^2}{4} + \frac{\alpha^2\beta^2}{16} + \frac{\beta^2}{4})\right] = 0$$

Thence

$$\phi_1 = 1$$

and

$$(\phi-1)^2 + (\phi+1)^2A^2 = 0$$

with

$$A^2 = \left[\frac{\alpha^2}{4} + \frac{\alpha^2\beta^2}{16} + \frac{\beta^2}{4}\right]$$

We find again

$$\phi_2 = \frac{1+iA}{1-iA} \qquad \phi_3 = \frac{1-iA}{1+iA} \quad (= \frac{1}{\phi_2})$$

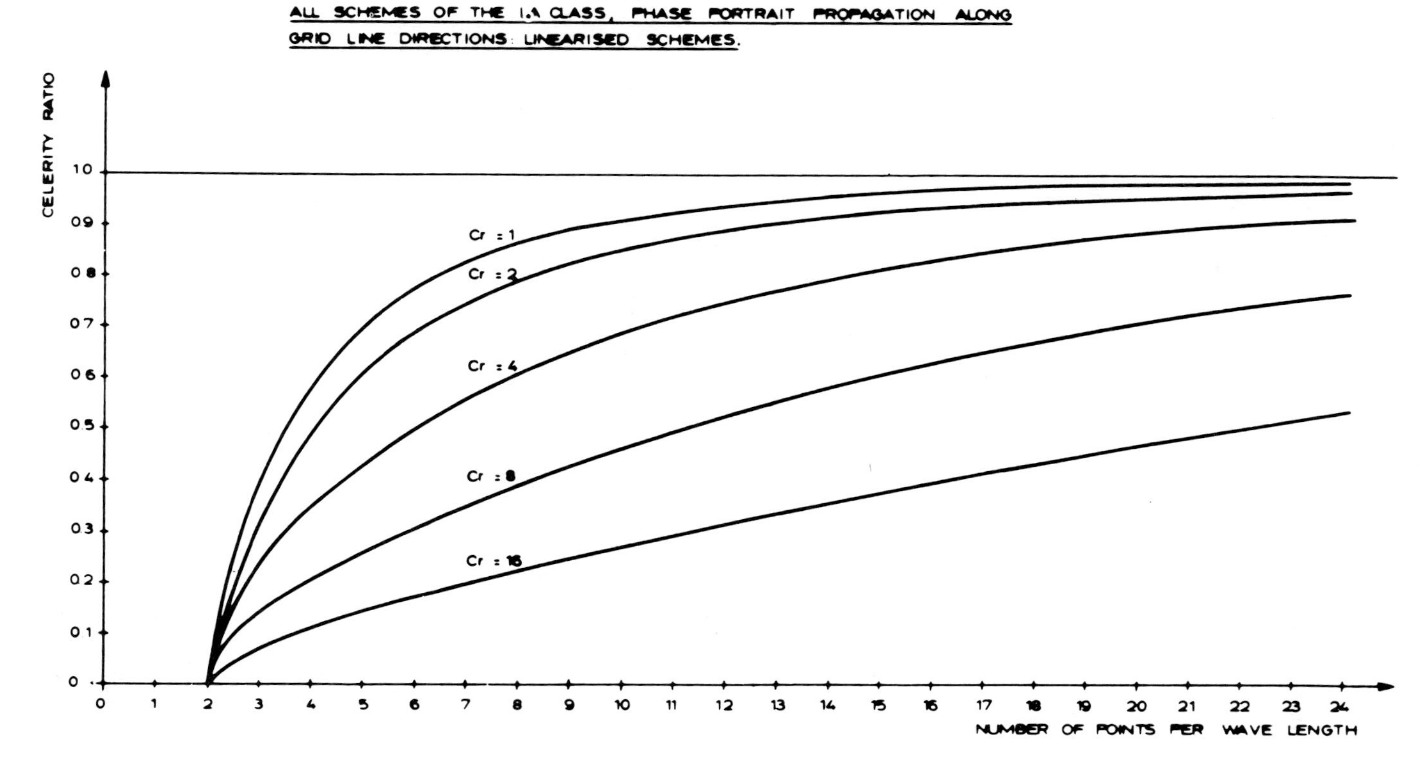

Figure 1. Phase portrait for propagation along grid line directions, linearized ideal scheme, and all schemes of that class.

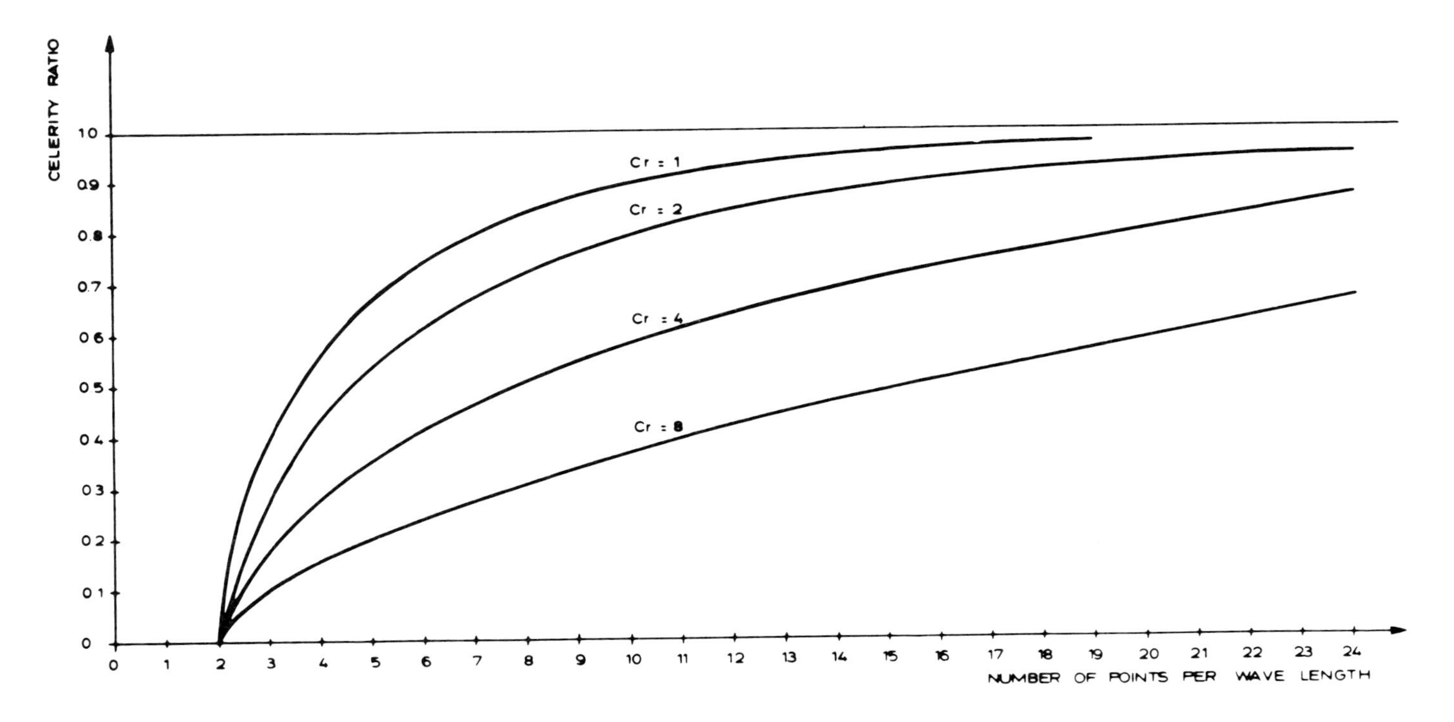

Figure 2. Phase portrait for propagation along directions at 45° to grid lines, linearized ideal scheme.

Then

$$| \phi_2 | = | \phi_3 | = 1$$

and

$$\phi_2 = \frac{(1 - A^2) + i2A}{1 + A^2}, \quad \phi_3 = \frac{(1 - A^2) - i2A}{1 + A^2}$$

For propagation along the grid direction, say, the x-direction, then $\beta = 0$ and $A^2 = \alpha^2/4$ or $A = \alpha/2 = C_r/2 \sin 2\pi/N_x$ and we verify only that the phase portrait is that of the class of schemes. (See Figure 1.)

For propagation along a line at 45° to the grid direction, $\beta = 0$ and

$$A^2 = \frac{\alpha^2}{2}(1 + \frac{\alpha^2}{8}) \quad \textit{or} \quad A = \frac{\alpha}{\sqrt{2}}(1 + \frac{\alpha^2}{8})^{1/2}$$

The celerity ratio then becomes

$$Q = \frac{\arctan(\frac{2A}{1-A^2})}{\frac{\sqrt{2} \cdot 2\pi C_r}{N_x}}$$

with

$$A = \frac{\alpha}{\sqrt{2}}(1+\frac{\alpha^2}{8})^{1/2}$$

and

$$\alpha = C_r \sin\frac{2\pi}{N_x}$$

The corresponding phase portrait is given in Fig. 3.

4) The S21 two-level form provides the Fourier transform

$$\begin{bmatrix} (\phi-1) & \begin{matrix}(\frac{gh_0\Delta}{4\Delta x})(2i\sin\sigma_1\Delta x)(\phi+1) \\ -(\frac{(gh_0)^2\Delta t^3}{64\Delta y^2\Delta x})(-4\sin^2\sigma_2\Delta y)(2i\sin\sigma_1\Delta x)(\phi+1)\end{matrix} & 0 \\ \frac{\Delta t}{4\Delta x}(2i\sin\sigma_1\Delta x)(\phi+1) & (\phi-1)+\frac{gh_0\Delta t^2}{16\Delta y^2}(-4\sin^2\sigma_2\Delta y)(\phi+1) & \frac{\Delta t}{4\Delta y}(2i\sin\sigma_2\Delta y)(\phi+1) \\ 0 & \frac{gh_0\Delta t}{2\Delta y}(2i\sin\sigma_2\Delta y)\phi & (\phi-1) \end{bmatrix} \begin{bmatrix} p^* \\ h^* \\ q^* \end{bmatrix} = 0$$

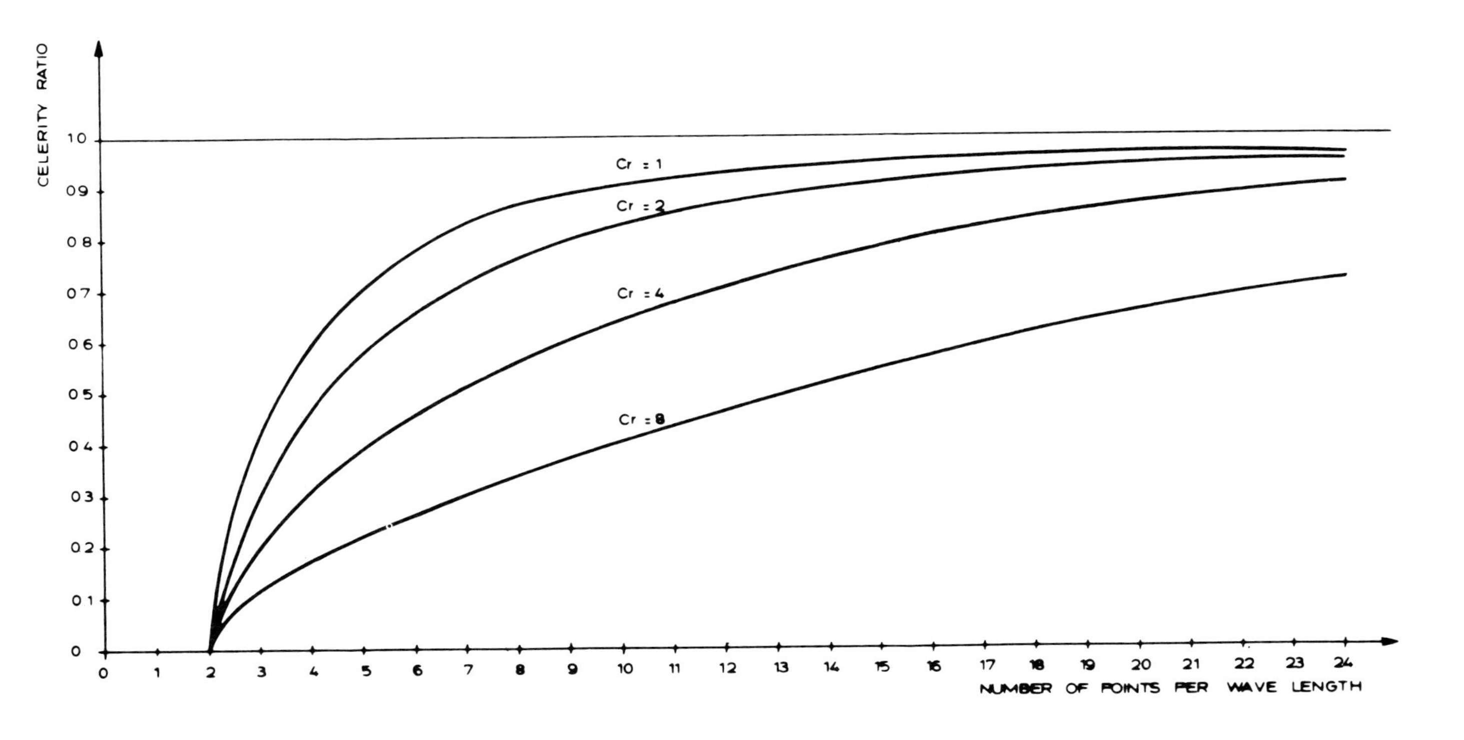

Figure 3. Phase portrait for propagation along directions at 45° to grid lines linearized. Leendertse (1967), Abbott (1968), and S21 schemes.

The transform determinant is

$$(\phi-1)\left[\phi-1\{(\phi-1) - (\frac{gh\Delta t^2}{16\Delta y^2}\,4\sin^2\sigma_2\Delta y)(\phi+1)\}\right.$$
$$\left. + \frac{gh\Delta t^2}{8\Delta y^2}\cdot 4\sin^2\sigma_2\Delta y \cdot (\phi+1)\phi\right]$$
$$+ (\phi-1)\left[+ (\frac{gh\Delta t^2}{16\Delta x^2}4\sin^2\sigma_1\Delta x)\right.$$
$$\left. + (\frac{(gh)^2\Delta t^2}{256\Delta x^2\Delta y^2}\cdot 4\sin^2\sigma_1\Delta x \cdot 4\sin^2\sigma_2\Delta y)\right](\phi+1) = 0$$

Setting

$$\frac{gh\Delta t^2}{\Delta x^2}\sin^2\sigma_1\Delta x = \alpha^2$$
$$\frac{gh\Delta t^2}{\Delta y^2}\sin^2\sigma_2\Delta y = \beta^2$$
$$A^2 = \frac{\alpha^2}{4} + \frac{\alpha^2\beta^2}{16} + \frac{\beta^2}{4}$$

we find

$$\phi^2(1+A^2) + \phi 2(-1+A^2) + (1+A^2) = 0$$

so that

$$\phi^2 + \frac{2(1-A^2)}{1+A^2}\,\phi + 1 = 0$$

or

$$\phi = -\left(\frac{A^2-1}{A^2+2}\right) \pm \sqrt{\left(\frac{A^2-1}{A^2+1}\right)^2 - 1}$$

We see that, since A is always real, both amplification factor moduli are always 1 and

$$\frac{\mathrm{Im}(\phi)}{\mathrm{Re}(\phi)} = \pm\,\frac{i2A}{1-A^2}$$

exactly as for schemes (2) and (3).

By virtue of the definition of the class we know *a priori* that all four schemes considered will have identical propagation characteristics along the grid directions but we now also find that all three algorithmically tractable schemes have identical propagation characteristics for any given angle to the direction of the grid. Thus, from the point of view of their propagation characteristics, the algorithmically tractable schemes considered here are indistinguishable. Of (2) and (3) it can be said that one is just a rewriting of the other, in that they have identical two-level forms, but this equivalence is by no means so clear in the case of (4).

4. CHARACTERICATION OF SCHEMES THROUGH THEIR VORTICITY PROPERTIES

Consider now the behavior of the discrete vorticity in the case of the four representative schemes. It can be shown that the scaling factor existing between the discrete vorticity and the continuum vorticity can be entirely accounted for by the conventional truncation error.*

4.1 Ideal Scheme. The x-Euler equations

$$\frac{(u_j^{n+1} - u_j^n)_{k+1}}{\Delta t} + g\left[\frac{(h_{j+1} - h_{j-1})_{k\pm1}^{n+1}}{4\Delta x} + \frac{(h_{j+1} - h_{j-1})_{k\pm1}^{n}}{4\Delta x}\right] = 0$$

imply that

$$\left[\frac{(u_{k+1} - u_{k-1})_j^{n+1} - (u_{k+1} - u_{k-1})_j^n}{2\Delta y \Delta t}\right] + g\left[\frac{(h_{j+1} - h_{j-1})_{k+1}^{n+1} - (h_{j+1} - h_{j-1})_{k-1}^{n+1}}{8\Delta x \Delta y} + \frac{(h_{j+1} - h_{j-1})_{k+1}^{n} - (h_{j+1} - h_{j-1})_{k-1}^{n}}{8\Delta x \Delta y}\right] = 0$$

Similarly the y-Euler equations

$$\frac{(v_k^{n+1} - v_k^n)_{j\pm1}}{\Delta t} + g\left[\frac{(h_{k+1} - h_{k-1})_{j\pm1}^{n+1}}{4\Delta y} + \frac{(h_{k+1} - h_{k-1})_{j\pm1}^{n}}{4\Delta y}\right] = 0$$

imply that

$$\left[\frac{(v_{j+1} - v_{j-1})_k^{n+1} - (v_{j+1} - v_{j-1})_k^n}{2\Delta x \Delta t}\right]$$

* *Editor's note:* A proof of this statement is available from the authors, but was deleted from this volume due to space requirements.

$$+ g \left[\frac{(h_{j+1} - h_{j-1})_{k+1}^{n+1} - (h_{j+1} - h_{j-1})_{k-1}^{n+1}}{8\Delta x \Delta y} + \frac{(h_{j+1} - h_{j-1})_{k+1}^{n} - (h_{j+1} - h_{j-1})_{k-1}^{n}}{8\Delta x \Delta y} \right] = 0$$

Thence, by subtraction,

$$\frac{(u_{k+1} - u_{k-1})_j^{n+1}}{2\Delta y} - \frac{(v_{j+1} - v_{j-1})_k^{k+1}}{2\Delta x} = \frac{(u_{k+1} - u_{k-1})_j^{n}}{2\Delta y} - \frac{(v_{j+1} - vj-1)_k^{n}}{2\Delta x}$$

so that

$$\omega_{j,k}^{n+1} = \omega_{j,k}^{n}$$

We observe that this result remains unaltered if $\partial(h+H)/\partial x$ is used instead of $\partial h/\partial x$ and $\partial(h+H)/\partial y$ is used instead of $\partial h/\partial y$, so as to describe a variable bed elevation H that is measured, positive upwards, from a reference horizontal plane.

2, 3) For the Leendertse (1967) and Abbott (1968) schemes we expand the Euler equations to

$$\frac{(u_{k+1} - u_{k-1})_j^{n+1}}{4\Delta s} - \frac{(u_{k+1} - u_{k-1})_j^{n}}{4\Delta s} + \frac{g\Delta t}{4\Delta s} \left[\left\{ \frac{(h_{j+1} - h_{j-1})_{k+1}^{n+1}}{4\Delta s} - \frac{(h_{j+1} - h_{j-1})_{k-1}^{n+1}}{4\Delta s} \right\} + \left\{ \frac{(h_{j+1} - h_{j-1})_{k+1}^{n}}{4\Delta s} - \frac{(h_{j+1} - h_{j-1})_{k-1}^{n}}{4\Delta s} \right\} \right]$$

$$+ \frac{gh\Delta t^2}{32\Delta s^3} \left[\left\{ (v_{k+2} - 2v_k + v_{k-2})_{j+1}^{n+1} - (v_{k+2} - 2v_k + v_{k-2})_{j-1}^{n+1} \right\} - \left\{ (v_{k+2} - 2v_k + v_{k-2})_{j+1}^{n} - (v_{k+2} - 2v_k + v_{k-2})_{j-1}^{n} \right\} \right] = 0$$

and

$$\frac{(v_{j+1} - v_{j-1})_k^{n+1}}{4\Delta s} - \frac{v_{j-1} - v_{j-1})_k^{n}}{4\Delta s}$$

$$\frac{g\Delta t}{4\Delta s}\left[\left\{\frac{(h_{k+1}-h_{k-1})_{j+1}^{n+1}}{4\Delta s}-\frac{(h_{k+1}-h_{k-1})_{j-1}^{n+1}}{4\Delta s}\right\}\right.$$
$$\left.+\left\{\frac{(h_{k+1}-h_{k-1})_{j+1}^{n}}{4\Delta s}-\frac{(h_{k+1}-h_{k-1})_{j-1}^{n}}{4\Delta x}\right\}\right]=0$$

to form the discrete vorticity relation

$$\omega_{j,k}^{n+1}=\omega_{j,k}^{n}-\frac{gh\Delta t^2}{32\Delta s^3}\left[\left\{(v_{k+2}-2v_k+v_{k-2})_{j+1}^{n+1}\right.\right.$$
$$\left.-(v_{k+2}-2v_k+v_{k-2})_{j-1}^{n+1}\right\}-\left\{(v_{k+2}-2v_k+v_{k-2})_{j+1}^{n}\right.$$
$$\left.\left.-(v_{k+2}-2v_k+v_{k-2})_{j-1}^{n}\right\}\right]=0$$

or

$$\omega_{j,k}^{n+1}=\omega_{j,k}^{n}-0\left(\frac{gh\Delta t^3}{4}\frac{\partial^4 v}{\partial x\partial y^2\partial t}\right)$$

With the truncation error shifted to the y-Euler equation this relation would then become, by symmetry

$$\omega_{j,k}^{n+1}=\omega_{j,k}^{n}+0\left(\frac{gh\Delta t^3}{4}\frac{\partial^4 u}{\partial x^2\partial y\partial t}\right)$$

and on average

$$\omega_{j,k}^{n+1}=\omega_{j,k}^{n}+0\left[\frac{gh\Delta t^3}{4}\frac{\partial^3}{\partial x\partial y\partial t}\left(\frac{\partial u}{\partial x}-\frac{\partial v}{\partial y}\right)\right]$$

Introducing $(h+H)$ instead of h now has the effect of changing the size of the truncation error. .

4) S 21 Mark 6 Scheme. We find

$$\left(\frac{(p_{k+1}-p_{k-1})_j^{n+1}}{2\Delta y}-\frac{(p_{k+1}-p_{k-1})_j^{n}}{2\Delta y}\right)$$
$$+\frac{gh\Delta t}{8\Delta x\Delta y}\left[\left\{(h_{j+1}-h_{j-1})_{k+1}-(h_{j+1}-h_{j-1})_{k-1}\right\}^{n+1}\right.$$

$$+ \left\{(h_{j+1} - h_{j-1})_{k+1} - (h_{j+1} - h_{j-1})_{k-1}\right\}^n\Bigg]$$

$$- \frac{(gh)^2\Delta t^3}{128\Delta y^3 \Delta x}\Bigg[\Bigg\{(h_{k+3} - 3h_{k+1} + 3h_{k-1} - h_{k-3})_{j+1}$$

$$- (h_{k+3} - 3h_{k+1} + 3h_{k-1} - h_{k-3})_{j-1}\Bigg\}^{n+1}$$

$$- \Bigg\{(h_{k+3} - 3h_{k+1} + 3h_{k-1} - h_{k-3})_{j+1}$$

$$- (h_{k+1} - 3h_{k+1} + 3h_{k-1} - h_{k-3})_{j-1}\Bigg\}^n\Bigg] = 0$$

and

$$\left\{\frac{(q_{j+1} - q_{j-1})_k^{n+1}}{2\Delta x} - \frac{(q_{j+1} - q_{j-1})_k^n}{2\Delta x}\right\}$$

$$+ \frac{gh\Delta t}{4\Delta x \Delta y}\left\{(h_{k+1} - h_{k-1})_{j+1} - (h_{k+1} - h_{k-1})_{j-1}\right\}^{n+1} = 0$$

so that, when expressed in terms of the discrete vorticity ω, the change over one time step is

$$h\omega_{j,k}^{n+1} = h\omega_{j,k}^n - u\left(\frac{(h_{k+1} - h_{k-1})_j^{n+1}}{2\Delta y} - \frac{(h_{k+1} - h_{k-1})^n}{2\Delta y}\right)$$

$$+ v\left(\frac{(h_{j+1} - h_{j-1})_k^{n+1}}{2\Delta x} - \frac{(h_{j+1} - h_{j-1})_k^n}{2\Delta x}\right)$$

$$+ \frac{(gh)^2\Delta t^3}{128\Delta y^3\Delta x}\Bigg[\Bigg\{(h_{k+3} - 3h_{k+1} + 3h_{k+1} - h_{k-3})_{j+1}$$

$$- (h_{k+3} - 3h_{k+1} + 3h_{k-1} - h_{k-3})_{j-1}\Bigg\}^{n+1}$$

$$+ (h_{k+3} - 3h_{k+1} + 3h_{k-1} - h_{k-3})_{j-1}\Bigg\}^n\Bigg]$$

$$- (h_{k+3} - 3h_{k+1} + 3h_{k-1} - h_{k-3})_{j-1}\Bigg\}^n\Bigg]$$

$$+ \frac{gh\Delta t}{8\Delta x \Delta y}\left[\left\{(h_{k+1} - h_{k-1})_{j+1} - (h_{k+1} - h_{k-1})_{j-1}\right\}^{n+1} - \left\{(h_{k+1} - h_{k-1})_{j+1} - (h_{k+1} - h_{k-1})_{j-1}\right\}^{n}\right]$$

so that

$$\omega_{j,k}^{n+1} = \omega_{j,k}^{n} + O\left(-\frac{u\Delta t}{h}\frac{\partial^2 h}{\partial y \partial t} + \frac{v\Delta t}{h}\frac{\partial^2 h}{\partial x \partial t} + \frac{(gh)^2 \Delta t^3}{4h}\frac{\partial^4 h}{\partial x \partial y^3} + \frac{g\Delta t^2}{2}\frac{\partial^3 h}{\partial x \partial y \partial t}\right)$$

If, however, we use the depth-integrated discrete vorticity Ω as a measure, so that

$$\Omega = \frac{(p_{k+1} - p_{k-1})_j^n}{2\Delta y} - \frac{(q_{j+1} - q_{j-1})_k^n}{2\Delta x}$$

we obtain

$$\Omega_{j,k}^{n+1} = \Omega_{j,k}^{n} + 0\left(\frac{(gh)^2 h}{\partial x \partial y^3} + \frac{gh\Delta t^2}{2}\frac{\partial^3 h}{\partial x \partial y \partial t}\right)$$

Shifting the truncation error to the y-Euler equation would then give

$$\Omega_{j,k}^{n+1} = \Omega_{j,k}^{n} - 0\left(\frac{(gh)^2 \Delta t^3}{4}\frac{\partial^4 h}{\partial x^3 \partial y} + \frac{gh\Delta t^2}{2}\frac{\partial^3 h}{\partial x \partial y \partial t}\right)$$

and on average

$$\Omega_{j,k}^{n+1} = \Omega_{j,k}^{n} + 0\left(\frac{(gh)\Delta t^3}{4}\frac{\partial^3}{\partial x \partial y \partial t}\left(\frac{\partial p}{\partial x} - \frac{\partial q}{\partial y}\right)\right)$$

Introducing $(h+H)$ instead of h again changes the size of the truncation error whereby, if rapid changes of bathymetry occur, this error can become substantial.

5. THE INFLUENCE OF THE CONVECTIVE TERMS ON THE PHASE PORTRAIT: THE CENTERED ONE-DIMENSIONAL DESCENT

The eigenvalues of the amplification matrix are given in this case of one-dimensional nearly-horizontal flow, with convective terms retained, by Abbott (1979, p. 180), as follows

$$\lambda = \frac{a^2 - (ib \pm c)^2}{(a + ib)^2 - c^2} = \frac{(a \pm c) - ib}{(a \pm c) + ib}$$

We first take the -ive case, so that

$$\lambda = \frac{(a - c) - ib}{(a - c) + ib}$$

where we use

$$a = \frac{u\Delta t}{\Delta x} \cdot \cos\left(\frac{2\pi}{N_x}\right)$$

$$b = \frac{(u^2 + gh)\Delta t^2}{2\Delta x} \cdot \sin\left(\frac{2\pi}{N_x}\right)$$

$$c = -\frac{(gh)^{1/2}\Delta t}{\Delta x}$$

or

$$a = Fr.\bar{C}r.\cos\left(\frac{2\pi}{N_x}\right)$$

$$b = (Fr^2 - 1)\frac{\bar{C}r^2}{2} \cdot \sin\left(\frac{2\pi}{N_x}\right)$$

$$c = -\bar{C}r$$

where

$$\bar{C}r = \frac{(gh)^{1/2}\Delta t}{\Delta x} \cdot \textit{the } \text{"Mean Courant Number"}$$

Since we can write

$$\lambda = \frac{(a - c)^2 - i2b(a - c) - b^2}{(a - c)^2 + b^2}$$

we have

$$-\frac{\mathrm{Im}(\lambda)}{\mathrm{Re}(\lambda)} = \frac{2(a-c)b}{(a-c)^2 - b^2}$$

$$= \frac{[(Fr\cdot\cos(2\pi/N_x)+1)\ \bar{C}r\ ]\ [(Fr^2-1)\,\bar{C}r^2\sin(2\pi/N_x)]}{(Fr\cdot\cos(2\pi/N_x)+1)^2\bar{C}r^2 - (Fr^2-1)^2\,\dfrac{\bar{C}r^4}{4}\,\sin^2(2\pi/N_x)}$$

When $Fr = 0$

$$-\frac{\mathrm{Im}(\lambda)}{\mathrm{Re}(\lambda)} = \frac{\bar{C}r\ [(-1)\,\bar{C}r^2\sin(2\pi/N_x)]}{Cr^2 - \dfrac{\bar{C}r^4}{4}\,\sin^2(2\pi/N_x)} = \frac{-\bar{C}r\cdot\sin(2\pi/N_x)}{1-\dfrac{\bar{C}r^2}{4}\sin^2(2\pi/N_x)}$$

as obtained earlier, this being the linear case.

More generally

$$-\frac{\mathrm{Im}(\lambda)}{\mathrm{Re}(\lambda)} = \frac{2\alpha\beta}{\alpha^2-\beta^2}$$

with

$$\alpha = (Fr\cdot\cos(2\pi/N_x)+1)\,\bar{C}r$$
$$\beta = (Fr^2-1)\,(\bar{C}r^2/2)\cdot\sin(2\pi/N_x)$$

We next take the positive case, so that

$$\lambda = \frac{(a+c)-ib}{(a+c)+ib}$$

and

$$-\frac{\mathrm{Im}(\lambda)}{\mathrm{Re}(\lambda)} = \frac{2(a+c)b}{(a+c)-b} = \frac{2\alpha\beta}{\alpha^2+\beta^2}$$

with

$$\alpha = (Fr\cdot\cos(2\pi/N_x)-1)\,\bar{C}r$$
$$\beta = (Fr^2-1)\,\frac{\bar{C}r^2}{2}\,\sin(2\pi/N_x)$$

Some typical phase portraits are given in Figure 4.

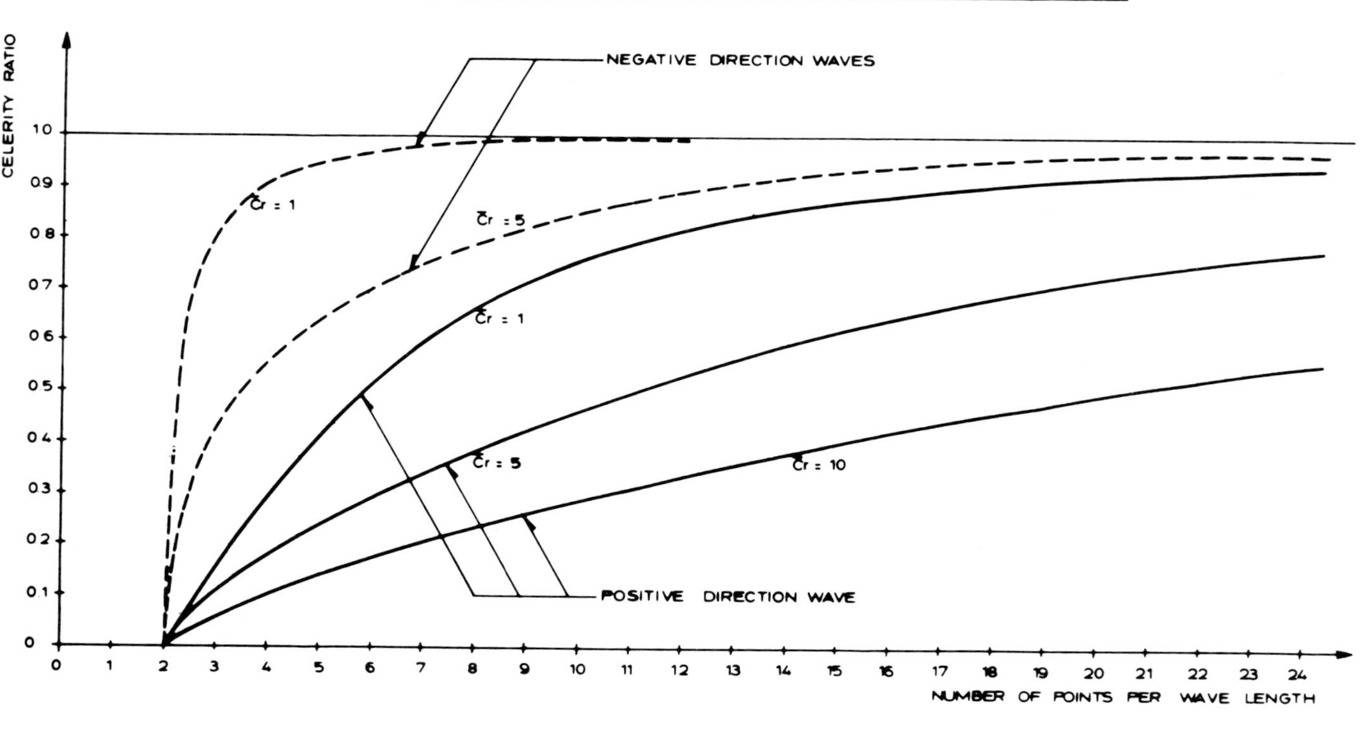

Figure 4. Phase portrait for propagation along grid line directions, when *Fr.* = 0.5 and flow is in the positive direction. Centered one-dimensional descent.

6. INTEGRAL REPRESENTATIONS OF THE CONVECTIVE TERMS OF THE EULER EQUATIONS

Consider the primitive x-Euler equation

$$\frac{\partial u}{\partial t} + \frac{\partial u}{\partial x} + v\,\frac{\partial u}{\partial y} = -g\,\frac{\partial h}{\partial x} \tag{6.1}$$

The left-hand side represents the transport of the x-velocity measure u with the resultant of the x- and y-velocities. The integral form of the transport equation, which corresponds to an exact solution of the differential form, is

$$u(x,y,t_2) = u\left(x-\int_{t_1}^{t_2} \bar{u}dt\ ,\ y - \int_{t_1}^{t_2} \bar{v}dt,\ t_1\right)$$

where, then, the filtered velocities $\bar{u}$ and $\bar{v}$ are given by

$$\bar{u} = \frac{1}{t_2-t_1}\int_{t_1}^{t_2} udt \quad \text{and} \quad \bar{v} = \frac{1}{t_2-t_1}\int_{t_1}^{t_2} vdt$$

This may be written, for one time step Δt, as

$$u(j\Delta x,k\Delta y,(n+1)\Delta t) = u(j\Delta x - \bar{u}\Delta t,k\Delta y - \bar{v}\Delta t,n\Delta t) \tag{6.2}$$

or

$$u_{j,k}^{n+1} = u_{j-Cr_x,k-Cr_y}$$

where

$$Cr_x = \frac{\bar{u}\Delta t}{\Delta x}\ ,\quad Cr_y = \frac{\bar{v}\Delta t}{\Delta x}$$

Equation (6.2) can as well be written as

$$u_{j,k}^n + \left(\frac{\partial u}{\partial t}\right)_{j,k}^n \Delta t + \left(\frac{\partial^2 u}{\partial t^2}\right)_{j,k}^n \frac{\Delta t^2}{2!} = u_{j,k}^n - \left(\frac{\partial u}{\partial x}\right)_{j,k}^n (\bar{u}\Delta t)$$
$$- \left(\frac{\partial u}{\partial y}\right)_{j,k}^n (\bar{v}\Delta t) + \left(\frac{\partial^2 u}{\partial x^2}\right)_{j,k}^n \left(\frac{\bar{u}\Delta t}{2!}\right)^2$$

$$+ 2\left(\frac{\partial^2 u}{\partial x \partial y}\right) \frac{\overline{uv}(\Delta t)^2}{2!}$$

$$+ \left(\frac{\partial^2 u}{\partial y^2}\right) \frac{(v\Delta t)^2}{2!} + \text{H.O.T.}$$

or

$$\left(\frac{\partial u}{\partial t} + \bar{u}\frac{\partial u}{\partial x} + \bar{v}\frac{\partial u}{\partial y}\right)^n_{j,k} - \left(\frac{-\partial^2 u}{\partial t^2} \cdot \frac{\Delta t}{2} + \frac{\partial^2 u}{\partial x^2} \cdot \frac{\bar{u}^2 \Delta t}{2}\right.$$
$$\left. - \frac{\partial^2 u}{\partial x \partial y} \overline{uv}\Delta t + \frac{\partial^2 u}{\partial y} \cdot \frac{\bar{v}^2 \Delta t}{2}\right)^n_{j,k}$$
$$= \text{H.O.T.} \tag{6.3}$$

It is then clear that, as $\Delta t \to 0$, so $\bar{u} \to u$, $\bar{v} \to v$ and the differential form (6.1) is retained. In a difference scheme, which comprehends the integration of a differential equation over a finite time Δt, there is no purpose in discretizing the left-hand bracketed quantity beyond second order accuracy unless the right-hand bracketed quantity is also included in the discretization. Thus, in a third-order accurate scheme, for example, the entire right-hand side of (6.3) must be represented. When the representation is centered at $(n+1)\Delta t$, (6.3) becomes

$$\left(\frac{\partial u}{\partial t} + \frac{\partial u}{\partial x} + \bar{v}\frac{\partial u}{\partial y}\right)^{n+1}_{j,k} - \left(\frac{\partial^2 u}{\partial t^2} \frac{\Delta t}{2} + \frac{\partial^2 u}{\partial x^2} \frac{\bar{u}^2 \Delta t}{2}\right.$$
$$\left. + \frac{\partial^2 u}{\partial x \partial y} \cdot \overline{uv}\Delta t + \frac{\partial^2 u}{\partial x \partial t} \bar{u}\Delta t + \frac{\partial^2 u}{\partial y \partial t} \bar{v}\Delta t\right)^{n+1}_{j,k}$$
$$= H.O.T. \tag{6.4}$$

so that in this case there are six "correction terms" in the second bracket. The terms of second and higher order in (6.3) and (6.4) are obtained simply through the centering of the finite scheme and, accordingly, must vary as the centering varies. In particular, if the scheme is centered at $(n+\frac{1}{2})\Delta t$, the even-order terms are eliminated.

The second space-derivatives in the second bracketed terms of (6.3) and (6.4) have the effect of introducing stresses in the fluid system, and these stresses may introduce a damping or an excitation of the system, depending upon the signs of their coefficients.

The first two of these second-derivative correction terms, which were investigated by C. H. Rassmussen using a quite other methodology, are

$$\frac{\partial^2 u}{\partial x^2} \cdot \frac{\bar{u}\Delta t}{2} \quad \text{and} \quad \frac{\partial^2 u}{\partial y^2} \cdot \frac{v\Delta t}{2}$$

These may be either dissipative or destabilizing depending upon the direction of the velocities. They are easily introduced into the double-sweep algorithm of a scheme of the given class in such a way as to allow $u_{j,k}$ to influence the correction, as shown in Figure 5.a. Such an introduction corrects the truncation error for the two quantities, allowing for a better approximation to the integral term (6.2)

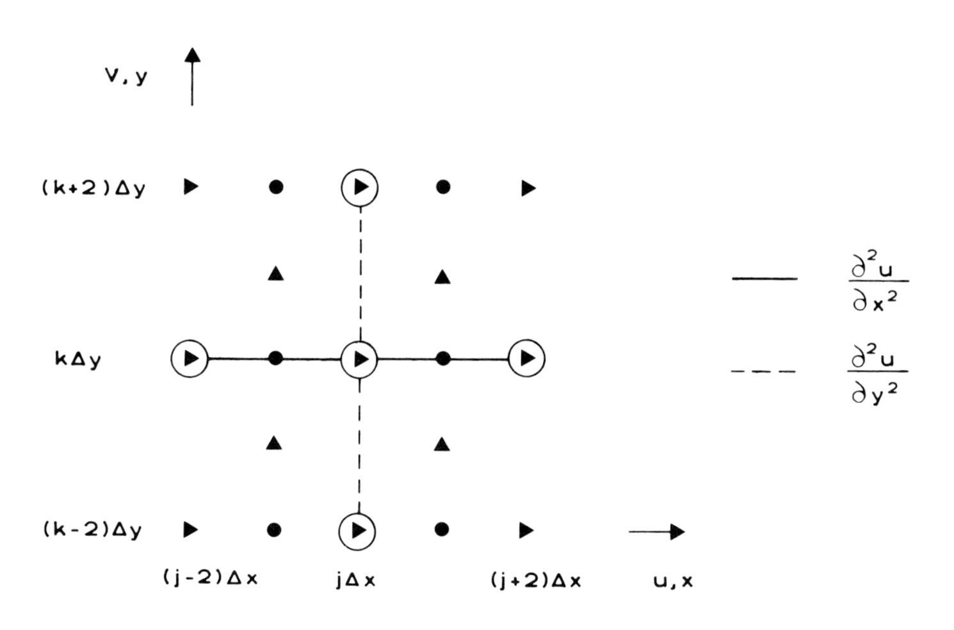

The term, also investigated by Rassmussen,

$$\frac{\partial^2 u}{\partial x \partial y}$$

can be represented as shown in Figure 5.b, but this does not usually allow the appearance of $u_{j,k}$ into the correction term and so is insensitive to $u_{j,k}$. Thus the representation of this term in the manner of Figure 5.b makes it insensitive to the typical double-solution "zig-zagging" structure of a centered scheme, it being quite unable to resolve any variation at highest resolvable wave number.

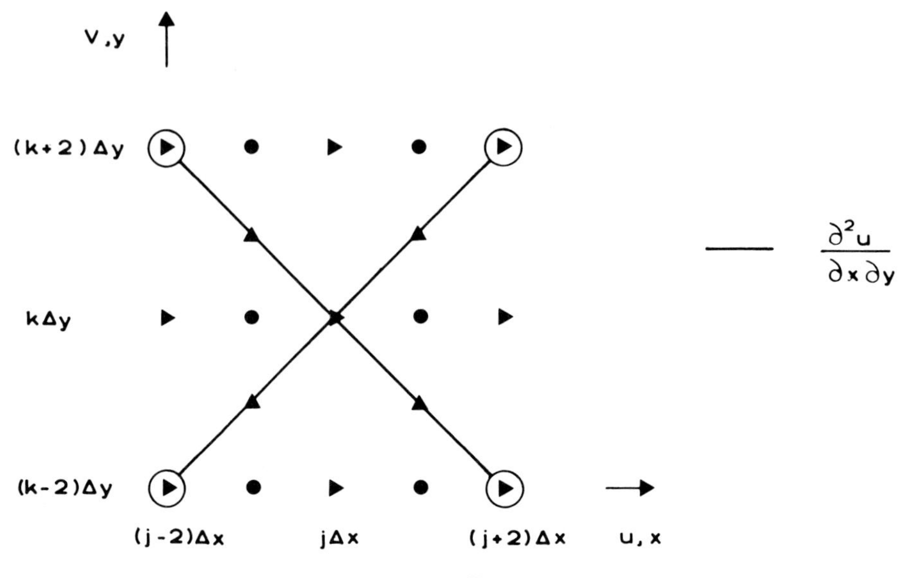

Figure 5b.

7. CORRECTION OF THE ONE-DIMENSIONAL LINEARIZED DESCENT TO THIRD ORDER TRUNCATION ERROR

The scheme is consistent with the forms

$$\frac{\partial p}{\partial t} + gh\,\frac{\partial h}{\partial x} = 0$$

$$\frac{\partial h}{\partial t} + \frac{\partial p}{\partial x} = 0$$

and the class of schemes considered read as

$$\frac{p_j^{n+1} p_j^n}{\Delta t} + gh\left(\frac{h_{j+1}^{n+1} - h_{j-1}^{n+1}}{4\Delta x} + \frac{h_{j+1}^{n} - h_{j-1}^{n}}{4\Delta x}\right) = 0$$

$$\frac{h_j^{n+1} - h_j^n}{\Delta t} + \left(\frac{p_{j+1}^{n+1} - p_{j-1}^{n+1}}{4\Delta x} + \frac{p_{j+1}^n - p_{j-1}^n}{4\Delta x} \right) = 0$$

We expand in Taylor's series to obtain

$$\frac{p_j^{n+1} - p_j^n}{\Delta t} = \left(\frac{\partial p}{\partial t}\right)_j^{n+1/2} + \left(\frac{\partial^3 p}{\partial t^3}\right)_j^{n+1/2} \frac{(\Delta t/2)^3}{3!} + H.O.T.$$

$$\left(\frac{h_{j+1}^{n+1} - h_{j-1}^{n+1}}{4\Delta x} + \frac{h_{j-1}^n - h_{j-1}^n}{4\Delta x} \right) = \left(\frac{\partial h}{\partial x}\right)_j^{n+1/2} + \left(\frac{\partial^3 h}{\partial x^3}\right)_j^{n+1/2} \frac{\Delta s^2}{3!} + \left(\frac{\partial^3 h}{\partial x \partial t^2}\right)_j^{n+1/2} \frac{(\Delta t/2)^2}{2!} + H.O.T.$$

Thence the truncation error is

$$\left(\frac{\partial^3 p}{\partial t^3}\right)_j^{n+1/2} \frac{\Delta t^2}{24} + gh \left[\left(\frac{\partial^3 h}{\partial x^3}\right)_j^{n+1/2} \frac{\Delta x^2}{6} + \left(\frac{\partial^3 h}{\partial x \partial t^2}\right)_j^{n+1/2} \frac{\Delta t^2}{8} \right] + H.O.T.$$

But

$$\frac{\partial^2}{\partial t}\left(\frac{\partial p}{\partial t}\right) = \frac{\partial^2}{\partial t^2}\left(-gh_0 \frac{\partial h_0}{\partial x}\right) = -gh_0 \frac{\partial^2}{\partial t \partial x}\left(\frac{\partial h}{\partial t}\right) = gh_0 \frac{\partial^3 p}{\partial t \partial x^2}$$

$$\frac{\partial^3 h}{\partial x^3} = \frac{\partial^2}{\partial x^2}\left(\frac{\partial h}{\partial x}\right) = \frac{\partial^2}{\partial x^2}\left(-\frac{1}{gh_0}\frac{\partial p}{\partial t}\right) = -\frac{1}{gh_0}\frac{\partial^3 p}{\partial t \partial x^2}$$

$$\frac{\partial^3 h}{\partial x \partial t^2} = \frac{\partial^2}{\partial x \partial t}\left(\frac{\partial h}{\partial t}\right) = -\frac{\partial^3 p}{\partial x \partial t^2}$$

So that the truncation error becomes

$$\left(\frac{gh\Delta t^2}{24} - \frac{\Delta x^2}{6} - gh\frac{\Delta t^2}{8} \right) \frac{\partial^3 p}{\partial t \partial x^2} = -\left(gh\frac{\Delta t^2}{12} + \frac{\Delta x^2}{6} \right) \frac{\partial^3 p}{\partial t \partial x^2}$$

The corrected momentum difference equation then reads

$$(p_j^{n+1} - p_j^n) + gh\frac{\Delta t}{4\Delta x}\left[(h_{j+1} - h_{j-1})^{n+1} + (h_{j+1} - h_{j-1})^n \right]$$

$$+ \frac{\Delta t}{4\Delta x^2 \Delta t} (gh \frac{\Delta t^2}{12} + \frac{\Delta x^2}{6}) \Big[(p_{j+2}2p_j + p_{j-2})^{n+1}$$
$$- (p_{j+2} - 2p_j + p_{j-2})\Big] = 0$$

The mass equation is corrected in the same manner, i.e., the truncation error is

$$(\frac{\partial^3 h}{\partial t^3})_j^{n+½} \frac{\Delta t^2}{24} + \left[\left(\frac{\partial^3 p}{\partial x^3}\right)_j^{n+½} \frac{\Delta x^2}{6} + \left(\frac{\partial^3 p}{\partial x \partial t^2} rigth)_j^{n+½} \frac{\Delta t^2}{8}\right] + H.O.T.$$

But

$$\frac{\partial^3 h}{\partial t^3} = \frac{\partial^2}{\partial t^2} (\frac{\partial h}{\partial t}) = \frac{\partial^2}{\partial t^2} (-\frac{\partial p}{\partial x})$$
$$= \frac{\partial^2}{\partial t \partial x} (-\frac{\partial p}{\partial t}) = \frac{\partial^2}{\partial t \partial x} (gh_0 \frac{\partial h}{\partial x}) = gh_0 \frac{\partial^3 h}{\partial t \partial x^2}$$

$$\frac{\partial^3 p}{\partial x^3} = \frac{\partial^2}{\partial x^2} (\frac{\partial p}{\partial x}) = \frac{\partial^2}{\partial x^2} (-\frac{\partial h}{\partial t}) = - \frac{\partial^3 h}{\partial t \partial x^2}$$

$$\frac{\partial^3 p}{\partial x \partial t^2} = \frac{\partial^2}{\partial x \partial t} (\frac{\partial p}{\partial t}) = \frac{\partial^2}{\partial x \partial t} (- gh_0 \frac{\partial h}{)} = - gh_0 \frac{\partial^3 h}{\partial t \partial x^2}$$

so that the truncation error in this case becomes

$$\left(\frac{gh\Delta t^2}{24} - \frac{\Delta x^2}{6} - \frac{gh\Delta t^2}{8}\right) \frac{\partial^3 h}{\partial t \partial x^2} = - \left(\frac{gh\Delta t^2}{12} + \frac{\Delta x^2}{6}\right) \frac{\partial^3 h}{\partial t \partial x^2}$$

and the corrected mass equation reads

$$(h_j^{n+1} - h_j^n) + \frac{\Delta t}{4\Delta x} \Big[(p_{j+1} - p_{j-1})^{n+1} + (p_{j+1} - p_{j-1})^n\Big]$$
$$+ (\frac{Cr^2}{48} + \frac{1}{24}) \Big[(h_{j+2} - 2h_j + h_{j-2})^{n+1} - (h_{j+2} - ‘2h_j + h_{j-2})^n\Big]$$

The Fourier transform now provides

$$Det \begin{bmatrix} (\phi-1) + (\frac{Cr^2}{48} + \frac{1}{24})(2\cos 2\sigma-2)(\phi-1) & \frac{\sqrt{gh}\,Cr}{4}(i2\sin\sigma)(\phi+1) \\ \frac{\Delta t}{4\Delta x}(2\sin\sigma)(\phi+1) & (\phi-1) + (\frac{Cr^2}{48} + \frac{1}{24})(2\cos 2\sigma-2)(\phi-1) \end{bmatrix} = 0$$

or

$$(\phi-1)^2(1+\beta)^2 + \alpha^2(\phi+1)^2 = 0$$

where, as before,

$$\beta = (\frac{Cr^2}{24} + \frac{1}{12})\ (\cos 2\sigma-1) \quad \cdot \quad \alpha = \frac{Cr}{2}\sin\sigma$$

but now

$$(\phi-1) = \pm i\gamma(\phi+1) \qquad \text{with} \qquad \gamma = \frac{\alpha}{1+\beta}$$

or

$$\phi_{\pm} = \frac{(1-\gamma^2) \mp 2i\gamma}{1+\gamma^2}$$

Some typical phase portraits are shown in Figures 6 and 7.

8. ANALYSIS OF THE RELATIVE INFLUENCES OF THE BOUSSINESQ TERM AND THE TRUNCATION ERROR CORRECTION TERMS

The correction term applied to the momentum equation is

$$(\frac{Cr^2}{48} + \frac{1}{24})\left[(p_{j+2} - 2p_j + p_{j-2})^{n+1} - (p_{j+2} - 2p_j + p_{j-2})^n\right]$$

while the Boussinesq term (Abbott, Skovgaard and Peterson, 1978) is

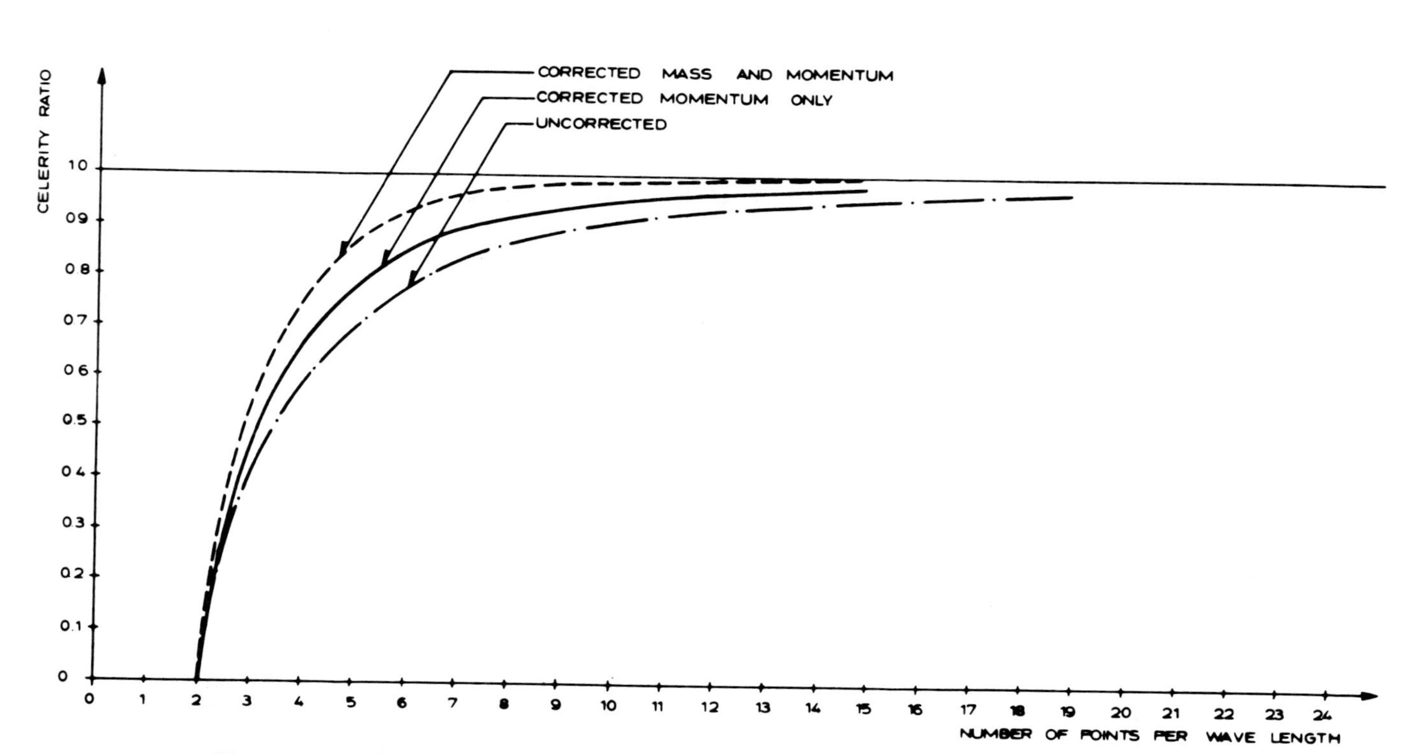

Figure 6. Phase portrait for propagation along grid line directions, when Cr = 1. One-dimensional descent, effects of correction.

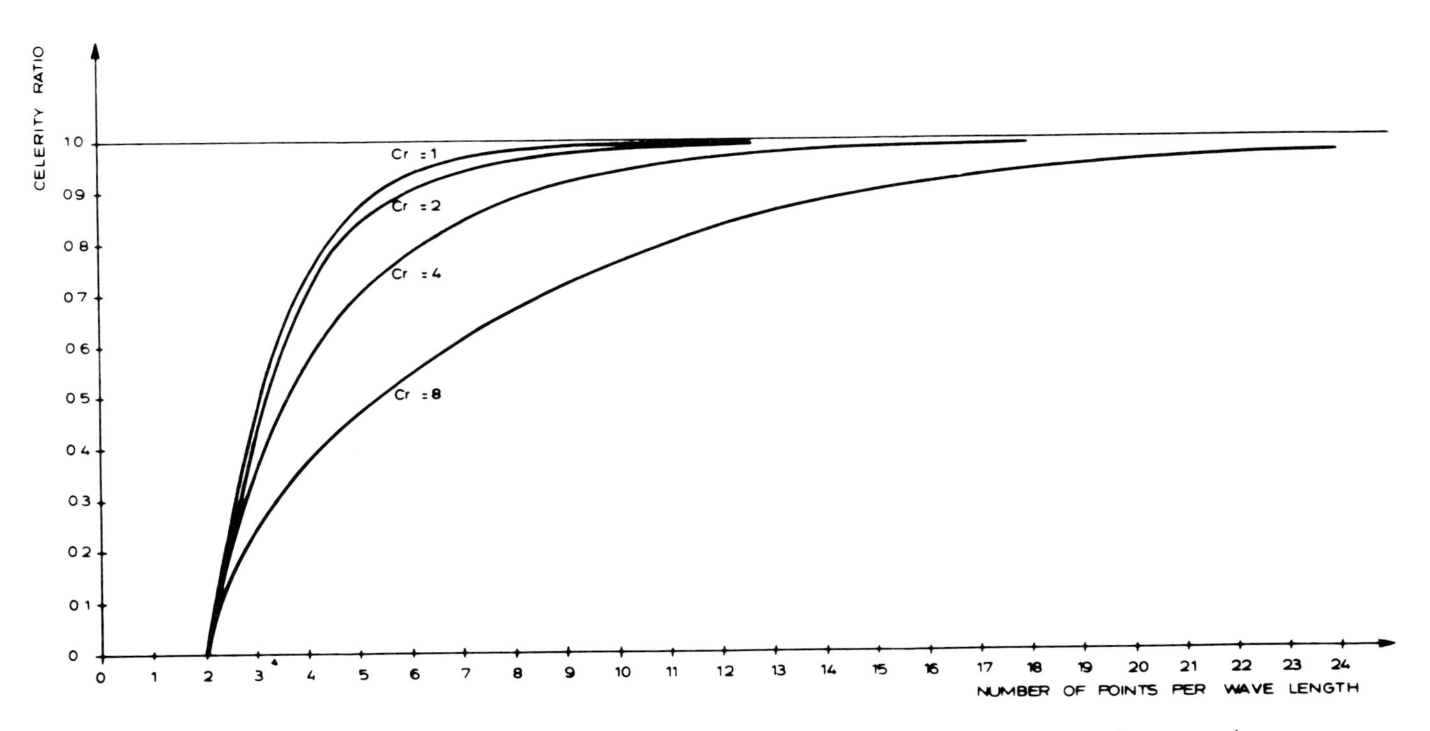

Figure 7. Phase portrait for propagation along grid line directions. One-dimensional descent, corrected on mass and momentum.

$$-\frac{h^2}{3}\cdot\frac{\partial^3 p}{\partial x \partial t}\cdot\Delta t = -\frac{h^2}{12\Delta x^2}\Big[(p_{j+2}-2p_j+p_{j-2})^{n+1} - (p_{j+2}-2p_j+p_{j-2})^{n}\Big]$$

A measure of the influence of the Boussinesq terms in this representation is the dimensionless quantity $A = (h/\Delta x)$. The relative "weights" of the Boussinesq term and the correction terms in this scheme is the dimensionless number

$$A' = \frac{h^2/\Delta x^2}{(\frac{Cr^2}{4}+\frac{1}{2})} = \frac{4h^2}{(Cr^2+2)\Delta x^2}$$

In the common situation whereby $A \approx 1$, $Cr \approx 1$,

$$A' \approx \frac{4}{3}$$

or, the weight of the Boussinesq term is to the weight of the correction terms as is the ratio of 4 to 3.

9. SUBGRID-SCALE MODELING

In view of the inclusion of the chapter of K. W. Bedford (Chapter 6) in this volume, it is convenient to discuss these subgrid-scale processes in terms of the theory of filters. In computational hydraulics, the starting point for this theory is the conceptual model of a discretization net or grid that is successively refined by successively doubling the number of net-points at each stage of the refinement.

In any one computation, a discretization scale or grid size is set and all components with a wave length less than twice this grid size cannot be resolved. Some components are aliased (Abbott, 1979) while others, like those shown in Fig. 8, are eliminated. This is to say that, in any grid-resolved description, these components are "filtered-out" of the description, completely or partially. Then if $f = f(x,y,t)$ is a function of bounded support in the $x-y$ plane defined (by linear interpolation, for example) on a grid the grid points of which will, in the limit, be so dense as to have as their set of limit points the entire support of f, then we define the resolved part of f, written as $\bar{f}$, by

$$\bar{f} = \int_{-\infty}^{\infty} d\xi \int_{-\infty}^{\infty} d\zeta \, [G(x-\xi, y-\zeta) f(\xi,\zeta)] \overset{\text{def}(*)}{=\!=} (G * f)$$

The functions $G(x,y)$ have unit area and are called filter functions. The filters most commonly used for eliminating the unresolvable wave numbers are shown, using one coordinate direction, in Fig. 9. The second of these, b, is most commonly used. The filters can, of course, also attenuate the resolved components. For example, the influence of filter a of Fig. 9 is to reduce the kth Fourier cosine component to

$$\frac{1}{\Delta} \int_{x-\Delta/2}^{x+\Delta/2} \cos\left(\frac{2\pi k \xi}{2l}\right) d\xi = \frac{2l}{2\pi k \Delta} \left[\sin\left(\frac{2\pi k (x+\Delta/2)}{2l}\right) - \sin\left(\frac{2\pi (x-\Delta/2)}{2l}\right) \right]$$

$$= \frac{4l}{2\pi k \Delta} \sin\left(\frac{2\pi k \Delta}{4l}\right) \cos\left(\frac{4\pi k x}{4l}\right)$$

Supposing, now, that we assume continuity of solutions, we can start out from the Navier-Stokes equations and the continuity equation for an incompressible fluid

$$\frac{\partial u_i}{\partial t} + \frac{\partial (u_i u_j)}{\partial x_j} = \frac{1}{\rho} \frac{\partial p}{\partial x_i} + \nu \nabla u_i^2 \tag{9.2}$$

$$\frac{\partial u_i}{\partial x_j} = 0 \tag{9.3}$$

We remark that the neutral element of the set of filtering operators, i.e. the element whose application leaves f unchanged, is the Dirac δ-distribution. It is then necessary that the class of filter functions be of the class of distributions. It may then be shown, under certain very general conditions which are satisfied in the present case (see, e.g., Guelfand and Chilov, 1962, p. 103) that

$$D(f^*g) = (Df^*g) = (f^*Dg) \tag{9.4}$$

where D is any differential operator.

Applying Eq. (9.1) and (9.4) to (9.2) and (.3) provides

$$\frac{\partial \bar{u}_i}{\partial t} + \frac{\partial (\overline{u_i u_j})}{\partial x_j} = -\frac{1}{\rho} \frac{\partial \bar{p}}{\partial x_i} + \nu \nabla^2 \bar{u}_i \tag{9.5}$$

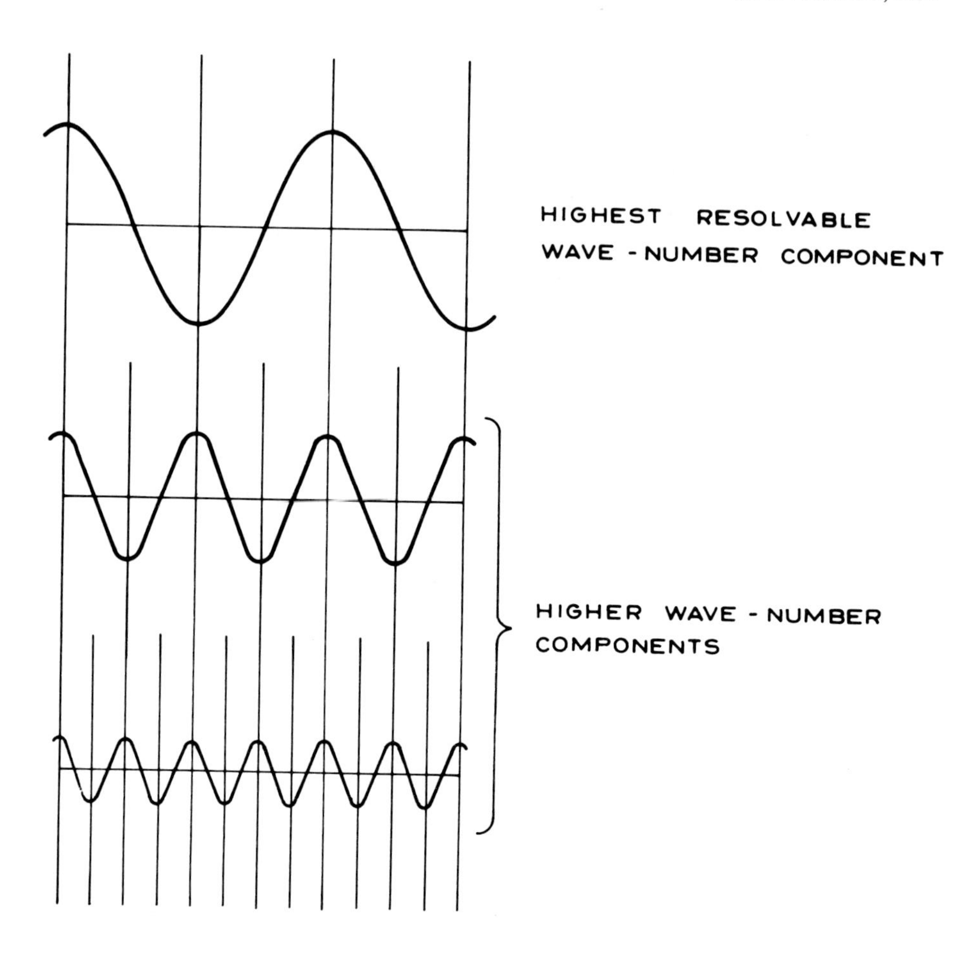

$$\frac{\partial \bar{u}_i}{\partial x_i} = 0 \tag{9.6}$$

We note that, *if* it was possible to replace $\overline{u_i u_j}$ by $\bar{u}_i \bar{u}_j$, the solution of a problem governed by Eq. (9.5) and (9.6) would differ from one governed by (9.2) and (9.3) "only" by virtue of the filtering of the initial (including boundary) conditions. We say "only", since it is known of solutions of these equations (filtered as well as unfiltered) that small differences in their initial conditions can produce large differences in their solutions. In fact it has been hypothesized by Welander (1955) that a finite line element may become of unbounded length in a finite time in a finite-velocity field, in analogy with the constructions of classical physics.

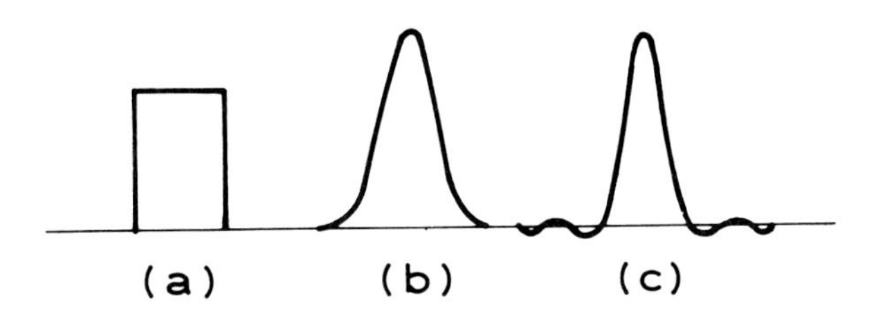

Figure 9.

Now when we integrate in the vertical over the fluid depth, we commonly assume steady-state conditions, whereby

$$G(x,y) = \epsilon(x,y)$$

where $\epsilon(x,y)$ is the ϵ-distribution defined by

$$\int_0^{+\infty} \epsilon(x,y,t)\,w(x,y,t) \overset{\text{def}(\epsilon)}{=\!=} \overline{w(x,y)}^{\,t}$$

where $\overline{w(x,y)}^{\,t}$ is time-average value of $w(x,y)$ and $w(x,y,t)$ is any test function satisfying $\overline{w(x,y)}^{\,t} < \infty$. Here $w(x,y,t)$ is interchangeable with the function $f(x,y,t)$.

We recall that the integration over depth provides, at least for a water density that is constant over depth, a "bed-resistance", defined in terms of the fluid velocity, the fluid depth and such bed properties as small-scale roughness, large scale features (dunes, flat bed, antidunes) and, perhaps, suspended load. This resistance enters as a sink into the two momentum-conservation equations and thence through non-linear transformations in conjunction with the mass-conservation equation, into a sink term in the energy-conservation equation and a source or sink term in the vorticity-conservation equation (e.g., Kuipers and Vreugdenhil, 1973).

The integration over depth provides, inseparably from the resistance law, a definite velocity distribution. This has the effect of moving fluid with different velocities at different elevations in the fluid, but the fluid so moved subsequently mixes by virtue of the vertical turbulence to provide a further horizontal redistribution of fluid and a corresponding filter in the horizontal. This process is schematized for a redistribution with the x-velocity in Fig. 10. In the event that we follow the momentum carried with the fluid, the corresponding filters are further sharpened by a weighting over depth with the velocities themselves (Elder,

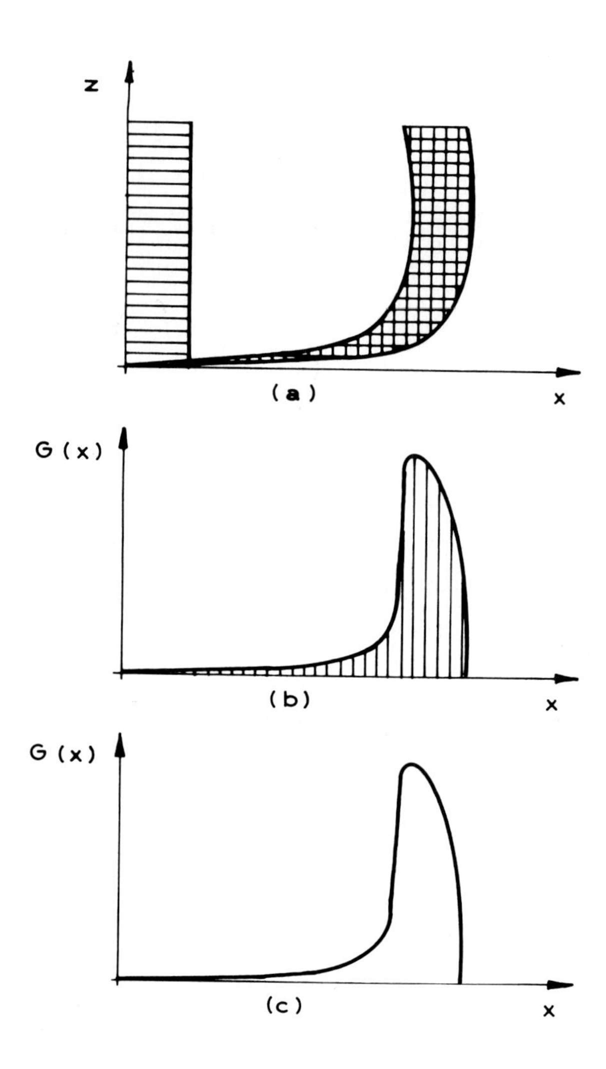

Figure 10.

1959; Abbott and Rassmussen, 1977). A closely related redistribution occurs due to rotation of the fluid about a vertical axis and a filter can be constructed in much the same way. It is common practice to replace the convolution of $f(x,y)$ with filters of the form shown in Fig. 10c by an approximating pure transport to the mass-center of the filter function and a pure diffusion providing the same variance as that of the filter function. Since the diffusion process is irreversible, energy is lost to the resolved system, even though the conservation of mass and momentum is maintained. We remark that a diffusion of mass leads to a loss of potential energy in free surface flows, without any intercession of resolved kinetic energy.

Although it is not usually possible to set $\overline{u_i u_j} = \bar{u}_i \bar{u}_j$, we can write Eq. (9.5) as

$$\begin{aligned}\frac{\partial \bar{u}_i}{\partial t} + \frac{\partial(\bar{u}_i \bar{u}_j)}{\partial x_j} &= -\frac{1}{\rho}\frac{\partial \bar{p}}{\partial x} + n\nabla^2 \bar{u}_i - \left(\frac{\partial \overline{u_i u_j}}{\partial x_j} - \frac{\partial \bar{u}_i \bar{u}_j}{\partial x_j}\right) \\ &= -\frac{1}{\rho}\frac{\partial \bar{p}}{\partial x} + \nu\nabla^2 u_i \\ &\quad - \left(\frac{\partial \overline{\bar{u}_i \bar{u}_j}}{\partial x_j} + \frac{\partial \overline{\bar{u}_i u'_j}}{\partial x_j} + \frac{\partial \overline{u'_i \bar{u}_j}}{\partial x_j} + \frac{\partial \overline{u'_i \bar{u}_j}}{\partial x_j} - \frac{\partial \bar{u}_i \bar{u}_j}{\partial x_j}\right)\end{aligned}$$

where $u'_i = u_i - \overline{u_i}$.

The terms

$$\frac{\partial \overline{\bar{u}_i \bar{u}_j}}{\partial x_j} - \frac{\partial \bar{u}_i \bar{u}_j}{\partial x_j} \tag{9.7}$$

that were the principle concern of Leonard (1974) are seen to be expressed entirely in terms of resolved quantities. Leonard proposed to express these terms, through Taylor series expansion, in the form

$$(\overline{\bar{u}_i \bar{u}_j}) - (\bar{u}_i \bar{u}_j) = \frac{\Delta^2}{4\gamma}\frac{\partial}{\partial x_k}\frac{\partial}{\partial x_k}(\bar{u}_i \bar{u}_j) + \text{H.O.T.} \tag{9.8}$$

where γ is the free parameter and Δ is the effective filter width in x, usually taken as $\Delta = 2\Delta s$, and $\gamma = 6$ in the error function filter, (b) of Fig. 9. These terms then appear in the form of third derivatives in the momentum conservation equations of resolved functions, again as shown in Chapter 6. They have accordingly a dispersive influence, providing a radiation of energy. Some of this radiation will then be resolved directly while the rest will be resolved indirectly through aliasings into the lower wave numbers. At the level of the momentum equation, the various dispersions may appear as radiation stresses. The differences on the left of Eq. (9.8) are commonly called the "Leonard stresses."

Applying a similar Taylor expansion to products of means and variations provides

$$\overline{\overline{u}_i u'_j} = -\ \overline{u}_i \frac{\Delta^2}{4\gamma}\ \frac{\partial^2 u_j}{\partial x_k \partial x_k} + \text{H.O.T}$$

Clark, *et al.*, (1977) suggested approximating this by

$$\overline{\overline{u}_i u'_j} = -\ \frac{\Delta^2}{4\gamma}\ \overline{u}_i\ \frac{\partial^2 \overline{u}_j}{\partial x_k \partial x_k} + \text{H.O.T.} \tag{9.9}$$

The remaining, Reynold's stress terms

$$\frac{\partial \overline{u'_i u'_j}}{\partial x_j}$$

can also be treated by Taylor series expansion, to provide the approximation

$$\overline{u'_i u'_j} = \frac{\Delta^4}{96\gamma}\left(\frac{\partial^2 \overline{u}_i}{\partial x_k \partial x_k} \cdot \frac{\partial^2 \overline{u}_j}{\partial s_l \partial s_l}\right) + \text{H.O.T.}$$

They are more conventionally treated since Smagorinsky (1965), however, by introducing a diffusion term in u_i into each of the by introducing a diffusion term in u_i into each of the the same in both x and y directions and given by

$$\boldsymbol{\nu}_x = \boldsymbol{\nu}_y = \frac{(c\Delta)^2 \overline{D}}{\sqrt{2}} \tag{9.10}$$

with $\overline{D}$ either the modulus of the local grid-resolved deformation rate,

$$\left[\frac{\partial \overline{u}_j}{\partial x_i}\left(\frac{\partial \overline{u}_i}{\partial x_j} + \frac{\partial \overline{u}_j}{\partial x_i}\right)\right]^{1/2} \tag{9.11}$$

or the equivalent modulus of the local grid-resolved vorticity. In two-dimensions, Eq. (9.11) then reads, for the class of schemes considered earlier (Deardorff, 1971)

$$
\begin{aligned}
\nu_{j,k} = (c\Delta x \Delta y)^2 \Bigg\{ & 2\left[\frac{(u_{j+1/2} - u_{j-1/2})_k}{\Delta x}\right]^2 + 2\left[\frac{(v_{k+1/2} - v_{k-1/2})_j}{\Delta y}\right]^2 \\
& + \frac{1}{4}\left[\frac{(v_{j+1} - v_j)_{k+1/2}}{\Delta x} + \frac{(u_{k+1} - u_k)_{j+1/2}}{\Delta y}\right]^2 \\
& + \frac{1}{4}\left[\frac{(v_j - v_{j-1})_{k+1/2}}{\Delta x} + \frac{u_{k+1} - u_k)_{j-1/2}}{\Delta y}\right]^2 \\
& + \frac{1}{4}\left[\frac{(v_j - v_{j-1})_{k-1/2}}{\Delta x} + \frac{(u_k - u_{k-1})_{j-1/2}}{\Delta y}\right]^2 \\
& + \frac{1}{4}\left[\frac{(v_{j+1} - v_j)_{k-1/2}}{\Delta x} + \frac{(u_k - u_{k-1})_{j+1/2}}{\Delta y}\right]^2 \Bigg\}^{1/2}
\end{aligned}
$$

Critical reviews of this approach, together with estimates of values of c, have been written by Ferziger (1977) and Leslie and Quarini (197() and a further discussion is provided in Chapter 6.

Alongside this work on sub-grid-scale turbulence, considerable work has been done upon the "semi-empirical" modeling of turbulent flows in which all turbulent energy, both resolved and sub-grid, is considered simultaneously. The most widely known of these models is the so-called "k-ϵ" model (Rodi, 1978) which traces the transport, diffusion, generation and decay of the turbulent energy level k and the transport, diffusion, generation and decay of a characteristic length scale, represented by the rate of turbulent energy loss to heat energy, ϵ. Although this type of model did not originally distinguish subgrid-scale processes in the above sense, it has been possible to modify it to comprehend subgrid-scale processes only (see also Leendertse and Liu, 1977).

The relative importance of the various sub-grid scale modeling components depends upon the area of application of the model. In many situations, this can be characterized simply by the dimensionless number $h/\Delta x$. When this number is very small, it appears that flow processes, including circulations, are mainly controlled by the distributions of the bathymetry, the convective momentum, the resistance and the wind stress, apart from the obvious initial and boundary conditions. The subgrid modeling is both relatively unimportant and rather poorly defined. However, as the number $h/\Delta x$ approaches unity, so the importance of the subgrid modeling increases rapidly. Moreover, in nearfield modeling, the transport and diffusion of turbulence can play a significant role, so that either a k-ϵ model has to be used or this model must be modified so as to comprehend only subgrid-scale processes. However, as $h/\Delta x$ increases, so a proper resolution of the flow necessitates that account be taken of variations from the hydrostatic pressure distribution. The simplest means of doing this is through the use of the Boussinesq equations. In rapidly rotating flows, the equivalent to $h/\Delta x$ is $R/\Delta x$, where R is the radius of curvature of the flow in the $x-y$ plane.

All of the above descriptions devolve around the assumption of continuity and strong consistency and convergence. However, observation of numerical solutions suggests that circulations at highest resolvable wave number persists with finer discretization, so that the limiting solution, with a countable infinity of grid points dense in the support of the function f in Eq. (9.1), may well be non-analytic (Abbott, 1979, p. 248). It will then not allow of a (strong) differential description. This possibility has hardly been explored at all, to date, but is an interesting area of research.

10. APPLICATIONS OF TWO-DIMENSIONAL FREE SURFACE FLOW MODELS

Moving on from the mainly theoretical considerations given above, this final section of the chapter is devoted to the engineering application of two-dimensional free surface flow models. The examples come from recent investigations carried out at the Danish Hydraulic Institute and have been selected to illustrate some of the main features of this type of model.

10.1 The North Sea Model

The North Sea Model, described more fully by Abbott (1979), was originally set up to predict storm surge levels along the southwest coast of Denmark. Since then it has been used in a number of investigations including a hindcasting study of the Bravo oil spill.

The model uses the MK6-3 version of the System 21 which can simultaneously compute over nested sub-domains, each having successively finer grid scales. In this way it is possible to focus the computation onto an area of direct interest, while maintaining a two-way exchange of information at the subdomain boundaries.

In addition to the results given by Abbott, the accuracy of the original model is further demonstrated by the surge level comparisons given in Fig. 11 at List and Hoejer for the January 3, 1976 storm. This area behind the west coast barrier islands was resolved in the second finer grid sub-domain of the model, with a mesh size of 2059 m.

More recently the North Sea Model has been applied to the problem of predicting currents, under the combined action of storm and tide, for the design of a gas pipeline to the Danish west coast. In order to provide an adequate description of the current conditions along the nearshore section of the pipeline, a total of four sub-domains, as shown in Fig. 12, were used. At each change of scale the degree of resolution was increased by a factor of three, so reducing the 18,580 m (10 nautical mile) grid size in SD0 down to 686 m in SD3.

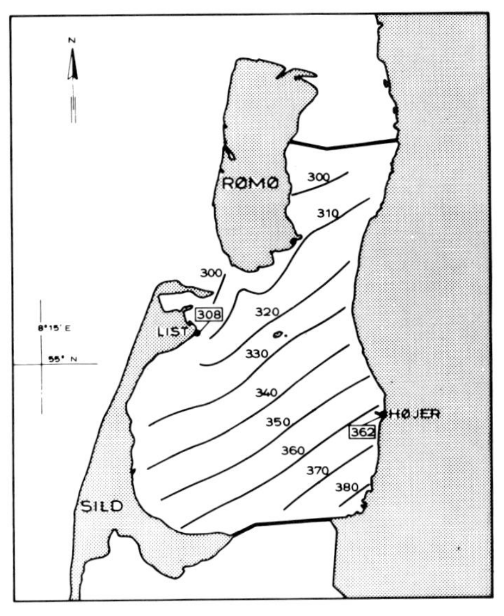

15 GMT.

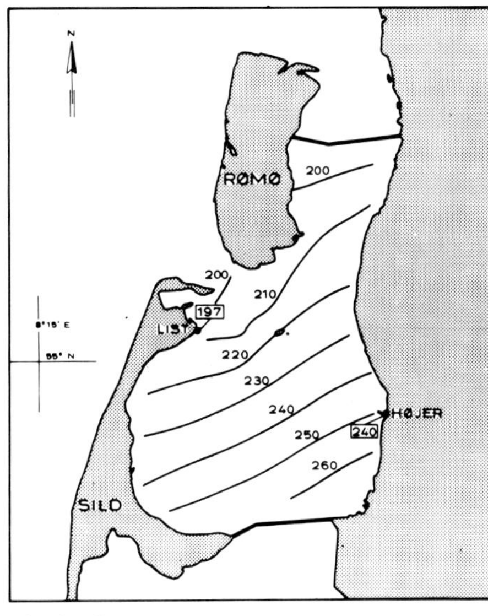

18 GMT.

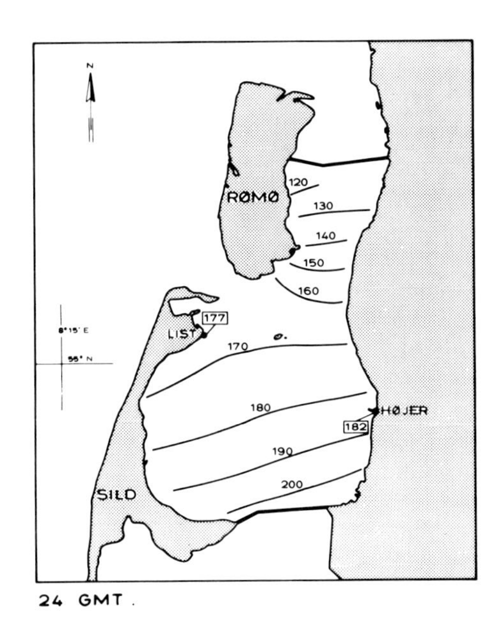

24 GMT.

NORTH SEA MODEL
SYSTEM 21

COMPUTED AND MEASURED WATER LEVELS IN A COASTAL EMBAYMENT. 3-1-76

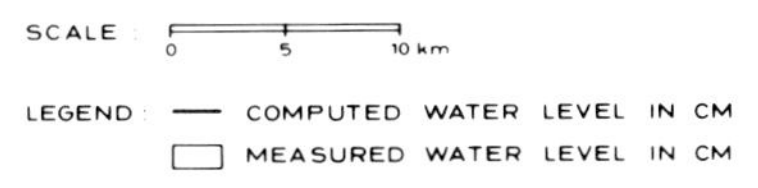

Figure 11. Water level comparisons in Listerdyb--January 3, 1976 storm.

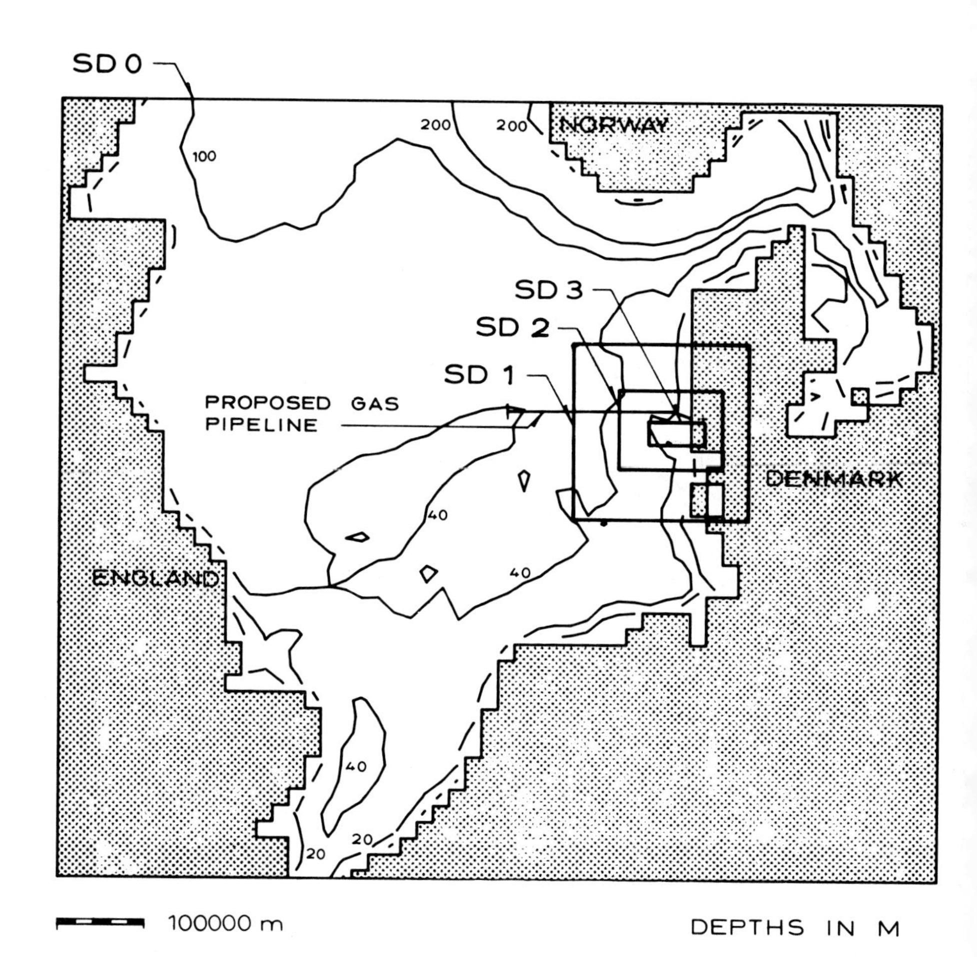

Figure 12. The North Sea Model with four subdomains, SD0, SD1, SD2, and SD3.

Although not directly in the region of interest, the finest grid SD3 was initially included to provide higher resolution of the currents through the complex bathymetry of Horns Reef shown in Fig. 13. However, during model calibration it was found that the effect of these currents, on the currents at the pipeline, could be adequately resolved in the coarser grid SD2. Sub-domain SD3 was therefore dropped from later runs resulting in a substantial saving in computational costs.

Water level comparisons at Graadyb Bar (the entrance to the port of Esbjerg) obtained during two calibration storms are shown in Fig. 14. The first storm (a)

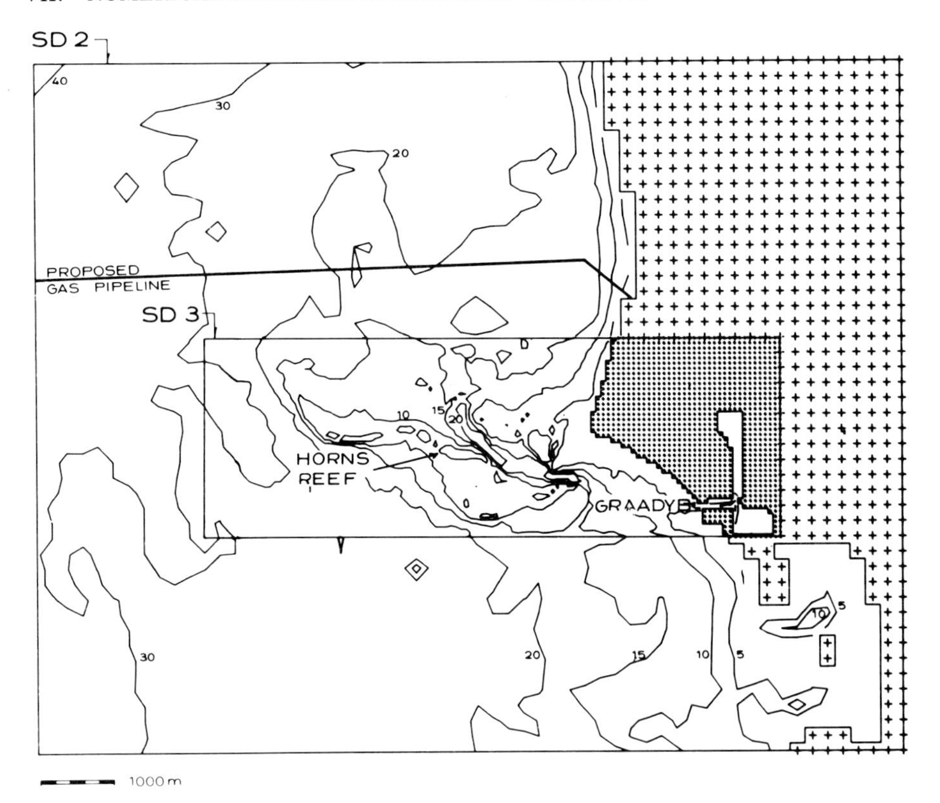

Figure 13. The bathymetry of subdomains SD2 and SD3.

corresponds to a surge level at Esbjerg having a return period of once in about 1.5 years, whereas the lower surge levels of the second (b) could be expected approximately eight times per year.

As part of the general calibration of the model, the velocity flow fields obtained in the overall coarse grid SD0 were compared against the North Sea current patterns given in *Atlas der Geseitenströmen für die Nordsee, der Kanal and die Irische See,* (1968). Comparisons of currents obtained at four different stages of the tide are given in Figs. 15-18.

10.2 Ho Bay

Both hydrodynamic and transport-dispersion modeling was carried out for an environmental study of Ho Bay, on the west coast of Denmark, The model area shown in Fig. 19 is characterized by deep narrow channels and extensive regions of tidal mudflats. This required the use of a relatively fine (250 m) grid System 21 model, which included a special flood and dry routine.

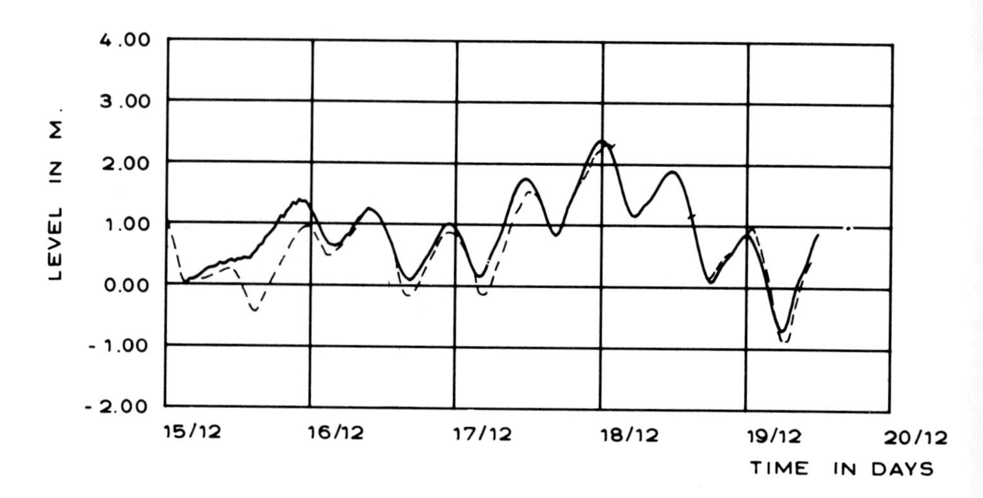

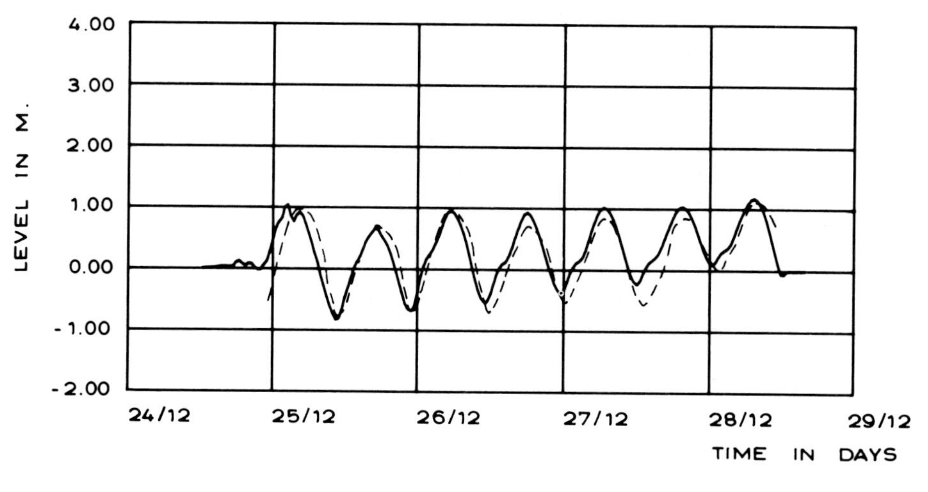

Figure 14. Water level comparisons at Graadyb for two storms in December 1979.

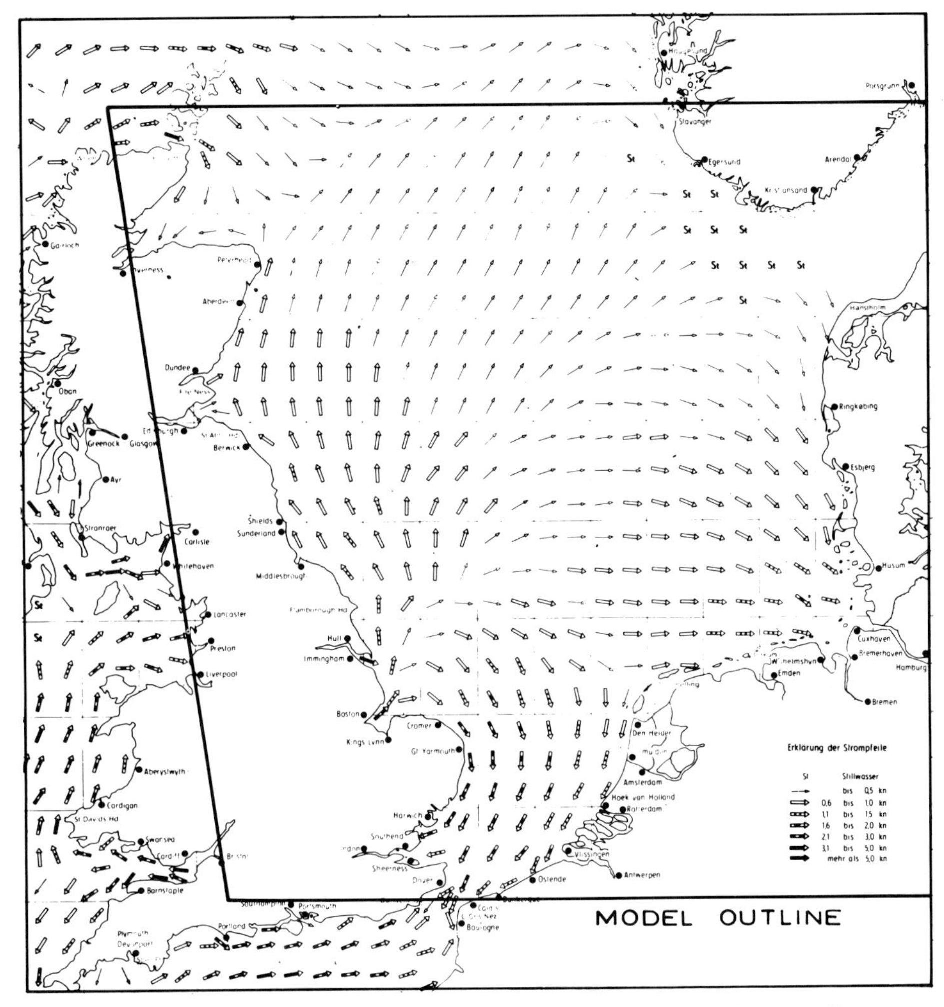

Figure 15. North Sea flow comparisons, two hours 30 minutes before high water at Dover.

In order to correctly describe the net circulations around the island of Fanoe, the hydrodynamic model was extended so as to include the large embayment to the south of Esbjerg. Water level variations were given as boundary conditions at the two main entrances, Graadyb and Knudedyb, with discharges given at the Varde River. The extensive flooding and drying of mudflats which occurs throughout the tidal cycle, is readily seen in the velocity fields of Fig. 20.

The water level and velocity results from the hydrodynamic model were used as input data to a more local high accuracy model for computing the transport, dispersion and decay of pollutants (see Hinstrup, Kej, and Kroszynski, 1977).

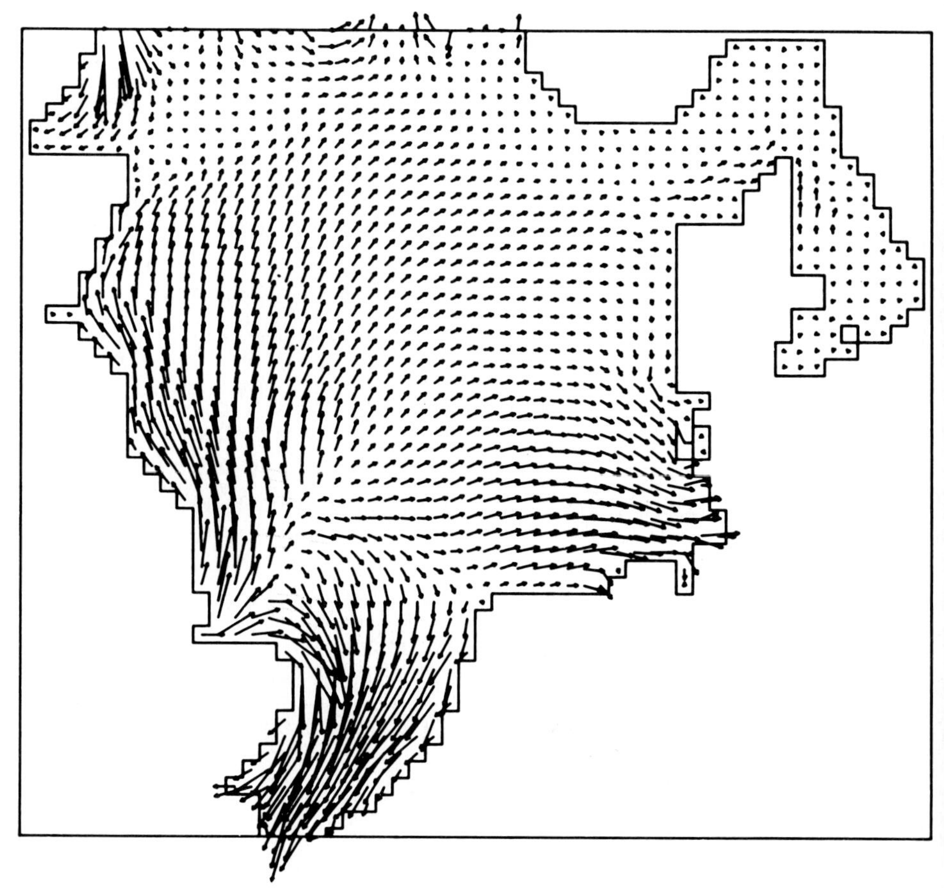

Figure 15, continued.

This transport dispersion model was calibrated against salinity measurements taken throughout the area. Comparisons of measured and computed salinities at high and low water are shown in Fig. 21.

10.3 Arhus Bay

The importance of specifying correct boundary conditions can be seen from the results of a model study of Arhus Bay, on the east coast of Jutland, in Denmark. The model area shown in Fig. 22 is generally stratified and for this reason

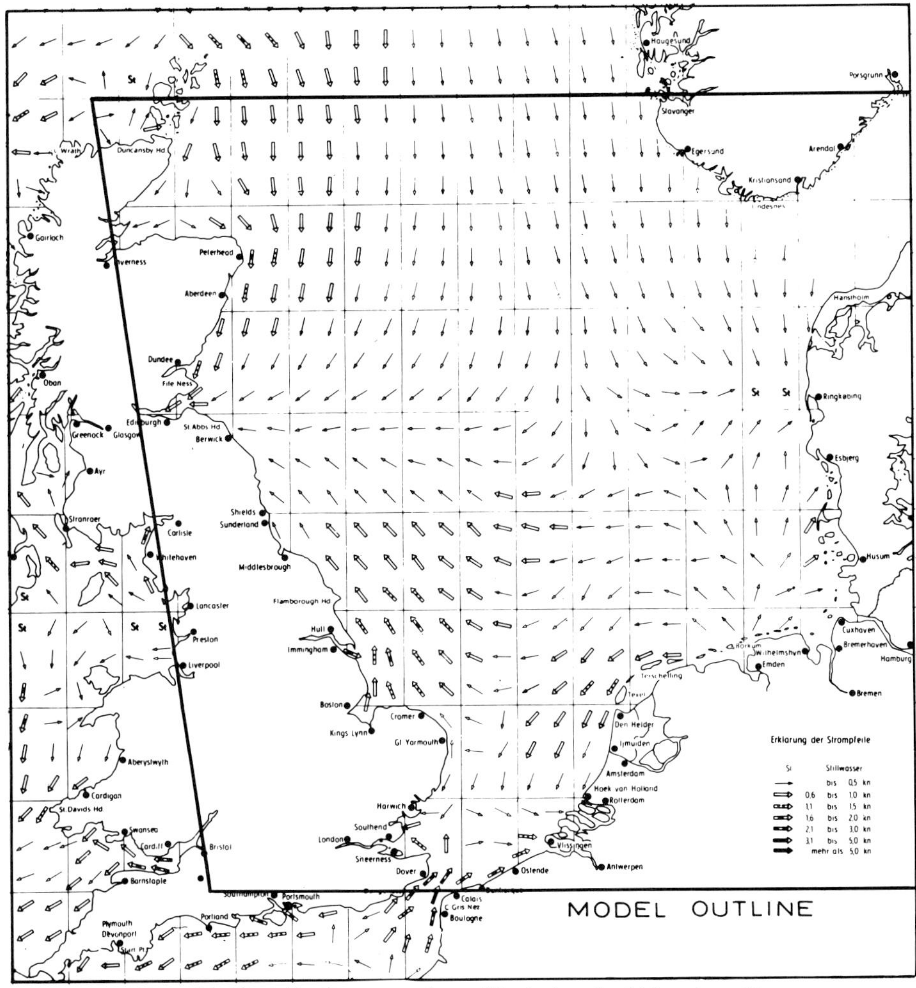

Figure 16. North Sea flow comparisons, 30 minutes after high water at Dover.

DHI's two dimensional, two-layer modeling system, System 22, was used. This system required surface levels, interface levels, and upper and lower layer densities to be given as boundary data along the southern boundary of the model.

Suitable interface and density boundary conditions were provided by long term measurements of temperature and salinity profiles taken at two representative positions between Norsminde and Helgenaes.

Initially the surface level boundary conditions were taken directly from tide recordings at Norsminde. However, in addition to the tidal and longer term trends, these measurements were found to contain small components with periods

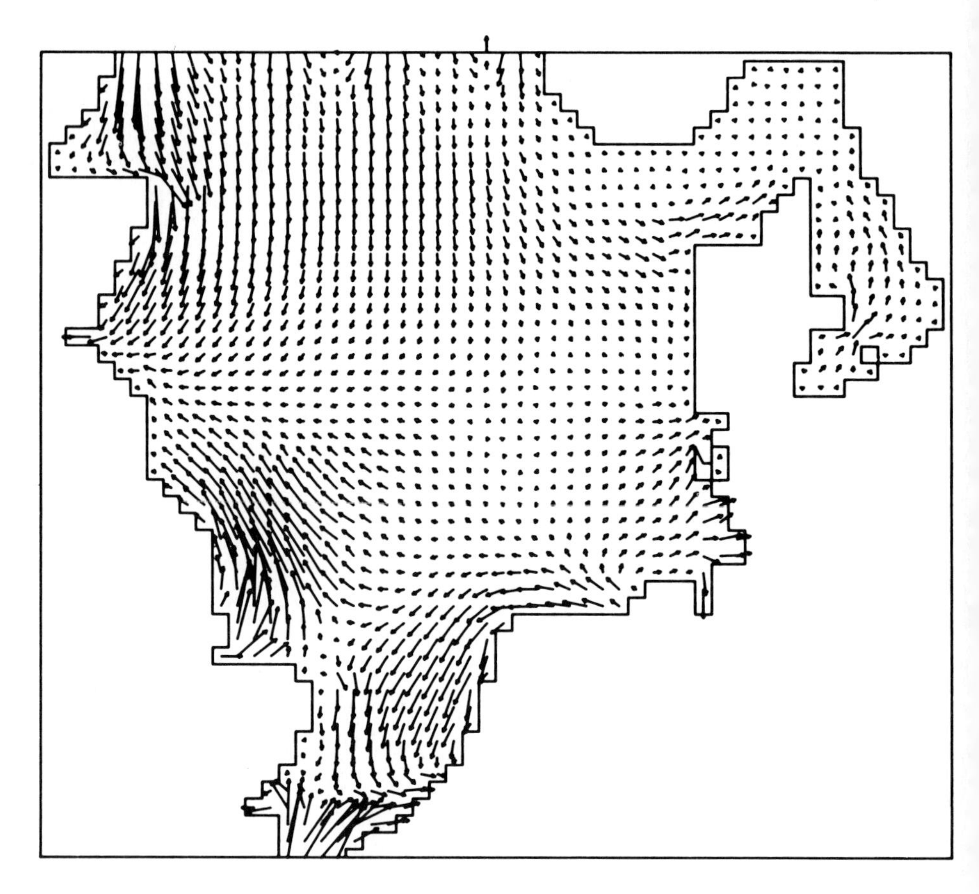

Figure 16, continued.

of about four hours. These corresponded closely to the natural frequency of Arhus Bay, and resulted in an artificial resonance of the model, as shown in the water level and discharge comparisons of Figs. 23(a) and 23(d) respectively.

In preference to using a radiation boundary condition to overcome this problem, it was considered more appropriate to filter the small amplitude components in the boundary data which were exciting the resonance. This was achieved by decomposing the boundary data into its Fourier components, and then damping all components with periods of 6 hours or less, to 20 percent of their original values. In this way, Figs. 23(c) and 23(e) show that it was possible to obtain satisfactory water levels and discharges at the entrance to Kaloe Vig, while the effect on the original boundary data shown in Fig. 23(b) is almost unnoticeable.

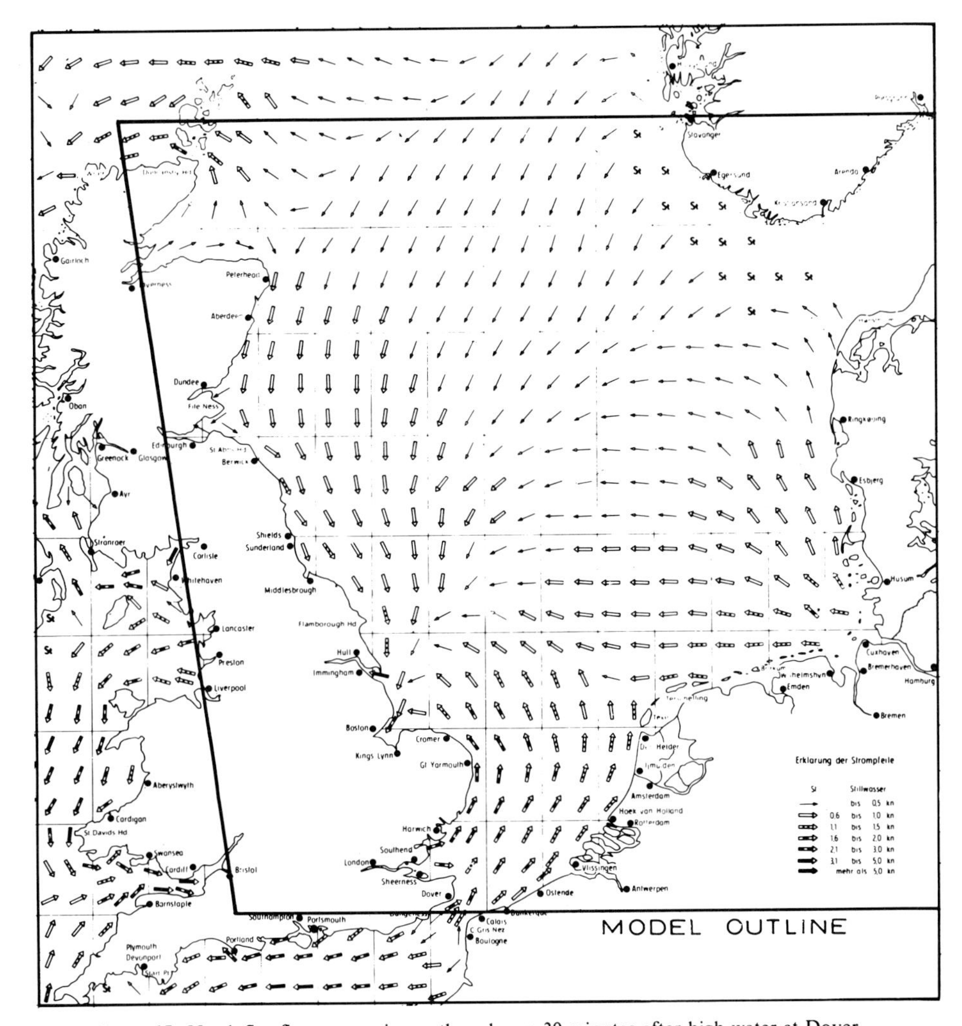

Figure 17. North Sea flow comparisons, three hours 30 minutes after high water at Dover.

10.4 Vendsysselvaerket

Two-dimensional hydrodynamic and transport-dispersion modeling was carried out to determine the likely environmental and recirculation effects of cooling water from a proposed 750 MW power station on the Limfiord, near Aalborg in Denmark.

In this investigation the one-dimensional, System 11, Limfiord model, described by Abbott (1979), was used to provide hydrodynamic boundary data to

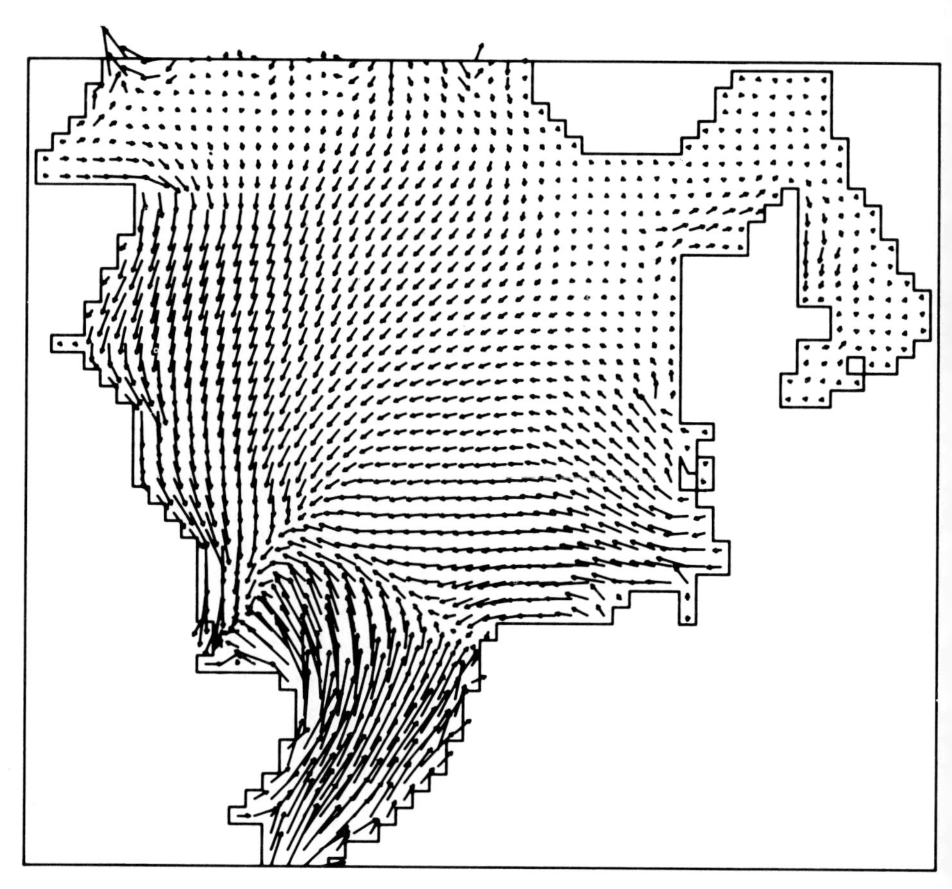

Figure 17, continued.

the more detailed, local System 11 model shown in Fig. 24(a). This local model was in turn used to provide both hydrodynamic and temperature field boundary data for a two-dimensional System 21 model of the area of direct interest, shown in Fig. 24b.

For the two-dimensional model, a 125 m grid size was initially chosen in order to provide the required degree of resolution for the temperature field calculations. However, a sensitivity analysis was carried out using other grid sizes, to ensure that the model would give a satisfactory description of the flow fields

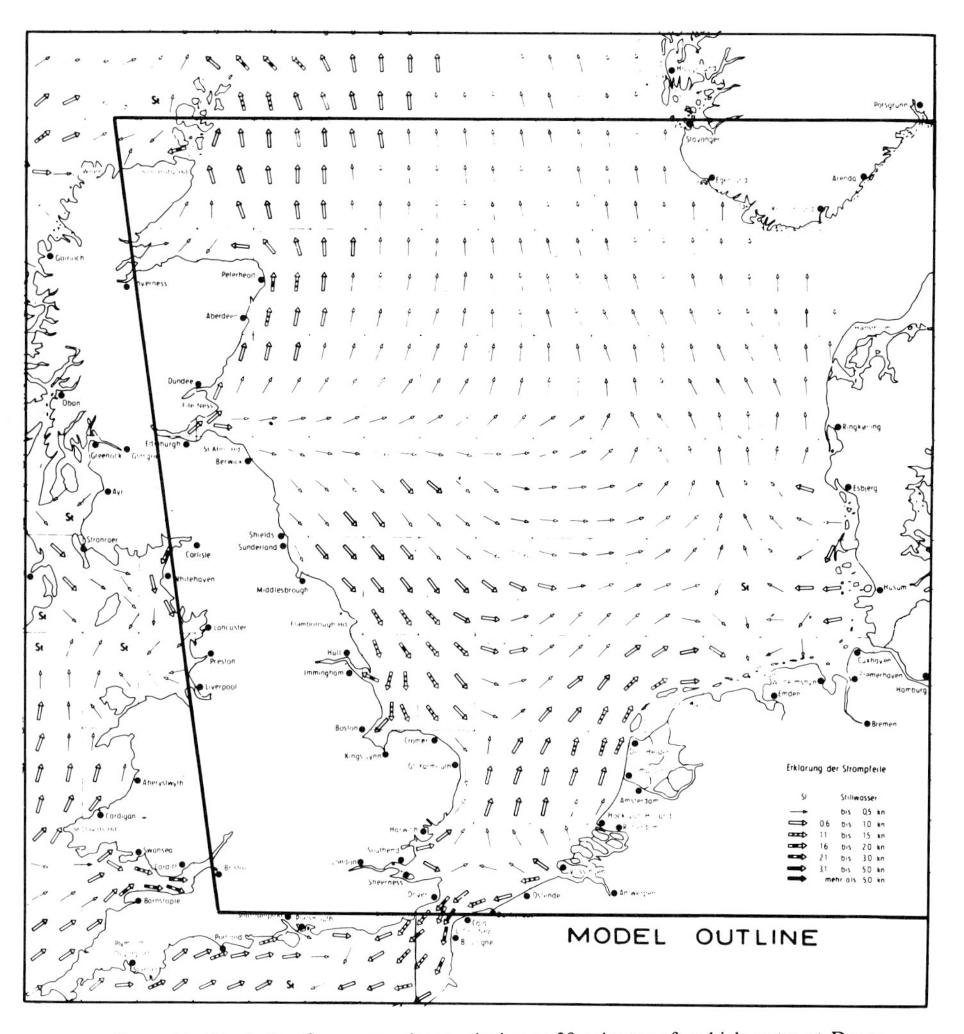

Figure 18. North Sea flow comparisons, six hours 30 minutes after high water at Dover.

around the bends of the relatively narrow main channel. Figure 25 shows a comparison of the flow fields and water level isolines obtained with (a) a 62.5 m grid (only 1 in 4 points shown), and (b) the 125 m grid. The results show only small variations and the 125 m grid model was used for subsequent computations.

10.5 Misurata

The usefulness of Boussinesq equations and the high accuracy MK8 version of the System 21, for describing short wave motions, has been discussed in length

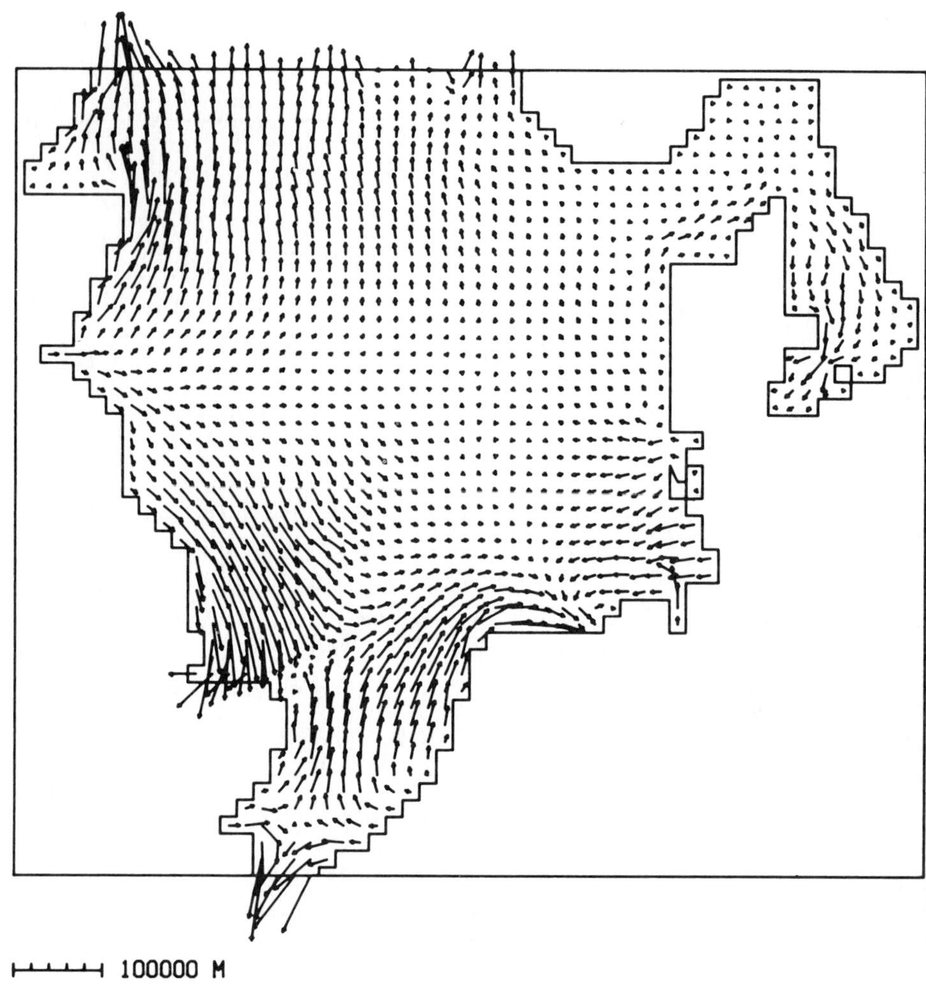

Figure 18, continued.

by Abbott, Petersen and Skovgaard (1978). However, after the discussion of Section 8 above, it seems appropriate to conclude this section of the chapter with a brief description of an application of this model.

Wave disturbance tests were carried out for a steel port at Misurata on the Mediterranean coast of Libya. Figure 26(a) shows the bathymetry used as input to the model and Fig. 26(b) shows resulting isolines of wave attenuation factors obtained for periodic 14 second north easterly waves. The ability of the model to reproduce the combined effects of refraction, diffraction and shoaling can be readily seen in Fig. 26(c).

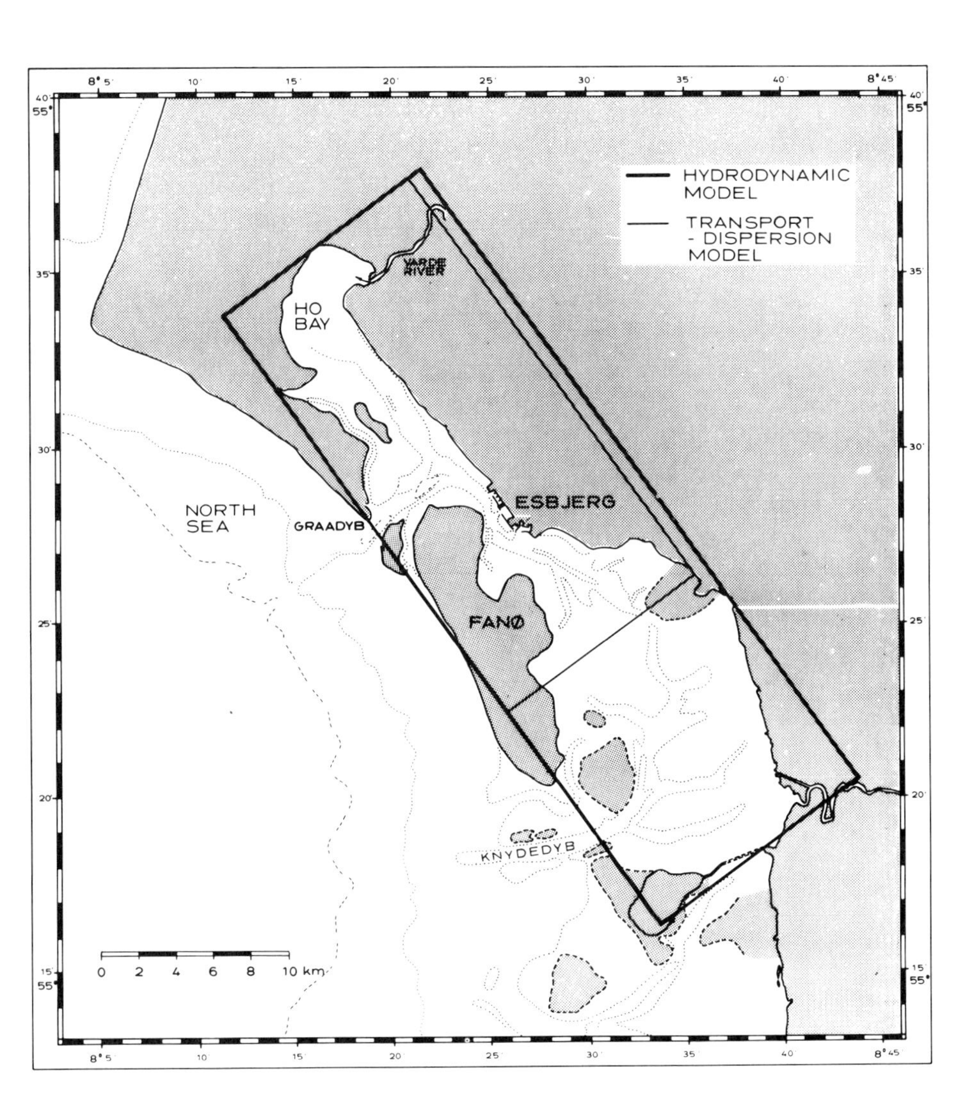

Figure 19. Model areas for the Ho Bay environmental study.

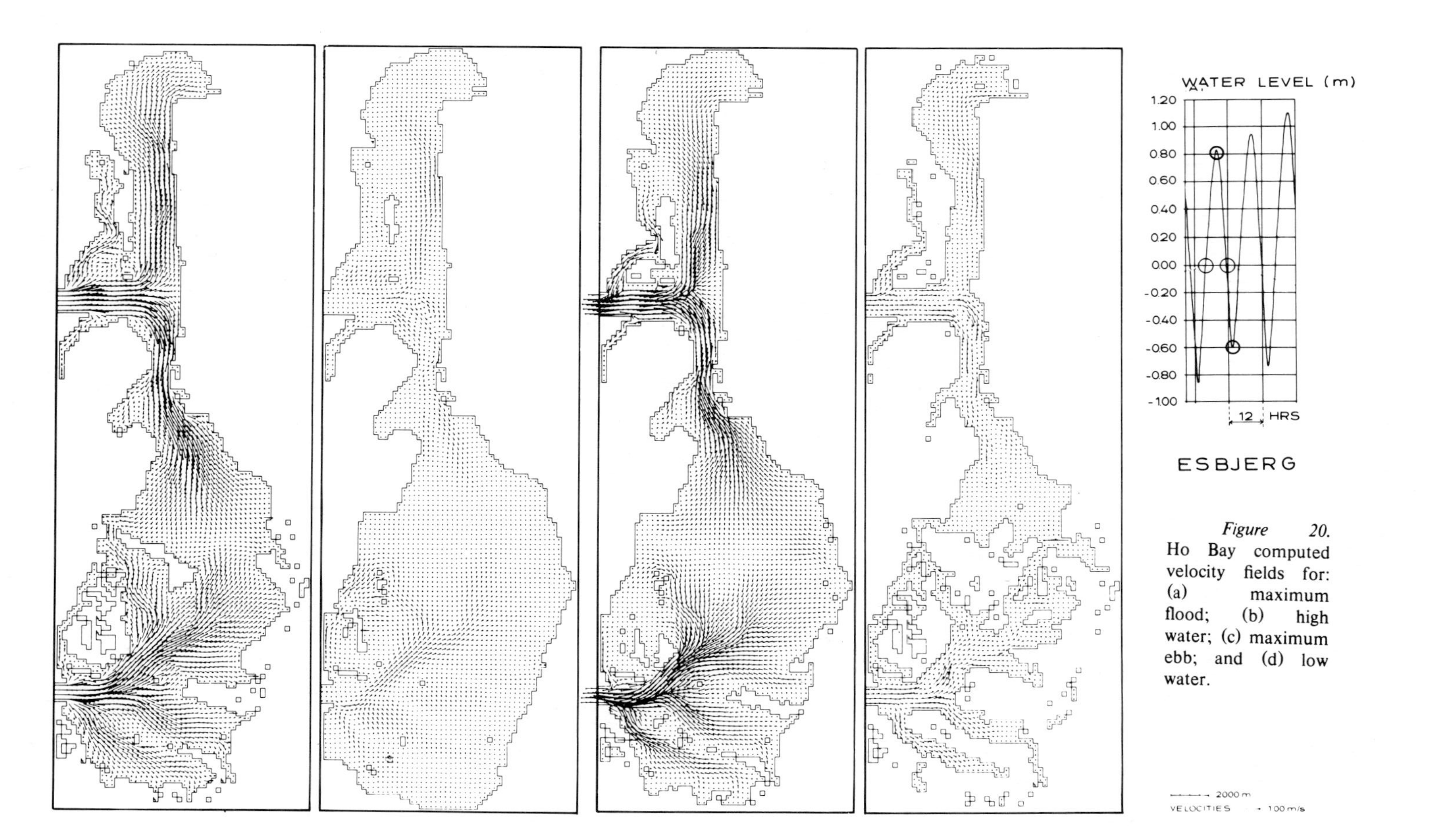

Figure 20. Ho Bay computed velocity fields for: (a) maximum flood; (b) high water; (c) maximum ebb; and (d) low water.

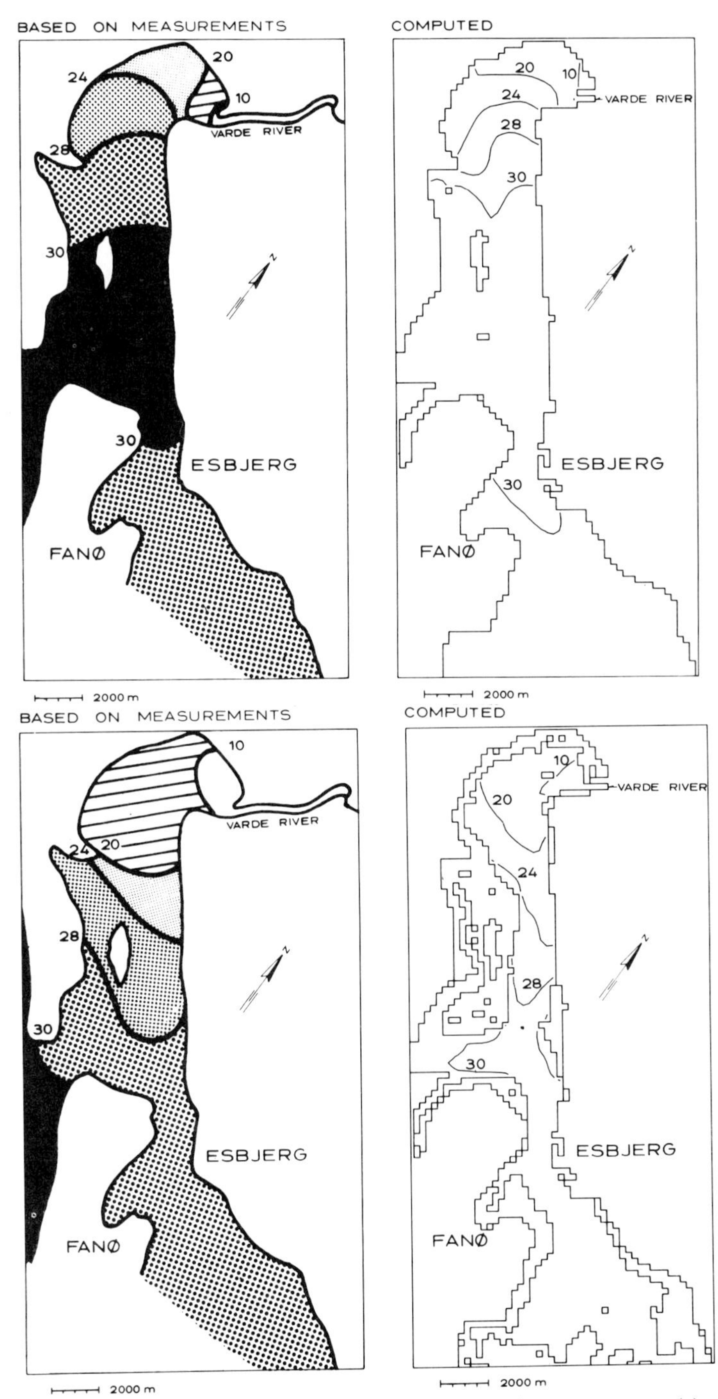

Figure 21. Ho Bay, measured and computed salinities for: (a) high water and (b) low water.

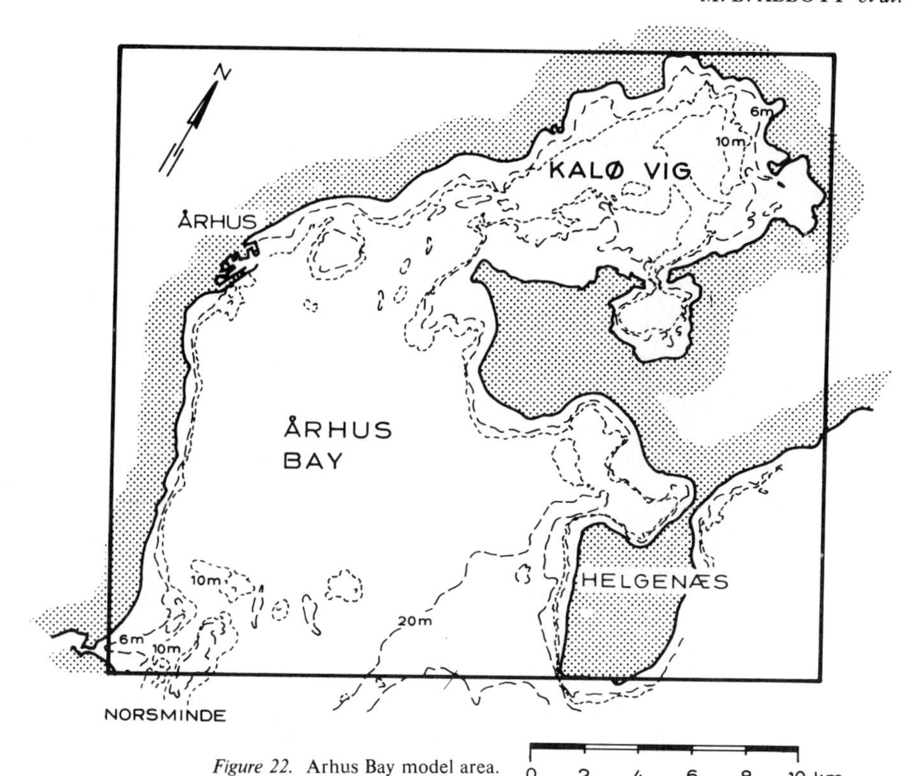

Figure 22. Arhus Bay model area.

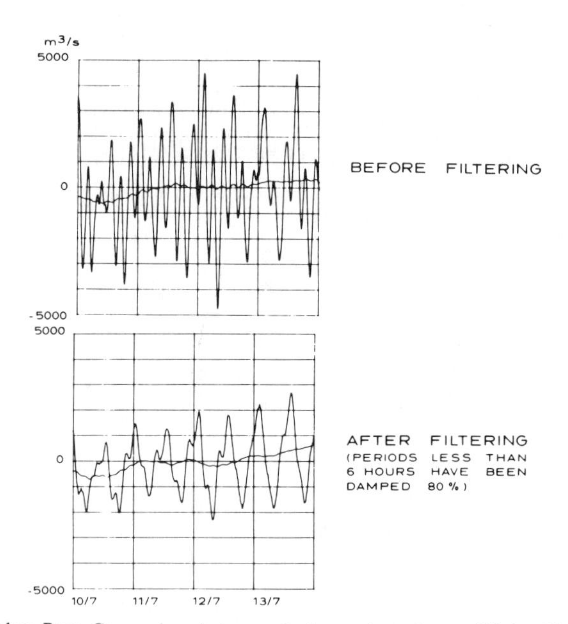

Figure 23a. Arhus Bay: Comparison between discharges in and out of Kaloe Vig calculated with System 22 before and after filtering of boundary data.

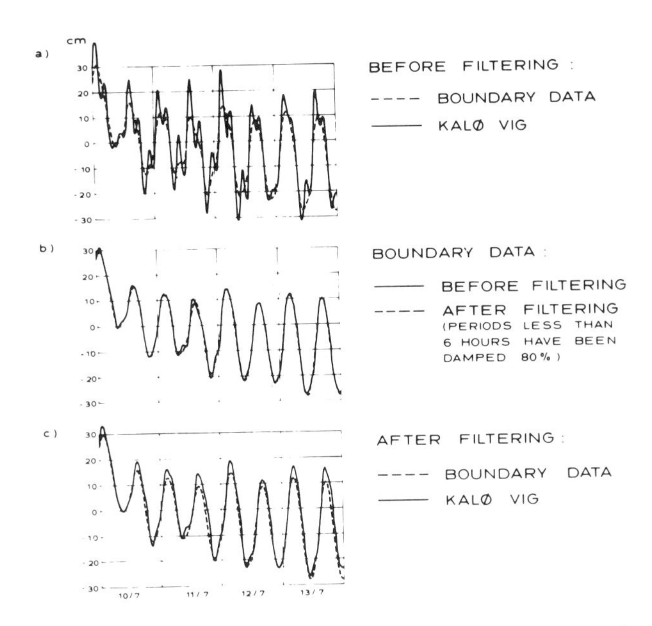

Figure 23b. Arhus Bay: Comparison between water levels in Kaloe Vig calculated with System 22 before and after filtering of the boundary data.

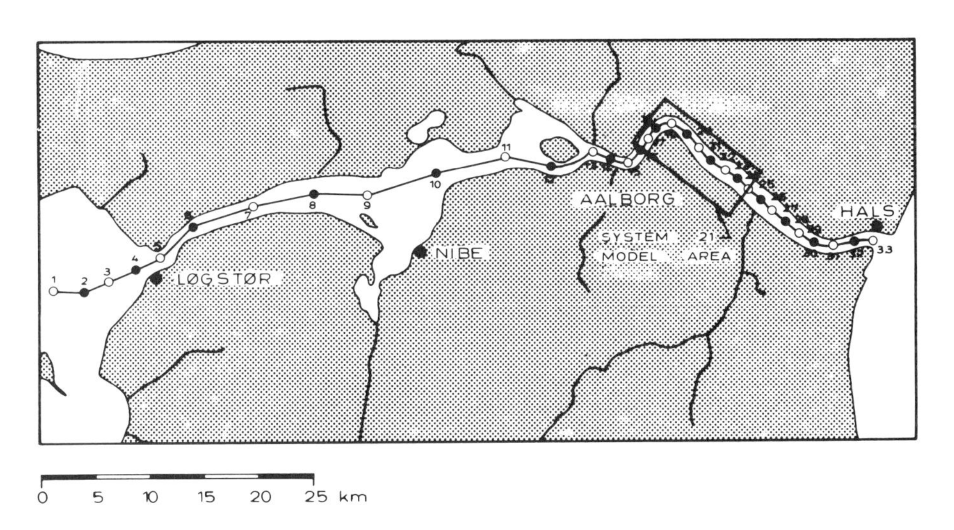

Figure 24. Vendsysselvaerket model areas.

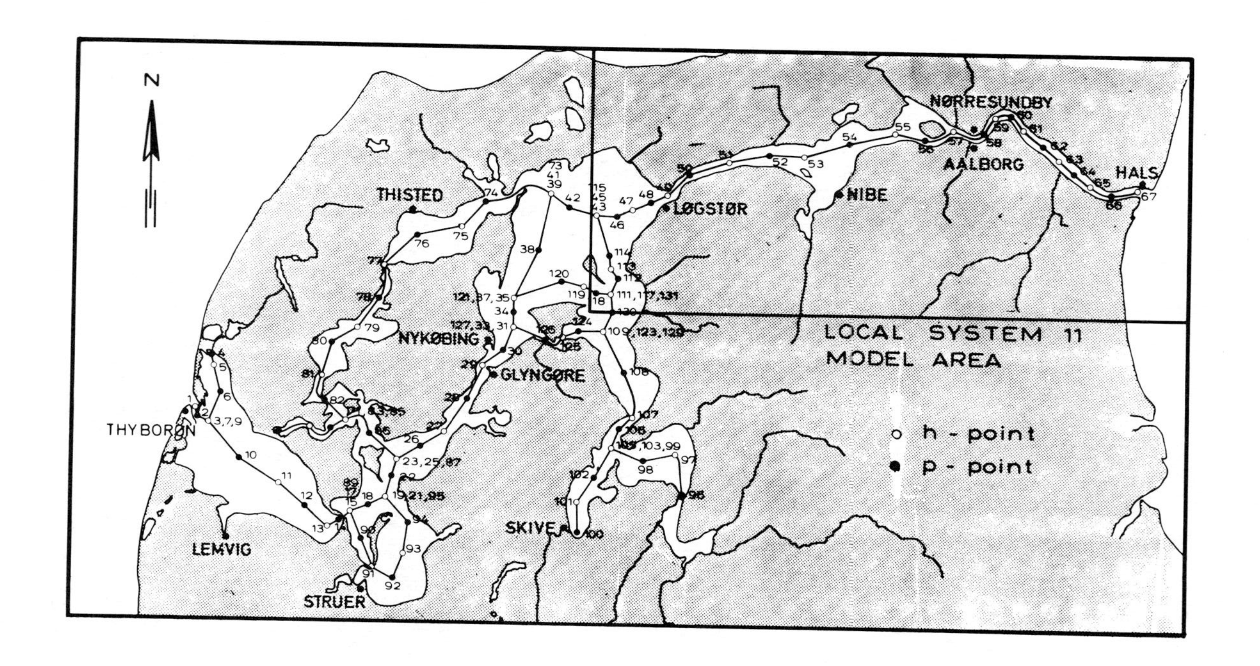

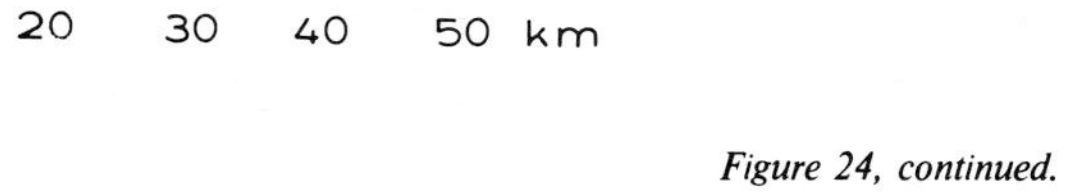

Figure 24, continued.

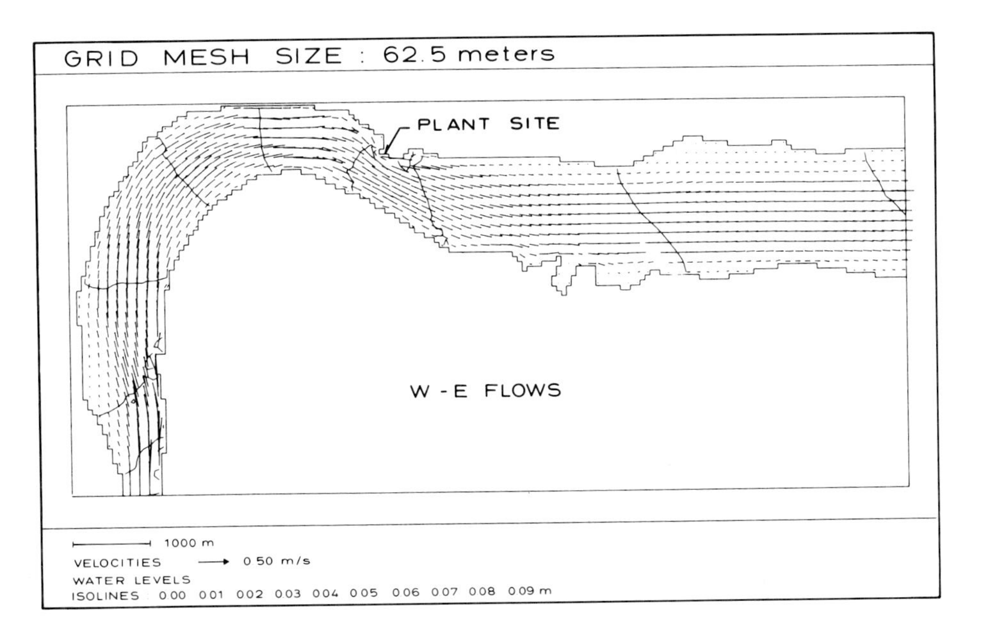

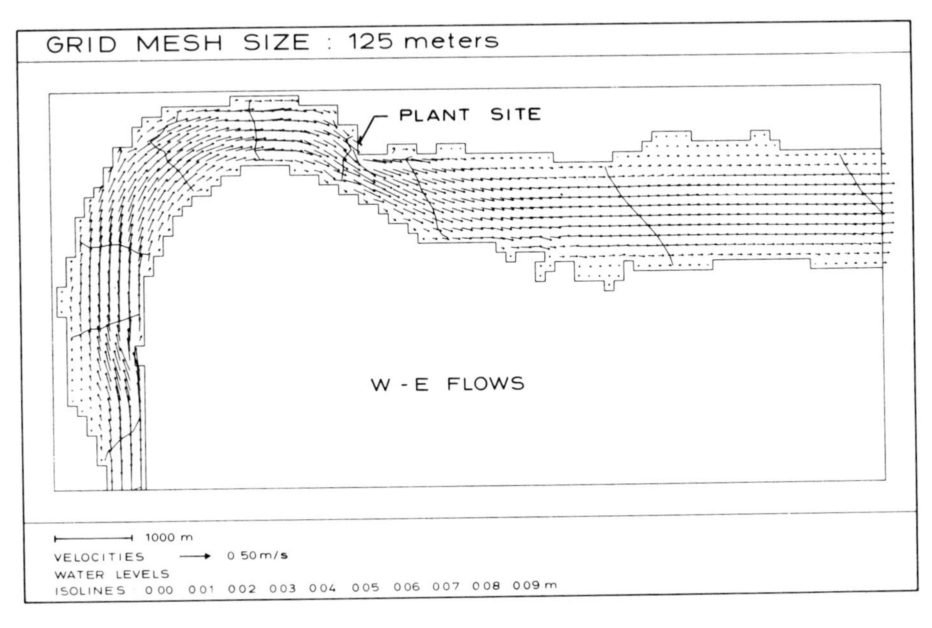

Figure 25. Vendsysselvaerket flow field comparisons.

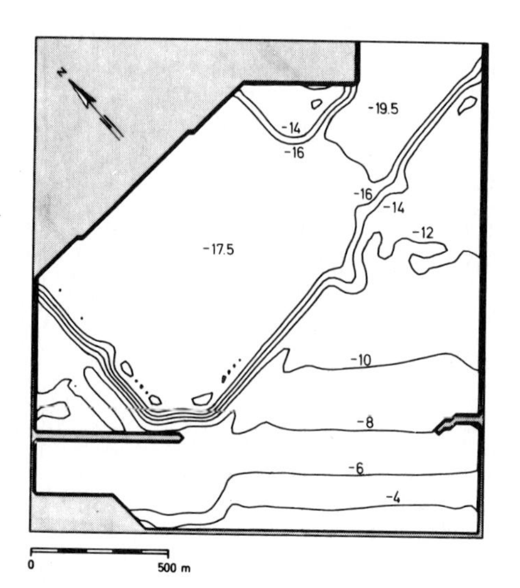

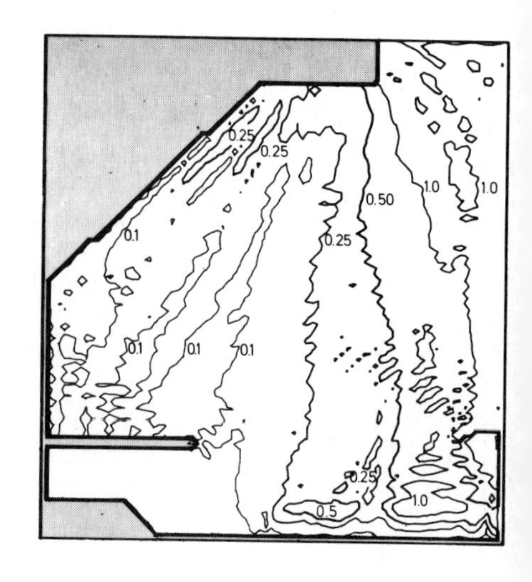

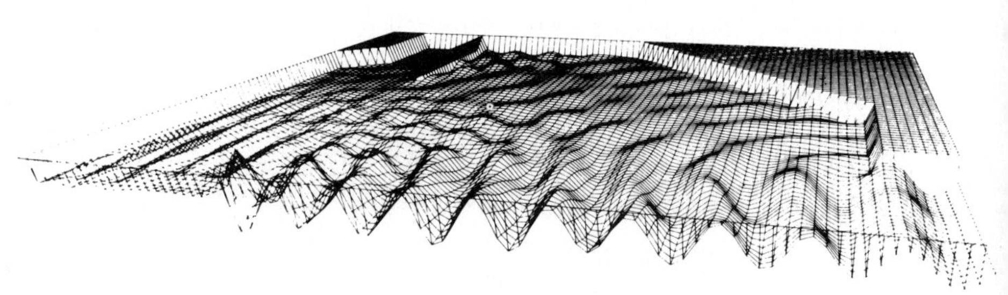

Figure 26. Misurata, System 21 Mk8. Short wave simulations T = 14 seconds. (a) Bathymetry. (b) Attenuation factors. (c) Perspective plot.

REFERENCES

Atlas der Gezeitenströmen für die Nordsee, der Kanal and die Irische See, (1968). Seehydrographische Dienst der D.D.R., Zweite Verbesserte Auflage, Rostock.

Abbott, M. B. (1979). Computational Hydraulics: Elements of the Theory of Free-Surface Flows. Pitman, London.

Abbott, M. B., Damsgaard, A. and Rodenhuis, G. S. (1973). System 21, Jupiter, A design system for two-dimensional nearly-horizontal flows. *J. Hyd. Res.,* **1,** 1-28.

Abbott, M. B., Petersen, H. M. and Skovgaard, O. (1978). Numerical modelling of short waves in shallow water. *J. Hyd. Res.,* **16,** 173-203.

Clark, R.A., Ferziger, J.H., and Reynolds, W.C. (1977). Evaluation of subgrid-scale turbulence models using a fully simulated turbulent flow. Thermosciences Division, Department of Mechanical Engineering, Stanford University, Rep. No. TF-9.

Deardorff, J. W. (1971). On the magnitude of the subgrid scale eddy coefficient. *J. Comp. Phys.,* **7,** 120-133.

Ferziger, J.H. (1977). Large eddy numerical simulations of turbulent flows. *AIAA Journal,* **15,** No. 9, 1261-1267

Guelfand, I. M. and Chilov, G. E. (1962). "Les Distributions." Dunod, Paris. (Translated by G. Rideau.)

Hinstrup, P. I., Kej, A., and Kroszynski, U. (1977). A high-accuracy two-dimensional transport dispersion model for environmental applications. *Proc. Congr. Int. Assoc. Hydraul. Res. 17th* **B17,** 129-137.

Kuipers, J. and Vreugdenhil, C. B. (1973),. Calculations of Two-dimensional Horizontal Flow. Delft Hydraulics Laboratory Report S163, Part I.

Leendertse, J J. and Lin, S.-K. (1977). A Three-Dimensional Model for Estuaries and Coastal Seas: Vol. IV, Turbulent Energy Computation. The Rand Corp., R-2187-OWRT.

Leendertse, J. J. (1967). Aspects of a Computational Model for Long Water Wave Propagation. Rand Corp., RH-5299-RR, Santa Monica, California.

Leonard, A. (1974). Energy Cascade in Large-Eddy Simulations of Turbulent Fluid Flows. *Adv. in Geophys. A.,* **18,** 37-248.

Leonard, B. P. (1979). A survey of finite differences of opinion on numerical muddling of the incomprehensible defective confusion equation. *Proc. Am. Soc. Mech. Eng., Winter Annual Meeting,* Publ. No. AMD-34.

Leslie, D. C. and Quarini, G. L. (1979). The application of turbulence theory to the formulation of subgrid modelling procedures. *J. Fluid Mech.* **91,** 5-91.

Millar, W. A. and Yevjevich, V. (1975). "Unsteady Flow in Open Channels." Vol. III, Bibliography. Water Resources Publications, Fort Collins, Colorado.

Rodi, W. (1978). "Turbulence Models and their Application in Hydraulics. A State of the Art Review." SFB 80/T/127.

Sobey, R. J. (1970). Finite Difference Schemes Compared for Wave-Deformation Characteristics, etc. Tech. Memorandum No. 32, U. S. Army Corps of Engineers, Coastal Engineering Research Center, Washington, D.C.

Smagorinsky, J. S. (1963). General circulation experiments with the primitive equations. I: The basic experiment. *Mon. Weath. Rev.,* **91,** 99.

Welander, P. (1955). Studies on the general development of motion in a two-dimensional, ideal fluid. *Tellus* **2,** 141-156.

A THREE-DIMENSIONAL MODEL FOR TIDAL AND RESIDUAL CURRENTS IN BAYS

Kim-Tai Tee

Atlantic Oceanographic Laboratory,
Bedford Institute of Oceanography,
Dartmouth, Nova Scotia, Canada

1. INTRODUCTION

Simulations of three-dimensional tidal currents in a sea or basin with irregular coastlines have usually involved a three-dimensional numerical model (i.e., Leendertse and Liu, 1975; Heaps, 1972). However, this type of computation is expensive and time-consuming and has poor vertical resolution because of limitations in computer time and storage. In this paper, a method for computing the three-dimensional structure of the first-order oscillating current and second-order tidally induced residual current is presented. The method, in comparison to the three-dimensional numerical model, is relatively simple and efficient, does not involve a large amount of computer time and storage, and provides accurate solutions, especially near the bottom.

The advantage of the simple model is that it can be applied by those who do not have access to a large-capacity computer, or who wish to avoid developing complicated computer programs, and also it is much more efficient in exploring the dynamics of a current system than the three-dimensional model. However, certain simplifications of the model may require input from numerical models and experimental observations. For example, the model cannot be applied to a highly non-linear system because of the linearization of the equation of motion; the vertical eddy viscosity coefficient is assumed to be time-independent; and the computation is applicable only for a homogeneous water column. Extension of the model to include stratification is being examined. The effects of the other simplifications on the predictability of the model will be discussed in Section 4.

Detailed descriptions of the model and the dynamics of the tidal current and tidally-induced residual current have been reported or submitted for publication elsewhere (Tee, 1979, 1980, 1981). In the following sections, we discuss only briefly some of the results, with particular emphasis on the method of computation and the predictability of the simple model. In Section 2, a method of computation

ISBN: 0-12-25852-0

is described. An example of the computation, which gives the first-order tidal current and second-order residual current for tides propagating perpendicularly to a straight coast, is given in Section 3. The predictability of the simple model is discussed in Section 4.

2. METHOD OF COMPUTATION

2.1 The Governing Equations

For a Cartesian system with coordinates x, y, and z, where x and y are the horizontal coordinates and z is the vertical coordinate measured vertically upward from the mean sea level, the equations describing the tidal motion in the homogeneous basin are

$$\frac{\partial \underline{u}}{\partial t} + \underline{u} \cdot \nabla \underline{u} + w \frac{\partial \underline{u}}{\partial z} + \underline{f} \times \underline{u} = -g\nabla\zeta + \frac{\partial}{\partial z} N \frac{\partial \underline{u}}{\partial z} + A_h \nabla^2 \underline{u}$$

$$\frac{\partial \zeta}{\partial t} + \nabla \cdot \int_{-D}^{\zeta} \underline{u} dz = 0 \tag{1}$$

where $\underline{u}$ is the horizontal velocity vector with components u and v in coordinates x and y; w is the vertical velocity component; d is the depth of the bottom below the mean sea level; ζ is the height of the water surface above the mean sea level; $\underline{f}$ is the Coriolis parameter; N is equal to $N(x,y,z)$ is a vertical eddy viscosity coefficient; A_h is a horizontal eddy viscosity coefficient; g is gravity; ∇ is the horizontal gradient operator; and ∇^2 is equal to $\partial^2/\partial x^2 + \partial^2/\partial y^2$ is the horizontal Laplacian operator.

Non-dimensionalizing (1) through the following transformation,

$$\begin{aligned} &(x,y) = L_l(x_n,y_n); \ \ z = D_l z_n; \ \ D = D_l D_n; \\ &t = \sigma^{-1} t_n; \ \ f = \sigma f_n; \ \ N = N_l N_n; \\ &\zeta = \zeta_l \zeta_n; \ \ \underline{u} = U_l \underline{u}_n; \ \ w = (U_l D_l / L_l) w_n \end{aligned} \tag{2}$$

where L_l is a horizontal length scale which is chosen to be $\sigma^{-1}(gD_l)^{1/2}$ (wavelength/2π), σ (tidal frequency) a frequency scale, N_l a vertical eddy viscosity scale, D_l a depth scale, ζ_l a tidal amplitude scale, and U_l the horizontal velocity scale which is equal to $(gD_l^{-1})^{1/2}\zeta_l$, we obtain

$$\frac{\partial \underline{u}_n}{\partial t} + \epsilon \left(\underline{u}_n \cdot \nabla \underline{u}_n + w_n \frac{\partial \underline{u}_n}{\partial z} \right) + \underline{f}_n \times \underline{u}_n$$

$$= -\nabla\zeta_n + E_1 \frac{\partial}{\partial z} N \frac{\partial \underline{u}_n}{\partial z} + E_2 \nabla^2 \underline{u}_n$$

$$\frac{\partial \zeta_n}{\partial t} + \nabla \cdot \int_{-D_n}^{\epsilon\zeta_n} \underline{u}_n dz_n = 0 \tag{3}$$

where E_1 is equal to $N_l(\sigma D_l^2)^{-1}$, E_2 is equal to $A_h(\sigma L_l^2)^{-1}$ and ϵ is equal to ζ_l/D_l (aspect ratio). With the frequency scale of σ instead of f, E_1 and E_2 are equivalent respectively to the Ekman Number and the horizontal Ekman Number. For typical values of $D_l \sim 20$m, $\zeta_l \sim 1$m, $\sigma \sim 10^{-4}\text{sec}^{-1}$ (semi-diurnal and diurnal tides), $N_l \sim 2\times10^{-2}\text{m}^2\text{sec}^{-1}$ (Tee, 1979) and $A_h \sim 10^2\text{m}^2\text{ sec}^{-1}$, we obtain $\epsilon \sim 0.05$, $E_1 \sim 0.5$ and $E_2 \sim 5\times10^{-5}$. The horizontal diffusion terms are thus negligible in a shallow sea, except in the coastal boundary layer. For the vertical diffusion, as E_1 is in the order of 1, we can simplify Eq. (3) by choosing the depth scale D_l is equal to $(N_l/\sigma)^{1/2}$ so that E_1 is equal to 1.

The current $\underline{u}$ has two components: the first is the tidal current which includes the oscillating current of various tidal constituents and, through non-linear interactions, induced shallow water and residual components; and the second is the wind driven current. In the tide-dominated estuary considered in this paper, the governing equations for the two components of the current are separable because the non-linear terms and the vertical eddy viscosity coefficient (linear form) involve only the first-order oscillating current. Thus, without losing any generality on the computation of the tidal current, we assume in this paper that wind forcing is equal to zero.

For simplicity, the subscript (n) of the non-dimensional variables is omitted in the following computation. Applying a perturbation to the order ϵ

$$\underline{u} = \underline{u}_1 + \epsilon \underline{u}_2; \quad \zeta = \zeta_1 + \epsilon\zeta_2$$

where the subscript (1) denotes the first-order oscillating component, and subscript (2) the second-order residual component, we obtain from (3) the first-order equations

$$\frac{\partial \underline{u}_1}{\partial t} + \underline{f} \times \underline{u}_1 = -\nabla\zeta_1 + \frac{\partial}{\partial z} N \frac{\partial \underline{u}_1}{\partial z} + E_2\nabla^2\underline{u}_1 \tag{4a}$$

$$\frac{\partial \zeta_1}{\partial t} + \nabla \cdot D\underline{U}_1 = 0 \tag{4b}$$

and the second-order equations

$$\overline{\underline{u}_1 \cdot \nabla \underline{u}_1} + \overline{w_1 \frac{\partial \underline{u}_1}{\partial z}} + \underline{f} \times \underline{u}_2 = -\nabla\zeta_2 + \frac{\partial}{\partial z} N \frac{\partial \underline{u}_2}{\partial z} + E_2\nabla^2\underline{u}_2 \tag{5a}$$

$$\nabla \cdot (D\underline{U}_2 + \overline{\zeta_1 \underline{u}_{1S}}) = 0 \tag{5b}$$

where the overbar ($\overline{\quad}$) denotes the time average over a tidal period, $\underline{u}_{1S}$ is the value of $\underline{u}_1$ at surface ($z = 0$), and $\underline{U}_1$ and $\underline{U}_2$ are the depth averaged values of $\underline{u}_1$ and $\underline{u}_2$

$$\underline{U}_{1,2} = <\underline{u}_{1,2}> = \frac{1}{D} \int_{-D}^{0} \underline{u}_{1,2}\, dz \tag{6}$$

The procedure of solving $\underline{u}_1$ and $\underline{u}_2$ is sketched in Fig. 1, and described in the following sections.

2.2 The Solutions of the Oscillating Current

To simplify some of the formulations, the following notations are used here

$$\Gamma_0 = \int_{-D}^{0} \int_{-D}^{z'} \frac{1}{N}\, dz dz' \tag{7}$$

$$\Gamma_1(s) = \frac{1}{D} \int_{-D}^{0} \int_{-D}^{z'} \frac{s}{N}\, dz dz' \tag{8}$$

and

$$\Gamma_2(s) = \frac{1}{D} \int_{-D}^{0} \int_{-D}^{z''} \frac{1}{N} \int_{z'}^{0} s\, dz dz' dz'' \tag{9}$$

where s is a variable which can be a scalar or a vector.

From Eq. (4a), the governing equation for the first-order oscillating current in the offshore area can be written as

$$i(1 + f)L_+ = -A_+ + \frac{\partial}{\partial z} N \frac{\partial L_+}{\partial z} \tag{10}$$

and

$$i(1 - f)L_- = -A_- + \frac{\partial}{\partial z} N \frac{\partial L_-}{\partial z} \tag{11}$$

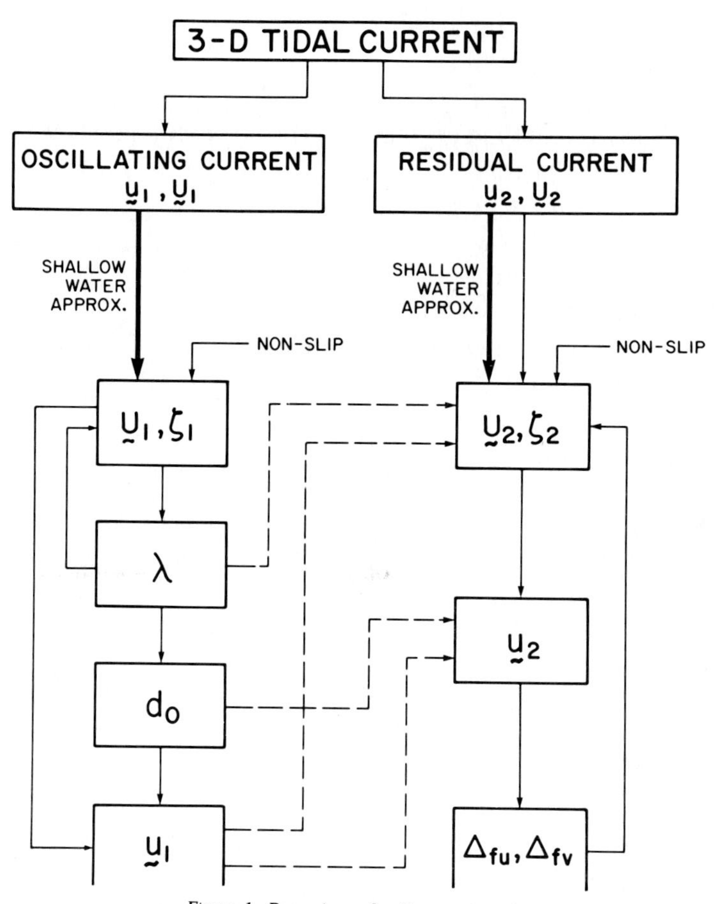

Figure 1. Procedures for the computation.

where $L_{\pm} = \hat{u}_1 \pm i\hat{v}_1$, $A_{\pm} = \frac{\partial \hat{\zeta}_1}{\partial x} \pm i\frac{\partial \hat{\zeta}_1}{\partial y}$ and a cavet such as $\hat{u}_1$ denotes a complex function in the transform

$$u_1 = \text{Re}\ (\hat{u}_1\ e^{it}) \tag{12}$$

Integrating Eq. (10) from $z = 0$ to $z = z'$, and applying the boundary condition $N\ \partial \underline{u}_1/\partial z = 0$ at $z = 0$, we obtain

$$N\ \frac{\partial L_+}{\partial z'} = A_+ z' - i(1 + f) \int_{z'}^{0} L_+ dz \tag{13}$$

Integrating Eq. (13) twice from $z' = z''$ to $-D$, and from $z'' = 0$ to $z'' = -D$, we obtain

$$\mathbf{L}_+ = A_+\alpha - i(1 + f)\Gamma_2(L_+) \tag{14}$$

where

$$\alpha \equiv \Gamma_1(z) \tag{15}$$

$\mathbf{L}_+ = \langle L_+ \rangle = 1/D\int_{-D}^{0} L_+ dz$ is the depth-average of L_+, which corresponds to $\hat{U}_1 + i\hat{V}_1$ where $\hat{U}_1$ and $\hat{V}_1$ are the x and y components of the depth-averaged velocity $\underline{\hat{U}}_1$. Substituting A_+ from Eq. (14) to (13), and setting z' in Eq. (13) to $-D$, we obtain

$$\frac{1}{D} N \left.\frac{\partial L_+}{\partial z'}\right]_{-D} = -\frac{1}{\alpha}\mathbf{L}_+ - i(1 + f)\mathbf{L}_+\Delta_+ \tag{16}$$

where

$$\Delta_+ = -\left(1 + \frac{1}{\alpha}\frac{\Gamma_2(L_+)}{\mathbf{L}_+}\right) \tag{17}$$

Substituting Eq. (16) back into Eq. (13) for $z' = -D$, we obtain

$$(1 + \Delta_+)i(1 + f)\mathbf{L}_+ = -A_+ + \frac{1}{\alpha}\mathbf{L}_+ \tag{18}$$

Similarly, we obtain the momentum equation for L_- as

$$(1 + \Delta_-)i(1 - f)\mathbf{L}_- = -A_- + \frac{1}{\alpha}\mathbf{L}_- \tag{19}$$

where

$$\Delta_- = -\left(1 + \frac{1}{\alpha}\frac{\Gamma_2(L_-)}{\mathbf{L}_-}\right) \tag{20}$$

From Eqs. (18) and (19), and the relationships

$$\hat{U}_1 = \frac{1}{2}(\mathbf{L}_+ + i\mathbf{L}_-) \quad \text{and} \quad \hat{V}_1 = \frac{1}{2i}(\mathbf{L}_+ - i\mathbf{L}_-) \tag{21}$$

The depth-averaged momentum equation of $\underline{U}_1$ can be written as

$$(1 + \Delta_\sigma) i\underline{\hat{U}}_1 + (1 + \Delta_c)\, \underline{f} \times \underline{\hat{U}}_1 = -\nabla\hat{\zeta}_1 + \frac{1}{\alpha}\, \underline{\hat{U}}_1 \tag{22}$$

where

$$\Delta_c = \frac{1}{2}\left[(\Delta_+ + \Delta_-) + \frac{1}{f}\,(\Delta_+ - \Delta_-)\right] \tag{23}$$

and

$$\Delta_\sigma = \frac{1}{2}[(\Delta_+ + \Delta_-)] + f(\Delta_+ - \Delta_-)] \tag{24}$$

From (22) and the continuity equation (4b), we can see that the depth-averaged velocity $\underline{U}_1$ can be calibrated if Δ_σ and Δ_c are known. Although the values of Δ_σ and Δ_c cannot be obtained easily because they depend intrinsically on the vertical variation of the tidal current, it will be shown later that the computation of $\underline{U}_1$ can still be carried out because $(1+\Delta_\sigma)$ and $(1+\Delta_c)$ can be approximated to unity.

To compute the vertical profile of the tidal current, we differentiate (10) and (11) with z and obtain the equation for the offshore area as

$$i(1 + f)L'_+ = \frac{\partial^2}{\partial z^2} N\, L'_+ \tag{25}$$

and

$$i(1 - f)L'_- = \frac{\partial^2}{\partial z^2} N\, L'_- \tag{26}$$

where $L'_\pm \equiv \partial L_\pm / \partial z$.

Four forms of N were considered in solving the first-order oscillating currents (Fig. 2): (i) N is independent of z; (ii) N varies parabolically with depth, e.g.,

$$N = N_m\{R_1 + 4(R_1 - 1)\eta + 4(R_1 - 1)\eta^2\} \tag{27}$$

where $\eta \equiv z/D, N_m$ is the maximum vertical eddy viscosity at $\eta = -½$, and R_1 is the ratio of N at the surface or bottom to N_m; (iii) N increases rapidly from a laminar sublayer at the bottom to a uniform eddy coefficient N_m in the turbulent layer, i.e.

$$N = \begin{cases} \nu_0\{1 + R_2 D(\eta + 1)\}^2 & \text{for } \eta \leqslant \eta_z \\ N_m = \nu_0(1 + R_2\delta_z)^2 & \text{for } \eta \geqslant \eta_z \end{cases} \tag{28}$$

where $\nu_0 = (1.4 \times 10^6 \text{m}^2\text{s}^{-1}/N_l)$ is the non-dimensionalized molecular eddy viscosity, $\eta_z = [(\delta_z/D) - 1]$ is the thickness of the transition layer between laminar and turbulent flow, and R_2 is a parameter. (iv) N has the combined characteristics of (ii) and (iii); that is, for $\eta < \eta_z$, N increases rapidly according to (28) and above that it increases to N_m at $\eta = \frac{1}{2}\eta_z$ according to (27).

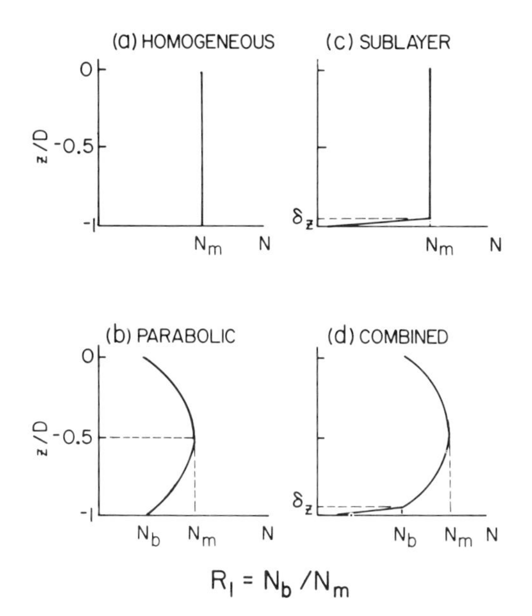

Figure 2. Various forms of the vertical eddy viscosity: (a) homogeneous model, (b) parabolic model, (c) sublayer model, and (d) combined model.

Cases (i), (ii), (iii) and (iv) are denoted as the homogeneous model, parabolic model, sublayer model and combined model, respectively. Equations (25) and (26) can be solved analytically for the homogeneous and sublayer models, and numerically for the parabolic and combined models. In order to resolve the rapid variation near the bottom, the numerical method involves using 110 levels with grid spacing increasing parabolically from the bottom. Detailed descriptions of the computation are given in Tee (1979). Using the boundary conditions $L'_{\pm} = 0$ at $=0$, $L_{\pm} = 0$ at $z = -D$, and $< L_{\pm} > = \boldsymbol{L}_{\pm}$ (6), it was found that $L_{\pm}$ could be written as (Tee, 1979)

$$L'_{+} = \frac{\boldsymbol{L}_{+}}{D} F_{+n} \quad \text{and} \quad L'_{-} = \frac{\boldsymbol{L}_{-}}{D} F_{-n} \tag{29}$$

where $F_{\pm n}$ is the function describing the vertical variation of the tidal current; it depends only on the non-dimensional parameter,

$$d_o(x,y) = D/(2N_m)^{1/2} \tag{30}$$

where N_m is defined as the maximum value of N at a vertical column (Figure 2). From (29), L_+ and L_- can be calculated as

$$L_{+} = \boldsymbol{L}_{+} F_{+I} \quad \text{and} \quad L_{-} = \boldsymbol{L}_{-} F_{-I} \tag{31}$$

where $F_{\pm I} = < F_{\pm n} >$. The three dimensional tidal current, $\underline{\hat{u}}$ is then

$$\hat{u}_1 = \frac{1}{2} [(\hat{U} + i\hat{V}) F_{+I} + (\hat{U} - i\hat{V}) F_{-I}]$$

$$\hat{v}_1 = \frac{1}{2i} [(\hat{U} + i\hat{V}) F_{+I} - (\hat{U} - i\hat{V}) F_{-I}] \tag{32}$$

For a given f and form of N, since $F_{\pm n}$ is only a function of d_o, we can see from (32) that $\underline{u}_1$ can be determined if d_o and the depth-averaged tidal current $\underline{U}_1$ are known. We now proceed to determine d_o and $\underline{U}_1$.

Using (31), Δ_+, and Δ_- become (17, 20)

$$\Delta_{+} = [1 + \frac{1}{\alpha}\Gamma_2 (F_{+I})] \quad \text{and} \quad \Delta_{-} = -[+ \frac{1}{\alpha}\Gamma_2 (F_{-I})] \tag{33}$$

For a given f and form of N, since $F_{\pm n}$ is only a function of d_o, we can see from (33), and the definition of Γ_2 (9), α (15), and $F_{\pm I}$, that Δ_+ and Δ_-, and thus Δ_σ and Δ_c are only functions of d_0. Figure 3 shows the amplitudes and phases of $(1 + \Delta_\sigma)$ and $(1 + \Delta_c)$ versus d_o for the sublayer model with $\delta_z \sim 0.1\ m/D_l$ We can see that $(1 + \Delta_\sigma)$ and $(1 + \Delta_c)$ deviates only slightly from unity. Similar solutions are found for the parabolic model of $R_1 > 0.1$, and combined model of $\delta_z = 0.07\ m/D_l$ and $R_1 = 0.5$. The values of $(1 + \Delta_\sigma)$ and $(1 + \Delta_c)$ for the homogeneous model which is considered unrealistic (Ianniello, 1977; Tee, 1979; also see Section 4), deviate more significantly from unity. For the known value of

D and N_m, we can calculate d_o from (30), and thus determine the parameters $(\frac{1}{\alpha})$, $(1 + \Delta_\sigma)$ and $(1 + \Delta_c)$ in the depth-averaged momentum equation (22) at the offshore area. Since $(1 + \Delta_\sigma)$ and $(1 + \Delta_c)$ are close to 1, we can approximate (22) to the shallow water equation

$$i\underline{\hat{U}}_1 + \underline{f}x\underline{\hat{U}}_1 = -\nabla\hat{\underline{\zeta}}_1 - \lambda\underline{\hat{U}}_1 \tag{34}$$

where

$$\lambda = -\frac{1}{\alpha} \tag{35}$$

is the linear friction coefficient. The value of N_m is chosen such that (35) is satisfied. The value of d_o can then be calculated from (30). The linear friction coefficient λ can be estimated by linearizing the quadratic bottom stress $\underline{\tau}_{b1} = \rho r \mid \underline{U}_1 \mid \underline{U}_1 \mid$ to a linearized form

$$\underline{\tau}_{b1} \approx \rho\,\lambda\,D\,\underline{U}_1 \tag{36}$$

which gives λ as proportional to the current amplitude U_m and inversely proportional the depth D, i.e., s.p

$$\lambda(x,y) = K\,\frac{\gamma U_m}{D}$$

where γ is the bottom friction coefficient usually taken to be 0.002 to 0.003, and K is a constant which is equal to $8/3\pi$ for a rectilinear flow and equal to 1 for the flow with similar amplitudes of major and minor axis. Figure 4 shows the variation of λ versus d_o for the sublayer with $\delta_z = 0.1\ m/D_l$. Similar results can also be obtained from the other models.

Using a two-dimensional (x,y) numerical or analytical model, $\underline{U}_1$ can be computed from the shallow water equation (34) and the continuity equation (4b). The three-dimensional tidal current $(\underline{u}_1)$ can then be calculated from (32).

To obtain a final solution for the closed or semiclosed basin, the tidal current in the coastal boundary layer remains to be investigated. As there is no simple method of making the computation in this layer where the horizontal mixing terms cannot be ignored, a three-dimensional numerical model with very fine grid spacing near the coast will probably have to be used.

In fact, in most of the numerical models, the coastal boundary layer is not resolved and the velocity inside the layer is not required. In this paper, the velocity structure in the coastal boundary layer is not studied. However, in the numerical evaluation of the depth-mean velocity, we include the horizontal mixing term in the momentum equation (34), i.e.,

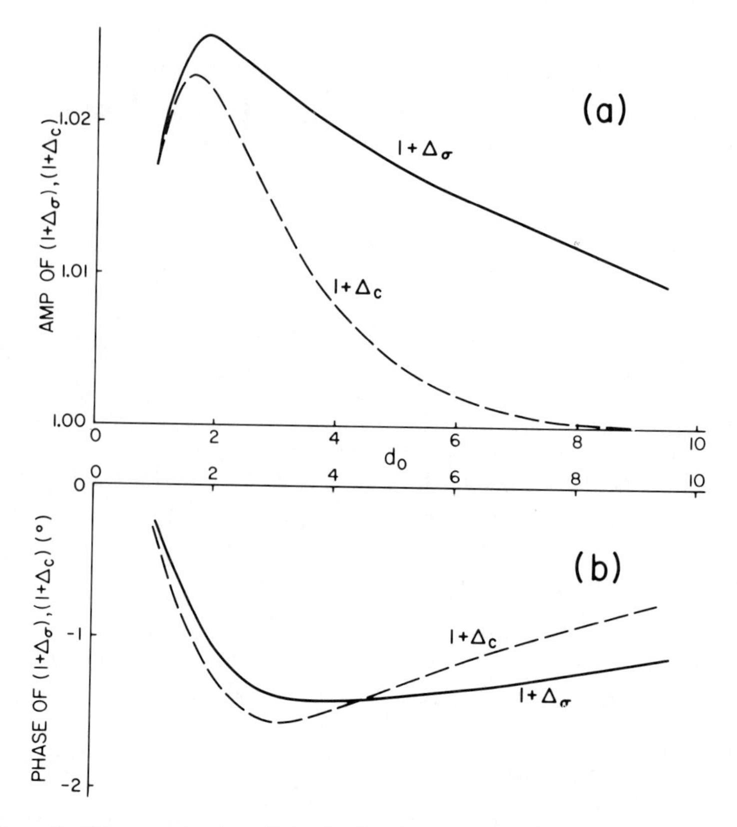

Figure 3. The approximation of the depth-averaged momentum equation for the first-order oscillation current, Eq. (22), to the shallow water equation (34). The amplitudes (a) and phases (b) of $(1 + \Delta_\sigma)$ and $(1 + \Delta_c)$.

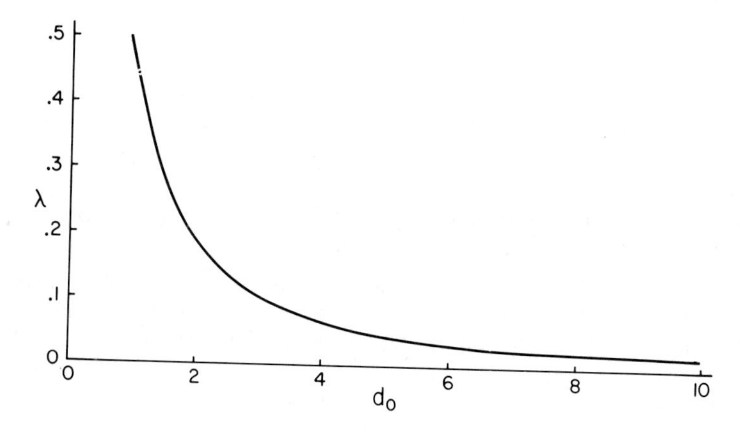

Figure 4. The linear frictional coefficient λ versus d_0.

$$i\underline{\hat{U}}_1 + \underline{f} \times \underline{\hat{U}}_1 = -\nabla\hat{\zeta}_1 - \lambda\underline{\hat{U}}_1 + E_2\nabla^2\underline{\hat{U}}_1 \tag{34a}$$

This inclusion of the horizontal mixing term implies a non-slip boundary condition at the coast and thus generates the correct vorticity values which, as indicated by Tee (1976, 1977), cannot be generally neglected in areas of strong tidal currents and irregular coastlines.

In summary, we first compute, using the given values of U_m and D, the horizontal distribution of λ. The depth-averaged tidal current, $\underline{U}_1$, can then be computed from the shallow water equations (34a) and (4b). The computation can be iterated by using the computed value of current amplitude, U_m, to update the value of λ. The equilibrium state is reached fairly quickly because a slight variation of λ has a negligible effect on the value of U_m. From the values of λ, the non-dimensional parameter d_o can be obtained from Fig. 4, and the functions describing the vertical variation of the tidal current (F_{+I}, F_{-I}) can then be computed. Using the computed depth-averaged tidal current, $\underline{U}_1$, and the functions F_{+I} and F_{-I}, the three-dimensional tidal current $\underline{u}_1$ can then be calculated from (32). Note that, although the tidal current in the coastal boundary layer is not solved, the three-dimensional structures of the tidal currents in the offshore area can still be predicted because the depth-averaged velocity U_1, on which u_1 depends, can be correctly computed from (34a) and (4b).

2.3 The Solutions of the Residual Current

To solve the residual current, we apply the depth-averaged operator to the momentum equation (5a) and obtain

$$\overline{<\underline{u}_1 \cdot \nabla \underline{u}_1>} + \overline{<w_1 \frac{\partial \underline{u}_1}{\partial z}>} + \underline{f} \times \underline{U}_2 = -\nabla\zeta_2 + \frac{\underline{\tau}_{s2} - \underline{\tau}_{b2}}{\rho D} + E_2\nabla^2\underline{U}_2 \tag{35}$$

where < > is the depth-averaged operator defined in (6)

$$\underline{\tau}_{b2} = \rho\left[N\frac{\partial \underline{u}_2}{\partial z}\right]\Bigg|_{z=-D} \quad \text{and} \quad \underline{\tau}_{s2} = \rho\left[N\frac{\partial \underline{u}_2}{\partial z}\right]\Bigg|_{z=0} \tag{36}$$

are the residual component of the bottom stress and the stress at $z = 0$. Note that although the stress at $z = \zeta$ is equal to zero in this study, the stress at $z = 0$ is equal to

$$\underline{\tau}_{s2} = -\frac{\partial}{\partial z} N \frac{\partial}{\partial z} \overline{\zeta_1 \underline{u}_1}\Bigg|_{z=0} \tag{37}$$

The result, which was derived by Dr. J. Ianniello, can be obtained from the expansion

$$\overline{N\frac{\partial u}{\partial z}}\bigg|_{z=\zeta} = \overline{N\frac{\partial u}{\partial z}}\bigg|_{z=0} + \overline{\zeta\frac{\partial}{\partial z}N\frac{\partial u}{\partial z}}$$

and the boundary conditions

$$\overline{N\frac{\partial u}{\partial z}}\bigg|_{z=\zeta} = 0$$

and

$$\overline{\frac{\partial \underline{u}_1}{\partial z}}\bigg|_{z=0} = 0$$

$\underline{\tau}_{s2}$, although it is not a stress on the surface $z = \zeta$, is present as a surface stress in the depth-averaged momentum equation (13). For convenience we therefore denote $\underline{\tau}_{s2}$ as a residual component of the surface stress in the following discussions.

The computation of $\underline{u}_2$ from (38) and (5b) is prevented by the unknown bottom stress $\underline{\tau}_{b2}$. To compute this parameter, we consider an area outside the coastal boundary, where the horizontal diffusion term can be neglected, and integrate (5a) from $z = 0$ to $z = z'$

$$\int_{z'}^{0} [\,\overline{\underline{u}_1 \cdot \nabla \underline{u}_1} + \overline{w_i \frac{\partial \underline{u}_1}{\partial z}}\,]dz + \underline{f} \times \int_{z'}^{0} \underline{u}_2 dz = -\nabla\zeta_2(-z') + \frac{\underline{\tau}_{s2}}{\rho} - N\frac{\partial \underline{u}_2}{\partial z} \tag{38}$$

Dividing all the terms in (38) by N and integrating from $z' = z''$ to $z' = -D$, we obtain

$$\begin{aligned}\underline{u}_2(z'') = {} & \nabla\zeta_2\,[\int_{-D}^{z''} \frac{z'}{N}\,dz'] \\ & - \int_{-D}^{z''} \frac{1}{N} \int_{z'}^{0} [\,\overline{\underline{u}_1 \cdot \nabla \underline{u}_1} + \overline{w_1 \frac{\partial \underline{u}_1}{\partial z}}\,]\,dz\,dz' \\ & + \frac{\underline{\tau}_{s2}}{\rho} \int_{-D}^{z''} (\frac{1}{N})\,dz' - \underline{f} \times \int_{z'}^{0} \underline{u}_2\,dz\,dz'\end{aligned} \tag{39}$$

The depth-average of (39) is

$$\underline{U}_2 = \nabla\zeta_2\,\Gamma_1(z) - \Gamma_2[\overline{\underline{u}_1 \cdot \nabla\underline{u}_1} + \overline{w_1\frac{\partial \underline{u}_1}{\partial z}}] + \frac{\underline{\tau}_{s2}}{\rho D}\,\Gamma_0 - \underline{f} \times \Gamma_2(\underline{u}_2) \quad (40)$$

Substituting $\nabla\zeta_2$ from (40) into (39) and setting z' in (41) to $-D$, we obtain the components of the bottom stress

$$\frac{\tau_{b2x}}{\rho} \equiv \frac{1}{D} N \frac{\partial u_2}{\partial z}\bigg|_{z=-D} = \lambda U_2 - \Delta_s \frac{\tau_{s2x}}{\rho D} + \Delta_u < \overline{\underline{u}_1 \cdot \nabla u_1} + \overline{w_1 \frac{\partial u_1}{\partial z}} > - \Delta_{fv}\, f V_2$$

$$\frac{\tau_{b2y}}{\rho} \equiv \frac{1}{D} N \frac{\partial v_2}{\partial z}\bigg|_{z=-D} = \lambda V_2 - \Delta_s \frac{\tau_{s2y}}{\rho D} + \Delta_v < \overline{\underline{u}_1 \cdot \nabla v_1} + \overline{w_1 \frac{\partial v_1}{\partial z}} > + \Delta_{fu} f U_2 \quad (41)$$

where τ_{s2x} and τ_{s2y} are the x and y components of the surface stress $\underline{\tau}_{s2}$

$$\Delta_u = -\left[1 + \frac{1}{\alpha}\,\frac{\Gamma_2\,(\overline{\underline{u}_1 \cdot \nabla u_1} + \overline{w_1 \frac{\partial u_1}{\partial z}})}{< \overline{\underline{u}_1 \cdot \nabla u_1} + \overline{w_1 \partial u_1/\partial z} >}\right] \quad (42)$$

$$\Delta_v = -\left[1 + \frac{1}{\alpha}\,\frac{\Gamma_2\,(\overline{\underline{u}_1 \cdot \nabla v_1} + \overline{w_1 \frac{\partial v_1}{\partial z}})}{< \overline{\underline{u}_1 \cdot \nabla v_1} + \overline{w_1 \partial v_1/\partial z} >}\right] \quad (43)$$

$$\Delta_{fv} = -\left[1 + \frac{1}{\alpha V_2}\,\Gamma_2\,(v_2)\right] \quad (44)$$

$$\Delta_{fu} = -\left[1 + \frac{1}{\alpha U_2}\,\Gamma_2\,(u_2)\right] \quad (45)$$

$$\Delta_s = -\,[1 + \Gamma_o/\alpha] \quad (46)$$

and $\lambda = -1/\alpha$ (35). Substituting the bottom stress (41) into (38), we obtain

$$[\overline{<\underline{u}_1 \cdot \nabla u_1>} + \overline{<w_1 \frac{\partial u_1}{\partial z}>}]\,(1 + \Delta_u) - fV_2\,(1 + \Delta_{fv}) =$$
$$- \frac{\partial \zeta_2}{\partial x} + \frac{\tau_{s2x}}{\rho D}\,(1 + \Delta_s) - \lambda U_2$$
$$[\overline{<\underline{u}_1 \cdot \nabla v_1>} + \overline{<w_1 \frac{\partial v_1}{\partial z}>}]\,(1 + \Delta_v) + fU_2\,(1 + \Delta_{fu}) =$$
$$- \frac{\partial \zeta_2}{\partial y} + \frac{\tau_{s2y}}{\rho D}\,(1 + \Delta_s) - \lambda V_2 \qquad (47)$$

In order to impose the non-slip boundary condition, which is found to be necessary for the computation of three-dimensional tidal current (Tee, 1979), the horizontal diffusion term is added to (47), i.e.,

$$\left[\overline{<\underline{u}_1 \cdot \nabla u_1>} + \overline{<w_1 \frac{\partial u_1}{\partial z}>}\right](1 + \Delta_u) - fV_2\,(1 + \Delta_{fv}) =$$
$$- \frac{\partial \zeta_2}{\partial x} + \frac{\tau_{s2x}}{\rho D}\,(1 + \Delta_s) - \lambda U_2 + E_2 \nabla^2 U_2$$

$$\left[\overline{<\underline{u}_1 \cdot \nabla v_1>} + \overline{<w_1 \frac{\partial v_1}{\partial z}>}\right](1 + \Delta_v) + fU_2\,(1 + \Delta_{fu}) =$$
$$- \frac{\partial \zeta_2}{\partial y} + \frac{\tau_{s2y}}{\rho D}\,(1 + \Delta_s) - \lambda V_2 + E_2 \nabla^2 V_2 \qquad (47a)$$

More discussion of the coastal boundary problem is given later.

The deviations of the surface stress (Δ_s) and the advection (Δ_u and Δ_v) can be easily included in the bottom stress because they can be computed from the known variables of the first-order solutions, such as the oscillating tidal current ($\underline{u}_1$ and W_1) and the vertical eddy viscosity coefficient (N). To determine the deviation of the Coriolis effect (Δ_{fu}, Δ_{fv}), the vertical variation of the residual current must be included, and thus cannot be done easily. An iteration method to carry out this computation is thus described.

We first set $\Delta_{fu} = \Delta_{fv} = 0$ in (47a), and denote the solution by a superscript (°). The depth-averaged residual current ($\underline{U}_2{}^{\circ}$) and the mean surface pressure gradient ($\nabla \zeta_2{}^{\circ}$) can be computed by solving the momentum equation (47a) and the continuity equation (5b) from a two-dimensional (x,y) model. The computation can be carried out by deriving the stream function Ψ from the continuity equation (5b)

$$DU_2 + \overline{\zeta_1 U_{1s}} = - \frac{\partial \Psi}{\partial y} \quad \text{and} \quad DV_2 + \overline{\zeta_1 V_{1s}} = - \frac{\partial \Psi}{\partial x} \qquad (48)$$

By taking the curl of the momentum equation (47a) to eliminate the surface elevation, and substituting $\underline{U}_2$ from (48), the resulting equation then involves only the unknown variable of the stream function, and can thus be solved numerically.

To compute the three-dimensional residual current ($\underline{u}_2{}^{\circ}$), we consider the area outside the coastal boundary, and let $S = u_2 + iv_2$, and

$$P = \left[\frac{\partial \zeta_2}{\partial x} + \overline{\underline{u}_1 \cdot \nabla u_1} + \overline{w_1 \frac{\partial u_1}{\partial z}} \right] + i \left[\frac{\partial \zeta_2}{\partial y} + \overline{\underline{u}_1 \cdot \nabla v_1} + \overline{w_1 \frac{\partial v_1}{\partial z}} \right]$$

(4a) can be reduced to

$$\frac{\partial}{\partial z} N \frac{\partial S}{\partial z} - i\, f\, S = P \tag{49}$$

where P is the forcing function which can be calculated from the known values of $\nabla \zeta_2{}^{\circ}$, $\underline{u}_1$ and w_1. Using the surface boundary condition of $\partial \underline{u}_2{}^{\circ}/\partial z$ (39) and $\underline{u}_2{}^{\circ} = 0$ at bottom ($z = D$), S can be solved numerically through a matrix decomposition. In order to resolve the rapid variation of the tidal currents near the bottom, we divide the water column into 110 levels with the grid spacing Δ_z increasing parabolically from the bottom. The residual current $u_2{}^{\circ}$ and $v_2{}^{\circ}$ can be obtained from the real and imaginary parts of the computed S.

After solving for $\underline{u}_2{}^{\circ}$, the values of Δ_{fu} and Δ_{fv} (denoted as $\Delta_{fu}{}^{\circ}$ and $\Delta_{fv}{}^{\circ}$) can be computed from (44) and (45). The new bottom stress, denoted by the superscript "1", becomes (41)

$$\left[\frac{1}{D} N \frac{\partial u_2}{\partial z} \bigg|_{z=-D} \right]^1 = \lambda U_2^1 - \Delta_s \frac{\tau_{s2x}}{\rho D} + \Delta_u < \overline{\underline{u}_1 \cdot \nabla u_1} + \overline{w_1 \frac{\partial u_1}{\partial z}} > - \left[\Delta_{fv}{}^{\circ} f V_2{}^{\circ} \right]$$

and

$$\left[\frac{1}{D} N \frac{\partial v_2}{\partial z} \bigg|_{z=-D} \right]^1 = \lambda V_2^1 - \Delta_s \frac{\tau_{s2y}}{\rho D} + \Delta_v < \overline{\underline{u}_1 \cdot \nabla v_1} + \overline{w_1 \frac{\partial v_1}{\partial z}} > + \left[\Delta_{fu}{}^{\circ} f U_2{}^{\circ} \right] \tag{50}$$

Using this new bottom stress, the improved solution of the residual current, denoted as $\underline{u}_2^1$, can be derived. The computation can be iterated by computing Δ_{fu}, Δ_{fv}, and thus the bottom stress from the improved solution of $\underline{u}_2$. An example of this computation is given in the next section.

As for the computation of the first-order oscillating current, we now discuss the coastal boundary problem in the computation of the second-order residual current. Although the vertical variation of the tidal current and residual current are not computed in the layer, the residual current in the offshore area can still be predicted because $\underline{u}_1$, $\underline{U}_1$ and $\underline{U}_2$ in the offshore area, on which $\underline{u}_2$ depends, can be correctly predicted from (22), (32) and (47). Note that in the computation of the depth-averaged residual current $(\underline{U}_2)$, the advective term (since it may not be approximated to $\underline{U}_1 \cdot \nabla \underline{U}_1$ in the rectilinear flow (Tee, 1979), may require the vertical structure of the oscillating current in the layer. However, this unavailable information for $\underline{u}_1$ in the coastal boundary layer is not likely to produce any difficulty in the actual computation of $\underline{U}_2$ because the thin coastal boundary layers, where the horizontal mixing terms cannot be ignored, usually are not resolved in the numerical computation of depth-averaged tidal and residual currents. The importance of introducing the horizontal mixing terms is to allow us to apply the non-slip boundary condition. The difficulties of resolving the coastal boundary layers in numerical computations do not appear to impose any serious restriction because the amount of vorticity in the coastal boundary layer can be correctly predicted and the computed depth-averaged residual currents have been found to agree very well with the observations (Tee, 1976, 1977).

In summary, using the known value of $\underline{u}_1$, the depth-averaged residual current $(\underline{U}_2)$ and the mean surface pressure gradient $(\Delta\zeta_2)$ can be solved from (47) where the deviations Δ_{fu} and Δ_{fv} can be included by using the iteration method (50). Using the known values of $\nabla\zeta_2$, the vertical variation of the residual current can be computed from (49).

The computation will be further simplified if the deviations are neglected so that the governing equation for the depth-averaged residual current becomes

$$<\overline{\underline{u}_1 \cdot \nabla \underline{u}_1}> + <\overline{w_1 \frac{\partial \underline{u}_1}{\partial z}}> + \underline{f} \times \underline{U}_2 =$$

$$- \nabla\zeta_2 + \frac{\underline{\tau}_{s2}}{\rho D} - \lambda \underline{U}_2 + E_2 \nabla^2 \underline{U}_2 \tag{51}$$

Eq. (51) is the shallow water equation for the depth-averaged residual current, which, in conjunction with the shallow water equation for the first-order oscillating current (34), is usually applied in the two-dimensional (x,y) tidal models. The accuracy and advantage of using these shallow water equations is discussed in the next section.

3. SOLUTION FOR TIDES PROPAGATING PERPENDICULARLY TO A STRAIGHT COAST

In order to illustrate the result of computation, and to examine the effects of various deviations from the shallow water equations, we apply here the general method given in the previous section to the simple case of a tidal wave propagating normal to a straight coast (in the x-direction, Fig. 5). All the variables are independent in the longshore direction ($\partial/\partial y = 0$). The simple case which exists in many coastal areas (U.S. Oceanographic Office, 1965) was applied by Johns and Dyke (1971, 1972) who computed the three-dimensional tidal and residual currents using the boundary layer approximation, and considered the case to be applicable in the Liverpool Bay area. The comparison of the residual currents between our results and those of Johns and Dyke will be discussed. A detailed description of the computation is given in Tee (1980).

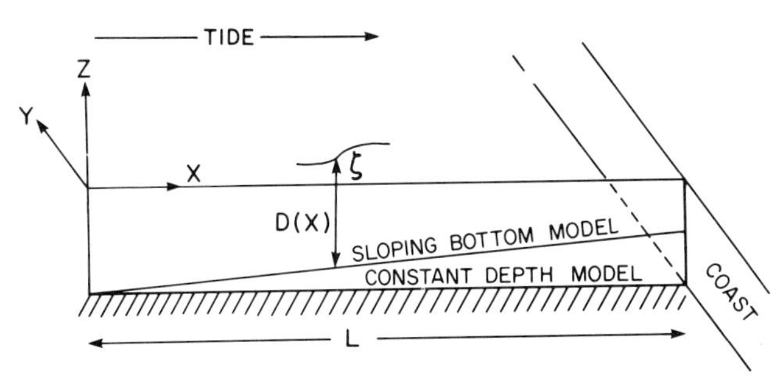

Figure 5. Schematic tidal model.

The tidal model is shown in Fig. 5. The tide is specified at $x = 0$ and the coastline located at $x = L$, where L is taken to be 1.26 which corresponds to 1/5 of the wavelength in the dimensional scale. Two models were considered (Tee, 1979): the constant depth model where the depth D is constant and equal to -1, and the sloping bottom model where $D = -1$ at $x = 0$ and increases linearly to -0.5 at $x = L$ (Fig. 5). Here, only the constant depth model is discussed.

As the solution of the parabolic model, the sublayer model and the combined model are quite similar (Tee, 1979, 1981), only one of the models is selected here to compute the second-order residual current. In this study, the sublayer model is used because the baroclinic velocity can be solved analytically. The vertical eddy viscosity N is given in (28) with $\delta_z = 0.1$ m$/D_l$. The value of 0.1 m$/D_l$ is chosen for δ_z because, as will be shown in the next section, the computed vertical variation of the tidal currents using this value is found to agree reasonably well with the observations. The frictional coefficient λ is calculated (37) by using $\gamma = 0.0021$ and $K = 1$. Other parameters applied in this study are: $U_l = 0.6$ m/sec, $D_l = 20$ m, $\sigma = 1.405 \times 10^{-4}$ sec^{-1} (semi-diurnal tide), $f = 1.169 \times 10^{-4}$ sec$^{-1}/\sigma$ (corresponding to the latitude of the Liverpool Bay), ζ_l $(= U_l\,(gD_l^{-1})^{-1/2}) = 0.86$ m, L_l $(= \sigma^{-1}\,(gD_l)^{1/2}) = 99$ km, N_l $(= \sigma D_l^2) = 5.6 \times 10^{-2}$ m^2/sec, and ϵ $(= \zeta_l/D_l) = 0.043$.

The vertical variations of the semi-major axis, the semi-minor axis (negative value means the tidal ellipse has a clockwise circulation), the inclination of the

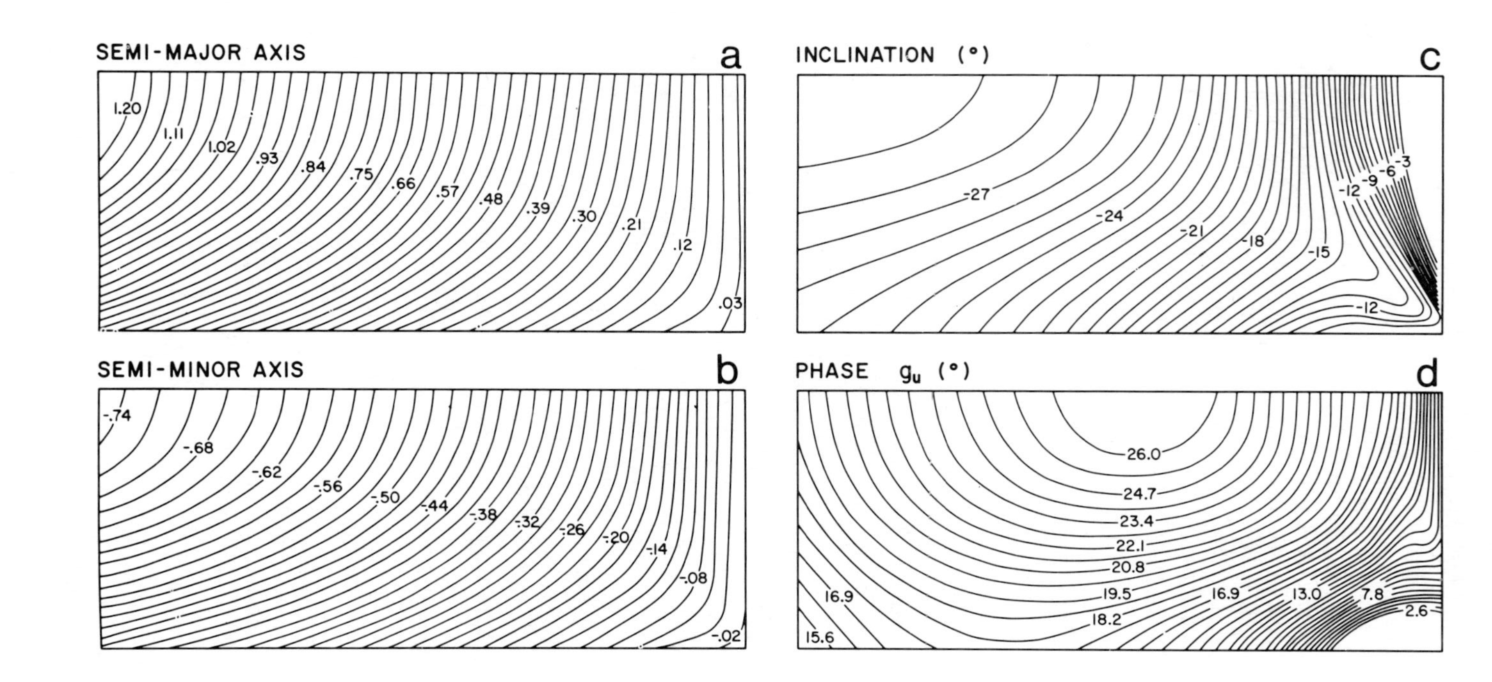

Figure 6. The oscillating tidal currents: (a) semi-major axis, (b) semi-minor axis, (c) inclination to the x-axis (°) and (d) phase (°).

semi-major axis to the x-axis (negative value means the axis rotate clockwise from the x-axis), the phase are shown in Fig. 6.

The solution of the residual current after the first iteration (after including $\Delta_{fu}{}^{\circ}$ and $\Delta_{fv}{}^{\circ}$ in 50) is very close to the equilibrium solution in most of the areas except near the coast. The result can be seen from Fig. 7 which shows the insignificant value of Δ_{fu} and Δ_{fv} after the first iteration in most of the area. In fact, as can be seen from the figure, the values of Δ_{fu} and Δ_{fv} are almost equal to zero between $x = 0$ and $\chi = 3/4L$. The computation of the second iteration, as expected, has a negligible effect on the solution of the first iteration except very close to the coast ($x > 0.95L$) where all the forcings and residual currents are very small (more than an order of magnitude smaller than those values near $x \sim 0$).

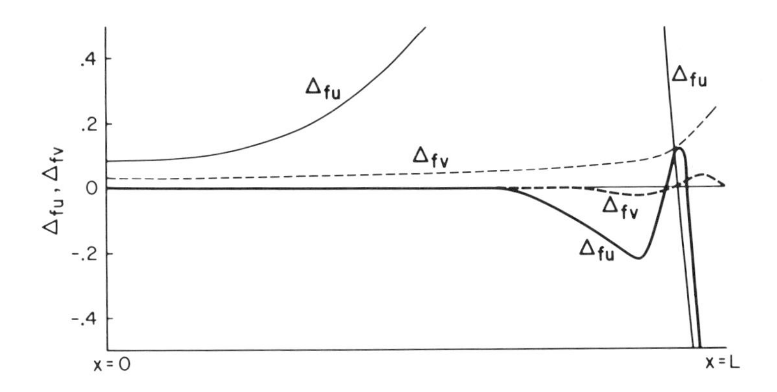

Figure 7. The deviation Δ_{fu} and Δ_{fv} before (thin curves) and after (thick curve) the first iteration.

Figure 8 shows the horizontal variations of U_2, V_2, and $\partial\zeta_2/\partial x$. The U_2 is in the offshore direction and its magnitude decreases toward the coast. One of the interesting results obtained is the generation of the relatively strong longshore residual current which is in the negative y-direction and has maximum strength near $x = 0$. The longshore Lagrangian residual current (Eulerian residual current + Stokes' drift velocity) is found to be in the same direction (Tee, 1980). This direction of the Lagrangian residual current is found to be opposite to that of a similar study of Johns and Dyke (1972) who unreasonably simplified the computation by applying the bottom boundary layer approximation, and assuming that there is no residual current in the frictionless layer. For the surface elevation, the result shows the set down on all x with the maximum gradient located at $x \sim 1/3L$. The result for the sloping bottom has also been investigated and found to produce similar results (Tee, 1980).

Figure 9 shows the distribution from $x = 0$ to $x = L$ of the three-dimensional residual current. The u_2 currents are mostly in the offshore direction except near the bottom at $x \sim L$ where there is a small onshore current. The

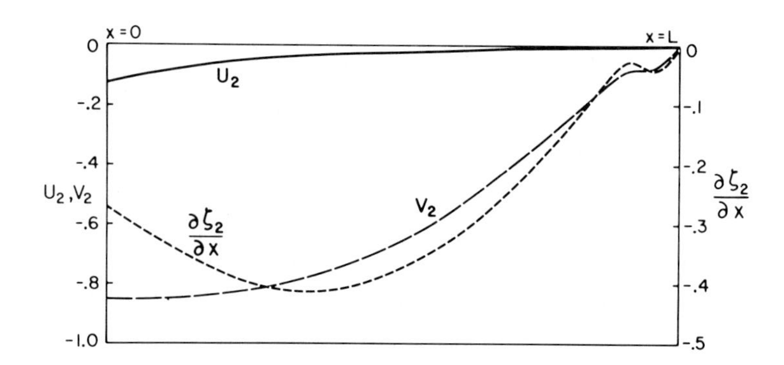

Figure 8. The horizontal and vertical variation of U_2, V_2, and $\partial\zeta_2/\partial\xi$.

magnitude of the offshore current increases from the surface and forms a maximum near the water column. The depth of this maximum in deeper water toward the coast. For the V_2 component, the magnitudes decrease with depth and have a maximum on the surface. The location of the maximum corresponds to that of the depth-averaged v_2 residual currents (at $x \sim 0$). Strong vertical variations of the v_2 component can be seen at the location of the maximum surface currents. The dimensional value of the residual current is $\epsilon \underline{U}_2 U_l$ which, for $\epsilon = 0.043$ and $U_l = 0.6$m/sec, gives the maximum value of the longshore residual current of about 0.026 m/sec (V_2 maximum of ~ 1.0).

The effect of the deviations (Δ_s, Δ_u, Δ_v, Δ_{fu}, and Δ_{fv}) on the residual currents is shown in Fig. 10 for the variation of V_2. We can see that including all the deviations in the depth-averaged momentum equations for the residual current improved the solutions by about 20 to 30 percent. However, there is an advantage in using the shallow water approximation to solve the depth-average residual current (Eq. 51) because the depth-averaged components of the first-order oscillating current and the second-order residual current may be solved together using the non-linear shallow water equation. The circulation of three-dimensional residual current obtained by neglecting all the deviations are generally the same as those shown in Fig. 9.

4. THE PREDICTABILITY OF THE MODEL

The advantage of the proposed method of computation is that the barotropic currents (depth-averaged) and baroclinic currents (vertical velocity gradient) can be solved separately. This separability, not possible in the three-dimensional numerical model, permits simpler programming and a highly efficient computation. The drawbacks are that the model is applicable only in a weakly non-linear system, and the coefficients of the vertical eddy viscosity, and thus the bottom friction are assumed to be time-independent. We now examine how these drawbacks affect the predictability of the model in a coastal environment.

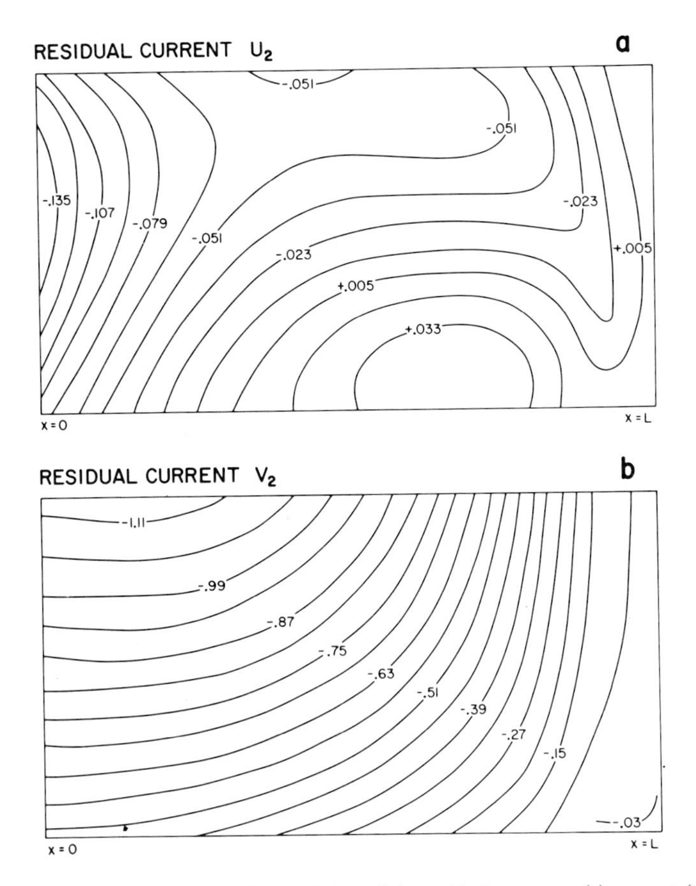

Figure 9. The horizontal and vertical variation of the residual currents: (a) u_2, and (b) (v_2).

The simplification of weak non-linearity of the tidal motion usually holds if the ration of tidal amplitude to the water depth is small, or the strength of the residual current is small compared to that of the first-order oscillating current. In many tidal models, it has been shown that, even in areas of complex coastal geometry, the first-order oscillating currents are quite insensitive to the non-linear effect and the residual currents are generally an order of magnitude smaller compared to the oscillating current (Tee, 1976, 1977; Pingree and Maddock, 1977). Thus, the assumption of weak non-linearity is probably reasonable in most coastal areas, except in a very turbulent and shallow area.

For the vertical eddy viscosity, the exact values and forms of the coefficient remain fairly uncertain in the present time. The linear forms of N are found to be adequate in reproducing the vertical variation of the first-order oscillating current.

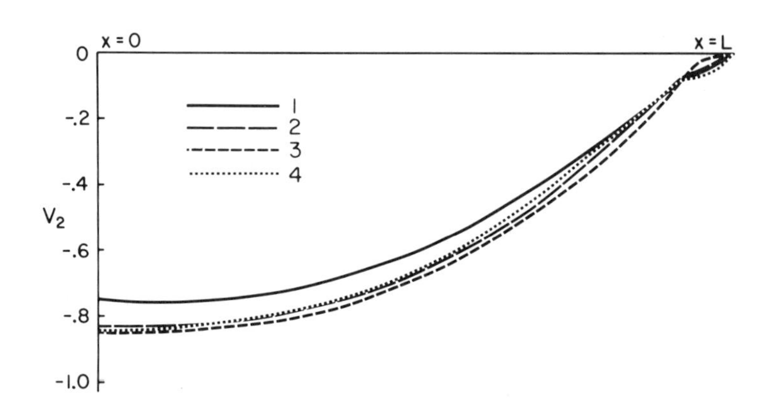

Figure 10. The horizontal variation of V_2 for varies correction to the bottom stress. Curve 1, (—): no correction; curve 2 (— —): includes Δ_s; curve 3 (- - -): includes Δ_s, Δ_u, and Δ_v; and curve 4 (·····): includes all the corrections.

This result can be seen from Fig. 11 which shows the comparison on the semi-major axis (u_J/U_J), semi-minor axis $((u_I - U_I)/U_J)$, phase lags $(g_J - G_J)$ and the inclination of the major axis of the model's results to the field observations taken at the entrance to Chignecto Bay at the head of the Bay of Fundy. The notations in Fig. 11 are that u_J and U_J are respectively the semi-major axes of $\underline{u}_1$ and $\underline{U}_1$, u_I, and U_I the semi-minor axes of $\underline{u}_1$ and $\underline{U}_1$, and g_J and G_J the phase lags of $\underline{u}_1$ and $\underline{U}_1$. The comparison shows that the theoretical computations of the sub-layer model with $\delta_z = 0.1\text{m}/D_l$, the parabolic model with $R = 0.1 \rightarrow 0.01$, and the combined model with $\delta_z = 0.07\text{m}/D_l$ and $R = 0.5$ agree reasonably well with observations. Detailed descriptions of the comparison are given in Tee (1981).

For the second-order residual current, the linear forms of N may exclude the portion of the residual current that is induced *directly* from the non-linear frictional term. The obvious non-linearity of N is that the eddy viscosity is larger during the period of higher velocities, such as those formulas given by the classical mixing length theory that gives N proportional to mixing length (time independent) and magnitude of vertical velocity gradient. This non-linearity indicates that the eddy viscosity has a strong M_4 component. However, this M_4 oscillation of N does not affect the second-order residual current because, by taking the time average of the vertical diffusion term, the combination of the M_4 component of N and the M_2 component of $\underline{u}_1$ produces zero residual component. The only contribution that can result from the non-linearity of N is the M_4 component of $\underline{u}$. However, the M_4 oscillation of $\underline{u}$ is considered to be the second-order component and the generated residual current is then in the third order, or higher order than the residual currents that are generated from the other non-linear terms in the equation of motion.

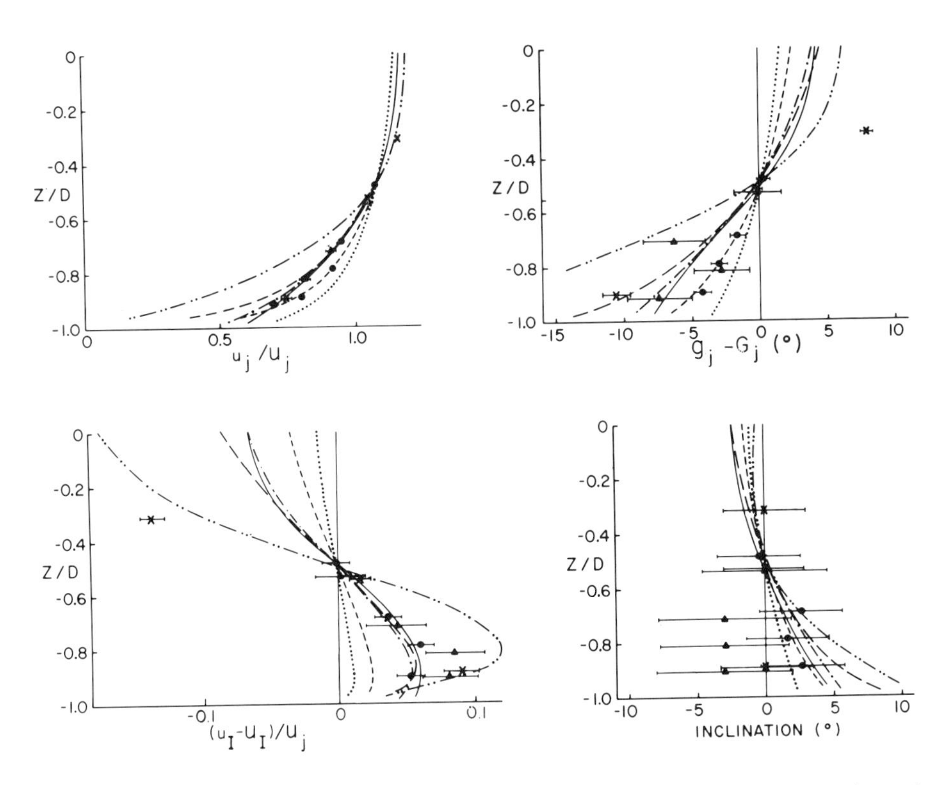

Figure 11. Vertical variation of the tidal current--comparison between theoretical and experimental results: (a) the semi-major axis u_J/U_J; (b) the semi-minor axis $(u_I - U_I)/U_J$; (c) the phase lags $(g_J - G_J)$; (d) the inclination of the major axis. (—··—··—) homogeneous model, (—) sublayer model, $\delta = 0.10$ m, combined model, $\delta = 0.07$m, $R = 0.5$, (— — —) parabolic model, $R = 0.1$, (-----) parabolic model, $R = 0.01$, (····) parabolic model, $R = 0.001$. The observed tidal currents are denoted as (●) for station 283, (▲) for station 306 and (×) for station 284.

A number of residual currents have been observed and computed. For example, it is well known that the residual current can be induced because of the frictional effect which deviates the amplitude of the reflecting wave from the incident wave, so that a pure standing wave does not exist in a semi-closed basin (Stokes' Drift effect, Longuet-Higgins, 1969). This type of residual current is expected to reproduced generally well in the model because the overall friction is well simulated there; that is, the linear bottom friction coefficient (λ) is proportional to the current amplitude and inversely proportional to the depth of the water column.

Many residual currents have been derived by using linear forms of N (corresponding to linear bottom friction), such as those of the channel models (Ianniello, 1977, 1979), or those generated by the topographic effect (Zimmerman,

1978). However, the applicability of the proposed method (Section 2) to those strong residual currents which have been found recently to exist in areas of strong tidal current and irregular coastline (Tee, 1976, 1977; Pingree and Maddock, 1977) requires some explanation. It was found (Tee, 1976; Pingree and Maddock, 1977) that these strong residual currents were induced mainly from the advection of the coastal vorticity, which is formed by the decreasing tidal current toward the coast. Two sources of vorticity can be generated near the coast: the first is a result of the formation of the coastal boundary layer, and the second is the decrease of the depth which then increases the bottom friction and thus reduces the tidal current. The former source of generation is included in the study because, by having the horizontal diffusion term, the non-slip boundary condition can be imposed, Eq. (4). The coastal boundary layers, although they cannot usually be resolved in numerical tidal models, can be approximated by imposing the non-slip boundary condition (Tee, 1976, 1977). The latter source of generation is also simulated in our linearized model because the frictional coefficient λ is proportional to the current amplitude and inversely to the depth. This source of generation can be important if the depth of the water column decreases very slowly toward the coast. However, if the bottom slope becomes steep, the vorticity generated from the coastal boundary layer becomes dominant. In any case, both mechanisms for generating the vorticity near the coast are included in the model.

From the above discussion, we can see that the model is applicable to many coastal areas.

5. SUMMARY AND CONCLUSIONS

A summary of the method of computation is shown in Fig. 1. The method includes the separate computation of the first-order oscillating current and the second-order tidally induced residual current. Using the shallow water approximation, the depth-averaged oscillating current ($\underline{U}_1$) and the surface deviation (ζ_1) can be solved from (34a) and (4b). The computation includes the iteration of the linear friction coefficient λ and the depth-averaged current $\underline{U}_1$. From the computed λ, the non-dimensional parameter d_o can be obtained from Fig. 3. Using these values of d_o and $\underline{U}_1$, the three-dimensional oscillating current $\underline{u}_1$ can be computed from (32).

From the known values of λ and $\underline{u}_1$, the depth-averaged residual current ($\underline{U}_1$) and mean surface elevation (ζ_2) can be computed from (47a) and (5b) by first neglecting the deviations Δ_{fu} and Δ_{fv}. The three-dimensional residual current ($\underline{u}_2$) can then be computed from the known values of ζ_2 and d_o (49). The final solution of the residual current can be achieved by a few numerical iterations, which involves updating the deviations Δ_{fu} and Δ_{fv} in the depth-averaged momentum equation (47) from the known values of $\underline{u}_2$ (44, 45).

The computation of the residual current will be much simpler if the shallow water approximation (indicated by the bold arrow in Fig. 1) is also applied in the depth-averaged momentum equation for the residual current (51). Although this approximation produces little effect on the oscillating current, it reduces the accuracy of the residual current by about 20 to 30 percent. Thus, the application of the shallow water approximation to the residual current depends on the required accuracy and simplification of the computation.

The computation is much simpler than the three-dimensional numerical model because, besides a few iteration between $\underline{U}_2$, $\underline{u}_2$, and Δ_{fu} and Δ_{fv}, the computation of the barotropic current (depth-averaged) and baroclinic current (velocity gradient) are separable.

The predictability of the simple model when applied to a coastal environment is discussed in Section 4, which shows that the model is acceptable in many coastal areas.

REFERENCES

Heaps, N.S. (1972). On the numerical solution of the three-dimensional hydrodynamical equations for tides and storm surges. *Mem. Soc. Roy. Sci. Liege,* **6,** 143-180.

Ianniello, J. P. (1977). Tidally-induced residual currents in estuaries of constant breadth and depth. *J. Mar. Res.* **35,** 755-786.

Ianniello, J. P. (1979). Tidally induced residual currents in estuaries of variable breadth and depth. *J. Phys. Oeanogr.* **9,** 962-974.

Johns, B. (1970). On the determination of the tidal structures and residual current system in a narrow channel. *Geophy. J. Roy. Astr. Soc.* **20,** 159-175.

Johns, B. and P. Dyke (1971). On the determination of the structure of an offshore tidal stream. *Geophys. J. Roy. Astr. Soc.* **23,** 287-297.

Johns, B. and P. Dyke (1972). The structure of the residual flow in an offshore tidal stream. *J. Phys. Oceanogr.* **2,** 73-79.

Leendertse, J. J. and Liu, S. K. (1975). A Three-dimensional Model for Estuaries and Coastal Seas, Vol. II: Aspects of Computation. The Rand Corporation, R-1764-OWRT, 123 pp.

Longuet-Higgins, M. S., (1969). On the transport of mass by time-varying ocean currents. *Deep Sea Res.* **16,** 431-447.

Pingree, R. D. and Maddock, L. (1977). Tidal residual in the English Channel. *J. Mar. Biol. Ass. U.K.* **57,** 339-354.

Tee, K. T. (1976). Tide-induced residual current, a 2-D non-linear numerical tidal model. *J. Mar. Res.* **34,** 603-628.

Tee, K. T. (1977). Tide-induced residual current--verification of a numerical model. *J. Phys. Oceanogr.* **7,** 396-402.

Tee, K. T. (1979). The structure of three-dimensional tide-generating current. Part I: oscillating current. *J. Phys. Oceanogr.* **9,** 930-944.

Tee, K.T. (1980). The structure of three-dimensional tide-generating current. Part II: residual currents. *J. Phys. Oceanogr.* (in press).

Tee, K. T. (1981). The structure of three-dimensional tide-generating currents: Experimental verification of the theoretical model. (Submitted to *Est. Coast. Mar. Sci.*).

U. S. Oceanograpic Office (1965). Oceanograpic Atlas of the Norh Atlantic Ocean, Sections I: Tides and Currents. No. 700, 75 pp.

Zimmerman, J. T. F. (1978). Topographic generation of residual circulation by oscillatory (tidal) currents. *Geophys. and Astrophys. Fluid Dyn.* **11,** 35-47.

A DYNAMIC RESERVOIR SIMULATION MODEL - DYRESM: 5

Jörg Imberger and John C. Patterson

University of Western Australia
Nedlands, Western Australia

1. INTRODUCTION

The dynamic reservoir simulation model, DYRESM, is a one-dimensional numerical model for the prediction of temperature and salinity in small to medium sized reservoirs and lakes. The model was constructed to form a suitable basis for a more general water quality model. The requirements imposed by this aim are made by Imberger (1978), who shows that for both substances with a fast turnover time such as phytoplankton and for conservative substances such as salt, the mixing mechanisms within the reservoir must be modeled accurately. Not only must a model be capable of predicting the net cumulative vertical transfer of mass induced by the mixing and the subsequent flow processes, but it must also be capable of resolving individual mixing regions.

This is a great departure from what has been assumed to date. Previously it has been generally accepted that once a particular reservoir model was capable of being tuned to correctly predict the vertical temperature profiles over a particular study period, that this then constituted a reliable model for the purposes of water quality modeling. However, as discussed in Fischer *et al.,* (1979) this is not the case since different parameters are dependent on different mixing processes for their distribution.

The problem confronting the modeler is thus difficult. On the one hand the model should be widely applicable and cheap to run, and on the other it must capture the mixing induced by the small scales of motion. The marriage of these seemingly conflicting aims requires careful consideration of not only the concept but also the architecture of a proposed model. Although increasing availability of computing power has led to an increased potential for application of two or three-dimensional reservoir models, the required resolution, down to centimeters and minutes, is not currently possible within the context of long term simulations, and a one-dimensional model appears to be the only realistic option.

The strong stratification normally found in small to medium sized reservoirs may be used to advantage. A strong stratification of the water inhibits vertical motions and reduces the turbulence to intermittent bursts. For such lakes the

ISBN: 0-12-25852-0

transverse and longitudinal direction play a secondary role and only the variations in the vertical enter the first order balances of mass, momentum and energy.

Departures from this state of horizontal isopycnals are possible, but these enter only as isolated events or as weak perturbations. In both cases the net effect is captured with a parameterization of their input to the vertical structure and a comparison of the model prediction and field data must thus be confined to periods of calm when the structure is truly one-dimensional.

The constraints imposed by such a one-dimensional model may best be quantified by defining a series of non-dimensional numbers. The value of the Wedderburn number

$$W = \frac{g'h}{u^{*2}} \cdot \frac{h}{L} \tag{1}$$

where g' is an effective reduced gravity across the thermocline, h the depth of the mixed layer, L the basin scale, and u^* the surface shear velocity, is a measure of the activity within the mixed layer. Spigel and Imberger (1980) have shown that for $W > 0(1)$ the departure from one-dimensionality is minimal and for $0(h/L) < W < 0(1)$ the departure is severe but may be successfully parameterized. For $W < 0(h/L)$ the lake overturns.

Inflows do not lead to severe vertical motions in the bulk of the reservoir provided the internal Froude number

$$F_i = \frac{U}{\sqrt{g'H}} < 1 \tag{2}$$

where U is the inflow velocity, g' the reduced gravity between the surface reservoir water and the inflow and H the reservoir depth.

The outflow dynamics is governed by a similar Froude number criterion

$$F_o = \frac{Q}{g'^{1/2}H^{5/2}} < 1 \tag{3}$$

where Q is the outflow discharge and g' the reduced gravity between the surface and bottom water.

Lastly, the earth's rotation will not lead to vertical motions if

$$\frac{g'\delta}{f^2B^2} < 1 \tag{4}$$

where g' is the effective reduced gravity over the depth δ, the scale of the velocity concentrations, f is the Coriolis parameter and B is the maximum width of the lake.

Within these constraints DYRESM yields an accurate description of the dynamics governing the vertical structure in a lake. It must, however, always be emphasized that superimposed on the structure predicted by DYRESM there are mixing processes which have only a second order influence on the vertical structures, but which are the first order mechanisms for transporting substances horizontally. These find their origins in enhanced heating and cooling at the edge of the lake, direct surface runoff, wind sheltering and shading by mountain ridges, and boundary mixing in the hypolimnion. Only the influence of these processes on the one-dimensional structure is modeled by DYRESM. A separate perturbation scheme is required to reveal their influence on horizontal variations in a reservoir. The first order vertical structure computed with DYRESM may in future investigations be used to compute such horizontal transport processes. These may be modeled with a simplified three-dimensional model using the one-dimensional vertical structure as the mean density field.

Even the one-dimensional problem is not simple. A uniform grid model which resolves these fine scales over the whole depth would require a large number of mesh points and computation would be slow, to say nothing about problems of stability and numerical diffusion. Even if local compaction of a wider grid were employed, which introduces its own problems, the model at any instant would have only an estimate of where the compaction was required for each process, and since several processes may occur at different positions simultaneously, little advantage would be gained.

With these considerations in mind, the model DYRESM was developed by Imberger *et al.* (1978) with a different approach to the task. Concentrating on parameterization of the physical processes rather than numerical solution of the appropriate differential equations, DYRESM makes use of a layer concept, in which the reservoir is modeled as a system of horizontal layers of uniform property. These layers move up and down, adjusting their thickness in accordance with the volume-depth relationship, as inflow and withdrawal increase and decrease the reservoir volume. In this way the problems of numerical diffusion associated with computation of vertical advection above the levels of inflow and withdrawal vanish.

There is, however, a far stronger reason for making use of adjustable layers as the model medium. Rather than being just a computational scheme, the layers may be given a functional character, and their dimension adjusted to suit the function expected of them. Thus the mixed layer, for example, may be modeled by a reasonably coarse layer structure in the epilimnion, fining down to very narrow layers in the transition zone. This adjustment of scale is performed by the algorithm itself, and thus only creates high resolutions when required. Further, by treating each of the dominant physical processes separately, only those layers affected by the process are operated on. In each of these process models, the proper range of scales may be accounted for and included explicitly.

The result of this concept is a model which is computationally very simple and yet conceptually more realistic than a fixed grid approach. Because of the computational simplicity, the algorithm is economical, and long term simulations which incorporate the finest scales become a realistic proposition.

2. DYRESM MAIN PROGRAM

The model DYRESM is constructed as a main program with subroutines which separately model each of the physical processes of inflow, withdrawal, mixed layer dynamics, and vertical transport in the hypolimnion. In addition, there are a number of service subroutines which provide maintenance of the layer system (volumes, positions, etc.) and provide calculations of physical properties which are frequently required such as density. The functions of the main program are therefore those of input/output, the calculation of fixed parameters, and control over timing of the calls to the various process subroutines.

The question of how frequently the process subroutines should be called naturally arises. For most purposes, a daily prediction of the thermal structure is sufficient, and the basic time step of the model is set to this value. Since the inflow and outflow volumes change only slowly, generally over a few days at least, this basic time step is also suitable for these processes, and the appropriate subroutines are called once daily. On the other hand, the mixed layer dynamics covers scales very much shorter than one day, and even though daily averaged meteorological inputs are mostly used, the mixed layer model subroutines need to be called more frequently to match the response of the diurnal deepening.

Thus the model incorporates two time steps; a fixed basic step of one day and a variable sub-daily time step for the mixing algorithm. The length of this sub-daily step is determined by the dynamics and ranges between one quarter hour and twelve hours if averaged data is used. This procedure allows small time steps when the dynamics so require; in less critical periods, the time step expands without loss in accuracy.

DYRESM is then organized in the following way (see Flowchart 2.1). The main program inputs the fixed data; physical dimensions, volume and area as a function of depth, physical properties of the inflowing streams, locations of the offtakes, an initial temperature an salinity profile, and output control parameters.

The initial profile, yielded from field data, may not be of sufficient resolution. This is expanded by interpolation, a density profile computed, and the initial layer structure formed. Each point in the expanded profile becomes the top of an individual layer of uniform property. The layer volumes and areas are computed, and their size checked against preset criteria.

The daily loop begins with the input of the current inflow, withdrawal, and meteorological data. After some output, the sub-daily loop commences. An index i_H is set to zero, representing the start of the first quarter hour of the daily loop. Subroutine HEATR is called to model the surface heat exchanges and to determine the length of the sub-daily time step. This routine is described in Section 3.1; briefly, the time step length is determined by considering the heat input and the mixed layer velocity increase in one quarter hour period. The time step length is set by limiting these parameters to preset maximum values. The routine returns

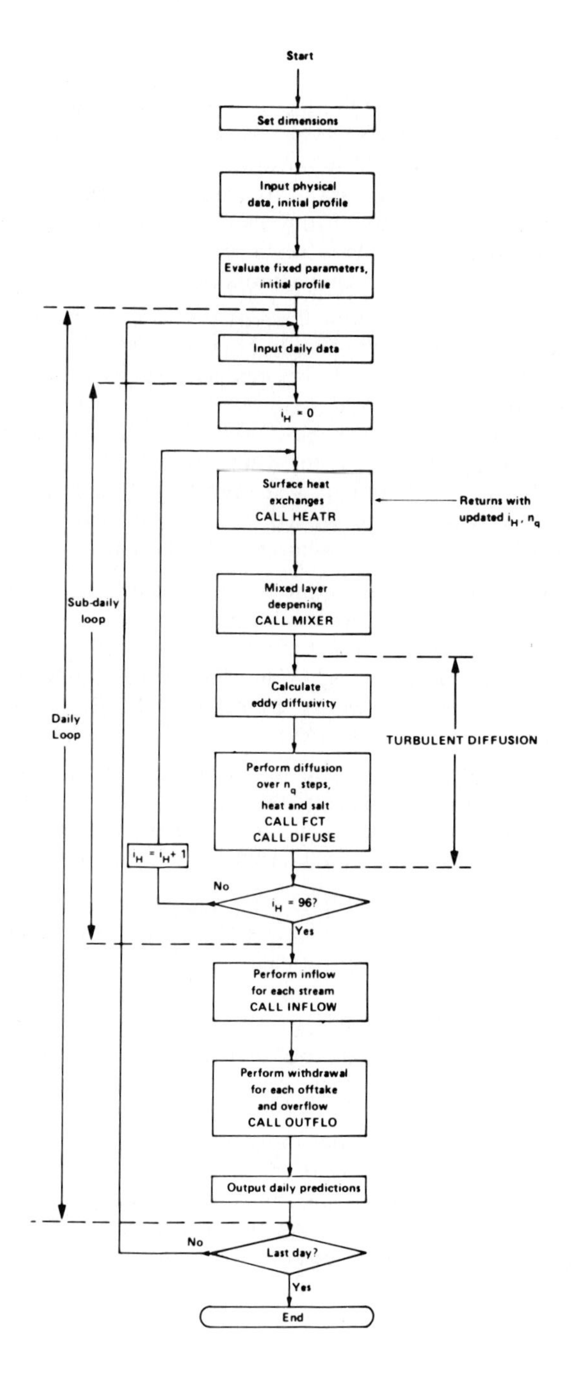

Flowchart 2.1. DYRESM Mainline.

with i_H increased by the number of quarter hours in this time, and the algorithm proceeds.

The mixed layer dynamics is modeled by a call to MIXER, which internally calls the Kelvin-Helmholtz billowing routine KH. These subroutines are described in Section 3.2. The eddy diffusivity described in Section 3.3 is then computed in the main program and the diffusion routines FCT and DIFUSE (Section 3.4) called in one quarter hour time steps for the sub-daily period for both heat and salt. The counter i_H, incremented in HEATR, is checked against one day; if a day has not elapsed, control is looped back and the process repeated.

After completion of the sub-daily loop, subroutine INFLOW is called for each inflowing river. This subroutine is described in Section 3.5. Subroutine OUTFLO (Section 3.6) models withdrawal from each submerged offtake, and if necessary, flow over the crest. The model permits overflow only if the predicted surface level is above both the measured level and the crest level; the quantity of overflow permitted is that which restores the predicted level without overflow to the measured level. This action is taken because of the unreliable nature of most overflow volume data.

At this stage the predicted temperature and salinity structure is output, along with other information, and the daily loop is complete. If the simulation has reached the required day, the computation ceases. Otherwise the entire procedure is repeated.

The service subroutines THICK, RESINT, DENSTY and SATVAP are described in Section 3.7. Execution of the program requires 55K (octal) words of memory, and for the 964 day simulation of Wellington Reservoir, required 2300 seconds of CP time on a Cyber 73 at the West Australian Regional Computer Centre.

3. DYRESM SUBROUTINES

3.1 Surface Heat and Mass Exchange

3.1.1 Theoretical background. Profile data of the air temperature, humidity and wind speed above the lake surface are rarely available and the modeler must rely on single elevation data at a single station on or near the lake.

This fact necessitates the use of the following bulk aerodynamic formulae for the assessment of the fluxes of momentum, τ/ρ, heat, $\tilde{H}$, and moisture E

$$\frac{\tau}{\rho} = u^*u^* = C_D U^2 \tag{5}$$

$$\frac{\tilde{H}}{\rho c_p} = -u^*\theta^* = -C_H U(\theta-\theta_s) \tag{6}$$

$$\frac{E}{\rho} = -u^* q^* = -C_W U (q - q_s) \tag{7}$$

where C_D, C_H and C_W are the respective bulk transfer coefficients. The wind speed U, air temperature θ, and the humidity q are measured at some reference height, and the subscript s refers to a surface measurement.

Equilibrium turbulence theory indicates that the bulk transfer coefficients are not constant, but are dependent on the stability of the air. Several authors (Deardorff, 1968; Carson and Richards, 1978) have investigated this dependency. Rayner (1980) has summarized their findings, and established a relationship between the actual coefficient and its value under neutral conditions in terms of a bulk Richardson number. When applying this correction care must, however, be taken to ensure that the single point sensors, from which the data are obtained, are all well within the developing equilibrium boundary layer of the lake.

Two versions of DYRESM are currently available. A standard version which assumes neutral values of the coefficients C_D, C_H and C_W dependent only on the wind speed and an experimental version, developed by Rayner (1980), which uses Hick's (1975) iterative procedure to correct the fluxes for the air column stability. The latter algorithm first estimates the fluxes using neutral values of the coefficient. The fluxes so computed are used to calculate a first estimate of the Monin-Obukhov length. The self similar profiles of the meteorological boundary layer theory are then used to correct the initial estimates of the fluxes. This leads to an iterative process which normally converges very rapidly.

Several authors provide estimates of the 10 m neutral transfer coefficients C_D, C_H and C_W arising from extensive field measurements. In DYRESM it is assumed that the value of $C_D = 10^{-3}$ for $u < 5\text{ms}^{-1}$ but rising linearly to 1.5×10^{-3} for wind speed of 15ms^{-1} (Hicks, 1972). The documented values of C_H and C_W range from 1.3×10^{-3} to 1.5×10^{-3} (Hicks, 1975; Pond *et al.,* 1974); a value of 1.4×10^{-3} is assumed.

In addition to the turbulent fluxes across the surface, heat transfers due to radiation must be considered. Long-wave radiation emitted from the water surface is given by (TVA 1972)

$$Q_{LR} = \epsilon \sigma T^4 \tag{8}$$

where ϵ is the emissivity (=0.96), σ the Stephan-Boltzman constant ($\sigma = 2.0411\times10^{-7}\text{kJ/m}^2\text{ hr } K^4$), and T the absolute temperature of the water surface. Some of this radiation is absorbed by atmospheric constituents and re-emitted back to earth. This atmospheric radiation is given by (Swinbank, 1963)

$$Q_{LA} = 0.937\times10^{-5} \sigma T_2^2 (1 - R_a) \tag{9}$$

where T_2 is the air temperature at 2 m height, and R the water surface reflectivity (=0.3). Field tests show this formula to be accurate to $\pm 12\, W\text{m}^{-2}$.

Clouds emit long-wave radiation and are accounted for by a factor in Eq. (9).

$$Q_{LAC} = (1 + 0.17C^2)\,Q_{LA} \tag{10}$$

where C is the part of the sky covered with clouds, measured in tenths of the total sky. It is assumed that Q_{LAC} is absorbed directly by the water in the upper surface slab.

On the other hand, short-wave solar radiation penetrates the surface and is absorbed within the water column, thereby heating it. Only the visible part of the solar spectrum (0.36 to 0.76 μ) penetrates below one meter and this serves as the definition of short-wave radiation. For depths greater than one meter the attenuation profile of this short-wave radiation is well described by Beer's Law

$$q(z) = Q_s e^{-\eta_1 z} \tag{11}$$

where Q_s is the solar radiation at the surface, z the depth, and η_1 the bulk extinction coefficient dependent on the turbidity of the water.

To describe attenuation over the full depth of water, account must be taken of the wavelength dependency of the extinction coefficient. The TVA (1972) report recommends use of three extinction coefficients, each associated with their own exponential decay. This option is included in DYRESM, but the actual attenuation profile and coefficient values used by DYRESM are dependent on the particular reservoir in question and must be specified by the user.

3.1.2 Surface heat and mass exchange algorithm. The heat exchanges through the surface are modeled by subroutine HEATR, which simulates radiation penetrative heating and the evaporative, conductive, and long-wave radiation exchanges at the surface.* These events are modeled by the bulk aerodynamic and radiation transfer formulae given by Eqs. (6) to (11).

The routine also sets the time step for itself and the following mixing and turbulent diffusion computations. As the short-wave radiation occurs only for half of the day, this time step is at most 12 hours. Further, to ensure that the turbulent velocity scale w^* and the mixed layer mean velocity U required by MIXER do not become excessively large, a further limit is imposed. This is set in the first case by limiting the change in surface temperature (before mixing) to a maximum of 3°C. This amounts to limiting the total $\tilde{H}^*$, and thus w^*, (defined in Section 3.2.1) for the total time step. In the second case, the mean velocity increase is limited to 1 cm/sec over the previous value, based on the previous mixed layer depth and the current value of u^*. The minimum time step set by these criteria is chosen down to a minimum of one quarter hour.

---

*Flow charts for all the subroutines mentioned in this section may be obtained by request to the authors.

Cooling at the surface or evaporation affect only the surface layer, and the density profile may be unstable following all the heat exchange occurring in a time step. HEATR makes no attempt to rectify this situation and merely passes the profile on to MIXER for stabilization.

The quarter hour time counter i_H is set to zero before the first call to HEATR. It is assumed that the short-wave radiation penetration occurs only for the second 12 hours of the day and thus is only applied if $i_H > 48$. The penetration law, Eq. (11), is applied, and the heat input to layer K evaluated as

$$\Delta Q_K = A_{K-1}\delta(Q_K) + (A_K - A_{K-1})Q_K \tag{12}$$

where A_K is the area of the layer, Q_K the radiation intensity arriving at the top of the layer determined by Eq. (11) and $\delta(Q_K) = Q_K - Q_{K-1}$

The evaporative, conductive and long-wave heat exchanges occur for all values of i_H and affect only the surface layer, layer NS. The heat input calculated from these effects is added to ΔQ_{NS}, and the temperature increment for each layer evaluated for a single quarter hour time step. Since the meteorological data responsible for these effects is averaged over a day, it is assumed that the temperature increment is a linear function of time and therefore that the 3°C change in surface temperature will occur in $n_q(T) = 3/\mid \Delta T \mid$ quarter hours, where ΔT is the surface change in one quarter hour, and n_q is rounded down, with a minimum value of one. Independently, the mean velocity criterion is given by $n_q(U) = 0.1h/u^{*2}$, where h is the mixed layer depth from the previous time step and u^* the current wind shear velocity; again n_q has a minimum value of one. The minimum of these two values is then taken as the number of quarter hours n_q for the current time step, i_H is incremented by n_q, cut off and n_q modified if the new i_H exceeds 48 or 96, and the total temperature increment for n_q quarter hours added into the temperature profile.

The salinity of the surface layer is adjusted for the effects of evaporation and rainfall, and finally a new density profile computed. Control is returned to the mainline ready for MIXER to be called.

3.2 Energetics of the Surface Layer

3.2.1 Theoretical background. Under the constraints given in the introduction the velocity and density distribution in the upper surface layer may be idealized as shown in Fig. 3.1. The velocity in the direction of the applied wind stress (ρu^{*2}) is characterized by a strong surface shear region near the water surface, with a thickness ranging from a few centimeters to one or two meters with increasing wind stress. Below this shear zone the velocity is relatively constant throughout the mixed layer. The mixed layer is defined as that surface water slab over which the water density is constant. Rayner (1980) has shown that this depth typically varies greatly within a day from zero to the depth of the seasonal thermocline. The momentum as well as the heat and mass introduced at the water surface are

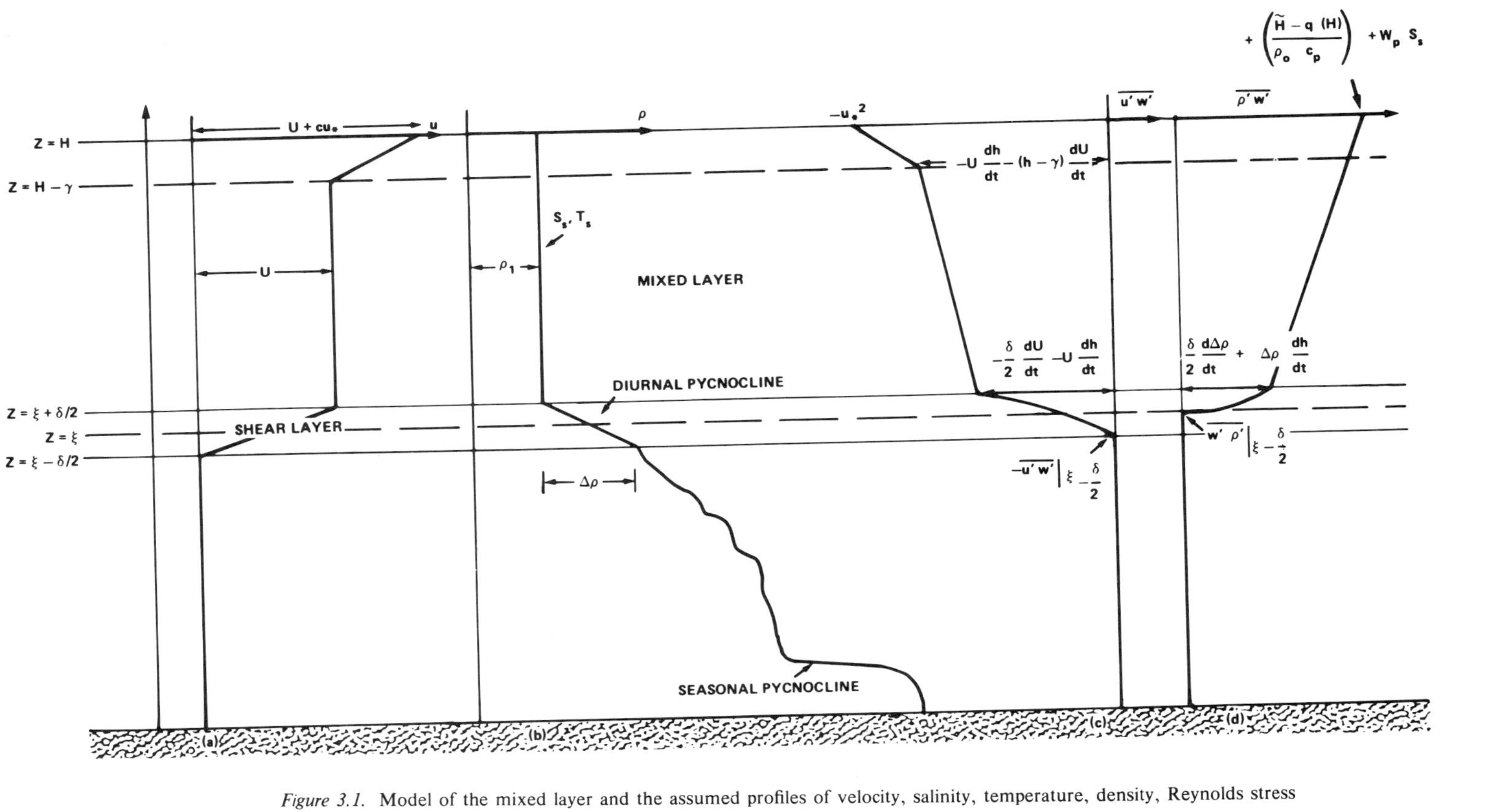

Figure 3.1. Model of the mixed layer and the assumed profiles of velocity, salinity, temperature, density, Reynolds stress and buoyancy flux.

continually being mixed or stirred into the surface mixed layer. The turbulent kinetic energy (T.K.E.) field will adjust to changes of the surface inputs in times of $0(h/u^*)$, where h is the mixed layer depth.

Consider first the behavior of this layer in the absence of horizontal pressure gradients which would normally be set up by the presence of the boundaries of the lake. With this simplification the momentum equation in any part of the profile reduces to (see Niiler and Kraus, 1977)

$$\frac{\partial u}{\partial t} = -\frac{\partial}{\partial z}\,(\overline{u'w'}) \tag{13}$$

Integration of Eq. (13) from below the shear layer to any part of the profile, using the velocity distribution shown Fig. 3.1 yields the distribution of Reynolds stress sketched in Fig. 3.1.

The density profile shown in Fig. 3.1 allows a similar calculation of $\overline{w'\rho'}$ from the mass equation

$$\frac{\partial \rho}{\partial t} = -\frac{\partial}{\partial z}(\overline{\rho' w'}) - \frac{\alpha}{c_p}\frac{\partial q(z)}{\partial z} \tag{14}$$

where q is the radiation heat flux at a depth z.

To obtain the mechanical energetics of the mean flow Eq. (13) may be multiplied by the velocity u and integrated over the mixed layer. The balance so obtained shows that during deepening $(dh/dt > 0)$ or billowing $(d\delta/dt > 0)$ events, a certain amount of mean kinetic energy is lost both in the surface shear layer and in the base shear layer. As is shown in Lumley and Tennekes (1972) this energy change is not a loss, but merely a transfer to turbulent kinetic energy and therefore it must be accounted for in the TKE budget.

The total transfer of energy accomplished via this adjustment is given by

$$\int_{\xi-\delta/2}^{\xi+\delta/2} \overline{u'w'}\frac{du}{dz}\,dz + \int_{H-\gamma}^{H} \overline{u'w'}\frac{du}{dz}\,dz = +\frac{\Delta U\delta}{6}\frac{d\Delta u}{dt} + \frac{1}{12}(\Delta U)^2\frac{d\delta}{dt} + \frac{(\Delta U)^2}{2}\frac{dh}{dt} + cu^{*3} + \gamma\frac{cu^*}{2}\frac{d\Delta U}{dt} \tag{15}$$

where ΔU is the velocity increment across the base of the mixed layer and c is a constant.

Similarly, the profile of $\overline{w'\rho'}$ shown in Fig. 3.1 leads to an adjustment of the mean potential energy of the mixed layer structure. During deepening and billowing the heavier fluid below the mixed layer is entrained into the base of mixed layer causing an adjustment of the profile shown in Fig. 3.1 with corresponding raising of the center of gravity of the water column and a rate of change of potential energy P.E. given by

$$\frac{d}{dt}\,\text{P.}\dot{\text{E}}. = -\frac{g}{2}\left\{2h\Delta\rho\,\frac{dh}{dt} + \frac{\delta\Delta\rho}{6}\,\frac{d\delta}{dt} + \left(h^2 + \frac{\delta^2}{12}\right)\frac{d}{dt}\,(\Delta\rho)\right\} \tag{16}$$

The above evaluations now allow an accounting of the turbulent kinetic energy budget expressed by the equation (see Denman, 1973)

$$\frac{1}{2}\,\frac{\partial\bar{E}}{\partial t} = -\,\overline{u'w'}\,\frac{\partial u}{\partial z} - \frac{\partial}{\partial z}\left|\overline{w'\left(\frac{p}{\rho_o} + \frac{E}{2}\right)}\right| - \frac{g\overline{\rho' w'}}{\rho_o} - \epsilon \tag{17}$$

where $\bar{E}/2 = \overline{(u'^2 + v'^2 + w'^2)}/2$ is TKE per unit mass, $\overline{u'w'}\partial U/\partial z$ is the shear production of TKE by Reynolds stresses $\overline{u'w'}$ and velocity shear of the mean flow $\partial u/\partial z$, $\partial/\partial z[\overline{w'(p'/\rho_o+E/2)}]$ is transport of TKE by vertical velocity fluctuations w', $g\overline{\rho' w'}/\rho_0 = g(-\alpha\overline{\theta' w'} + \beta\overline{s'w'})$ is the work done locally against buoyancy forces by TKE lifting heavier fluid (α,β are compressibilities for heat and salt; ρ',θ',s' are density, temperature and salinity fluctuations, respectively), and ϵ is the dissipation of TKE by viscosity. Equation (17) is valid at any point in the water column and may be integrated from the base of the thermocline $z = \xi - \delta/2$ to the water surface $z = H$ using the assumed profiles of Fig. 3.1, with a finite thermocline thickness δ.

Making use of the integrated equations for heat and mass as well, the integrated TKE budget may be written as

$$\underbrace{\frac{1}{2}\,\frac{\partial}{\partial t}\,(E_S h)}_{1} + \underbrace{\left\{\frac{gh\Delta\rho}{2\rho_0}\,\frac{dh}{dt} + \frac{g\delta^2}{24\rho_0}\,\frac{d}{dt}\,\Delta\rho + \frac{g\delta\Delta\rho}{12\rho_0}\,\frac{d\delta}{dt}\right\}}_{2} =$$

$$+\underbrace{\left\{\frac{\Delta U\delta}{6}\,\frac{d\Delta U}{dt} + \frac{1}{12}\,\Delta U^2\,\frac{d\delta}{dt} + \frac{\Delta U^2}{2}\,\frac{dh}{dt}\right\}}_{3} + \underbrace{\frac{c}{2}\,u^{*3}}_{4}$$

$$-\underbrace{\overline{w'\left(\frac{p'}{\rho_0} + \frac{E}{2}\right)}\Bigg|_H}_{5} + \underbrace{\frac{\alpha gh\tilde{H}^*}{2\rho_0 C_p}}_{6} + \underbrace{\frac{\beta gh\tilde{w}S_S}{2}}_{7} - \underbrace{\frac{g(h+\delta)\overline{\rho' w'}}{2\rho_0}\Bigg|_{\xi-\delta/2}}_{8}$$

$$+\underbrace{\overline{w'\left(\frac{p'}{\rho_0} + \frac{E}{2}\right)}\Bigg|_{\xi-\delta/2}}_{9} - \underbrace{\epsilon_s(h+\delta/2)}_{10} - \underbrace{\overline{u'w'}\Big|_{\xi-\delta/2}\Delta U}_{11} \tag{18}$$

Each term and its related scales are as follows:

1. Rate of change of depth integrated TKE, as determined by the various sources and sinks of TKE on the right hand side of the equations. E_S is the depth integrated TKE in the mixed layer. Since turbulence is generated by both wind and convective overturn, we introduce the scale q^* incorporating both these effects (see also Zeman and Tennekes, 1977)

$$q^{*3} = w^{*3} + \eta^3 u^{*3} \tag{19}$$

where now

$$w^{*3} = \frac{\alpha g h \tilde{H}^*}{\rho_o c_p} + \beta g h \tilde{W} S_S \tag{20}$$

is modified to include the effects of evaporation at the surface, which leaves a residue of heavier, saltier water at the surface. $\tilde{H}^*$ is net surface cooling corrected for the stabilizing effects of solar radiation absorption, as described below in connection with term 6. Hence, $E_S \sim q^{*2}$ and $\partial(E_S h)/\partial t \sim q^{*2} dh/dt$.

2. Rate of change of potential energy due to deepening and billowing. Billowing is due to shear instability at the base of the mixed layer.

3. Shear production within the lower shear layer.

4. Shear production within the surface shear layer.

5. Downward transport by vertical velocity fluctuations of TKE input at the free surface. Terms 4 and 5 both scale with u^{*3} and are the sources of TKE for the wind stirring mechanism.

6. TKE produced by convective overturn resulting from surface cooling, corrected for the effects of solar radiation. $\tilde{H}^*$ is defined by

$$\frac{\tilde{H}^*}{\rho_o c_p} = \overline{\theta' w'}(H) + q(H) + q(\xi) - \frac{2}{h} \int_{\xi}^{H} q(z)\, dz \tag{21}$$

$\overline{\theta' w'}(H)$ is the net heat exchange that occurs just at the water surface, including all turbulent and long-wave radiative transfers. $q(z)$ is the solar radiation flux in the water. The three solar radiation terms in Eq. (21) account for the stabilizing effect on the water column of the exponential attentuation of solar radiation with depth. Whereas $\overline{\theta' w'}(H)$ is a source of TKE for net cooling, $q(H) + q(\xi) - 2/h \int_{\xi}^{H} q(z)\, dz$ will be a sink for TKE if the $q(z)$ profile is concave downward, as is usually the case. This may be better understood by considering the thermal energy equation

$$\frac{\partial T}{\partial t} = -\frac{\partial \overline{\theta' w'}}{\partial z} - \frac{1}{\rho_o C_p} \frac{\partial q}{\partial z} \tag{22}$$

If $q(z)$ is a linear function of z, then $\partial q/\partial z=$ constant so that heating of the water column $\partial T/\partial t$ by solar radiation will be uniform with depth, with no net stabilizing or destabilizing effects produced. (Note that for $q(z)$ linear, the solar radiation terms in Eq. (21) sum to zero). As Rayner (1980) points out, the form of Eq. (21) emphasizes the importance of correct modeling of solar radiation absorption for mixed layer energetics. Finally, we note that term (6) can be a sink for TKE if $\tilde{H}^*$ is negative (net heating).

7. TKE produced by convective overturn due to salinity increases caused by evaporation. Terms 2, 6 and 7 result from the integral

$$\int_{\xi-\delta/2}^{H} \overline{w'\rho'}dz$$

8. Leakage of turbulent fluxes of heat and salt from the mixed layer into the hypolimnion

$$\left.\frac{\overline{\rho' w'}}{\rho_0}\right|_{\xi-\delta/2} = \left(-\alpha\overline{\theta' w'} + \beta\overline{s' w'}\right)\Big|_{\xi-\delta/2}$$

is a sink for mixed layer TKE.

9. Downward transport by vertical velocity fluctuations of TKE into the hypolimnion. As with

$$\left.\overline{u'w'}\right|_{\xi-\delta/2}\Delta U$$

term 9 may act as a source for internal waves in the hypolimnion and is a sink for mixed layer TKE.

10. Depth integrated dissipation in the mixed layer, $\epsilon_S \sim q^{*3}/h$.

11. Rate of working by the lower shear layer on the hypolimnion.

Using the scales given above we introduce the following parameterization

$$\frac{1}{2}\frac{d(E_S h)}{dt} = C_T q^{*2}\frac{dh}{dt} \tag{23}$$

$$\frac{cu^{*3}}{2} - \overline{w'\left(\frac{p'}{\rho_o} + \frac{E}{2}\right)}\Bigg|_H - \epsilon_S(h+\delta/2) = \frac{C_K}{2}\eta^3 u^{*3} \tag{24}$$

$$\frac{\alpha gh\tilde{H}^*}{2\rho_o c_p} + \frac{\beta ghWS_S}{2} = \frac{w^{*3}}{2} \tag{25}$$

$$\frac{gh\overline{\rho'w'}}{2\rho_0}\Bigg|_{\xi-\delta/2} + \overline{u'w'}\Big|_{\xi-\delta/2}\Delta U + \overline{w'\left(\frac{p'}{\rho_o} + \frac{E}{2}\right)}\Bigg|_{\xi-\delta} - \epsilon_S(h+\delta/2)$$

$$= \Lambda_L + (1-C_K)\frac{q^{*3}}{2} \tag{26}$$

Substituting Eq. (19) to Eq. (26) into the TKE budget Eq. (18), rearranging terms, and writing the result in finite difference form gives

$$\underbrace{\left(\frac{C_T}{2}q^{*2} +\right.}_{1} \underbrace{\left.\frac{\alpha\rho gh}{2\rho_0} + \frac{g\delta^2}{24\rho_0}\frac{d\Delta\rho}{dh} + \frac{g\Delta\rho\delta}{12\rho_0}\frac{d\delta}{dh}\right)}_{2}\Delta h$$

$$= \underbrace{\frac{C_K}{2}(w^{*3} +}_{3} \underbrace{\eta^3 u^{*3})\Delta t +}_{4}$$

$$\underbrace{\frac{C_S}{2}\left(\Delta U^2 + \frac{\Delta U^2}{6}\frac{d\delta}{dh} + \frac{\Delta U\delta}{3}\frac{d\Delta\delta}{dh}\right)\Delta h}_{5} - \underbrace{\Lambda_L\Delta t}_{6} \tag{27}$$

$\Delta\rho$ is now the density difference between the mixed layer and the next layer of thickness Δh. The left hand side of Eq. (27) gives the energy required in time Δt to entrain the next layer, while the right hand side gives the energy available for mixing. C_T, C_K, η and C_S are 0(1) coefficients related to the efficiencies involved in the mixed layer energetics; their values as determined by experiment are discussed in Section 4. For the solution of the problem Eq. (27) must be solved simultaneously with an equation for the velocity shear ΔU and the billow thickness δ.

The characteristics of the billowing have recently been documented in the laboratory by numerous investigators. Corcos and Sherman (1976) discuss an inviscid model and Thorpe and Hall (1977) have given evidence of their effect in Loch Ness. However, the processes involved during the latter stages of the instability and the subsequent mixing has still yet to be resolved. Much of this work is summarized in Corcos (1979) who suggests the time scale for billowing, collapse and re-establishment of a new mean density profile is given by

$$T_B = \frac{20\Delta U}{g'} \tag{28}$$

Even with severe billowing this time is quite short typically being of order 10 - 60 seconds. This means that an interface which has been sharped by the erosion from surface stirring energy source will retain a billowed interface with an interface width δ characterized by

$$\delta = \frac{0.3(\Delta U)^2}{g'} \tag{29}$$

Since the shear ΔU changes on a much slower time scale than this, the interface thickness will be modulated by the time scale of the shear and

$$\frac{d\delta}{dt} = \frac{0.6\Delta U}{g'}\frac{d(\Delta U)}{dt} - \frac{0.3(\Delta U)^2}{g'^2}\frac{dg'}{dt} \tag{30}$$

where the variation of g' can often be neglected over a time scale relevant for the shear.

Lastly, an equation for ΔU must be derived from the momentum Eq. (13), but with the pressure gradient added. In principle, this would involve a rather lengthy computation of the velocity field at each time step. In the construction of DYRESM it was felt that while it was important to include the parameterization of the shear production as this is responsible for the majority of the deepening during severe storms, it would be computationally too time consuming to solve for the full velocity field in the reservoir even when Coriolis accelerations are neglected.

A very simple parameterization of the integral of momentum equation for the mixed layer only presents itself. It basically consists of assuming that the integral to Eq. (13) at the center of the lake is given by

$$\Delta U = \frac{u^{*2}t}{h} \tag{31}$$

for a time before the internal waves generated by the end boundaries have propagated to the center of the lake and set up the pycnocline. In Fischer, *et al.*, (1979) it is shown that this occurs in a time of one quarter of the internal wave period. Beyond this time the shear across the interface will oscillate and decay to zero as dissipation takes effect. Thus a simple linear increase with a full cut off of the shear is conceptually the simplest representation of the actual behavior of the momentum. This idea was incorporated in DYRESM by Spigel (1978) and includes provision for a variable wind and an effective dissipation time.

3.2.2 Mixed layer dynamics algorithm. The mixed layer dynamics are simulated by subroutines MIXER and KH, which determine the mixed layer deepening and the Kelvin-Helmholtz billowing at the interface respectively. The routines are based on Eqs. (27), (30) and (31). The algorithm acts on the density profile generated by heat transport subroutine HEATR. As this profile is the result of both surface heat exchanges and penetrative heating from radiation, there may be a density instability present at the surface. Stabilization of this profile is equivalent to the calculation of the buoyancy flux and the total distributed heat flux $\tilde{H}^*$. Since evaporative losses are also included, w^* may be directly calculated from Eq. (20), or alternatively, from an equivalent mean potential energy formulation. If no surface cooling has occurred, the profile will not require stabilization, the mixed layer depth becomes the top model layer thickness, and $w^* = 0$.

The simplest possible solution of Eq. (27) would be an explicit evaluation of all of the available energies in a time step concerned with the mixing of a particular layer, followed by a comparison with the required energy, with the process applied layer by layer until the available energy is exhausted. The difficulty here is that in cases of high wind stress and small mixed layer depth, an excessively high shear velocity is produced in the fixed time step Δt. In these cases, the time step needs to be reduced below that set by HEATR. In the context of DYRESM, this is undesirable.

To avoid this problem, the balance described by Eq. (27) is applied in sections. Firstly, the deepening by stirring alone is applied layer by layer until there is insufficient energy available for further deepening. To the residual is added energy made available from shear production and billowing during this initial deepening after an adjustment of the shear velocity via Eq. (31). Deepening by shear production and billowing is then applied to further layers by again comparing the available energy, including the residual, to that required. At each incremental deepening, all of the variables are updated to their current values.

The resulting interface is then opened an amount corresponding to the shear instability for the time step by subroutine KH, which evaluates the size of the Kelvin-Helmholtz billows. If these meet certain size and time criteria, the sharp interface is relaxed over the billow thickness.

Thus the mixed layer dynamics is modeled in four distinct sections; deepening by convective overturn, deepening by stirring, deepening by shear production which includes continual readjustment of the shear velocity, and mixing of the pycnocline by Kelvin-Helmholtz billows.

The routine MIXER is called with the density profile stored in model layers 1 to NS, where layer NS is the surface layer, with the residual energy available for mixing from the previous step, and with the various parameters from the previous step required by the shear production algorithm.

Convective overturn. The density of layer NS is checked against that of layer NS-1; if an instability exists, the layers are mixed, a new density computed and compared with layer NS-2. The process is repeated until a stable gradient is achieved. Each such mixing releases a quantity of potential energy, and the corresponding turbulent velocity scale w^* is calculated. This potential energy formulation of w^* may be equated with the TKE form Eq. (20). If layers K, $K+1$, ---, NS have mixed, w^* is then given by

$$w^{*3} = \frac{g}{\rho_f \Delta t} \left\{ \sum_{i=K}^{NS} \frac{\rho_i (d_i - d_{i-1}) \, (d_i + d_{i-1})}{2} - \frac{(d_{NS} + d_{K-1})}{2} \sum_{i=K}^{NS} \rho_i (d_i - d_{i-1}) \right\} \tag{32}$$

where ρ_i is the density and d_i the height of the top of the ith layer, and ρ_f the density of the mixed layers. The mixed layer depth becomes $h = d_{NS} - d_{K-1}$. Should the mixed layer deepen to the bottom the time counter is incremented and control is returned to the main program with reinitialized velocity and residual energy.

Stirring. With the evaluation of w^* from the stabilization section complete, the total energy made available for mixing by stirring in a single time step Δt is calculated. To this is added any residual available energy from the previous time step. Thus the total available energy from stirring mechanisms becomes

$$E_a = \frac{C_K q^3 \Delta t}{2} + E_a \tag{33}$$

where q is given by Eq. (19)

The energy required to mix layer $K-1$ is given by

$$E_r = \left(\frac{\Delta \rho}{\rho} gh + C_T q^2\right) \frac{\Delta h}{2} \tag{34}$$

where $\Delta h = d_{K-1} - d_{K-2}$ the thickness of layer $K-1$ and $\Delta\rho/\rho = 2(\rho_{K-1} - \rho_f)/(\rho_{K-1} + \rho_f)$. The layer is mixed if $E_a \geqslant E_r$, in which case a new ρ_f is calculated, K decremented, E_a reduced by an amount E_r, and new values of $\Delta\rho/\rho$ and Δh established. The process is repeated until $E_a < E_r$, at which stage the mixed layer has deepened by a further amount $\tilde{h}$. Thus the total mixed layer depth is now

$$h = h + \tilde{h} + d_{NS} - d_{K-1} \tag{35}$$

Again, if deepening is vigorous enough for the mixed layer to reach the bottom, the algorithm is complete and control is returned after incrementing the time counter and reinitializing the velocity and energy residual.

Shear production. The deepening resulting from shear at the interface requires the solution of Eq. (31) for the mean velocity shear ΔU. The precomputation of mixing by the stirring mechanism ensures that in times of high wind stress the mixed layer depth will also be relatively large, and the resultant shear velocity remains at a sensible level. The procedure for determining the mean velocity is a complex one and is discussed fully by Spigel (1978). Briefly, the procedure is as follows.

Equation (31) yields a linear growth in ΔU with time for a fixed u^*, up to an effective cut off time t_{eff} at which time ΔU falls to zero. The cut off time assumes use only of the energy produced by shear at the interface during the first wave period T_i, and thus is defined by

$$\int_0^{t_{eff}} \left(\frac{u^{*2}t}{h}\right)^2 dt = \int_0^{T_i} u^2 dt = 4 \int_0^{T_i/4} \left(\frac{u^{*2}t}{h}\right)^2 dt \tag{36}$$

where T_i is calculated for an equivalent two layer profile and no damping is assumed. This yields

$$t_{eff} = 1.59\ T_i/4 \tag{37}$$

Spigel (1978) inserts a correction for damping to yield

$$t_{eff} = \frac{T_i}{4}\left[0.59\{1 - sech\left(\frac{T_d}{T_i} - 1\right)\} + 1\right] \tag{38}$$

where L_d is the damping time for the equivalent two layer system. Thus the shear velocity is effectively given by

$$\begin{aligned} \Delta U &= \frac{u^{*2}t}{h} + \Delta U_o \qquad & t \leqslant t_{eff} \\ \Delta U &= 0 & t > t_{eff} \end{aligned} \tag{39}$$

where ΔU_o is the last value of ΔU and, since t_{eff} may extend over more than one day, u^* may be a function of time. Once cut off has occurred, ΔU remains zero until a change in wind stress occurs.

Thus the algorithm requires the last value of ΔU from the previous time step and a time counter to determine the position of the current time relative to t_{eff}. The previous value of the slope of the velocity function u^{*2}/h is also required to determine a new start up time. The slope of the function may of course change within one period of t_{eff} as the wind stress changes from day to day.

With the procedure for establishing ΔU within a time step, the algorithm proceeds in the following manner. The shear velocity computed in the previous time step is adjusted for the new mixed layer depth from momentum considerations. This adjusted value is used as an initial value and the layer accelerated with acceleration u^{*2}/h over the time step Δt or, if cut off occurs before Δt, until t_{eff}

is reached. An average velocity for the time step is formed

$$U_{av} = \frac{1}{\Delta t} \int_0^t \Delta U \, dt \tag{40}$$

This velocity is used to evaluate the energy produced by shear and billowing during the stirring phase, given by

$$E_a = \frac{C_S}{2} \left[U_{av}^2 \tilde{h} + \frac{U_{av}^2 \Delta(\delta)}{6} + \frac{U_{av} \delta \Delta(U_{av})}{3} \right] + \left[g \frac{\Delta\rho}{\rho} \frac{\delta^2 \tilde{h}}{24h} - g \frac{\Delta\rho}{\rho} \frac{\delta\Delta(\delta)}{12} \right] \tag{41}$$

where $\Delta(U_{av})$ is the change in U_{av} from the previous time step, $\Delta(\delta)$ is the current billowing thickness change given by

$$\Delta(\delta) = \frac{0.6\ U_{av} \Delta(U_{av})}{g \Delta\rho/\rho} \tag{42}$$

To this energy is added any residual energy from the previous phase.

The algorithm now considers the mixing of the next layer, layer $K-1$. From shear production, additional energy

$$E_a = \frac{C_S}{2} \left[U_{av}^2 \Delta h + \frac{U_{av} \Delta(\delta)}{6} + \frac{U_{av} \delta \Delta(U_{av})}{3} \right] + \left[g \frac{\Delta\rho}{\rho} \frac{\delta^2 \Delta h}{24h} - g \frac{\Delta\rho}{\rho} \delta \frac{\Delta(\delta)}{12} \right] \tag{43}$$

is available, where $\Delta h = d_{K-1} - d_{K-2}$. The energy required is again given by

$$E_r = (g \frac{\Delta\rho}{\rho} + C_T q^2) \frac{\Delta h}{2} \tag{44}$$

Layer $K-1$ is mixed if $E_a \geqslant E_r$, in which case K is decremented, a new density computed, E_a reduced by an amount E_r, U_{av} adjusted for the deeper mixed layer, and δ and $\Delta\rho/\rho$ recalculated. The process is then continued, with $\Delta(U_{av})$ and $\Delta(\delta)$ being the changes in U_{av} and δ from the previous comparison. The loop concludes when $E_a < E_r$ for some layer, or the deepening has reached the bottom, in which case control is passed back to the main program, after incrementing the time counter and reinitializing the velocity and residual. Any residual E_a is stored for use in the following time step.

Billowing. The interface resulting from the mixing phases above is unstable to shear if its thickness has been reduced below that given by Eq. (29). In this case Kelvin-Helmholtz billows will be formed. This is accounted for by a call to subroutine KH, which evaluates a billow scale given by Eq. (29). If the billows are of sufficient thickness, at least six layers are formed over the shear zone, incorporating or splitting existing layers as required. These layers are then successively mixed from the interface out, yielding an approximately linear density gradient across the billow thickness. If the billows are of insufficient size, the interface is unaffected.

The final act of KH is to coalesce all the layers above the shear zone into a single large layer to facilitate computation in the following stages of the main program. Control then returns to MIXER, which runs a final check on the density profile stability before returning control to the main program.

3.3 Vertical Diffusion in the Hypolimnion

3.3.1 Theoretical background. The vigorous mixing in the epilimnion has been parameterized in Section 3.2, and this allows the calculation of the depth of the thermocline. The waters of the hypolimnion are protected by the thermocline from disturbances above. In addition, the density gradient in the hypolimnion, although weak, has a stabilizing effect, and it may be expected that vertical mixing in the hypolimnion is small, perhaps only on a molecular level.

There have been many studies in which the evolution within the hypolimnion of concentration distributions of natural or artificially introduced tracers have been measured. Fischer, *et al.*, (1979) summarize the results of these studies and conclude that there is evidence of relatively vigorous, but isolated, mixing in the hypolimnion. The only explanation for such patchy mixing is that although there is not sufficient kinetic energy to cause overall mixing, there are areas in the lake at any particular time where the energy density has been increased by some type of concentrating mechanism allowing a local breakdown and mixing of the structure.

The mixing mechanisms are most likely varied from one lake to the next, but it appears (see Garrett, 1979) that they fall into three major classes. First, internal wave interaction leads, under the right conditions, to a growth of one wavelength at the expense of the others until breaking occurs. These interactions may involve triads of internal waves, parametric instabilities or wave-wave shear interaction. Second, the local shear may be raised, by the combining of long and short internal waves, to such a level that Kelvin-Helmholtz billowing can take place. Thirdly, gravitational overturning can be induced by absorption of wave energy at critical layers.

Regardless of the mechanism responsible, mixing in a stratified fluid is characterized by an overturning motion with a scale of at most a few meters (see Fig. 3.2a). The energetics of the mixing has so far not been explained in any

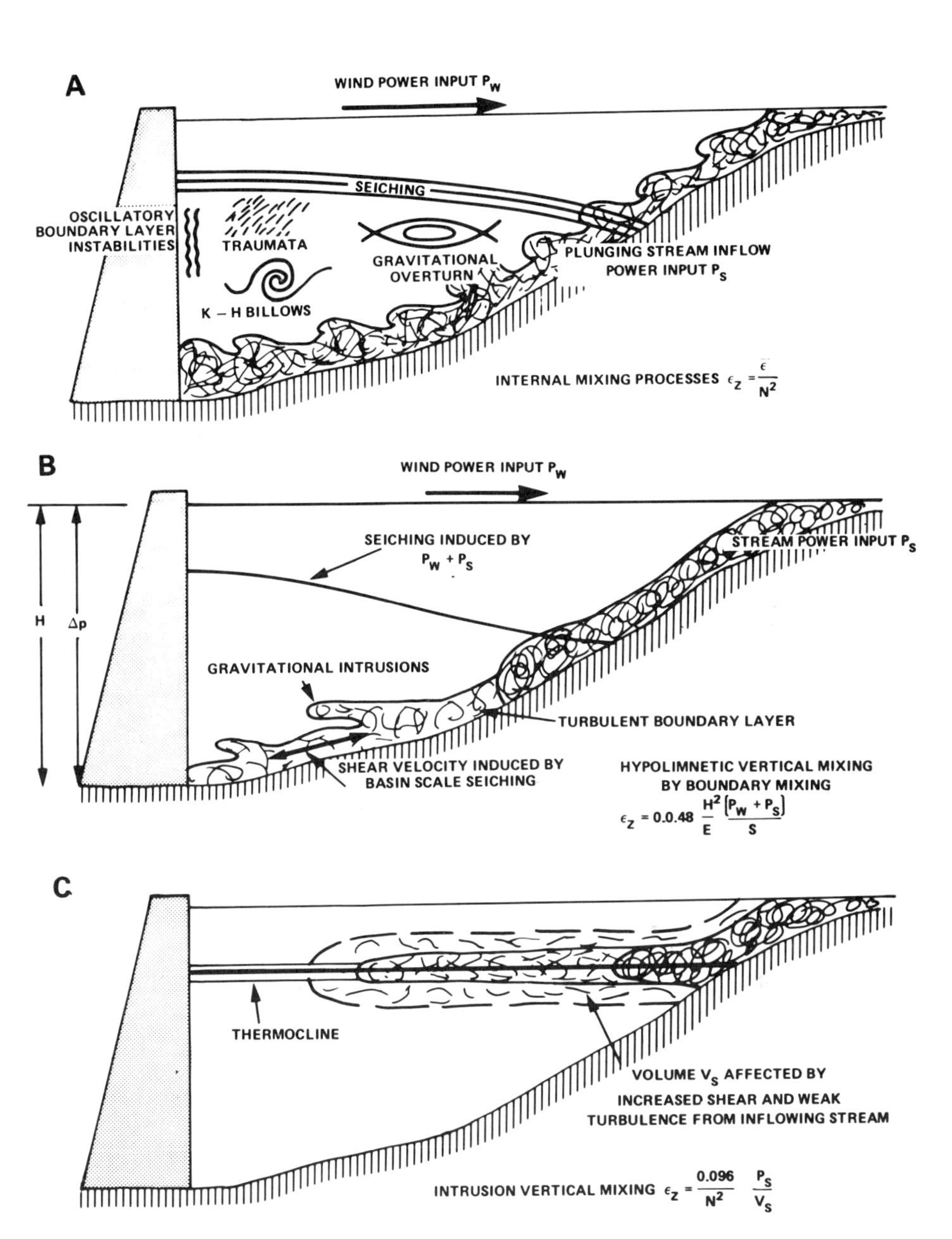

Figure 3.2. (a) Model of basin scale mixing in the hyplimnion of a lake. Wind and river inflow induce basin scale motions which in turn induce internal as well as boundary mixing. (b) Boundary mixing model. The vertical structure is actually adjusted via laminar horizontal intrusions which are driven by boundary mixing. (c) Local river induced mixing. A river intrusion often takes place near the seasonal thermocline, with little associated basin scale oscillation. Such intrusions do, however, induce mixing immediately below and above the intrusion.

detail, but in general the energy introduced by the wind at the lake surface or by the inflow in tumbling to a depth within the lake, induces mean motions and internal waves with scales ranging from the basin size to a few centimeters. These non-mixing motions interact in the hypolimnion until locally one of the above mechanisms leads to a local overturning event with subsequent mixing. The turbulent motions within the mixing patch lead to a readjustment of the potential energy. The remainder of the mean kinetic energy is either dissipated or radiated away by internal waves generated by the collapsing mixed patch. The local increase in potential energy caused by the mixing is in turn utilized to drive horizontal intrusions which are ultimately responsible for the adjustment of the mean density profile in the hypolimnion.

Garrett and Munk (1975) postulate a spectrum of internal waves truncated to allow the right amount of mixing for the particular external energy inputs. The presence of the external lake boundaries prohibits the direct application of the Garrett and Munk (1975) spectrum since, in lakes, one of the dominant mixing mechanisms appears to be the dissipation of basin scale seiching by turbulent boundary layers at the boundary of the lake (Fig. 3.2b). Here, the concept of equilibrium mixing related to the local gradients via the Cox number has only local significance. A simple dimensional argument may be used to derive an effective diffusion coefficient resulting from the general interaction between localized mixing and subsequent gravitational adjustment of the mixed patches.

Let P_W be the power introduced by the wind at the surface of the lake, P_S the power introduced by the inflowing streams and E the potential energy of the stratification in the whole lake. Then the global vertical diffusion coefficient ϵ_z may be written as

$$\epsilon_z = \alpha_1 \frac{H^2(P_W+P_S)}{ES} \tag{45}$$

where α_1 is a function depending on the basin shape, the stratification and the forcing history and S is a stability parameter $H/\Delta\rho \; d\rho/dz$ providing a local variation of ϵ_z with N^{-2} as indicated by experimental evidence. H is the depth of the lake and $\Delta\rho$ is the total density difference between the epilimnion and the hypolimnion. The formulation expressed by Eq. (45) has the advantage that it is simple, yet accounts for the energy input in a fashion similar to the more complicated spectral models and in addition the potential energy parameter E introduces a first order dependence on the basin shape.

Experience with DYRESM has indicated that α_1 may be taken as a constant (= 0.048) provided the lake hydrography is not too contorted. However, while Eq. (45) appears to describe the variation of the vertical transport coefficient of heat and mass due to mixing induced by basin scale motions, the simulation of Lake Argyle has indicated that the assumption that the power P_S is spread uniformly over the volume of the lake is not always valid. River inflows which do not plunge to great depth, but which intrude horizontally just above or below the thermocline, do not induce seiching, but rather merely cause local turbulence and shear within the intrusion itself (Fig. 3.2c).

Such a process suggests a diffusion coefficient formulation as proposed by Ozmidov (1965) and we may postulate a local diffusion coefficient

$$\epsilon_z = \frac{\bar{\epsilon}}{N^2} \tag{46}$$

where $\bar{\epsilon}$ is the local dissipation. Once again if equilibrium flow is assumed an estimate for $\bar{\epsilon} = \alpha_2 P_S / V_S$, where V_S is the volume of the reservoir effected by the stream intrusion.

At present DYRESM uses the diffusion coefficient given Eq. (45) for values of $\epsilon_z < 10^{-4}\mathrm{m}^2\mathrm{s}^{-1}$ and equal to $10^{-4}\mathrm{m}^2\mathrm{s}^{-1}$ for values greater than $10^{-4}\mathrm{m}^2\mathrm{s}^{-1}$. For the Wellington lake where most major inflows plunge to the bottom of the lake this procedure works very well. However, simulations on Lake Argyle and Kootenay Lake, British Columbia, have shown an intensification of mixing around the inflows. The algorithm is therefore being modified to include intensification of the diffusion coefficient by adding the effects of Eq. (45) and (46). Much more work is required to justify the arbitrary cut-off of $10^{-4}\mathrm{m}^2\mathrm{s}^{-1}$

3.3.2 Turbulent diffusion algorithm. The effective turbulent mixing occurring during a time step is simulated by subroutine DIFUSE, which calculates (separately) the redistribution of salt and heat governed by the diffusion equation. The form of the diffusion equation solved is the constant flux model

$$\frac{\partial y_i}{\partial t} = \frac{1}{\rho_i A_i} \frac{\partial}{\partial z}\left(A_i \rho_i \epsilon \frac{\partial y_i}{\partial z}\right) \qquad i = 1,\ NS \tag{47}$$

when y_i is the property (temperature or salinity), ρ_i the density, and A_i the area of the ith layer. The eddy diffusivity $\epsilon(z)$ is calculated from Eq. (45). This equation is solved explicitly in time steps of one quarter hour, time stepping until the current time step set by HEATR is reached. The entire term on the right hand side, including $\epsilon(z)$ is calculated by subroutine FCT for each layer.

The algorithm introduces a mixing time scale

$$T_M = \frac{E}{P_S + P_W} \tag{48}$$

where E is the potential energy locked in the stratification.

$$E = g\left(\rho_{av} \sum_{i=1}^{NS} V_i d_i - \sum_{i=1}^{NS} \rho_i V_i d_i\right) \tag{49}$$

and P_S and P_W are the rates of working of the inflows and wind respectively

$$P_S = g\ \Delta\rho\ D\ Q \tag{50}$$

for each stream, and

$$P_W = C\ A_{NS}\ U_3 \tag{51}$$

with ρ_{av} the average density, V_i the volume and d_i the height of the ith layer, $\Delta\rho$, D and Q the density jump, the level of neutral buoyancy, the river discharge (computed in this instance without entrainment), U the average wind speed, and C a parameter which incorporates the drag coefficient and a factor representing the degree of sheltering of the surface from wind.

Subroutine FCT is called with this value of T_M, the eddy diffusivity is computed, the molecular value for heat is added, and the remainder of the right hand side is evaluated for each layer. The redistribution of salt during one quarter hour period is evaluated by DIFUSE, which ensures that no reversals in gradient occur, the time counter checked against the current time step and the process repeated if necessary. The entire process is then repeated for temperature.

Following these diffusion calculations, the service subroutine THICK is called to split the completely mixed upper layers into a number of smaller layers in anticipation of the next action, either further heat transfer with HEATR or the inflow computations with INFLOW.

3.4 *Inflow Dynamics*

3.4.1 Theoretical background. The inflow process may be divided into three stages as shown in Fig. 3.3 As the stream enters the lake it will push the stagnant lake water ahead of itself until buoyancy forces, due to any density differences which may be present, have become sufficient to arrest the inflow. At this point the flow either floats over the reservoir surface if the inflow is lighter or it plunges beneath the reservoir water if it is heavier. If the entrance point is a well defined drowned river valley, as in most reservoirs, then the side of the valley will confine the flow and a plunge line will be visible across the reservoir at which point the river water submerges uniformly and travels down the channel in a one-dimensional fashion.

Once the inflowing river has negotiated the plunge line it will continue to flow down the river channel, entraining reservoir water as it moves towards the dam wall. Hebbert *et al.* (1979), generalizing the work of Ellison and Turner (1959), show that the entrainment coefficient E is given by

$$E = -\frac{h}{2u\Delta}\frac{D\Delta}{Dt} \tag{52}$$

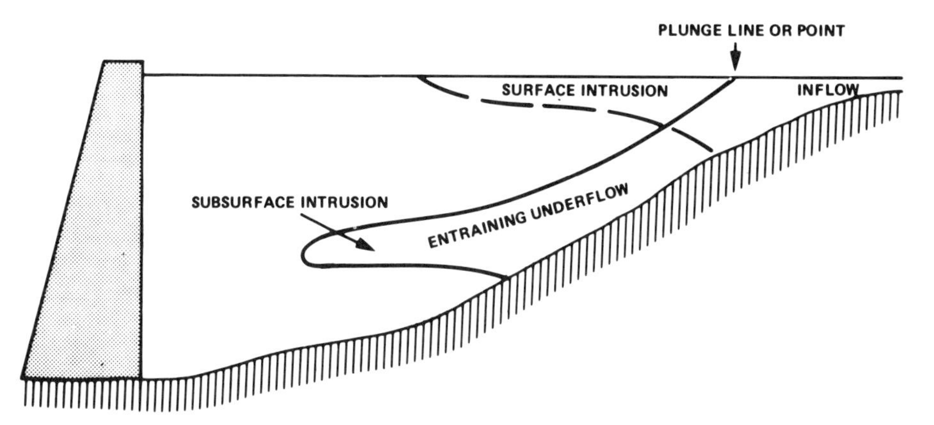

Figure 3.3. Schematic of the various inflow configurations parameterized in DYRESM.

in which Δ is the non-dimensional density anomaly $\Delta = (\rho_u - \rho_r)/\rho_r$, ρ_u and ρ_r the underflow and reservoir densities, u the mean velocity of the underflow, and $D\Delta/Dt$ the variation in Δ moving with the mean velocity u. Direct measurements of u, Δ and h yielded a value of $E = 1.9\times10^{-4}$ for the Collie River Valley.

By assuming that the discharge varies slowly in comparison to the internal adjustment time, and the valley section was approximately triangular, Hebbert *et al.* (1979) showed that the drag coefficient may be expressed as

$$C_D = \frac{\sin\alpha}{5}\left[\frac{5\sin\Phi}{F_i^2} - \frac{4}{3}E\left\{3 + \frac{2}{F_i^2}\right\}\right] \tag{53}$$

where F_i is the internal Froude number, $F_i = u/(g\Delta h)^{1/2}$, h is the hydraulic depth, and Φ the angle between the river bed and the horizontal. For normal flow, F_i is constant. With the above value of E, $\tan\Phi = 10^{-3}$, and $F_i = 0.24$, Eq. (52) implies a value of $C_D = 0.015$ for the Collie River Valley.

The corresponding Manning's n is 0.05, which is typical for such a natural stream.

Rewriting Eq. (53) for the entrainment E yields

$$E = \frac{3}{4}\left\{\frac{5\tan\Phi}{F_i^2} - \frac{5C_D}{\sin\alpha}\right\}\frac{F_i^2}{(3F_i^2 + 2)} \tag{54}$$

where F_i is determined by the slope and roughness of the river bed. The corresponding expression for the flowing depth h then follows. (See Hebbert *et al., 1979).*

Little data exist about the relationship between F_i, the slope of the bed and the bed roughness. The dynamics of the underflow are however characterized by the same processes as those described for the mixed layer and the TKE Eq. (27) may be applied here. For streams with very mild slopes all the entrainment will be induced by the bottom shear stress u^*, where

$$u^* = C_D^{1/2} u$$

Now since F_i is small, $C_T = 0$, and $w^* = 0$, Eq. (27) reduces to

$$E = \frac{\eta^3 C_k C_D^{3/2} F_i^2}{2} \tag{55}$$

On first thought $\eta^3 C_K$ should be the same as found for the mixed layer dynamics. However, the experiments of Hebbert, *et al.*, (1979) and Elder and Wunderlich (1972) suggest a considerably higher efficiency of 3.2 for mixing in a downflow. The difference is most probably due to the very different distribution of mean shear within the underflow.

Equating Eqs. (54) and (55) leads to the result

$$F_i^2 = \frac{\sin\alpha \tan\Phi}{C_D} (1 - 0.85\ C_D^{1/2} \sin\alpha) \tag{56}$$

Hence, once C_D is fixed F_i and E, can be determined.

The entrainment E leads to an increase ΔQ in the underflow volume Q given by conservation of volume

$$\Delta Q = Q\left(\left\{\frac{h}{h_o}\right\}^{5/3} - 1\right) \tag{57}$$

The flow may thus be routed down the river channel of a stratified reservoir by assuming the flow to be gradually varied and by applying Eqs. (54), (56) and (57) locally in each horizontal slab within which the density is assumed constant. At the transition from one slab to the next h_o is redefined so that Q and F_i are continuous across the transition. In this way the flow may be routed down the channel slope until neutral conditions are achieved, at which stage the horizontal penetration of the slug is assumed to commence.

To start the algorithm the initial h_o is required from a plunge point analysis. For a triangular cross section, Hebbert *et al.,* (1979) equated Eq. (56) to the entrance Froude number and showed that the initial flowing depth h_o is given by

$$h_o = \left\{\frac{2Q^2}{F_i g \Delta \tan^2\alpha}\right\}^{1/5} \tag{58}$$

This formulation applies only for streams of small bed slope; in the case of larger bed slopes, shear production and mixing at the plunge point must also be considered.

The entrainment given by Eq. (55) leads to a decrease in the density of the plunging inflow until at some level, the inflow and reservoir densities balance and the inflow penetrates the reservoir. This intrusion into a stratified water body has been studied by Imberger *et al.*, (1976) and is characterized by the nondimensional number $R = F_i Gr^{1/3}$, where Gr is the Grashof number $N^2 L^2/\epsilon_z^2$ and F_i is the Froude number BNL^2/Q, both being determined at the point of insertion. The length of the intrusion e is given by

$$e = 0.44\, LR^{1/2}\, t' \qquad t' \leqslant R \tag{59}$$

$$e = 0.57\, LR^{2/3}\, (t)^{5/6} \qquad R < t' \leqslant Pr^{5/6} \tag{60}$$

where $t' = tN/Gr^{1/6}$, Pr is the Prandtl number of the fluid, N the Brunt-Vaisala frequency and L the reservoir length at the level of insertion. Equation (59) describes an inertia-buoyancy balance and Eq. (60) a viscous-buoyancy balance. For $R \geqslant 1$ Eq. (59) is valid for the whole intrusion process; on the other hand, the smaller the value of R, the larger the applicability of Eq. (60). In the context of DYRESM, the process is split at $R = 1$, using Eq. (59) for $R \geqslant 1$, and Eq. (60) for $R < 1$, irrespective of the time of insertion. Given the two preceding expressions for the length of the intrusion as a function of time it is possible to compute an entrance thickness 2δ such that the inserted fluid is always in "static" equilibrium with the fluid in the slabs which it pushes ahead of itself. Thus

$$\delta = \frac{Q}{B}\left\{1 - \frac{e}{L}\right\} \tag{61}$$

in which B is the width at the entrance. The flow to be inserted during one day is then distributed over a thickness, 2δ, such that

$$u = u_{\max}\left\{\cos\frac{\pi Z}{\delta} + 1\right\} \tag{62}$$

3.4.2 Inflow dynamics algorithm. The dynamics of the river discharges entering the main body of the reservoir are modeled by the subroutine INFLOW, which is based on the equations described above. The subroutine is called separately for each river entering the reservoir.

The entrance of the river inflow into the reservoir is modeled by the insertion of the volume into a number of existing layers at the appropriate height. If the layer volumes become excessive, new layers are formed. The increased volume of these layers causes those above to move upwards, decreasing their thickness in accordance with the given volume-depth relationship, to accommodate the volume increase. The layers affected by inflow are those encompassed by the inflow layer thickness 2δ, centered at the level of neutral buoyancy. This level is determined by a comparison of the inflow density, modified by the entrainment from the layers already passed, with the current layer density. The apportionment of the total inflow volume (discharge + entrainment) is done in such a way that the inflow velocity takes a bell shaped profile.

In detail, the computational procedure is as follows. The inflow properties of volume Q, temperature T, salinity s, and density ρ are initialized to the river values. The inflow density ρ is compared with that of the top layer ρ_{NS}: if $\rho < \rho_{NS}$, or $NS = 1$, the total volume is added to the top layer, a new surface level and properties are computed, and the control is returned to the main program. If, however, $\rho > \rho_{NS}$, $NS \neq 1$, and underflow occurs, entrainment calculations are necessary.

The entrainment ΔQ from the layers adjacent to the underflow is computed from Eq. (55) and this quantity of water is added to the inflow volume Q. The properties T, S and ρ are then adjusted and ρ is compared with the next lowest layer. If the inflow density is the smaller, the level of insertion is taken to be the mid-point of current layer. If not, the process is repeated until a neutrally buoyant level is found or the bottom is reached. The intrusion layer thickness 2δ is calculated from Eq. (61), and the layer numbers n_B and n_T corresponding to the upper and lower limits of 2δ, centered on the level of insertion, are evaluated.

In the calculation of δ, an iterative procedure ensures that δ is less than the distance over which $d\rho/dz$ is calculated. Thus local density inequalities are prevented from yielding erroneously high δ values.

The total inflow volume is apportioned amongst layers n_B to n_T so that the required velocity profile is achieved. Since the inflow layer may not be precisely symmetric because of the differing model layer thicknesses or intersection of the inflow layer with the surface or bottom, upper and lower thicknesses δ_T, δ_B are formed. Thus the bell shaped profile becomes

$$\begin{aligned} u &= \frac{u_{\max}}{2}\left[\cos\left(\frac{\pi z}{\delta_B}\right) + 1\right] \qquad z \leqslant 1 \\ &= \frac{u_{\max}}{2}\left[\cos\frac{\pi z}{\delta_T} + 1\right] \qquad z > 1 \end{aligned} \tag{63}$$

where z is measured from the level of insertion. Since

$$\int_{-\delta_B}^{\delta_T} u\, dz = q;\; u_{\max} = \frac{2q}{\delta_T + \delta_B} \tag{64}$$

and the volume inserted into layer i, above the level of insertion is

$$dV_i = \frac{q}{\delta_T + \delta_B}\left[\frac{\delta_T}{\pi}\left\{\sin\frac{\pi}{\delta_T}(d_i - d_p) - \sin\frac{\pi}{\delta_T}(d_{i-1} - d_p)\right\} + d_i - d_{i-1}\right] \quad (65)$$

A similar expression is obtained for layers below the level of insertion.

3.5 Outflow Dynamics

3.5.1 *Theoretical background.* The outflow structures of a reservoir vary greatly and range from flow over a sharp crested spillway extending the whole width of the dam to single pipe inlets housed in an offtake tower situated in the central basin. Once again the single pipe outlets differ greatly in design from one lake to the next. However, there are basically two extreme idealizations of these structures which serve as suitable models for the development of the theory. First, the flow may be assumed to be two-dimensional and flowing into a line sink. Second, single pipe outlets may be approximated by point sinks in the stratified lake.

In both cases the flow is determined primarily by the discharge q (or Q for the three-dimensional flow) and the stratification gradient. During peak flows and periods of weak stratification the inertia forces associated with the outflows will be much larger than the buoyancy forces and the flow will not "feel" the stratification. Near the sink the flow will be radial with a potential flow dictated by the geometry in the remainder of the lake (Fig. 3.4a). Such flows are rare, since in most cases the stratification is severe enough to have a first order influence on the flow.

Consider now the extreme of very severe continuous stratification and a small discharge. In this case vertical motion is inhibited, but not horizontal motions or streamline curvatures. The flow under such conditions will be more as sketched in Fig. 3.4b with the fluid below the offtake remaining unaffected by the outflow and the fluid above the outlet falling vertically to make up the volume lost through the outflow but preserving the horizontal isopycnals.

In the potential flow shown in Fig. 3.4a the streamlines are such that the whole depth of reservoir is sampled equally. On the other hand with stratification dominating, the flow becomes selective and the quality of the water withdrawn is the average of an elongated volume extending a long way into the reservoir (Fig. 3.4b).

Often the stratification is not uniform, but characterized by a uniform epilimnion, a very strong pycnocline and only rather weak stratification in the hypolimnion. Withdrawal from an outlet in the hypolimnion (or the epilimnion) leads to three possible flows. First, for small discharges the density gradient in the hypolimnion is sufficient to constrain the vertical motion and the flow is selective. Second, at slightly larger discharges the hypolimnion density gradient is insufficient to prevent drawdown, but the density gradient at the pycnocline inhibits relative

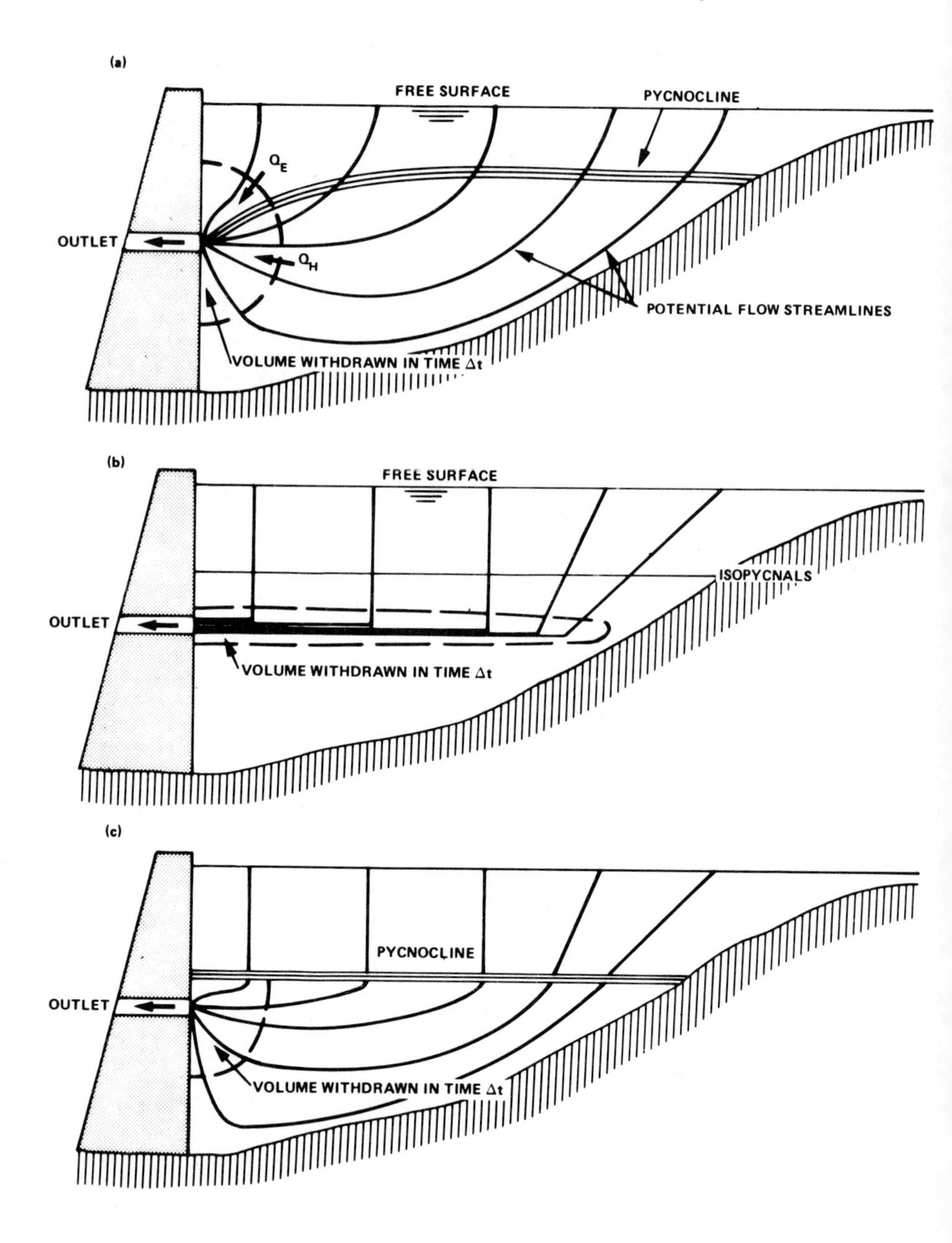

Figure 3.4. (a) Potential flow out through a strong discharge. (b) Selective withdrawal under very stable conditions. (c) Influence of a strong pycnocline on the outflow streamlines.

vertical motion and the pycnocline acts as the free surface in the first case. The water in the epilimnion once again is that in a container with a falling bottom (Fig. 3.4c). Third, as the discharge is raised to a critical value, the pycnocline will also be drawn down (or up for an outlet in the epilimnion) and the water withdrawn will be a mixture of epilimnion and hypolimnion water. The ratio of the mixture will tend increasing to that given by the potential flow solution (Fig. 3.4a) as the discharge is increased or the density difference across the pycnocline is decreased.

In practice, the density profile is as shown in Fig. 3.5. There is usually a well mixed layer containing a diurnal thermocline, a strong pycnocline and a hypolimnion stabilized with a definite density gradient. For weak outlet discharges, the linear stratification model is most applicable, but as the discharge is increased care must be taken that no contradiction arises between the application of this theory and the possibility of drawdown of the seasonal pycnocline. A complete review of the problems has recently been given by Imberger (1980) and we shall confine ourselves to a summary of the applicable formulae.

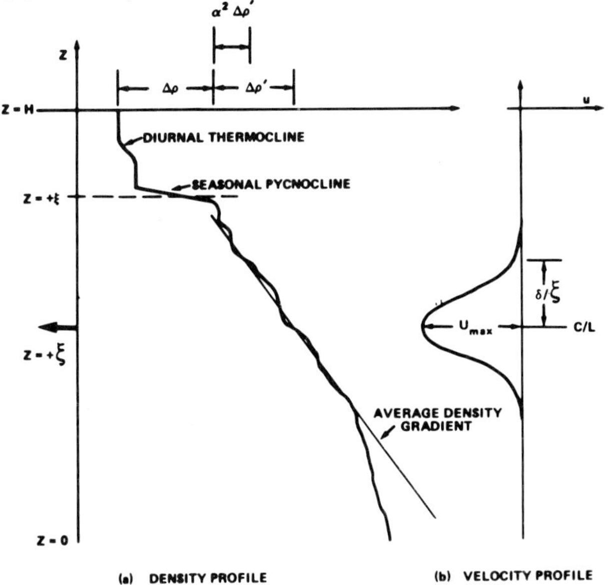

Figure 3.5. Definition sketch of the density and velocity profiles for the selective withdrawal calculations.

Consider first slot outlets, spillway flow or point outlets in very narrow reservoirs.

Let q be the discharge per unit width, N the average buoyancy frequency over the layer thickness, ϵ_z the effective vertical diffusion coefficient, L the length of the reservoir and

$$R_L = F_L Gr_L^{1/3} \tag{66}$$

where $F_L = \dfrac{NL^2}{q}$ and $Gr = N^2L^4/\epsilon_z^2$ then

$$\delta = 5.5\ L\ Gr_L^{-1/6}; \qquad R_L < 1 \tag{67}$$

with 64% of the layer lying above the sink centerline and

$$\delta = 4.0\ L\ F_L^{1/2}; \qquad R_L > 1 \tag{68}$$

The set up time for these two steady state solutions is $T_L = 0(N^{-1}Gr_L^{1/6})$ for $R_L < 1$ and $T_L = 0(N^{-1}F_L^{-1/2})$ for $R_L > 1$.

The velocity distribution associated with these flows is a complex integral of the equations of motion. However, a good approximation is again given by the equation

$$u = u_{\max}\left(1 + \cos\frac{2\pi z}{\delta}\right) \tag{69}$$

This neglects the weak back flow regions at the boundaries of the withdrawal layers, but in terms of a water quality model this is insignificant as these flows are internal and do not affect the outflow.

The withdrawal layer will intercept the seasonal pycnocline when $\delta/2 = 0(\xi - \zeta)$. The present versions of DYRESM retains the formulation given by Eqs. (67) and (68) and for large discharges any possible sharp density changes are incorporated into the computation by using the average of N. This is not strictly correct since once the discharge has reached the critical value where the layer intersects the pycnocline, the strong density gradients there inhibit the development of the withdrawal to a far greater degree than is accounted for by the average gradient.

The layer intercepts the pycnocline when

$$q = 0.35\ (\zeta - \xi)^{3/2}\left(\frac{\Delta\rho' g}{\rho_o}\right)^{1/2} \tag{70}$$

yet drawdown does not occur until

$$q = 2.6\ (\zeta - \xi)^{3/2}\left(\frac{\Delta\rho\ \alpha^2\rho'}{\beta}\ g\right)^{1/2} \tag{71}$$

For the two results to agree as $\Delta\rho \to 0$ requires $\alpha = 0.13$. The correct procedure thus is to test if a change in formulae is warranted or not.

In the case of $R_L < 1$ and the close proximity of an interface is best handled with the average gradient method. In practice the discharge is rarely larger than that given by Eq. (70). DYRESM makes no allowance for this case other than the average gradient method.

Second, consider the flow into single outlets of a dam wall or offtake structure where the lake is very wide and the outflow is essentially radial in the horizontal plane. We shall quantify this assumption after establishing the present formulae.

For very small discharges, where the withdrawal layer is governed by a viscous buoyancy balance there is no difference to that in the two-dimensional case, although Koh's work (1966) indicates that the layers could be about 20% thicker. For Q large enough for the dynamics to be a balance of buoyancy and inertia leads to a withdrawal layer thickness (Imberger, 1980) of

$$\delta = 1.58 \frac{Q^{1/3}}{N^{1/3}} \qquad R_{3,L} > 1 \tag{72}$$

where

$$R_{3,L} = F_{3,L}\, Gr_{3,L}^{1/2}; \quad F_{3,L} = \frac{Q}{NL^3} \text{ and } Gr_{3,L} = \frac{N^2 L^4}{\epsilon_z^2} \tag{73}$$

Hence, once again a critical discharge is reached when the layer intercepts the seasonal pycnocline. This will occur when

$$Q = \tilde{Q}_c = 1.00\,(\zeta - \xi)^{5/2}\left[\frac{\Delta\rho' g}{\rho_o}\right]^{1/2} \tag{74}$$

For the case of a sink situated in a dam wall (other cases can be considered) drawdown occurs when

$$Q = Q_c = 2.54\,(\zeta - \xi)^{5/2}\left[\frac{\Delta\rho + \alpha^2\Delta\rho'}{\rho_o}\, g\right]^{1/2} \tag{75}$$

Once again taking the limit $\Delta\rho \to 0$ requires $\alpha = 0.80$. For Q much larger than Q_c withdrawal will be dominated by the two layer flow and the ratio of epilimnion to hypolimnion draw can be obtained from the laboratory experiments of Lawrence and Imberger (1979). For value of $Q \sim Q_c$ uncertainty still remains. Version 5 of DYRESM has provisions only for two dimensional withdrawal.

The division between whether a particular flow should be considered to be described by three or two-dimensional formulae may be estimated by comparing the magnitude of the drawdown cone with the reservoir width. Since the drawdown cone is given by approximately $4(\zeta - \xi) = 2\delta$ for a drawdown situation (for very large discharge this may be larger) the three-dimensional drawdown must be checked first. If the cone intersects the edges of the reservoir then the two-dimensional formulae are applied. Offtakes in the corner of the reservoir must of course once again have a corrected Q.

The outflow subroutine, OUTFLOW is presently under review to incorporate three-dimensional flow and the possibility of drawdown. The current version uses Eqs. (67) and (68) by defining an equivalent two-dimensional discharge $q = Q/B$, where B is the width of the reservoir.

3.5.2 Withdrawal dynamics algorithm. The computational strategy is similar to that of the inflow model, with the level of withdrawal being fixed as the level of the current offtake. Thus a withdrawal layer is again constructed from Eqs. (67) and (68), depending on the withdrawal volume q in the daily time step and the stratification at the offtake level. In this case however, the upper and lower half layer thicknesses are computed separately as δ_T and δ_B as the withdrawal layer may be highly unsymmetric if located near the thermocline. In addition, the layers generated by withdrawal from the bottom offtake or by overflow are likely to intersect the reservoir bottom and surface respectively and δ_B and δ_T must therefore be restricted.

Once again the daily flow is apportioned over the model layers encompassed by the withdrawal layer $\delta_T + \delta_B$ in such a way that the same velocity profile results. After checking that this apportionment does not completely empty a model layer, the withdrawal volumes are removed from the affected layers, the layers above moved downward to accommodate the volume change, and the properties (temperature and salinity) of the withdrawal computed. The subroutine then returns control to the main program.

In the computational procedure, the layer numbers corresponding to each offtake are precalculated by subroutine RESINT; the layer number n corresponding to the current offtake is passed to OUTFLOW as an argument. This identifies the offtake, the flow volume Q and the offtake height h. The outflow properties are initialized and Q compared with zero. In the case of no withdrawal, the routine is complete and control is passed back to the main program.

If Q is non zero, the computation of the half layer thicknesses begins, after first checking that more than one model layer is present. If only one layer exists, all the withdrawal is taken from that layer. The calculation of δ_T and δ_B is performed by two passes through a loop. In either case, an iterative procedure is employed to ensure that the calculated δ is smaller than the distance dz over which the local density gradient is calculated, ensuring that local density gradients do not lead to erroneously high δ values. The bounds of the withdrawal layer $h - \delta_B$, $h + \delta_T$ are checked against the reservoir bottom and surface and the δ values changed if necessary.

In order to apportion the total volume Q amongst the layers, the layer numbers i_B, i_T corresponding to the lower and upper bounds of the withdrawal layer are determined. The apportionment of the volume to layers i_B to i_T is identical to the procedure for inflow and the volume taken from the ith layer is given by the same expression. In this case, each dV_i must be checked against the total model layer volume V_i to ensure that a layer is not completely emptied or attains a negative volume. Should this be the case, a small amount is left in the model layer and the remainder of the withdrawal component spread equally over the neighboring layers.

Subject to this adjustment, the dV_i are removed from the affected layers, the layer heights of these and the layers above adjusted to account for the reduced volume, and the properties of the outflow computed as a mixture of the dV_i from the respective model layers. The algorithm is now complete, and control is passed back to the main program.

3.6 Service Subroutines

A number of service routines which are called from the various segments of the main program and the dynamics subroutines complete the structure of DYRESM. These are THICK, which maintains the model layer volumes between specified limits, DENSITY, which calculates the density of water for given temperature and salinity, SATVAP, which evaluates the saturated vapor pressure of air corresponding to a given temperature, and RESINT, which provides an interpolation between depths, volumes and areas from the physical data input. Only brief details of these subroutines are given.

THICK: This subroutine ensures that the model layer volumes are maintained between specific limits and is called after any operation which may have affected layer volumes, with the exception of MIXER, which establishes the mixed layer as a single large layer to facilitate the diffusion computations. THICK is called after these computations to maintain good resolution for the heating and cooling calculations following.

The upper layer volume $V_{\max}$ is set to ensure that adequate resolution is maintained, and the lower limit $V_{\min}$ to ensure that an excessive number of layers is not required. However, $V_{\min}$ must also be sufficiently small to avoid problems of numerical diffusion through a large number of layer amalgamations. A good compromise has proved to be $V_{\min} = S/N$, where S is the capacity of the reservoir and N the maximum number of layers allowed, and $V_{\max} = 4\ V_{\min}$.

The routine checks each layer volume against $V_{\max}$ and $V_{\min}$. Any layer exceeding $V_{\max}$ is split into an appropriate number of smallest layers of identical properties, all layers renumbered, and new positions computed by RESINT. A layer smaller than $V_{\min}$ is amalgamated with its smaller neighbor, and all layers renumbered.

DENSITY: This subroutine evaluates the density ρ of water of temperature T and salinity S according to the formula proposed by Chen and Millero (1977).

The routine is called after any mixing of two layers, or any readjustment of a layer property.

SATVAP: In many cases the surface temperature is not available as daily data. Thus the saturated vapor pressure at the surface temperature, required by HEATR, is not able to be precalculated. Subroutines SATVAP evaluates this variable, using the predicted surface temperature T.

RESINT: Subroutine RESINT calculates the layer volumes and surface areas corresponding to given layer heights or conversely, layer heights and surface areas corresponding to given volumes. The routine is called following any operation which affects either volumes or heights.

In either case the calculation is an interpolation on the given depth-volume-area data for the reservoir, which is assumed to have the local functional form

$$\frac{V_i}{V_{i-1}} = \left[\frac{d_i}{d_{i-1}}\right]^{a_i} \tag{76}$$

$$\frac{A_i}{Ai-i} = \left[\frac{d_i}{d_{i-1}}\right]^{b_i} \tag{77}$$

where V_i,A_i and d_i are the ith volume, area, and height data value given.

An additional function of RESINT is to evaluate the model layers corresponding to the levels of the various offtakes and the crest. This is performed by comparing each layer height with the given height of each offtake.

4. DISCUSSION

The development of DYRESM and the associated algorithms described in Section 3 are part of a program designed to understand the physical mixing processes in a lake. The priority attached to the many processes operating in a lake is determined by the long standing aim to obtain a fuller understanding of the interaction between physical mixing processes and the biological kinetics of the lake. The initial construction and constant updating of DYRESM thus serves a twofold purpose. First, it is an organized way to evaluate the influence of a particular process to the overall mixing patterns in a lake, to validate a particular parameterization and to study the competition for available energy among the great host of mixing mechanisms. Second, DYRESM serves as a useful predictive tool for testing reservoir management strategies, for evaluating methods for the control of the lake stratification and for water quality simulations.

DYRESM has been tested on a number of different lakes, but its major evaluation and development has taken place with data from the Wellington Reservoir. This storage is situated about 160 km south of Perth in Western Australia and has a storage capacity of $185 \times 10^6 \text{m}^3$. The reservoir is approximately 30 m deep at the dam wall, extends some 24 km along the main Collie River Valley and has a surface area of approximately 16 sq. km at the crest level.

The catchment has been severely disturbed by clearing native forest for agricultural development. This has resulted in substantial increases in the stream salinities with the first winter flows possessing salinities as high as 3500 p.p.m. Salinities generally decrease during the late winter and spring. These high salinities, although extremely detrimental to the usefulness of the storage, are valuable for the critical evaluation of the model. In this lake the temperature and the conductivity of the water form two independent tracers with independent inputs.

The seasonal variability of the various inputs to the reservoir over the period from the Julian day 133 in 1975 to day 365 in 1977 are shown in Fig. 4.1a. Depicted are the wind speeds, the short-wave solar radiation as computed from cloud cover records, the salinity and temperature of the inflowing water and the total rate of inflow from the Collie River which contributes approximately 85% of the total inflow. The remaining inflow is included in the simulations, but is not shown in Fig. 4.1a.

Figures 4.1b and 4.1c show the field temperature and salinity structures length averaged along the Collie River Valley as a function of time over the period January 1975 to August 1978. The salinity data gathered between October 1977 and June 1978 is regarded as unreliable and is not shown. The broken lines in Fig. 4.1b indicate that no data was taken in the period covered and the thermal structure is interpreted from the structure before and after the period of interruption.

The yearly cycle is clearly evident in Figures 4.1a and 4.1c. The cold salty inflows lodge in the base of the homogeneous reservoir in the months of June, July and August. Summer stratification builds up until December, when surface winds begin to mix the surface layers and a thermocline forms, protecting the waters below. In early winter, the air temperature falls and the winds increase, with the result that the reservoir is completely mixed before the following inflows arrive. The marked difference in the thermocline structure between 1976 and 1977 was caused by a change in the withdrawal policy. In 1976 all the water was withdrawn from the offtakes at 15 m height, whereas in 1977 a large quantity of salty water was scoured through the offtake at the bottom of the dam wall.

It is also clear from Fig. 4.1b that the temperature regime of the reservoir is determined by the inflows and the surface heating and cooling. The bottom temperature of the reservoir for most of the year is determined by the temperature of the coldest inflows, whereas the surface temperature is determined by the meteorological forcing. This is typical of lakes in temperate regions, and is in contrast to tropical lakes, where the temperature regime is solely determined by the meteorological inputs.

The data set gathered at the Wellington Reservoir forms an ideal test for DYRESM or any other model. The data was collected at reasonably regular intervals and a series of process oriented field expeditions yielded further data on many of the details of individual events. This work is continuing and it is to be expected that further development of DYRESM will take place from these results.

(a)

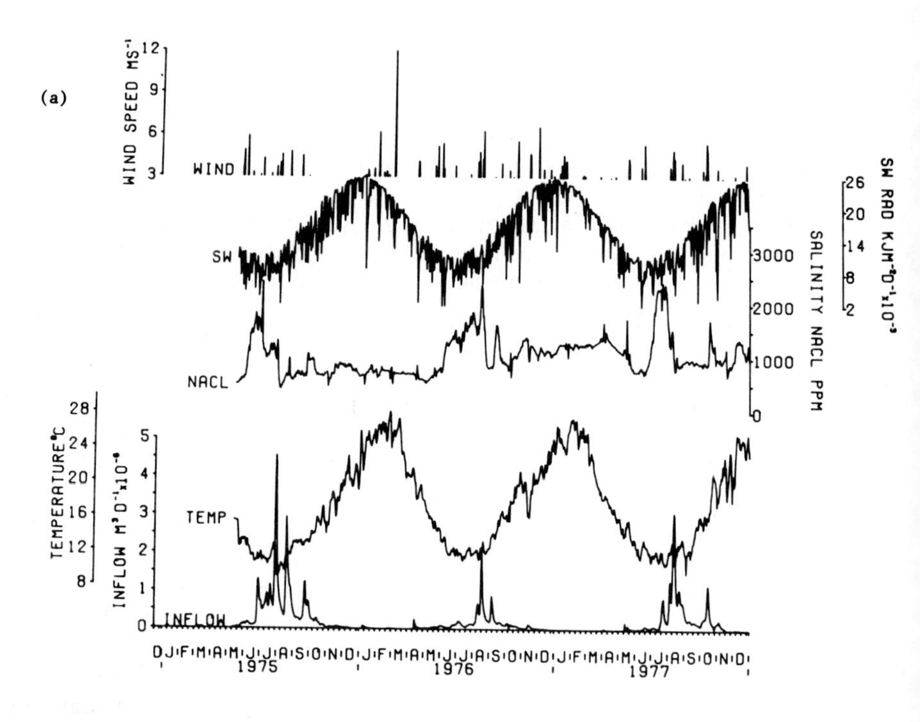

(b)

(c)

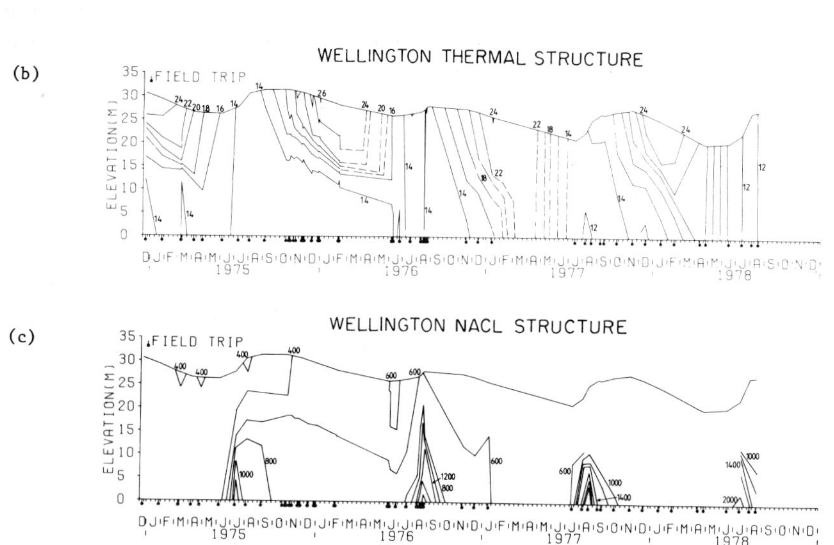

Figure 4.1. (a) Seasonal variation of wind speed, short-wave radiation, salinity and temperature of inflow and discharge for the simulation period of the Wellington Reservoir. (b) Field data showing the temperature structure in the Wellington Reservoir for the duration of the simulation period. (c) Field data showing the salinity structure in the Wellington Reservoir for the duration of the simulation period.

There are seven constants which must be specified by the user before applying DYRESM. Of these only one is truly adjustable--the others are related to well identified physical processes and are determined from experimental or field data. The constants are described below, together with experimentally determined values.

1. C_D is the drag coefficient for inflowing streams. C_D was determined independently of DYRESM in a field study described in Hebbert *et al.*, (1978). The value determined in that study, $C_D = 0.015$, is used here.

2. η_1 is an extinction coefficient for short-wave solar radiation penetrating the water. It relates the solar radiation received at the water surface, to that penetrating to a depth z. A single exponential decay formula was used as only limited field measurements were available. An average value of $\eta_1 = 0.35$ is taken from data presented by Hutchinson (1975), based on the fact that the Wellington is fairly clear in the summer months when surface heating is an important effect.

3. α_1 is a constant occurring in the expression for the diffusivity calculated for the deep hypolimnetic mixing. It represents the efficiency with which the power input from the surface wind and river inflows is converted to a gain in potential energy of the lake water due to vertical mixing. A value $\alpha_1 = 0.048$ determined in earlier calibrations of DYRESM over the 100 day period from day 133 to 233 has proven satisfactory throughout.

4. C_K is the coefficient measuring the stirring efficiency of convective overturn. Experimental results summarized by Fischer *et al.*, (1979) suggest an average value of $C_K = 0.125$.

5. η, in combination with C_K as $C_K\eta^3$, is a coefficient measuring the stirring efficiency of the wind. The value given by Wu (1973) is adopted here, since it was shown by Spigel (1978) that in his experiments stirring effects dominated the entrainment, with shear production, temporal effects, and internal wave radiation losses negligible. Wu's deepening law is $dh/dt = 0.23\ u^*/Ri^*$ giving $C_K\eta^3$, and thus $\eta = 1.23$.

6. C_S is the coefficient measuring the efficiency of shear production for entrainment. Values ranging from 0.2 to 1 are reported by Sherman *et al.*, (1978). $C_S = 0.2$ was chosen as representing a good estimate for most experimental results. A value of 0.5 was used before the energy released by billowing was separately included as is done in version 5 of DYRESM.

7. C_T is associated with temporal, or unsteady, non-equilibrium effects due to changes in surface wind stress or surface cooling. C_T is constructed by the requirement that for a turbulent front entraining into a homogeneous fluid, $dh/dt = 0.3u^*$ (Zenman and Tennekes, 1977) giving a value of $C_T = C_K\ \eta/0.3 = 0.510$.

These coefficient values were used for a 962 day simulation of the lake dynamics starting with initial profile data on day 75133. The results from this simulation are shown in Figs. 4.2a and 4.2b.

Comparison of Figs. 4.1b and 4.2a shows that the thermal structure is reproduced extremely well with all the stratifying and mixing regimes occurring at the

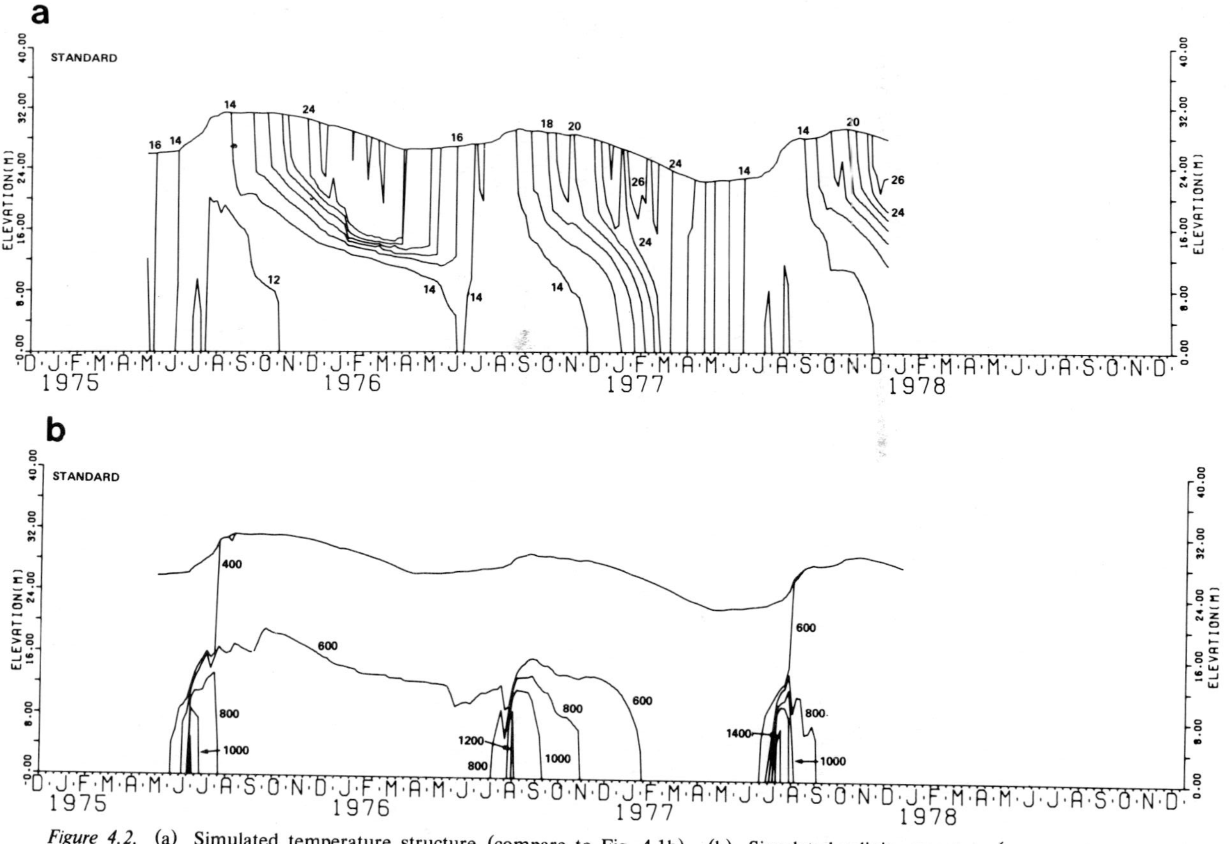

Figure 4.2. (a) Simulated temperature structure (compare to Fig. 4.1b). (b) Simulated salinity structure (compare to Fig. 4.1c).

correct times and of the correct magnitude. The slug of 12°C water predicted by the model in August to September 1975 does not appear in the field data. This slug is derived from the previous inflow and its presence may be due to errors in the inflow temperature data. In any case, the predicted temperature gradient in the bottom at this time is small, and the actual difference between field and predicted temperatures is less than 1°C.

The predicted and observed salinity variations also compare well, the most significant anomaly being the mismatch of the 600 p.p.m. line at the period of maximum inflow in 1976. The field data suggest somewhat more energetic surface mixing which may be due to small errors in the wind data. Additionally, the model does not reproduce the peak salinities of the inflows. This is due to the construction of the model layers in terms of layer volumes, hence fixed by the depth resolution in the upper part. The bottom three or four meters always appear as mixed, and any salinity slug occupying less than this will be mixed with the layer above. Thus, there is no error in terms of the salt load, but one of display of the structure, and it was accepted in order to save computer time. If greater resolution were required, this could be achieved by specifying smaller minimum slab sizes.

Overall, DYRESM appears to faithfully reproduce even very severe changes in the reservoir structure caused by such diverse forcing as large saline inflows, active scouring, strong wind deepening and winter convective cooling. Perhaps more importantly, the model correctly simulates two independent parameters, salinity and temperature, to a resolution equal to that of the field program. More accurate field data is required if more stringent tests are to be applied.

Profile data, as computed by the various components of MIXER are reproduced in Fig. 4.3 for day 75138. The result of each computational step is shown starting with Fig. 4.3a which depicts the result of HEATR adding both the surface heat losses and the subsurface heating due to solar radiation penetrating to some depth. A 3° C deficit due to surface cooling, was quickly achieved, leaving a convectively unstable density profile, Fig. 4.3b, with a density anomaly of nearly 0.5 kg m^{-3}. The potential energy stored in this unstable profile is released by relaxing the density profile until it and the mixed layer have reached a depth of 5 meters. The total amount of potential energy released yields the effective convective velocity scale w^* (Fig. 3.4c). The stirring energy $C_K/2\eta^3 u^{*3}\ \Delta t$ is then added to this pool of TKE causing a further deepening of the mixed layer of 3.6 meters (Fig. 3.4d). The velocity in the mixed layer is reduced due to the deepening, but there is still sufficient mean kinetic energy to cause further 2 meters of deepening by shear production as is seen in Fig. 3.4e. At this stage the algorithm leaves a structure characterized by a mixed layer and a sharp thermocline. The subroutine K-H smears this interface over a distance $0.3\ \Delta u^2/g'$ which in this case amounted to nearly 3 meters (Fig. 4.3f). This process is carried forward from one time step to the next.

On day 75138 all the deepening processes were additive. This is not always the case as is illustrated by the deepening on day 75228. In Fig. 4.4a the net surface heat flux is plotted as a function of time and in Fig. 4.4b the corresponding ΔU (final) for the mixed layer. The plot in Fig. 4.4c shows the resulting mixed

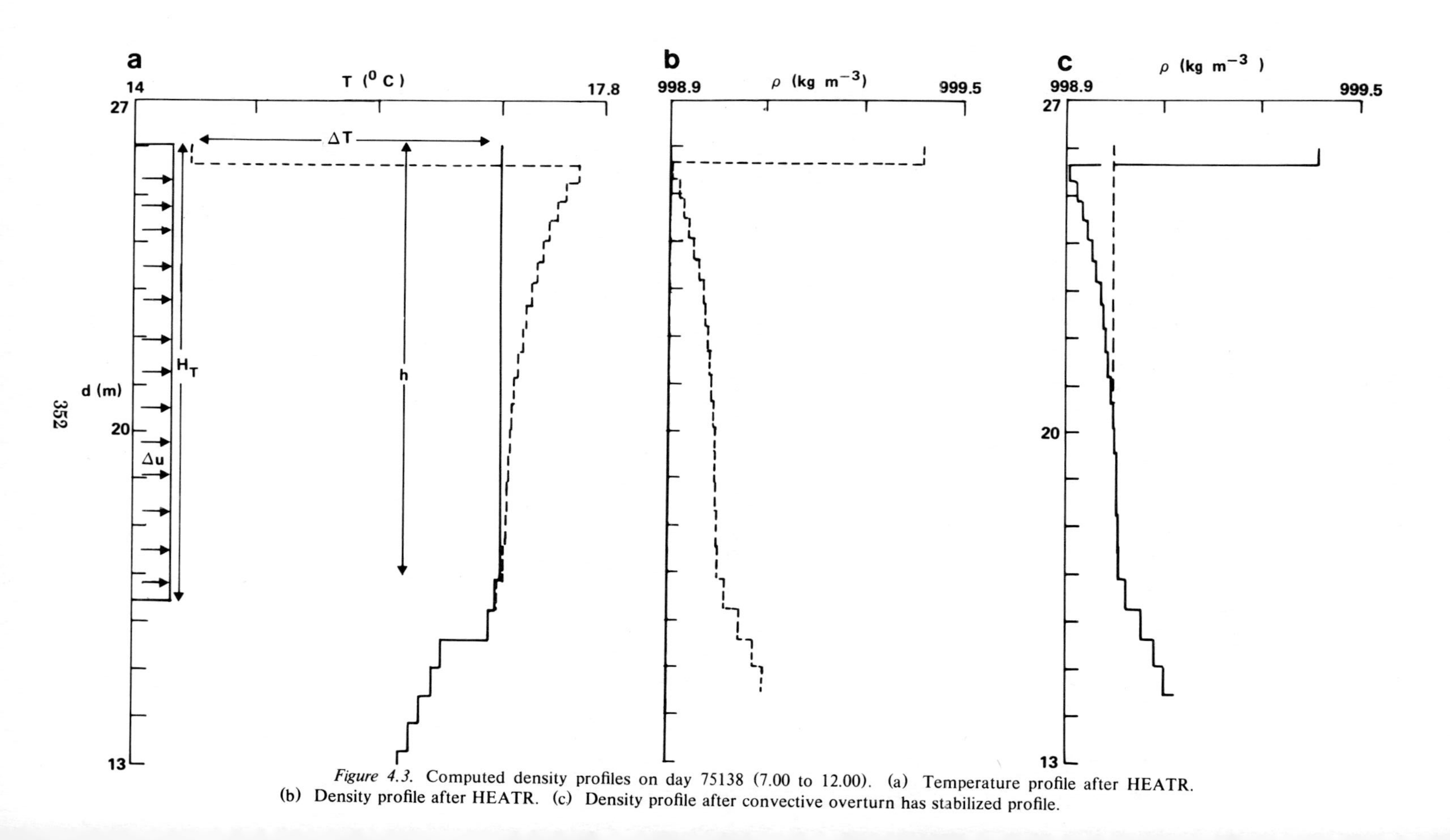

Figure 4.3. Computed density profiles on day 75138 (7.00 to 12.00). (a) Temperature profile after HEATR. (b) Density profile after HEATR. (c) Density profile after convective overturn has stabilized profile.

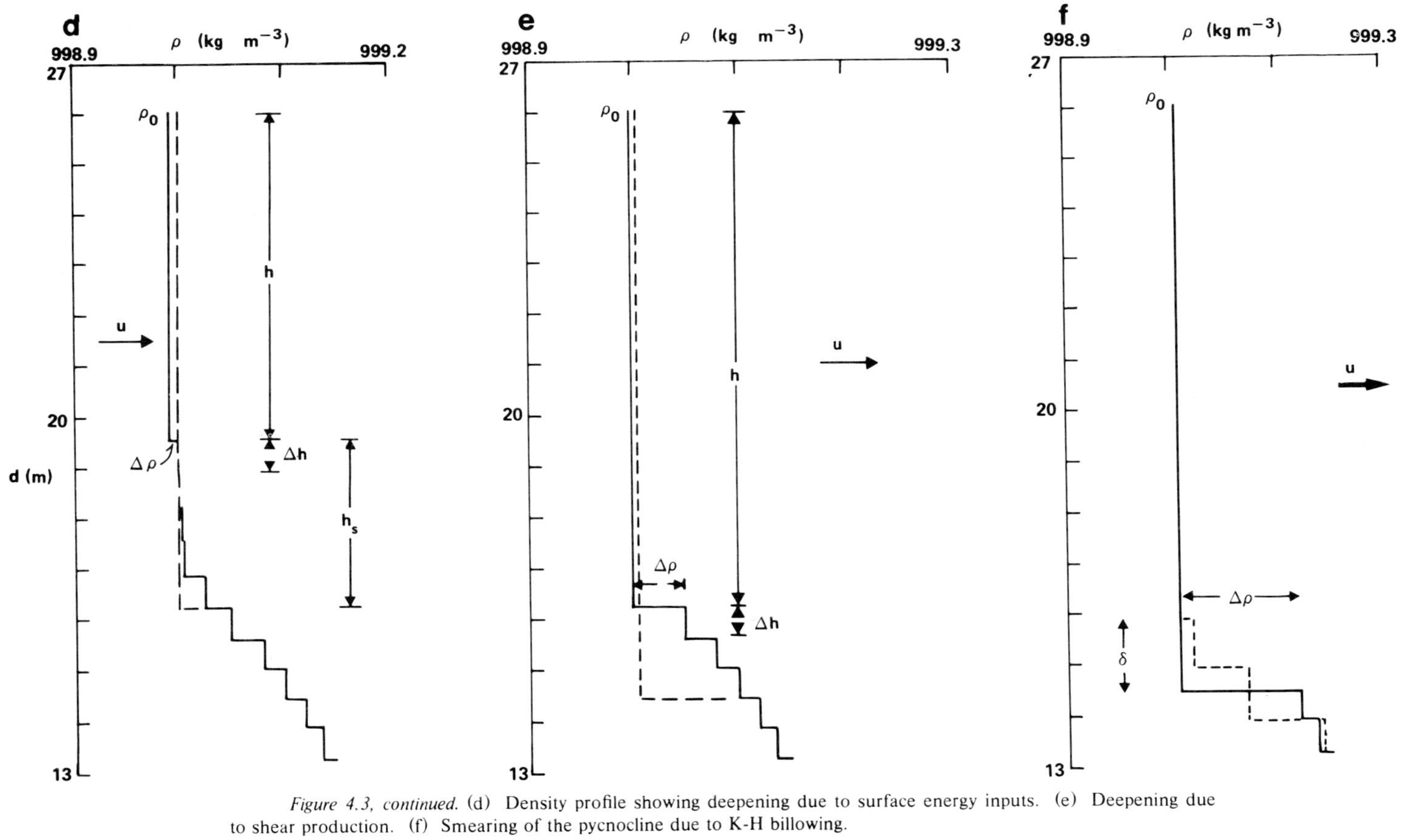

Figure 4.3, continued. (d) Density profile showing deepening due to surface energy inputs. (e) Deepening due to shear production. (f) Smearing of the pycnocline due to K-H billowing.

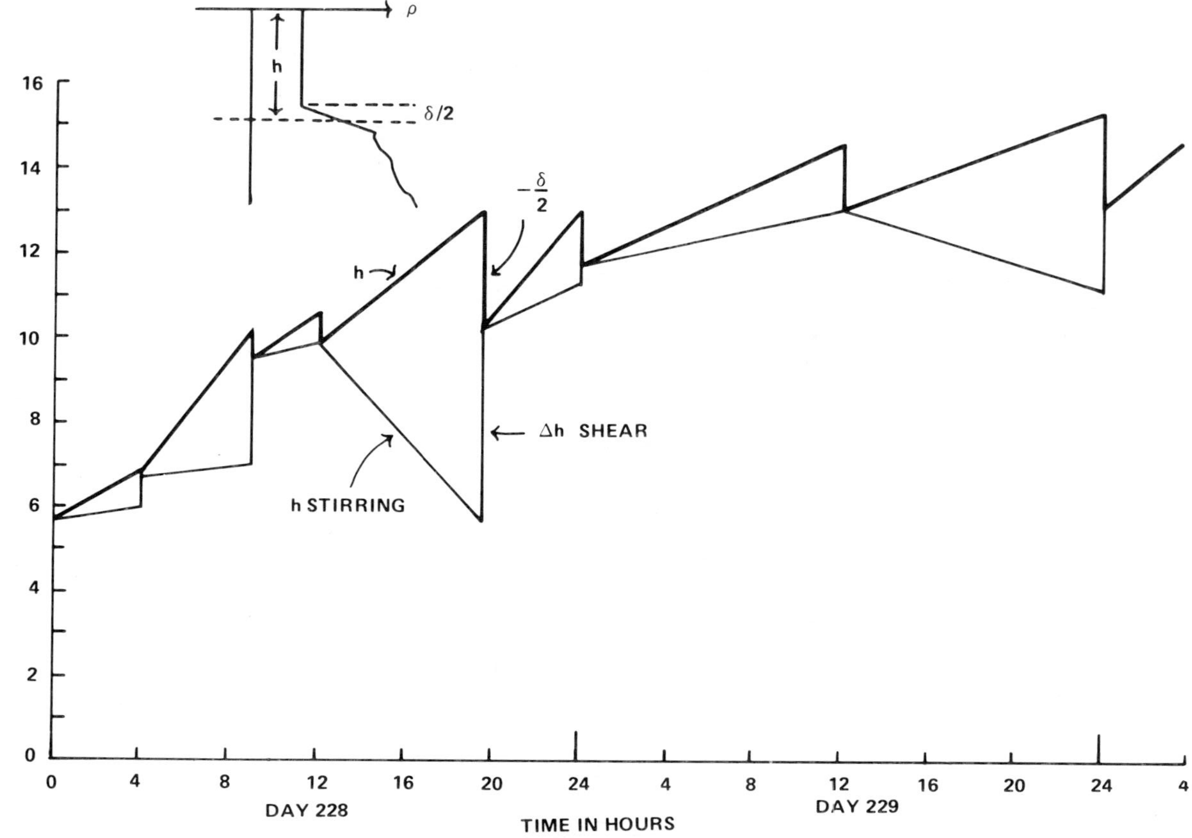

Figure 4.4. Deepening predictions on day 75228. (a) Thermal input at the surface. (b) Calculated mean speed of mixed layer. (c) Predicted deepening. Solid line is net depth of mixed layer. Thin line represents deepening only due to the stirring mechanism. Jumps at each time step account for K-H billowing. Notice the decrease in the depth during the 4th time step if only stirring energy is accounted for.

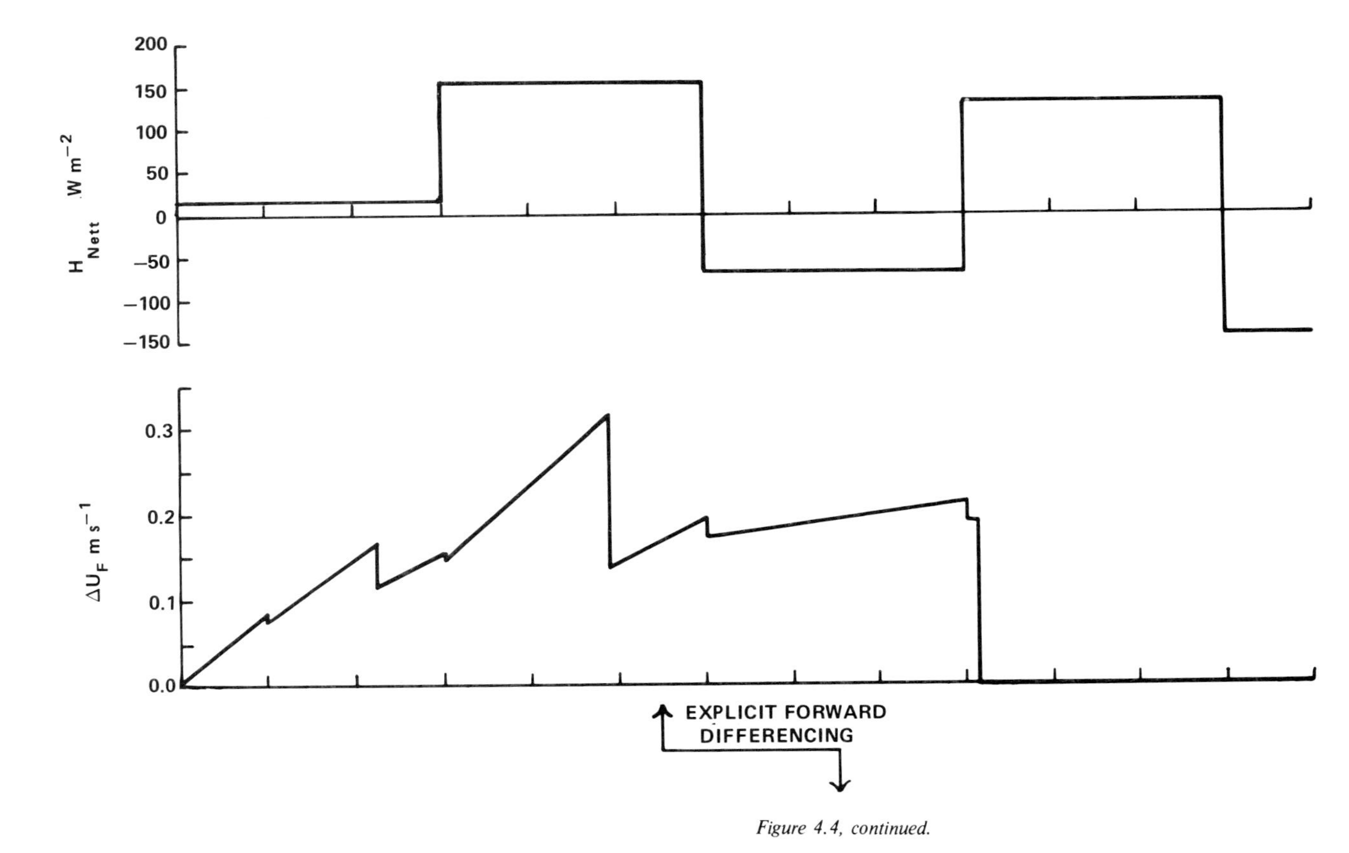

Figure 4.4, continued.

layer behavior. The solid line is the total depth after all the processes illustrated in Fig. 4.3 have acted. The thin line shows the deepening due to stirring only and the jump at the end of each time period is the result of K-H billowing decreasing the layer depth. In the fourth time period, at the beginning of severe surface heating the stirring mechanism actually predicted a decrease in layer thickness. This is explained simply by noting that the Monin-Obukhov length was decreased by the positive buoyancy flux due to the surface heating. The decreased depth however raised the velocity and greatly enhanced shear deepening.

Each day of the simulation is a variation of these two extremes and the depth of the mixed layer at the end of a season is merely the cumulative effect of individual diurnal deepening events.

The subroutine DIFUSE is called after the mixed layer dynamics calculations have been completed. Fig. 4.5 is a graph of the density profile of day 75138 and the associated vertical diffusion coefficient for heat. The value of 10^{-4}m sec^{-1} is the cut-off value introduced to simulate a fully turbulent condition. It is seen that as N increases ϵ_z decreases sharply leading to decreased mixing in stable regions.

The great advantage of DYRESM is that each mixing process may be identified and each constant has a well-defined influence on the simulated results. In Fig. 4.6b to 4.6d are shown three perturbations around the standard simulation reproduced in Fig. 4.6a.

First, increasing the shear production efficiency greatly increases the deepening during the wind events in January of 1976, but otherwise introduces little change. The structure recovers quickly from the overdeepening indicating that the dynamics of the Wellington Reservoir are strongly forced with little long term "ringing" of the structure. This is also reflected in the short seiching periods. This is not the case for a recent simulation of Kootenay Lake in British Columbia. Here T_i was close to 28 days and the structure tended to oscillate for long periods after a wind event. This necessitated a modification of MIXER routine which captured the wind direction as well as the wind speed.

The arbitrary cut-off of 10^{-4} m^2 sec^{-1} in the diffusion coefficient has only a minor influence on the temperature structure as is illustrated in Fig. 4.6c where the cut-off has been raised to 10^{-2} m^2 sec^{-1}. The salinity concentrations were however, appreciably altered with the increased diffusion reducing the peak salinity at the bottom. These simulations make clear that the diffusion in the hypolimnion is a two parameter process; the second parameter being the appropriate diffusion in the absence of stratification.

The importance of good input data is shown in Fig. 4.6d. A bad data point in February of 1976 was not discovered until the recent introduction of the billowing subroutine. Previous versions of DYRESM were insensitive to isolated short duration wind episodes, but as seen in Fig. 4.6d a 12 m sec^{-1} wind in February now had a dramatic deepening influence. A check of the wind record revealed an error in transcription.

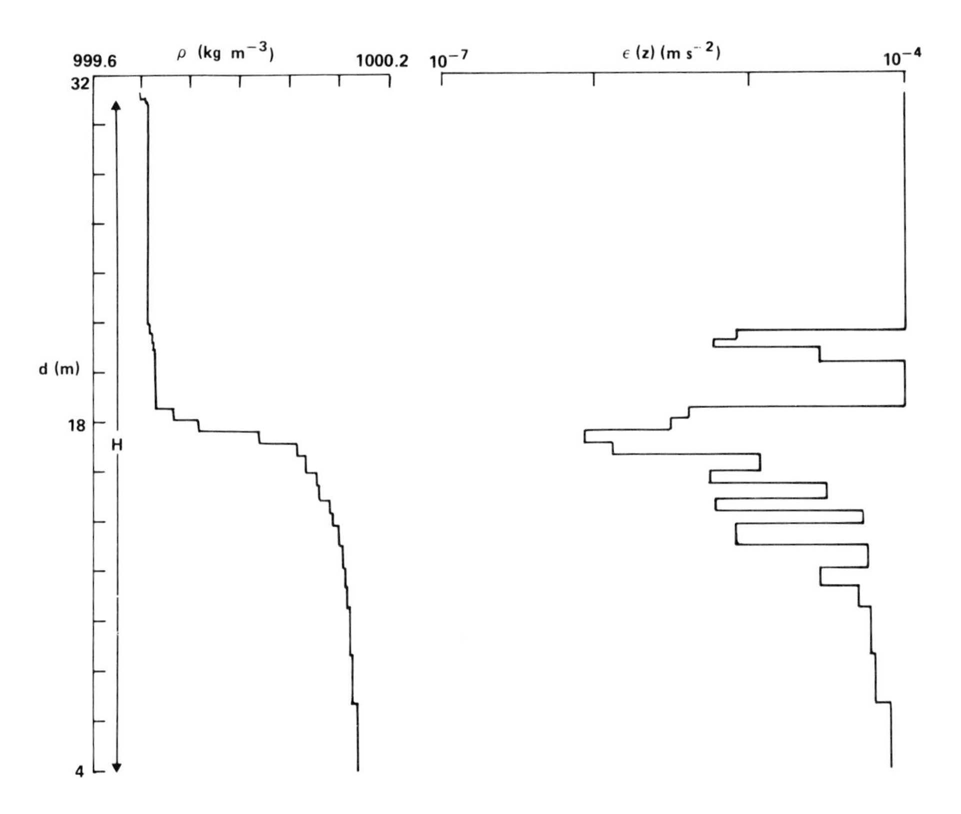

Figure 4.5. A plot of density and vertical diffusivity on day 75138.

5. CONCLUSION

The dynamic simulation model DYRESM appears to successfully model the dynamics of small and medium sized lakes. The model has only one calibratable coefficient and even the value of this is not expected to vary much from one lake to the next. The model has been validated in terms of temperature and salinity variations in 3 major lakes and two lakes with only poor data. It is now necessary to collect data specific to certain processes in the lake and validate the details of the algorithms MIXER and DIFUSE. Such a field program has been commenced in Western Australia.

Certain changes to version 5 are presently in progress. These include the provision for drawdown at the offtake structure, localizing mixing around a river intrusion, the effect of wind direction, a specific parameterization of boundary mixing and the influence of rotation on the shear production parameterization.

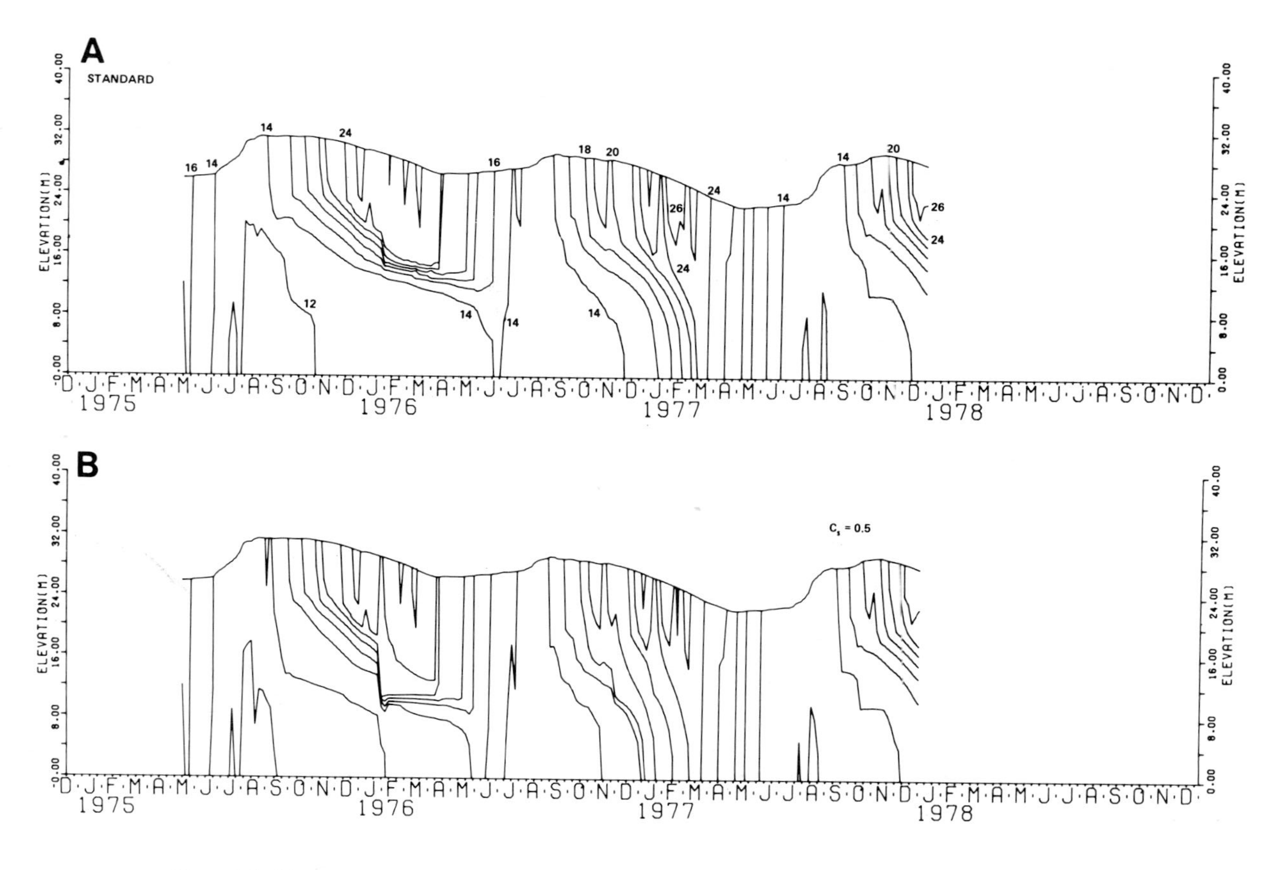

Figure 4.6. (a) Predicted temperature structure. (b) Predicted temperature structure with C_S increased to 0.5

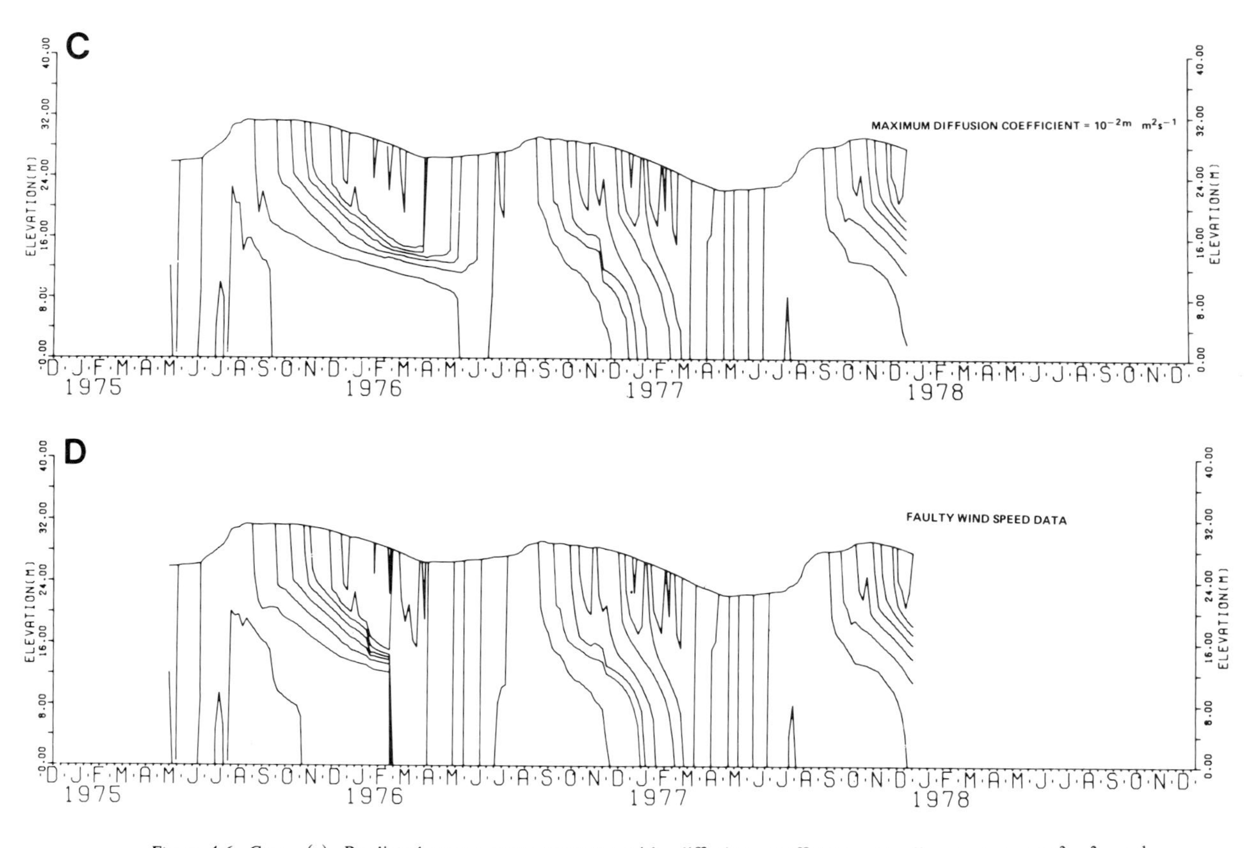

Figure 4.6, Cont. (c) Predicted temperature structure with diffusion coefficient cut-off raised to 10^{-2} m^2 sec^{-1}. (d) Predicted temperature structure showing the influence of an erroneous wind speed of 12 m sec^{-1} in February 1976.

Acknowledgements. DYRESM has been under development for over 6 years and numerous people have contributed to its construction. In particular the authors would like to thank Dr. Robert H. Spigel for his involvement in the MIXER subroutine.

The model was developed with funding from the Australian Water Resources Council, the Water Resources Center of the University California, the Australian Research Grants Committee, the Public Works Department of Western Australia and the Metropolitan Water Supply, Sewerage and Drainage Board of Western Australia. This assistance is gratefully acknowledged.

REFERENCES

Carson, D. J. and Richards, P. J. R. (1978). Modelling surface turbulent fluxes in stable conditions. *Boundary-Layer Meteorol.* **14,** 67-81.

Chen, C. T. and Millero, F. J. (1977). The use and misuse of pure water PVT properties for lake waters. *Nature* **266,** 707-708.

Corcos, G. M. (1979). The Mixing Layer: Deterministic Models of a Turbulent Flow. College of Engineering, University of California, Berkeley, Report No. FM-79:2.

Corcos, G. M. and Sherman, F. S. (1976). Vorticity concentrations and the dynamics of unstable shear layers. *J. Fluid Mech* **73,** 241-264.

Deardorff, J. W. (1968). Dependence of air-sea traner coefficients on bulk stability. *J. Geophys. Res.* **73,** 2549-2557.

Denman, K. L. (1973). Time dependent model of the upper cone. *J. Phys. Oceanogr.* **3,** No. 2, 173-184.

Dyer, K. R. (1974). The salt balance in stratified estuaries. *Estuarine Coastal Mar. Sci.* **2,** 273-281.

Elder, R. A. and Wunderlich, W. O. (1972). Inflow density currents in TVA reservoirs. *Proc. Int. Symp. Statrified Flow., Am. Soc. Civ. Eng.*

Ellison, T. H. and Turner, J. S. (1959). Turbulent entrainment in stratified flows. *J. Fluid Mech.* **6,** 423-448.

Fischer, H. B., List, E. J., Koh, R. Y. C., Imberger, J. and Brooks, N. H. (1979). "Mixing in Inland and Coastal Waters." Academic Press, New York.

Garrett, C. (1979). Mixing in the ocean interior. *Dyn. Atmos. Oceans* **3,** 239-256.

Garrett, C. J. R. and Munk, W. H. (1975). Space-time scales of internal waves: a progress report. *J. Geophys. Res.* **80,** 291-297.

Hebbert, B., Imberger, J., Loh, I. and Patterson, J. (1979). Collie River underflow into the Wellington Reservoir. *J. Hydraul. Div. Proc. Am. Soc. Civ. Eng.* **150,** HY5, 533-545.

Hicks, B. B. (1972). Some evaluations of drag and bulk transfer coefficients over water bodies of different sizes. *Boundary-Layer Meteorol.* **3,** 201-213.

Hicks, B. B. (1975). A procedure for the formulation of bulk transfer coefficients over water. *Boundary-Layer Meteorol.* **8,** 515-524.

Huber, D. G. (1960). Irrotational motion of two fluid strata towards a line sink. *J. Eng. Mech. Div. Proc. Am. Soc. Civ. Eng.* **53,** No. 2, 329-349.

Imberger, J. (1977). On the validity of water quality models for lakes and reservoirs. *Proc. Congr. Int. Assoc. Hydraul. Res. 17th* **6,** 293-303.

Imberger, J., Patterson, J., Hebbert, B. and Loh, I. (1978). Dynamics of reservoir of medium size. *J. Hydraulic. Div. Proc. Am. Soc. Civ. Eng.* **104,** No. HY5, 725-743.

Imberger, J. and Spigel, R. H. (1980). Billowing and its influence on mixed layer deepening. *Proc. Symp. Surf. Water Impoundments., Minneapolis.* In press.

Imberger, J., Thompson, R. and Fandry, C. (1976). Selective withdrawal from a finite rectangular tank. *J. Fluid Mech.* **78,** 389-512.

Koh, R. C. Y. (1966). Viscous stratified flow towards a sink. *J. Fluid Mech.* **24,** 555-575.

Lawrence, G. A. and Imberger, J. (1979). Selective Withdrawal through a Point Sink in a Continuously Stratified Fluid with a Pycnocline. Department of Civil Engineering, University of Western Australia, Report No. ED-79-002.

Lumley, J. L. (1965). On the interpretation of time spectra measured in high intensity shear flows. *Phys. of Fluids* **8,** 1056.

Niiler, P. P. and Kraus, E. B. (1977). One-dimensional models of the upper ocean. *In* "Modelling and Prediction of the Upper Layers of the Ocean," (Ed. E. B. Kraus). Pergamon Press, New York.

Ozmidov, R. V. (1965a). Some features of the energy spectrum of oceanic turbulence. *Dokl. Acad. Nauk. SSSR Earth Sci.* **160,** 11-18.

Ozmidov, R. V. (1965B). On the turbulent exchange in a stably stratified ocean. *Atmos. Oceanic Phys.,* Ser. 1, 853-860.

Pond, S., Fissel, D. B. and Paulson, C. A. (1974). A note on bulk aerodnamic coefficients for sensible heat and moisture fluxes. *Boundary-Layer Meteorol.* **6,** 333-339.

Rayner, K. N. (1980). Diurnal Energetics of a Reservoir Surface Layer. Joint Report: Environmental Dynamics, Department of Civil Engineering, University of Western Australia, Report ED-80-005, and Department of Conservation and Environment, Western Australia, Bulletin No. 75.

Sherman, F. S., Imberger, J., and Corcos, G. M. (1978). Turbulence and mixing in stably stratified waters. *An. Rev. Fluid Mech.* **10,** 267-288.

Spigel, R. H. (1978). Wind Mixing in Lakes. Unpublished doctoral thesis, University of California, Berkeley.

Spigel, R. H. and Imberger, J. (1980). The classification of mixed layer dynamics in lakes of small to medium size. *J. Phys. Oceanogr.* (in press).

Swinbank, W. C. (1963). Longwave radiation from clear skies. *Quart. J. Roy. Meteorol. Soc.* **89,** 339-348.

Thorpe, S. A. and Hall, A. J. (1977). Mixing in the upper layer of a lake during heating cycle. *Nature* **265,** 719-722.

Tennessee Valley Authority (1972). Heat and Mass Transfer between a Water Surface and the Atmosphere. Division of Water Resources Research, Tennessee Valley Authority, Report No. 14.

Wu, J. (1973). Wind induced entrainment across a stable density interface. *J. Fluid Mech.* **61,** 275-278.

Zeman, O. and Tennekes, H. (1977). Parameterisation of the turbulent energy budget at the top of the daytime atmospheric boundary layer. *J. Atmos. Sci.* **34,** 111-123.

MODELING OF HEATED WATER DISCHARGES ON THE FRENCH COAST OF THE ENGLISH CHANNEL

François Boulot

Laboratoire National d'Hydraulique,
Electricité de France, Chatou, France

1. INTRODUCTION

During the last decade, France launched an extensive nuclear plant development program. The search for appropriate sites for such plants, which would not have major impacts on the environment, has led to the undertaking of substantial studies. A certain number of sites along the sea-coast, in particular, along the English Channel (see Fig. 1.1), with an open-circuit cooling (i.e. once-through) system were specifically studied.

A substantial problem to be resolved was that of the dilution of the heated effluents discharged into the sea medium. Indeed, all ecological studies were subordinated thereto. This problem was mainly tackled by means of mathematical models. We propose herein to explain the methodology applied. We shall not go into the details of the modeling, particularly the numerical modeling, which is either conventional or which may be found in other papers which we have published elsewhere. We shall rather emphasize the concepts which guided such choice of method, and a certain number of points dealing with the difficulties encountered during the implementation of the modeling.

For illustration purposes, we have selected the Paluel site whenever possible, thereby enabling us to consider the overall study somewhat as a "case example." At present, we do not have measurements *in situ* giving the temperature rise of the sea, which may be compared to the results predicted by the codes, since the first plant, studied according to the methods herein described, is due to start up during 1980. However, we do have other measurements (such as sea currents, laboratory experiments) which allow for interesting comparisons with the calculations made; these we shall refer to in this paper.

The sites studied, among which the Paluel site is quite illustrative, are characterized by strong tides (4 to 8 m), inducing periodic and substantial currents

TRANSPORT MODELS FOR
INLAND AND COASTAL WATERS

ISBN: 0-12-25852-0

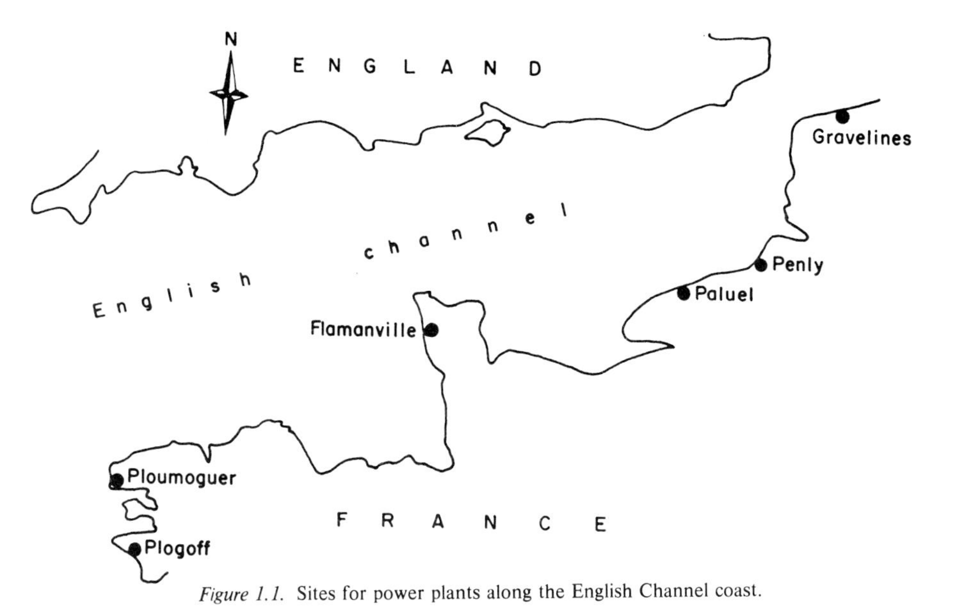

Figure 1.1. Sites for power plants along the English Channel coast.

(of the order of one meter per second). The constant and repeated action of such currents constitutes the predominate factor for the dilution of the heated water. Proper modeling thereof is therefore of primary importance. We shall see in Section 2 how such modeling has been carried out by examining, in particular, the problems of numerical solution and calibration of the model.

The action of the currents is two-fold. On the one hand, their periodic components induce a to-and-fro motion of water masses over a distance which we shall refer to as the "excursion", and, on the other hand, their "continuous" component (residual circulation) induces a generally weak drift, but particularly important by its effects, since it contributes to the renewal of the water in an area near the plant. Such residual circulation and the problems raised in its modeling shall likewise be discussed in Section 2.

Then, the dilution of the outfall shall be dealt with. We shall first examine, in Section 3, the far field where vertical differences of temperature are assumed to be small. A two-dimensional model may then be adopted using modeled current fields from Section 2. Such modeling raises the question of what should be the "diffusion"; we shall discuss it.

We shall also examine the problems raised by use of different boundary conditions of such a model, and those raised by the use of different sized grids in various regions of the calculational domain.

Finally, in the last section, we shall deal with the near field of the outfall structure. The flow and temperature fields are substantially three-dimensional. A model applicable to such a region and comparisons between the results calculated and measured in the laboratory will be presented.

2. MODELING OF TIDAL-INDUCED CURRENT FIELD

2.1 A Brief Account of the Difficulties Encountered during the Numerical Solution of Shallow Water Equations

In a pollution problem, knowledge of currents is essential. In tidal seas, in many cases, uniformity of the flow speed and direction over the water depth may be assumed. The equations governing such currents will then be the shallow water equations; these are sufficiently known and need not be restated in this paper.

When examining the tidal velocity field in the immediate vicinity of the coast, and the field encountered in the open sea, one is struck by the fact that the former may cause separated flows boundary layer on capes and large amplitude vortices (see Fig. 2.1, showing the Brest Roadstead), whereas in the open sea, in the Channel for instance, measurements show the propagation of several waves influenced by Coriolis acceleration giving rise to a current field which remains fairly simple.

It has been a long-established fact in fluid mechanics that in order to reproduce separated flows (creation of vorticity) a momentum diffusion operator need be introduced. Indeed, recirculation proceeds from the interaction of the advection terms with the diffusion terms (both of which need be properly represented if a pollution problem is to be dealt with). Consequently, an equation system is arrived at which is formally expressed as follows: (a) vertical mass flow + divergence of horizontal mass flow = 0; and (b) non-linear relative acceleration terms = pressure gradients + momentum diffusion + friction + Coriolis acceleration effects. A preliminary qualitative analysis shows that such equations represent the "superposition" of three operators: advection of momentum (non-linear); diffusion of momentum (linear); and propagation of momentum (non-linear). Without going into a highly complex analysis from a mathematical viewpoint, it should be pointed out that the existence of diffusion (which is an elliptic operator), changes the nature of the equations which, in its absence, would be purely hyperbolic. In the hyperbolic case, a full theory is available whereby analysis can be effected, and which, in particular, determines the number of boundary conditions required in order that the problem be properly posed. The existence of diffusion perturbs such analysis, and, as of this date, the problem of the number of necessary boundary conditions, although the subject of recent research work, does not appear to be fully clarified. This first, essentially mathematical, difficulty having been stated, let us now examine the difficulties encountered when one attempts an approximation of the solution, that is, when only a finite number of parameters (d° of freedom) is taken to describe the solution.

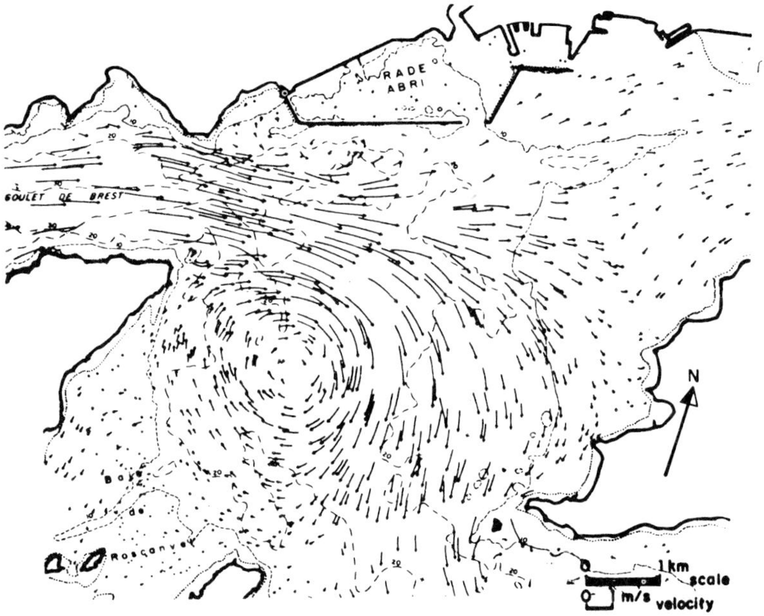

Figure 2.1a. Tidal currents in the Bay of Brest (Brittany) during flood-tide obtained from a physical model.

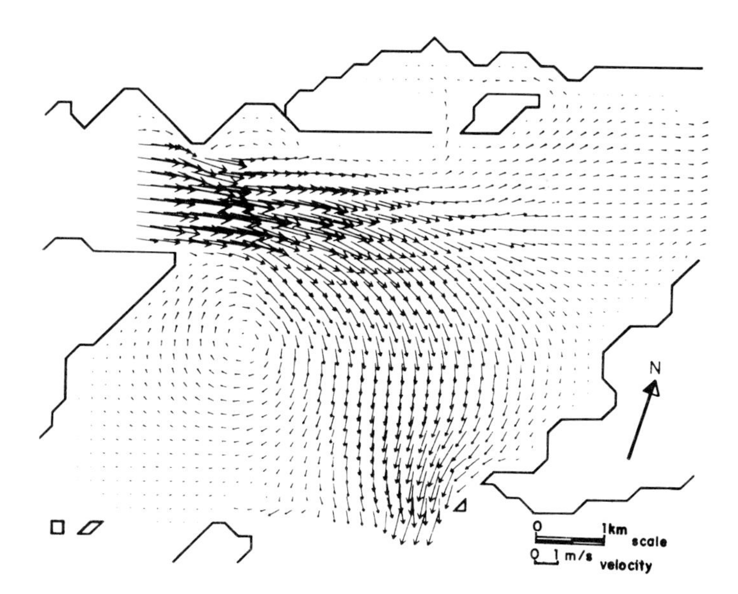

Figure 2.1b. Tidal currents in the Bay of Brest (Brittany) during flood-tide obtained from a mathematical model.

Since the problem to be solved is one in a determined area, boundary conditions need be introduced which will generate tides in the calculation domain. This point will be discussed later.

The first difficulty encountered when one deals numerically with a fluid mechanics problem resides in discretization of the non-linear *advection* operator (Benque, Ibler and Labadie, 1980). Such an operator is found in the shallow water equations, and is particularly important in the coastal area. Let us now examine the principal problems appearing in the processing thereof.

Whatever the methods of discretization (finite differences or finite elements), numerical analysts are well aware that such operators are physically responsible for the creation of shocks which are the causes of substantial difficulties. This is due to its one-directional hyperbolic nature. The equation written reflects the predominant influence of the upstream over the downstream to be introduced in discretization. To reflect such fact, off-centering upstream need be effected. This method is coupled with a loss of accuracy which, unless care is taken, may cause a numerical diffusion much more substantial than the physical diffusion which is also modeled. The simplest off-centering upstream differencing introduced a first order error, which is seen in the Taylor expansion of the advective term by a second order derivative influencing the velocity field. This is quite analogous to diffusion with a coefficient equal to the product of half the velocity amplitude and the space grid size: (i.e. $u - 1$ m/s, $\Delta x - 100$m, where the numerical diffusion coefficient is 50 m^2/s). This is generally unacceptable and a more sophisticated differencing scheme which introduces only a second-order error must be resorted to. In such a case, there generally occurs a substantial phase-lag caused by a third operator which remains in the Taylor expansion. This, in the presence of steep gradients, causes the well known "overshoot" phenomenon (which causes oscillations in the results from one mesh to the other). Generally, even-order errors must not be introduced and odd-order errors appear preferable. To do this one must search for third-order space discretization. The search for such an algorithm is not trivial, and in the case of the examples hereinafter set forth, was carried through on the basis of a construction method specific to first-order quasi-linear hyperbolic operators.

To clarify matters, let us take, for instance, a one-dimensional case, where the unknown function f is governed by:

$$\frac{\delta f}{\delta t} + u\,\frac{\delta f}{\delta x} = 0$$

where $u(x,t)$ is given as piecewise linear function. f is known at the time t_o at the different grid points $i\Delta x$. One seeks to know it at the same points at the time $t_o + \Delta t$. To do so, one finds the path coming from the point $i\Delta x, t_o + \Delta t$ along $dx/dt = u(x,t_o)$ (the characteristic curve) which can be analytically integrated if $u(x,t_o)$ is piecewise linear.

The transport equation to be solved gives f as being constant along a characteristic curve. The value of the function f at the foot of a trajectory (characteristic curve) is interpolated by means of a third degree polynomial. By using such a method, a rather high order of spatial precision (third order) can be achieved without great complications, and without problems of stability.

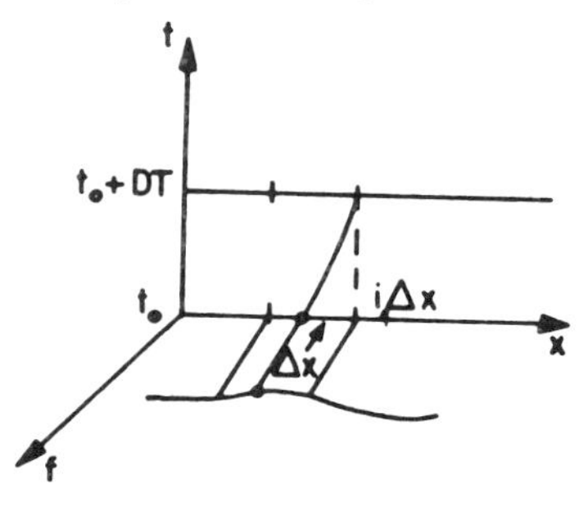

Figure 2.2.

However, it should be noted that, in the case of strong non-linearity, a loss of accuracy is introduced when the Courant Advection Number is greater than one. On the other hand, a comparison with the analytical solutions of the equation $\delta u/\delta t + u(\delta u/\delta x) = 0$ shows that the result obtained is very close to the analytical result, if the condition regarding the Courant Number is complied with.

This manner of proceeding, generalized to two-dimensions, yields highly accurate results in test cases. No instability is shown, and the solution is sensitive to the value of the physical diffusion coefficient introduced (thus proving that the numerical diffusion is an order of magnitude less than physical diffusion).

In order to estimate the effect of discretization the following non-linear test was made: let us consider a one-directional velocity field with the profile shown in Fig. 2.3 at the initial time. The exact solution has the same shape, with the slope between P_0 and P_1 becoming steeper as the time increases since the point P_0 moves with the velocity u_o and the point P_1 is at rest. The domain is a square with 15 × 15 nodes. The distance between two successive nodes is $\sqrt{2/2}$ meters, the time step is one second and the maximum velocity is $\sqrt{2\text{m/s}}$. Figure 2.3b shows the grid with the initial velocity profile. The method which is used in this test is a finite element method and the interpolation is quadratic. Figure 2.3c shows the exact and computed velocity profiles along the main diagonal. Although it is a rather severe test, the method shows good behavior. We note, in particular, the small damping of the numerical solution, which is observed.

Discretization of the *diffusion* operator is obtained by using discretization between second order derivatives and does not raise specific problems. In regard to tidal wave *propagation,* distinction needs to be made between the cases of a large domain problem (e.g. Continental Shelf, English Channel, North Sea), and a small domain problem (e.g. port, roadstead, bay). In the case of a large domain problem, let us first specify a few orders of magnitude. The grid size is approximately 10 km, velocity is approximately one meter per second and depth approximately 40 m, thus leading to a free surface wave velocity propagation of 20 m/s.

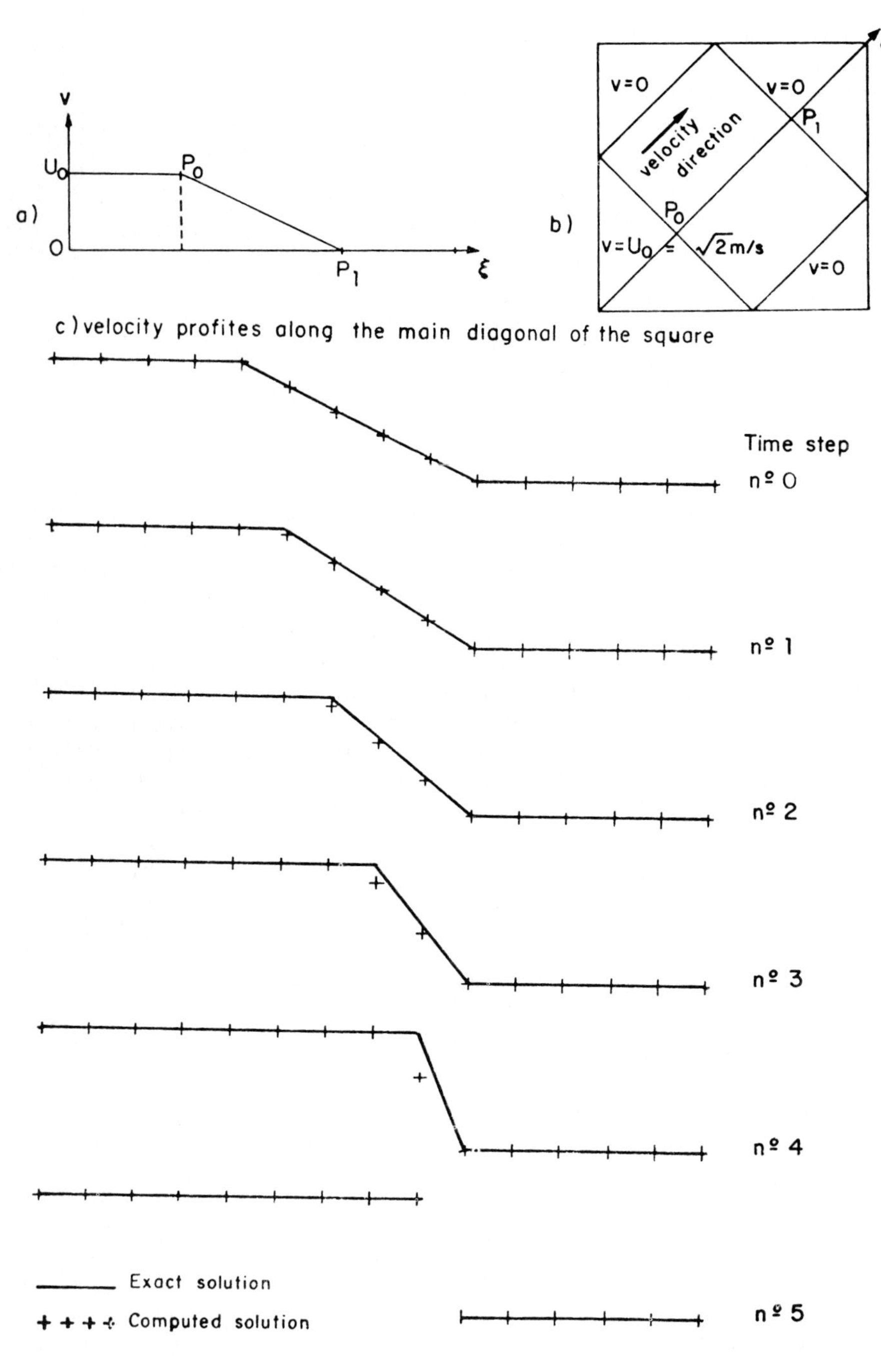

Figure 2.3. Non-linear advection test--quadratic *fe* approximation. (c) shows velocity profiles along the main diagonal of the square.

By choosing the time step so that during a step the waves should cross only one mesh, calculation of a tide requires 100 time steps, a reasonable value that permits any space centered discretization method to obtain good results. It may even be advantageous to use an explicit method requiring few calculations at each time step, and which becomes unstable if the "CFL" criterion is exceeded during the time step. This is what has been done for our English Channel model.

In the case of a small domain problem, such as a harbor, the situation is different: the size of the mesh may reach 50 m, and the depth is of the order of a few dozen meters. The same time step limitation as before would then lead us not to exceed a step of 5 seconds, which is hardly realistic if a twelve-hour phenomenon is to be described. In the case of a finite difference discretization, this problem can be solved by a central-differenced implicit method given by Preissmann (Cunge and Preissmann, 1961) and generalized to two space dimensions. The result to be reached differs from the previous case, in that no attempt will be made to represent the forward and backward motions of the tidal waves in the port, which are effected in a few tenths of a second, but an attempt will only be made to obtain detailed information on the velocity field which may be highly complex, the free surface remaining virtually horizontal. In regard to the dimensions of the area studied, this is a slow transient flow in which the natural period of the harbor installation is much less than the tidal period.

A detailed analysis of the numerical scheme has shown that the filtering effected by the latter when the time step was large (100 s), was absolutely compatible with the result sought; the scheme dampens the rapid forward and backward waves which do not greatly affect the flow and follows the variation of the velocity field with the tide.

When the solution algorithm is accurate and reliable, the accuracy of the result depends only on the algorithm with which the integration domain is described. In the case of shallow water equations, two difficulties appear: one often deals with a complex shape domain (very indented coast, existence of islands); and the geometry may be variable in time, with shoals awash at low tide.

The description of a complex calculation domain may be effected by three alternative methods: rectangular grids, curvilinear grids, and finite elements. If the finite difference method in curvilinear grids contributes a supplement as compared to conventional rectangular grids, it would appear that in the case of coastal geometrical shape, only a finite method would actually allow the currently existing gaps to be filled in. We are attempting to develop such a method. The concepts previously developed have been reviewed and adapted to implementation of the finite element method, which is more cumbersome and more costly than the finite difference method. The results to be presented in the following pages were obtained with rectangular grids, the only ones to be perfectly operational at the time when the calculations were performed.

2.2 Use of Nested Mathematical Models

2.2.1 Method. Depending on the case involved, the numerical reproduction of tide-induced flows requires accurate representation of either the levels (storm surges, tidal power plants operation) or of the currents (pollutant dispersion, removal of sediments, or navigation problems). The terms to be specified as boundary conditions on the limit of domain represented by the mode may affect the level of the free surface and the magnitude of the flow components per width unit. With regard to the nature of the equations solved, two conditions appear to be necessary, when the current flows into the domain, and when it flows out. Generally, the two flow components are specified when currents are of greater interest and in the opposite case the level and flow tangent component are specified.

In the former case, with which we are here concerned, the phenomenon studied is usually of a local nature, that is, the field of interest is small as compared to the tidal wave length. It may then appear natural to limit the domain of the numerical study on a small-sized model corresponding to the area concerned by the phenomenon studied. However, the calibration of such a model requires the knowledge of a large number of data obtained *in situ* at the cost of a large and expensive series of measurements.

It would appear preferable to undertake the study on a large-sized domain for which numerous data *in situ* are generally available (see the English Channel model). This is also desirable because the sensitivity of the results to the quality of the conditions imposed at the boundaries is then reduced. The results of the calculations in the area concerned depend less on the details of the boundary conditions, and more on the topography of the sea bottom and of the coast line, since most of the domain is remote from the boundaries.

Spacing of the boundaries appears to be inconsistent with the high accuracy required by the local study. However, two solutions allow for reconciliation of such requirements. In the first, the results of an initial calculation performed on a large area model provide the boundary conditions for a more detailed model. Different models of successively reduced size may thus be used in series, until sufficiently accurate results are obtained. To initiate such a process the overall model must first of all be calibrated by means of successive adjustments of the boundary conditions (which, at the outset, are only inaccurately known), and of the roughness of the depths, a parameter which becomes increasingly important as the size of the domain is increased. Such calibration is generally the most delicate. By way of example, Fig. 2.4 provides for the overall model of the English Channel, a comparison between the levels and currents measured and the levels and currents calculated at a few points.

The second solution is use of a refined mesh model. In such a model, the flow characteristics may be calculated at one pass with greater accuracy, by using successive mesh refinements, as one gets closer to the area involved.

The first method has an advantage in that the latter model high precision can be more reliably obtained through adjustments, if any, of intermediate boundary

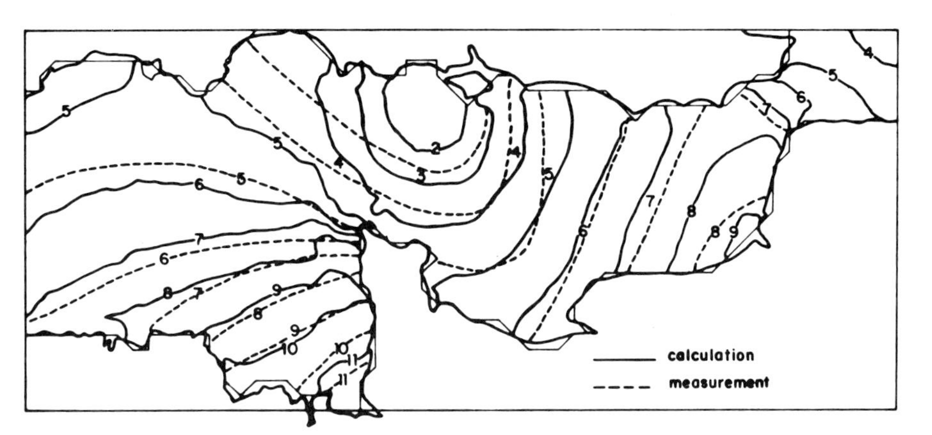

Figure 2.4a. Results from a mathematical model of tidal currents in the English Channel for a mean spring tide. Lines of equal tidal amplitude (in meters) are shown.

conditions and the computing time and the computer storage capacity requirements are more reduced since the models are decoupled. Moreover, an overall model may be used for several studies. However, on the other hand, by using the second method, more coherent results may be obtained in the various areas, an important fact when subsequent utilization of the results involves several of the areas concerned. Also in the overall domain, response to the various configurations represented in the detailed model (operation of a tidal power plant) may likewise be taken into account.

The first method, which is more adapted to local studies, has been dealt with in this paper from the viewpoint of the problems it may raise, and the results which may be obtained from it.

2.2.2 Problems posed by the use of nested models. Levels: Poor reproduction of water level is frequent with small-sized models, when only the flow conditions are imposed. Variation of the level of the free surface depends on the difference between in-flows and out-flows. When the boundaries are quite close together, such flows are of the same order. A small relative error on each may therefore involve a substantial error on the variation of the level. Such errors may be due to interpolations from which the boundary conditions are inferred, or to the trucation errors (if any) of the Fourier series expansion of the results of the previous model from which such boundary conditions are inferred, when using such models, in cases where the level is nearly uniform, by adjusting the balance of the flows introduced thereby requiring only a very slight change in the individual balance.

Currents: In the case of a complex geometry of the bottom, the natural flow will not be uniform. The overall wide mesh model strongly schematizes reality, and provides comparatively smooth results; the boundary conditions inferred therefrom for the detailed model then prove ill-adapted to such a geometry and their interpolation must be modified, and if need be the magnitude or allocation of

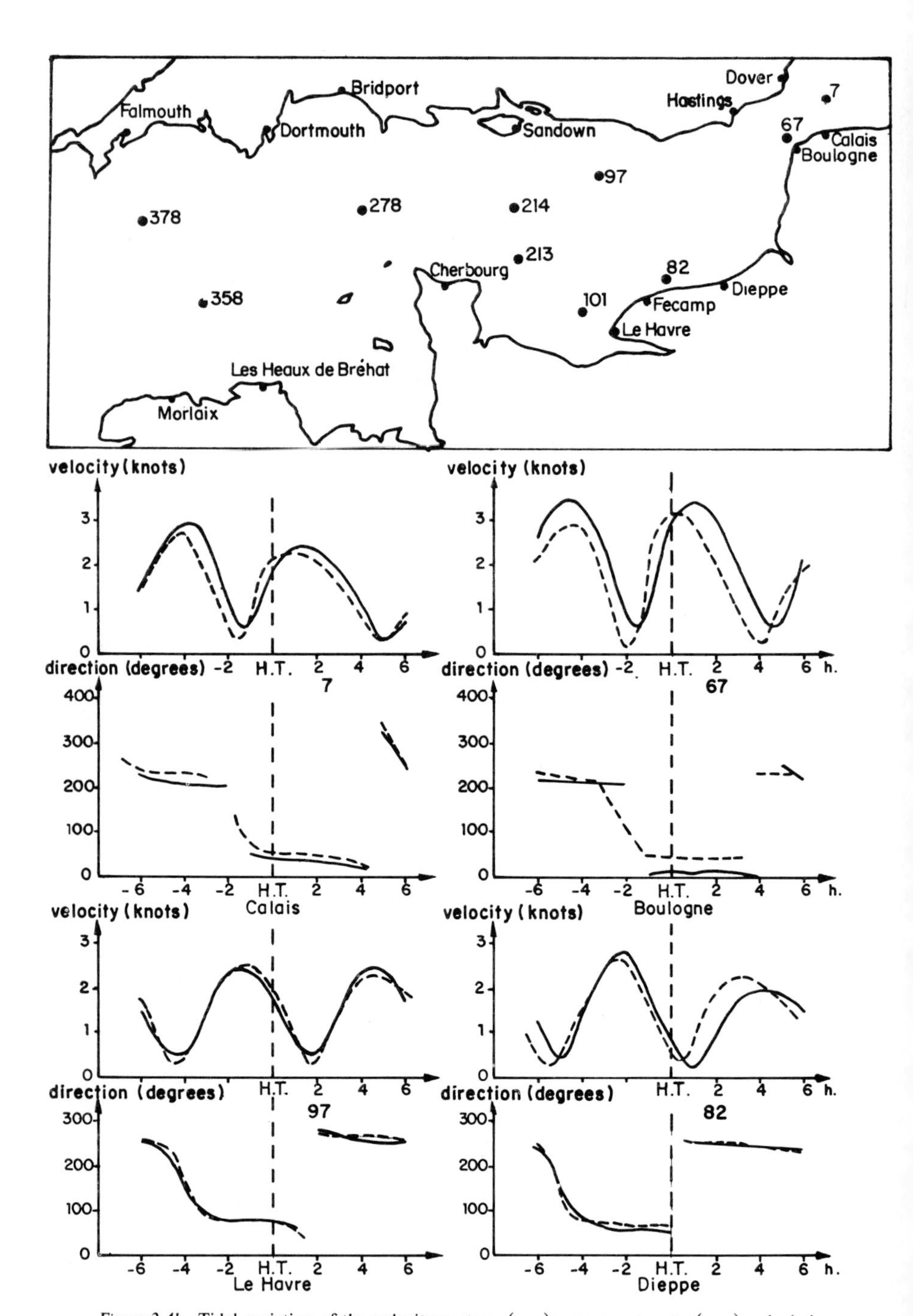

Figure 2.4b. Tidal variation of the velocity vector. (——): measurement. (-----): calculation.

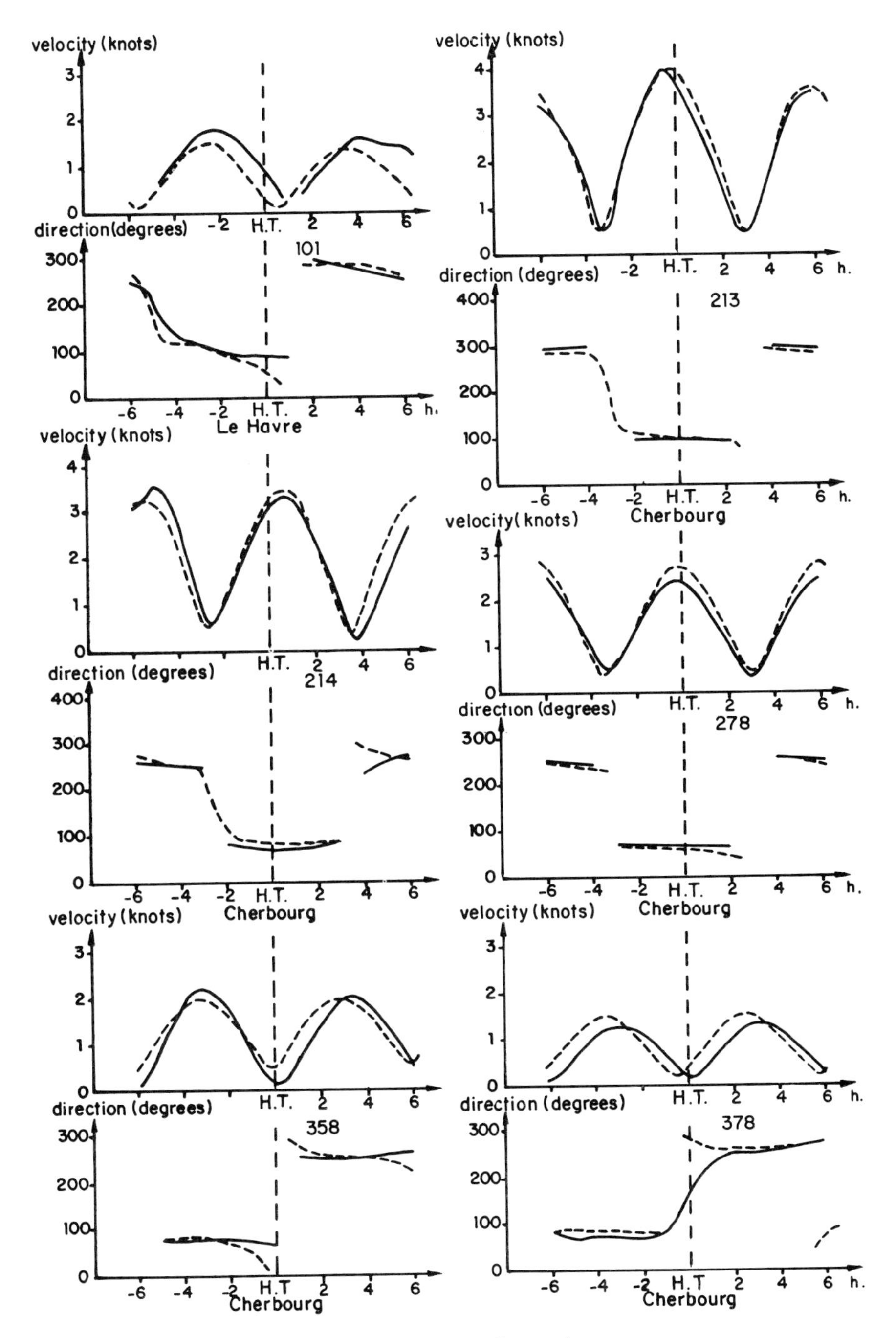

Figure 2.4b.--Continued.

such a flow must be locally verified. Such modifications may likewise prove necessary, near the coast in the "boundary layer" of the overall model, if the flow rates have been underestimated.

It should be noted that all such modifications do not in any case involve a full disruption of the boundary conditions.

2.2.3 A few examples. In practice the series determination of the flow in the nested models may operate without any modification of the boundary conditions or, on the contrary, may require substantial adjustments therein. The following two examples will illustrate such possibilities.

The Penly site: For the Penly site, located in the "Baie de Somme", three nested models were used successively: an overall model for the English Channel (10 km grid); a regional model for the "Baie de Somme" (2 km grid); a local model for Penly (500 m grid). The size of the latter two models is shown in Fig. 2.5.

Due to the geometrical regularity of the sea bottom and of coastal line, the tidal flow off Penly was determined step by step, without any adjustment in the intermediate boundary conditions, using a well-calibrated overall model. Figure 2.6 gives the two-point comparison between currents measured and currents calculated by the regional and local models.

New outport of Calais: Current and sediment studies relating to the extension of the port of Calais involved, in particular, implementation of four nested numerical models. Use of the regional Pas-de-Calais model (Fig. 2.8) after the overall model of the Southern part of the North Sea (Fig. 2.7), required considerable adjustments of the boundary conditions due to the following difficulties: the average results were obtained in that area from the overall model (which used an oversized mesh in relation to the width of the strait); the complexity of the configuration of the coast and of the sea-bottom (shoals), as shown by Fig. 2.8; the complexity of the flow arising at the meeting point of two tidal waves.

The boundary conditions which, in this instance, bore only on the flow, had to be adjusted, hour by hour, and boundary after boundary, using a trial and error method. Such modifications were effected only by means of multiplication coefficients applied to the imposed flow rates (using an identical coefficient for the two flow components), except along the Western boundary where the phase of the flow was shifted as a function of the ordinate. The values of such coefficients fluctuated here between 0.5 and 2. Figure 2.9 illustrates the variation of such coefficients along the various sea boundaries four hours after high tide at Brest.

Figure 2.10 shows the substantial improvements obtained by such adjustments in the boundary conditions given by the overall model.

2.2.4 Conclusion. The local and detailed configuration of tide-induced flows may be determined by means of a series of nested models operated successively, ranging from a model representing a large maritime domain to a model of a size adapted to the local problems studied. The results of each model are then used to furnish the boundary conditions of the following model, with more or less

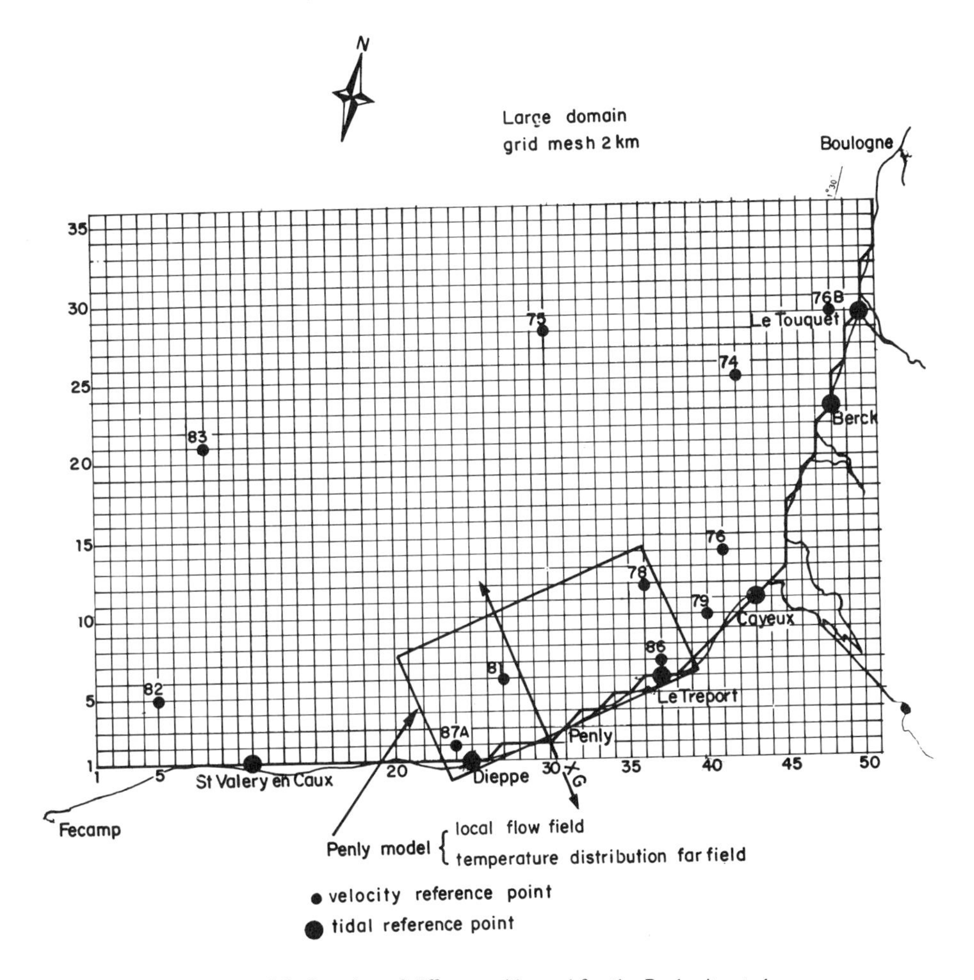

Figure 2.5. Location of different grids used for the Penly site study.

substantial adjustments, if any, according to the complexity of the case concerned, which, however, do not in any case involve complete disruption of such conditions.

By using such a process, one can spare oneself many measurements *in situ* while achieving a generally satisfactory final result. Limits on calculation time and costs involved with such a method, to a great extent, depend on having an easy transition from one model to the following model.

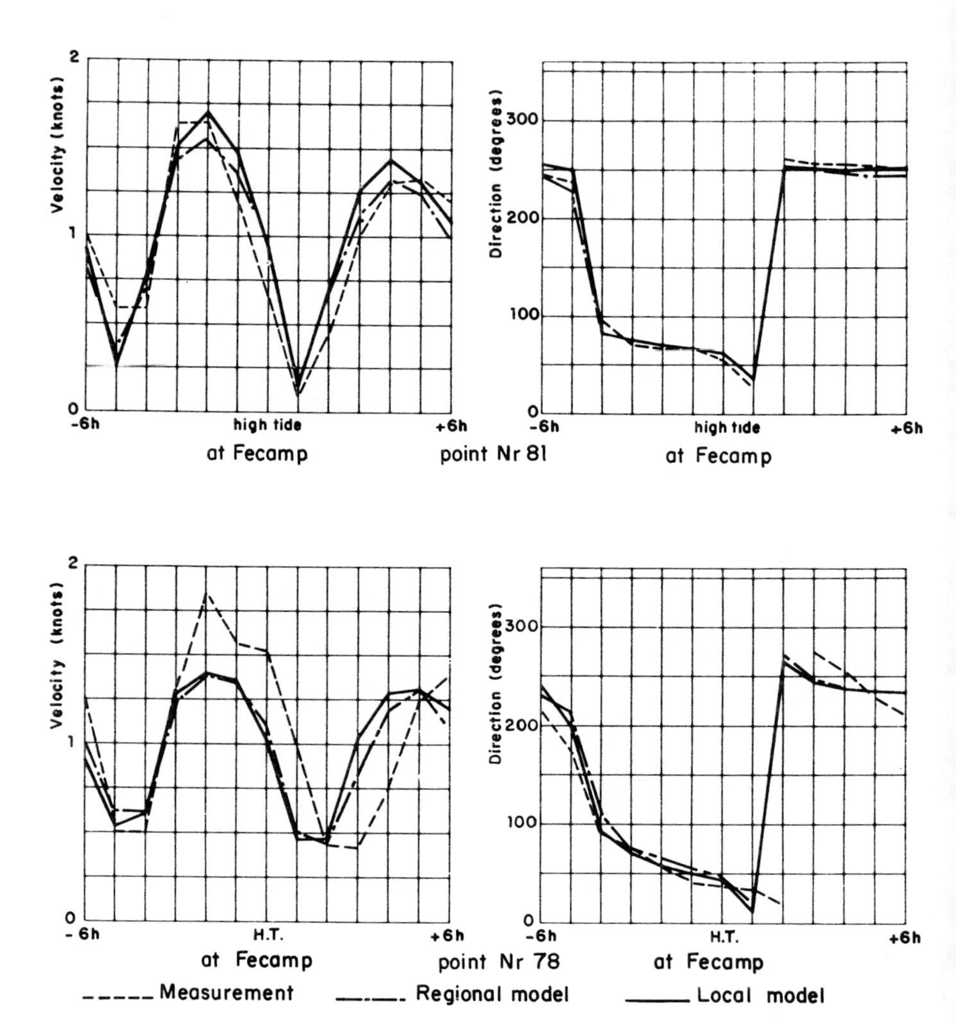

Figure 2.6. Comparison between measured and calculated velocities using the Bay of Somme regional model and the Penly local model.

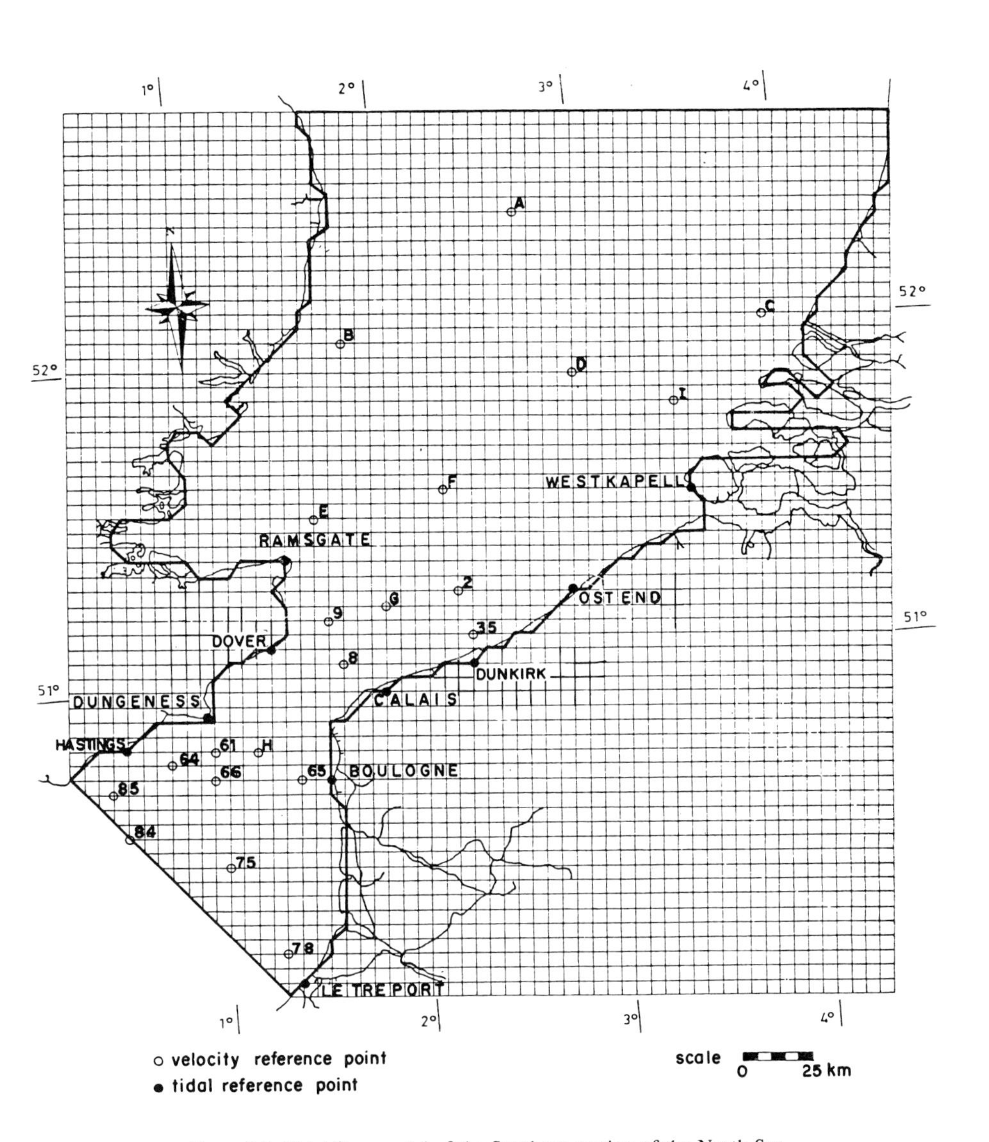

Figure 2.7. Tidal flow model of the Southern section of the North Sea.

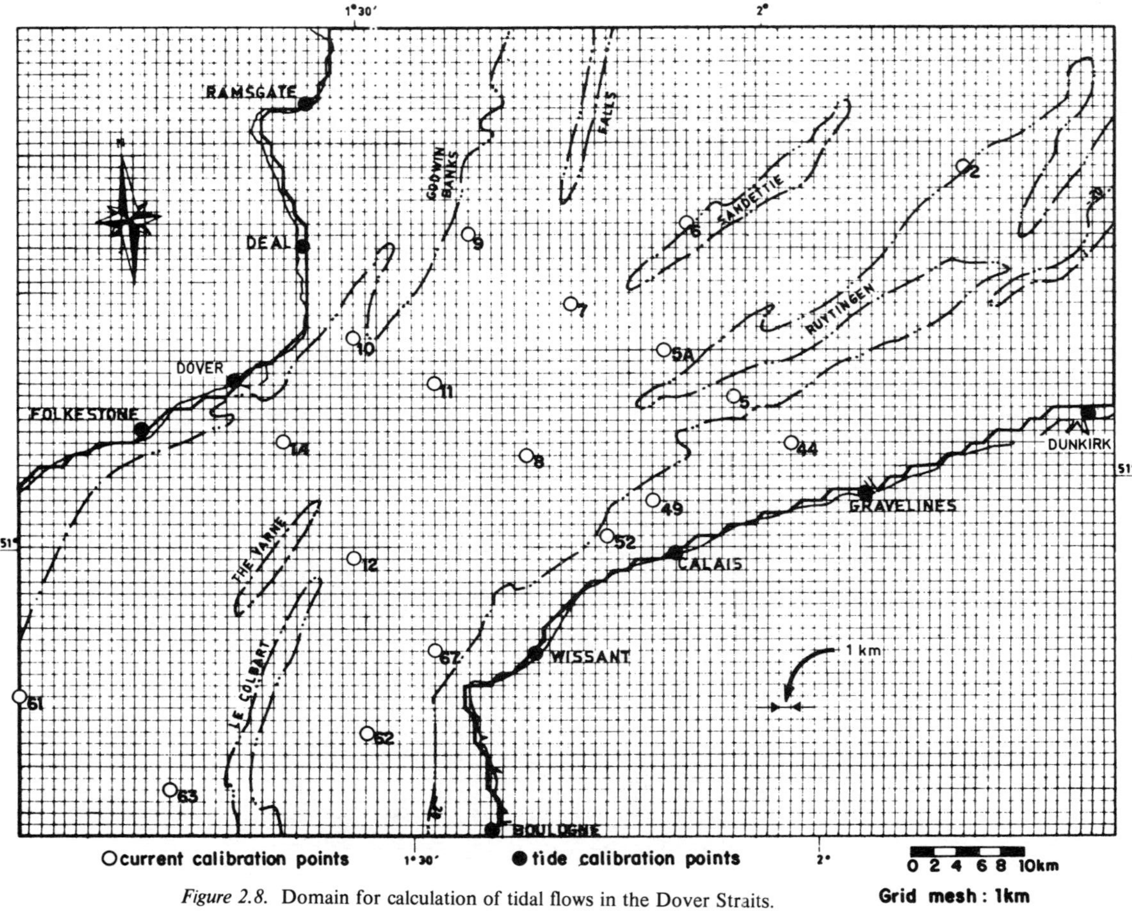

Figure 2.8. Domain for calculation of tidal flows in the Dover Straits.

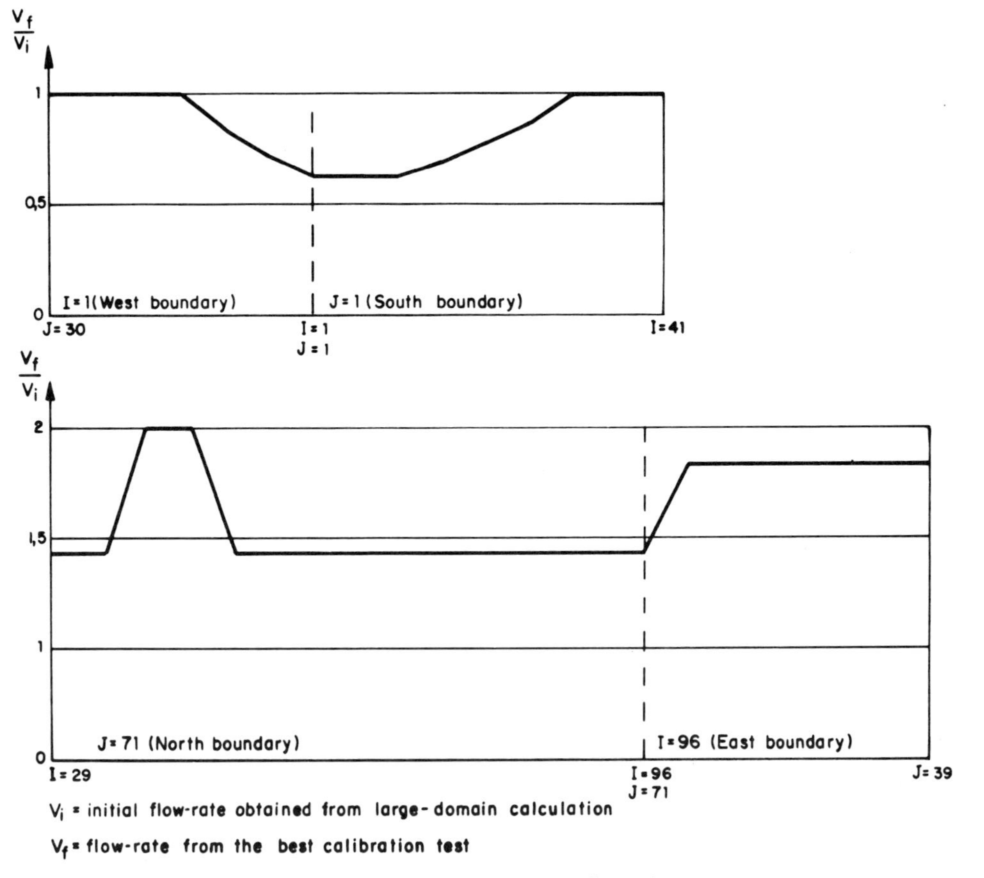

Figure 2.9. Results from the Dover Straits model, with effects of boundary conditions.

2.3 Residual Circulation in Tidal Seas

The evolution of long-lived pollutants (for example, some chemicals and radioactive substances), mainly depends on the residual circulation (after filtering of tidal currents) which renews the water in front of the outfall structure and will carry pollution away from it.

Such residual circulation even for a sea as well-studied as the Channel is usually poorly known. This is because the residual currents are very weak as compared to the tidal currents (a few centimeters per second as against a meter per second) and are thus difficult to measure. Residual circulation is highly complex, both spatially and through its time evolution. We shall not attempt here to deal in detail with this subject, but we shall merely show the difficulty thereof, the

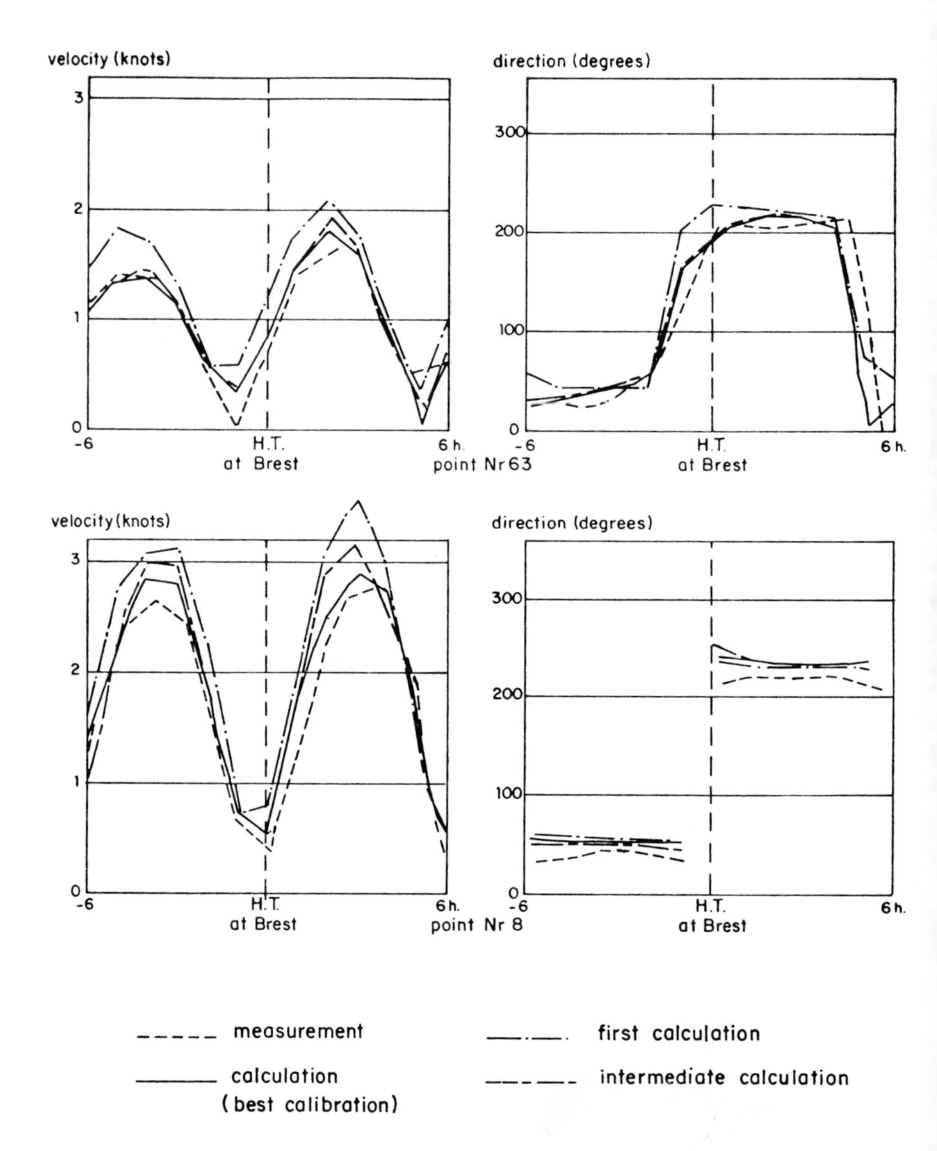

Figure 2.10. Comparison between measured and calculated velocities in three calibration calculations.

modeling attempts, and the manner in which this was taken into account, and the studies effected for power plant siting.

The origin of such circulation is mainly two-fold: the tide, on the one hand, and weather hazards on the other. The former comes from the fact that the sea behaves mechanically as a non-linear system, harmonically excited. The second accrues from the inhomogeneities of atmospheric pressure and ambient winds on the sea surface (storm surges).

Residual current field modeling is only at its beginning and what has hitherto been done has only succeeded in giving a very partial representation of such a highly complex reality. It was thought first that residual circulation could be calculated by filtering the velocity field obtained in a tidal current calculation (by a Fourier series expansion, for example). Unfortunately, it appears that the results thus obtained are rather inaccurate since the velocity sought is in a number of cases of the order of magnitude of the accuracy of the tidal current calculation. Another method was developed on the basis of the concepts of Nihoul and Ronday which consists of writing the equations governing the residual circulation, and solving them numerically (Lomer, 1978). In Fig. 2.11, we show the result of such a calculation, giving the residual current field in the Channel induced by the average spring tide (Coefficient. 95). For the same tide, the following figure gives an idea of the float trajectories induced by residual circulation only, on the one hand, and by tidal currents, on the other (Fig. 2.12).

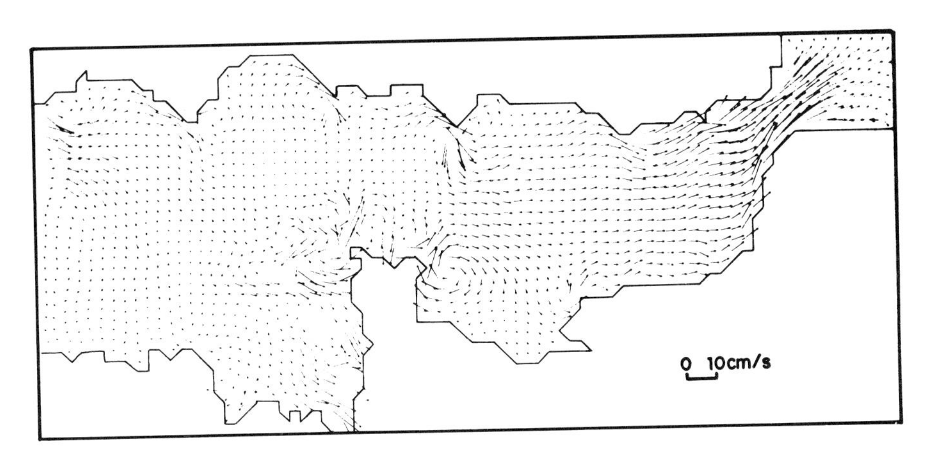

Figure 2.11. Calculated results for residual currents in the English Channel.

It appears that a purely Lagrangian component (that obtained from the harmonic component of the tide current at each point) is of the same order of magnitude as the Eulerian component. As for instantaneous currents, a residual circulation (obtained from a large-sized model, such as that of the Channel) may give

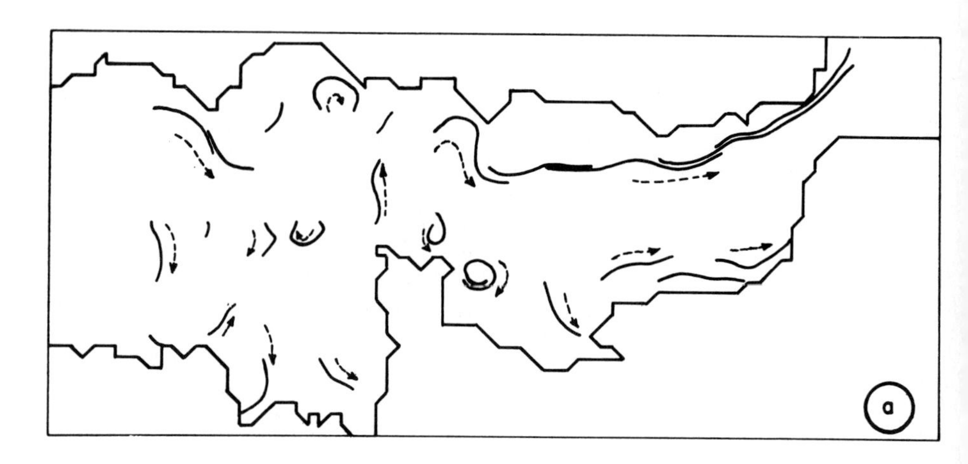

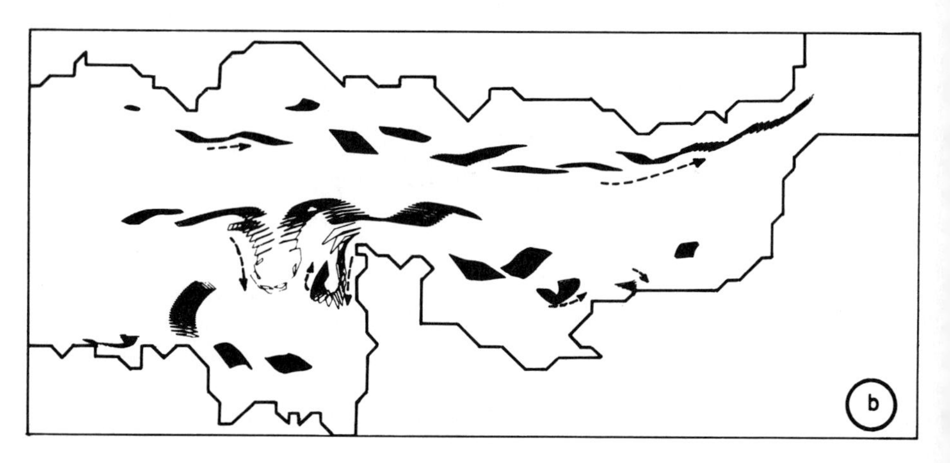

Figure 2.12. Float trajectories during 25 tidal cycles induced by (a) Eulerian residual currents only; (b) tidal currents only.

boundary conditions for a smaller sized model. Figure 2.13 gives the results of such a model as applied to the "Baie of the Seine".

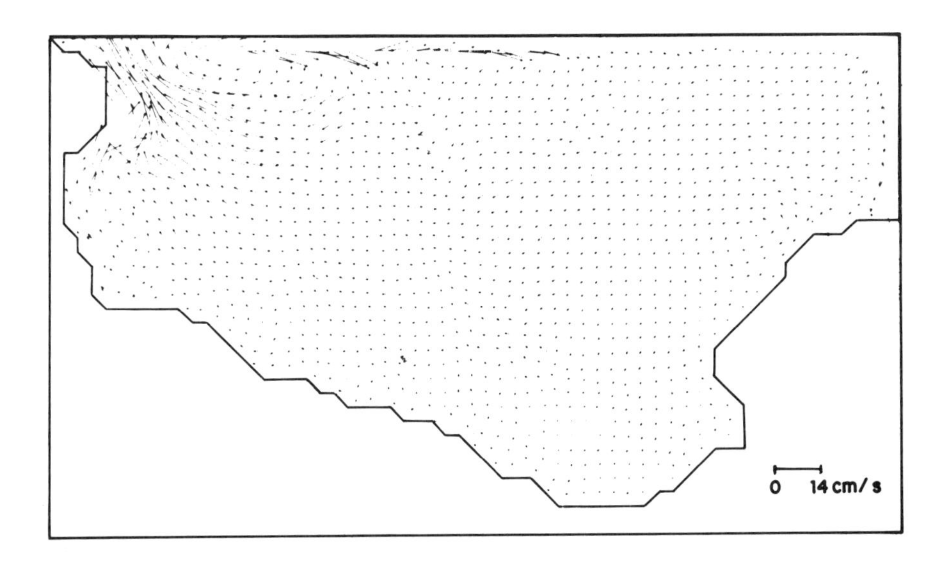

Figure 2.13. Tidal residual circulation in the Bay of the Seine.

Storms can generate a velocity field of the same order of magnitude as the tide-induced residual field. The model based on the shallow-water equations affords a certain idea of this. Figure 2.14 shows the results concerning the south of the North Sea influenced by a westerly wind. An important problem, however, is the boundary conditions to be imposed; in the case with which we are dealing here, the excess level is nil, and this obviously affects the velocity field obtained.

In the calculations performed for the dilution of thermal effluents discharged by power plants, residual circulation was taken into account; its value was determined on the basis of the analysis of the *in situ* current measurements achieved over very long periods of time. A parametric study of its effect was conducted in certain cases.

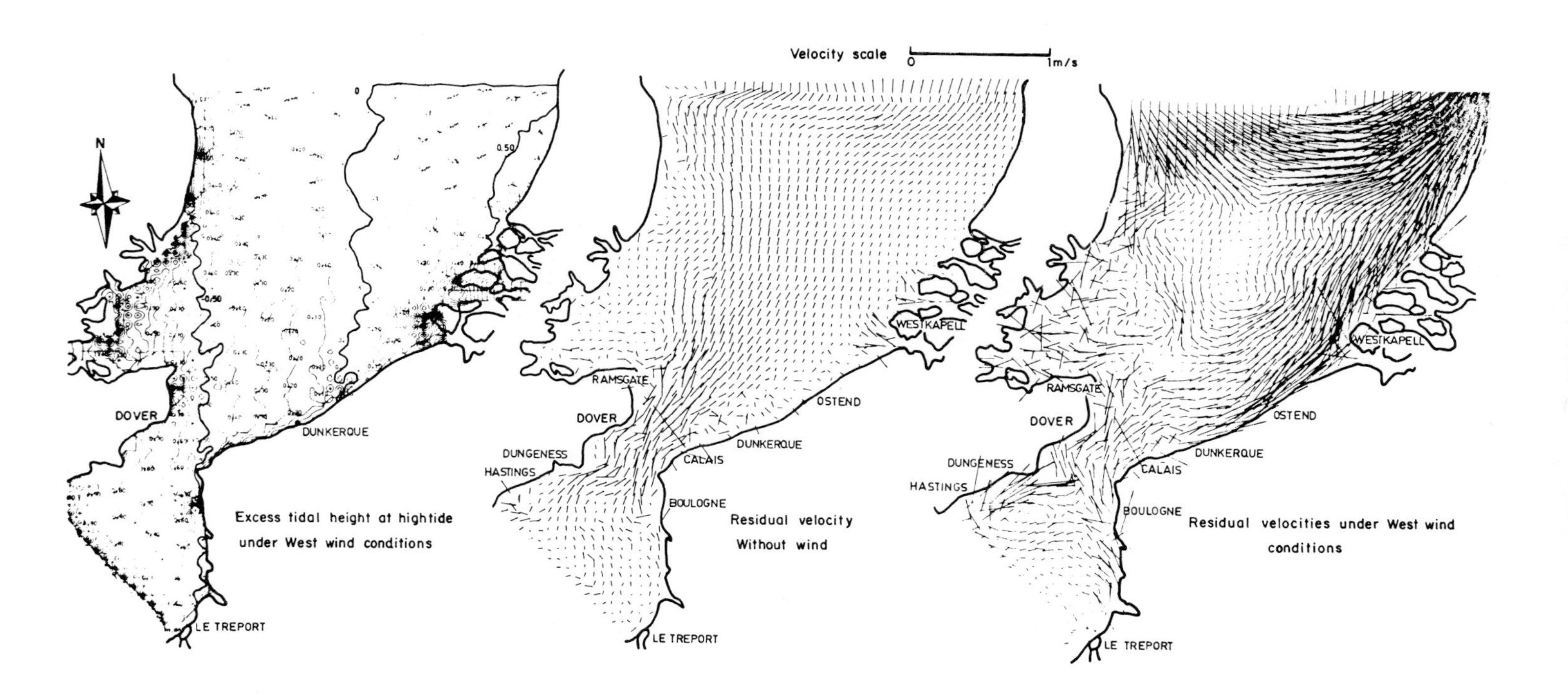

Figure 2.14. Residual circulation induced by a west wind (25 m/s) in the southern portion of the North Sea.

3. FAR-FIELD DILUTION

3.1 Dispersion Tensor

Within a small distance, of the order of 100 meters, from outfall structure, all the water column is affected by the effluent due to the strong tide-induced velocity, on the one hand, and to the small depths (about 10 meters) on the other hand. Thus, the averages over a depth are of interest. The depth-averaged thermal balance includes terms which express the interaction of local velocity and temperature variances. It can be shown (Daubert, 1974, Benque and Warluzel, 1979) that these terms can be modeled by means of a tensor, the so-called dispersion tensor, which is formally analogous to the diffusion tensor. Its physical nature and order of magnitude, however, are completely different.

Based on calculations using a vertical allocation of the turbulent diffusion coefficient, or of vertical velocity profile measurements, the elements of such a tensor and its variation during a tidal-cycle may be calculated. The findings show that dispersion is very great at the inversion time, a fact which is equally accounted for by the great vertical velocity difference at that time, and the low value of turbulent diffusion. To clarify ideas, let us examine the mean values of the elements of such a tensor over a tide cycle. With a 20 meter sea-bottom and tidal currents of the magnitude of one meter per second, a longitudinal coefficient (in the direction of the current) of the order of 100 m^2/s, a few m^2/s cross-wise, and 10 m^2/s crossed are found.

The most important coefficient for calculation of the dilution in the far field is the cross-wise coefficient. In our calculations of the temperature field, this coefficient was taken as ranging between 0.5 and 5 m^2/s. Parametric studies have demonstrated that over such a range the heated surface level was approximately changed by a factor of two.

On the Paluel site, where the coast is relatively straight and the flow alternates in direction, an estimate of the average dispersion tensor leads to the following values

$$D_{xx} = 60 \text{ m}^2/\text{s}, \quad D_{xy} = -7 \text{ m}^2/\text{s}, \quad D_{yy} = 2 \text{ m}^2/\text{s}$$

where x is the direction of the coast and y the perpendicular direction. The crossed coefficient effect is quite negligible. Diagonalization of the tensor does not significantly modify the diagonal terms and induces a very small angle variation. The important effect in such a tensor is indeed the large distortion between the direction of the average flow (60 m^2/s) and the normal direction (2 m^2/s).

Figure 3.1 shows the comparison of the results obtained at Paluel with the same current field, but with two dispersion tensors (the numerical method used will be described later), one having $-D_{xx} = 5$ m^2/s, $D_{xy} = 0$, and $D_{yy} = 5$ m^2/s, and the other the estimated tensor referred to above. In the second case, the results show a field "adhering" much more to the coast, and far closer to the experimental data which would be obtained typically, especially in the case of a river. More than the influence of the new values introduced, it would appear that it is actually the tensor distortion which affects the shape of the field, since the x component of the tensor spreads along the coast the heat which the y component does not disperse out to sea. The heated water then remains in a limited coastal area and does not reach further out to sea, where, in the case of Paluel, there exists a strong drift toward the east.

Such a difference in shape is coupled with a substantial reduction of the heated surface areas. The average surface of the heated areas, exceeding 1°C temperature increase, drops from 20 km^2 in the first case to 11 km^2 in the second case (see Fig. 3.1b). This is probably related to the coefficient parallel to the coast (60 sq.m/s), which is much less limiting than the 5 m^2/sec of the first calculation when the other coefficient is of the same order of magnitude.

The distortion effect of the dispersion tensor is therefore quitc substantial on the shape of the field since it may or may not lead the heated water toward areas which differ in their current-field characteristics. Moreover, the average dispersion effect (which may be characterized by the trace of the tensor) appears to influence directly the extent of the heated surface.

3.2 Numerical Solution of the Temperature Equation

The depth averaged temperature field obeys the following conservation equation

$$\frac{\partial hT}{\partial t} + \frac{\partial}{\partial x}(uTh) + \frac{\partial}{\partial y}(vTh) = \text{div}\,(\bar{\bar{D}} \nabla T) + \frac{A}{\rho Cp}(T - T_E) \tag{1}$$

where h is the instantaneous water height, u,v are the components of the depth-averaged velocity due to the tide (h,u,v being elsewhere determined as a decoupled problem), $\bar{\bar{D}}$ designates the second order dispersion tensor, and A is the atmosphere heat exchange coefficient. If h,u,v are obtained by means of a tidal current numerical model, they satisfy the continuity equation, and Eq. (1) is then simplified as follows

$$\frac{\partial T}{\partial t} + u\frac{\partial T}{\partial x} + v\frac{\partial T}{\partial y} = \frac{1}{h}\,\text{div}\,(\bar{\bar{D}} \nabla T) + \frac{A}{\rho Cph}(T - T_E) \tag{2}$$

We have solved this equation by a splitting finite difference method. Advection by u and v are computed by an absolutely stable explicit characteristic method, and dispersion in x and y by double sweeping. For boundary conditions it is assumed that the presence of the coast blocks lateral thermal exchanges ($\partial T/\partial n = 0$) and that at sea, far from the outfall point, the temperature is the equilibrium temperature ($T = T_E$).

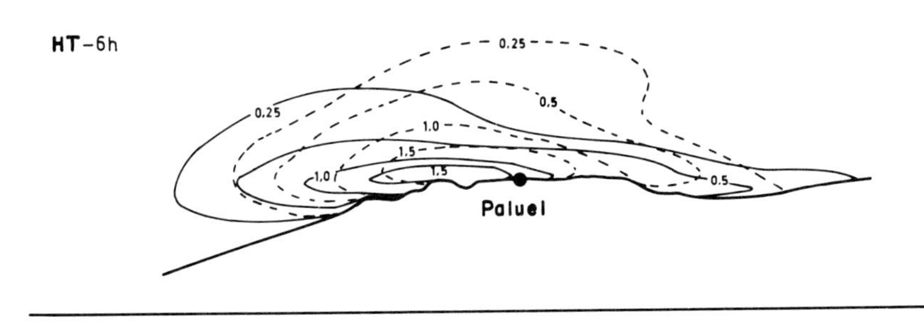

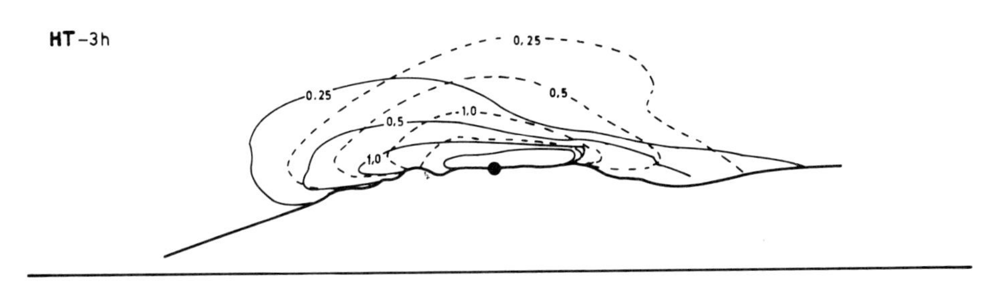

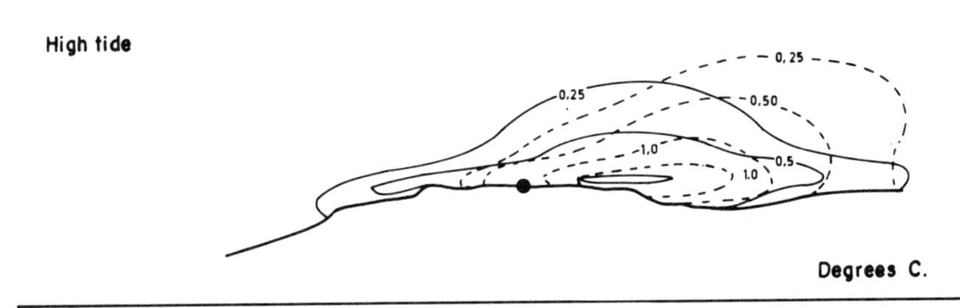

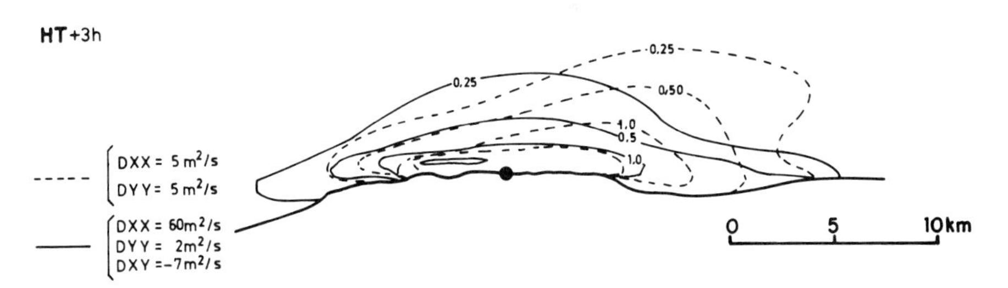

Figure 3.1a. Influence of the dispersion tensor value on the temperature field of the Paluel power plant. Excess temperature at different moments durng a tidal cycle.

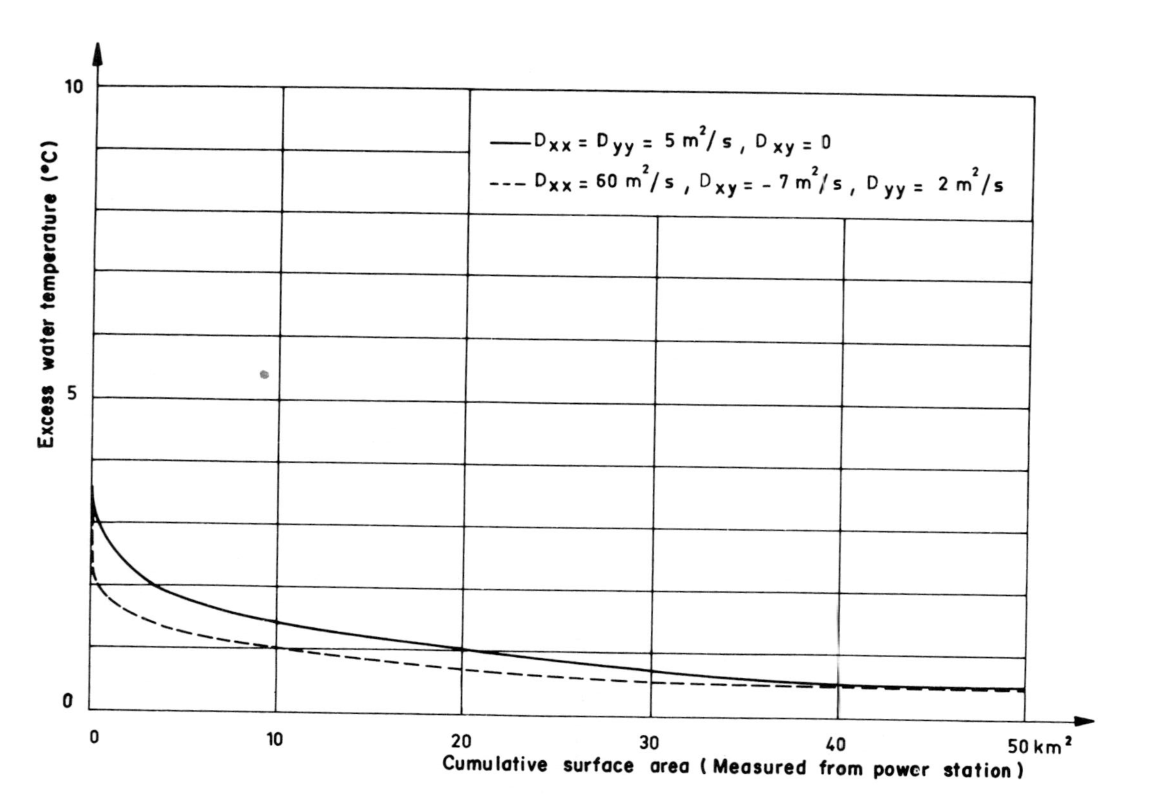

Figure 3.1b. Influence of the dispersion tensor value on the temperature field of the Paluel power plant. Cumulative excess temperature distribution near the power plant.

3.3 *Practical Problems Related to the Calculation of the Heated Outfall from a Power Plant*

In practice, the siting of power plants (5 GW), along the Channel posed the problem of determination of the temperature field, during a typical tide, assumed to be repeated and equal to itself, due to the continuous outfall of 180 cubic m/s of heated water at 15°C (11 GW). The tidal mixing causes the heated water to drop very rapidly to 1 or 2°C. Such water is then spread out over a large surface area (1 to 30 sq.km). Finally, and only then, the heat is significantly dissipated towards the atmosphere by atmospheric exchanges. The numerical modeling of such transfer must permit power injection to a small-sized area, followed by spread over a large surface area with minimum errors regarding the surface areas involved. The overall thermal balance may be grossly inaccurate unless special precautions are taken. Indeed, for such a large surface, on the order of 20 km × 30 km is involved, and an average 10 m depth, an average numerical error of 0.02°C per tide distorts the heat evaluation by a factor of 2 (which shows moreover the extent of care required). Different grid areas are required, small meshes near the outfall structure, and large meshes further off where the thermal gradients are smaller. It should be noted that the area very close to the outfall structure, where the velocity and temperature fields are strongly three-dimensional, is not properly modeled here. It is generally included within one calculation mesh where the important thing is to introduce an accurate thermal power. The modeling of this area will be described below.

Several different-sized models (see Fig. 3.2) are used in the orthogonal mesh finite difference calculations, with variable step and substantial distortion (limited, nevertheless, to a factor of 10). Several solutions may be envisaged for power injection and operation of which the following are important.

Injection: The heat rise (15°C) may be imposed on the outfall points considered as the boundary. In the case of an outfall not located on the coast, there exists a tidal-induced flow on the outfall which makes any control on the injected power very difficult ($P = \rho Cp \Delta T \times Q$). Alternately, the power due to outfall mixing may be predicted on a small surface S (four or five meshes) by adding at each time step an average heat rise over the five meshes. This alternate manner of proceeding makes for outfall transparency; an elementary schematic field carrying the required energy is added at each time step. The injected power is perfectly controlled.

Solution in several models: Even if the conservation equation (2) is numerically properly solved in each of the models (this already poses a certain number of numerical conservation problems), the handling of the two models may also give rise to errors in thermal power transfer. In the first place, one may proceed by the simultaneous solution of the various models with a magnifying glass effect on the smallest: solution in the large domain; solution in the small domain after injection with boundary conditions interpolated from the larger; interpolation of the result in the largest; etc. This type of solution also leads to a hazard in the conservation of energy, having regard to the tidal flows crossing the boundaries of the small domain. Indeed, perfect control of the heat transported through a boundary

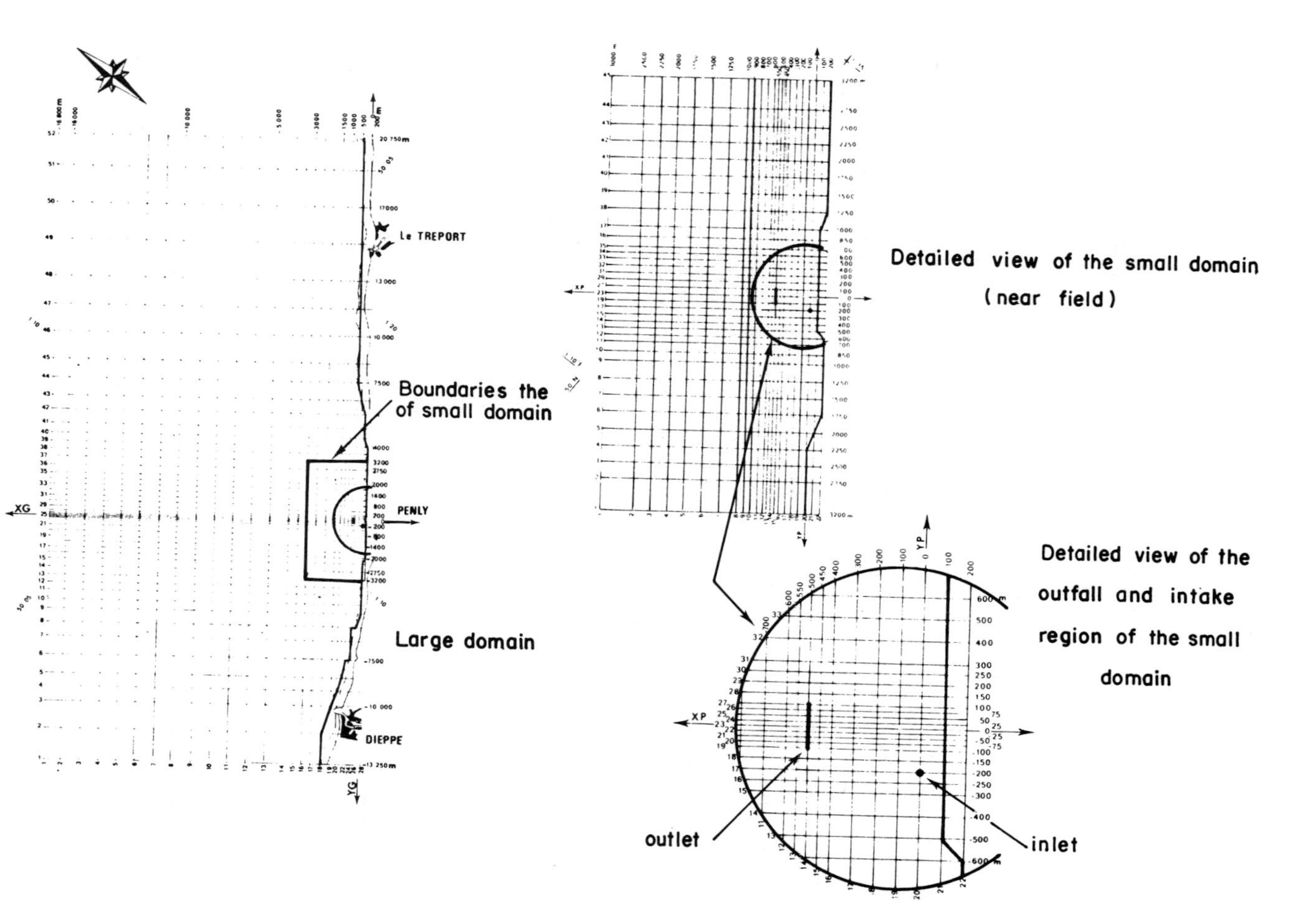

Figure 3.2. Calculational domain of the thermal mathematical model for the Penly power plant.

is difficult to achieve when the temperature is imposed therein. An alternate way of proceeding consists in solving each of the domains successively (multi-stage models used from the smallest to the largest) by availing oneself of the linear nature of equation (2). The dispersion calculations are conducted in the small domain with boundary conditions $T = T_E$ during a sufficiently small period of time (of the order of half an hour) to ensure that the heated water does not reach the boundaries. The result is stored, the heating put back at zero, and the operation is started over again during the time interval following the tide, and so forth. After a full tidal cycle, temperature fields corresponding to a first dilution of a half-hour injection are available. These are then injected into a larger model, every half-hour, and their advection-dispersion is calculated. In the meshing examples attached, two models were sufficient: the size of the model was adequate to ensure the spreading of the field and atmospheric heat transfer, while the losses at the boundaries of the model did not inordinately perturb the temperature field.

Most of our calculations (see Fig. 3.3) were determined about thirty tides after injection. In that period equilibrium is almost reached: approximately 6,500 Térajoules are stored in the domain and ensure such a level of temperature that during a tide, out of the 510 Térajoules injected by the plant, 50 Térajoules (10%) are still stored in the domain, 130 (25%) are lost at the boundaries of the model, and 330 (65%) are dissipated towards the atmosphere. These latter figures (50 and 130), compared with energy stored in the domain (6,500) lead us to believe that the heat rise level attained will hardly be modified later. In practice, this is what is ascertained. The marked heat rise (in excess of 1°C) no longer varies. Lower temperatures have not as yet fully stabilized, but this is not too much of a hindrance since it will hardly have any impact on the environment. In case of a greater heat loss at the model boundaries, the operation consisting of the storage, putting the heat rise back to zero, followed by the injection in a still larger model, should be started over again. A schematic chart of the multi-stage model technique is presented in Fig. 3.4. By using such injection techniques, together with perfectly conservative solution programs, storage followed by thermal transfer may be ensured by limiting as much as possible any errors in heat transfer evaluation. It has been found that given the same current fields and solution programs, the magnitude of error in the tidal power evaluation has dropped from 20 to 1 percent by decoupling of the small and large domains.

3.4 A Few Results Achieved in the Channel

During the past four years the National Hydraulics Laboratory ("Laboratoire National d'Hydraulique") has conducted studies on the thermal impact of the power plants along the English Channel, shown in Fig. 1.1. Except for the Gravelines site, which was studied when the Channel model was not as yet available (but with the results of a physical model for the new very close Dunkirk outport), the current fields were obtained by the multi-stage mathematical model technique previously described. All the temperature fields were obtained by means of the mathematical model solving the additional temperature equation (advection,

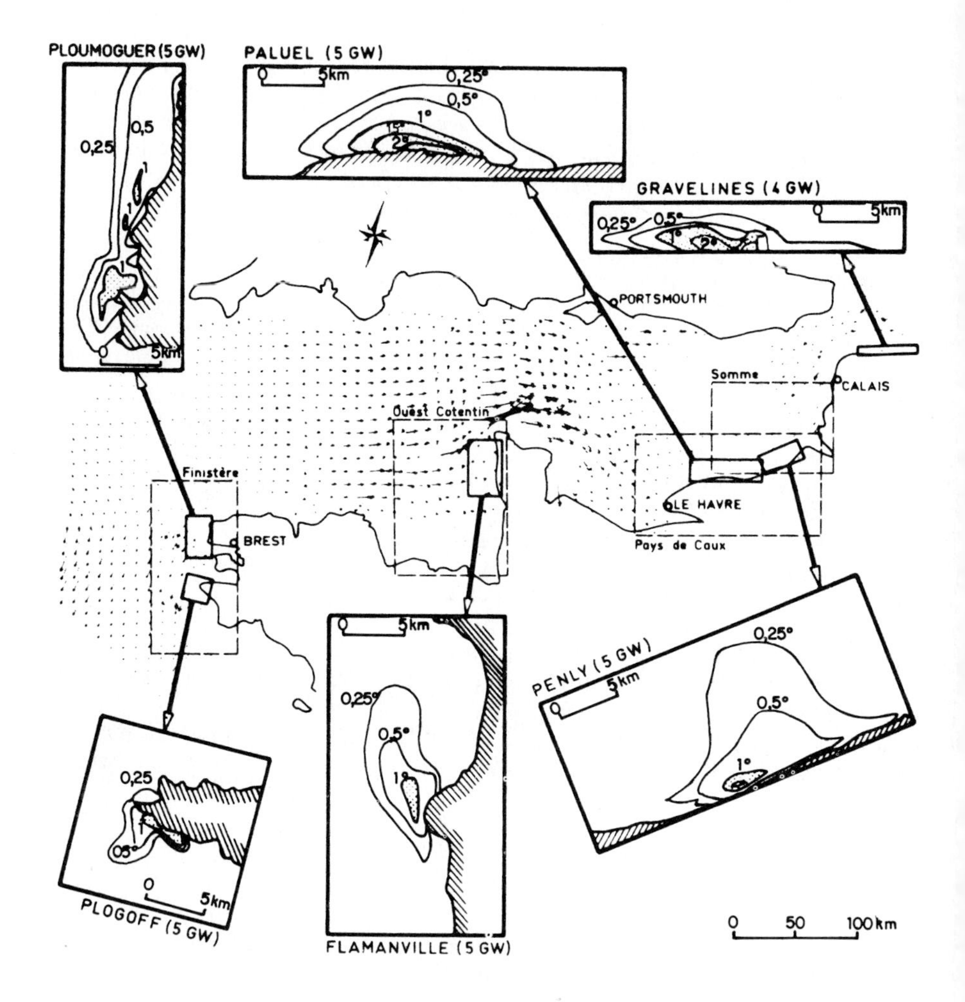

Figure 3.3. Surface isotherms of plumes computed for the nuclear power plants projected for siting on the French coast of the English Channel.

dispersion and atmospheric exchanges), and by using the current-field of a mean spring tide (coefficient 95) defined previously. Figure 3.3 shows the size of each of the thermal models, and the temperature field obtained at equilibrium for average drift, dispersion and atmospheric exchange conditions.

The overall results presented in Figs. 3.3 and 3.5 correspond to the same type of plant, involving four 1300 MWe units each. For this a total of approximately 5 GW is rejected, requiring a total cooling water flow of 180 m^3 per second, with a temperature rise of 15°C at the outlet of the condensers. Only the Gravelines plant, of a slightly different type, has been equipped for a total of 4 GW electric power and uses a 160 cubic meter/second flow, with a temperature rise of 12°C.

Gravelines site: Tidal currents, having a maximum velocity of 1 m/s, run parallel to the coast. The are driven east at high tide, and west at low tide. In a mean spring tide, the flood velocity substantially exceeds the ebb, leading to an eastward drift parallel to the coast of about 3 to 3.5 km per tide.

The atmospheric heat exchange coefficient was taken to be $A = 100$ W/sq.m°C. To investigate variations from average conditions, a result of which is shown in Fig. 3.3 (discharge flow rate W = 160 cubic m/s, dispersion coefficient $D = 5$ sq.m/s) two effects were studied (Benque and Boulot, 1975): the effect of the initial dilution of discharge Q was raised to 400 m^3/s (and the temperature rise at the outlet reduced to 5°C); and the effect of the dispersion coefficient was tested with a value of $D = 0.5 m^2/s$. The corresponding results, expressed in terms of average heated surface as a function of temperature, are shown in Fig. 3.5. Dilution of the discharge can be seen to have only a very localized effect on temperature. Away from the outfall the temperature rises are identical in the two cases, and depend only on the discharged power and on the local current field. The effect of the change in dispersion, however, is more marked.

Paluel site: On the site, the ebb currents flow in an east-west direction. However, the flood currents form a 10° angle towards the north from the east axis. This is due to the angle formed by the coast to the west of the site. The cape probably induces a slight easterly drift off to sea, at the site (measurements taken have given an order of magnitude of 1 km per tide), which diminishes near the coast, and may possibly be cancelled out or reversed according to hydrometeorological conditions.

On the site, three calculations were effected, two with $D = 5$ sq.m/s dispersion coefficient, and the third with the full tensor already referred to, and $A = 100$ $W/m^{2\circ}C$ atmospheric heat exchange coefficient. The first calculation used a schematic current field parallel to an assumed rectilinear coast, with a uniform drift of 1 km per tide parallel to the coast. The second and third used a field calculated by means of multi-stage models which allow for better reproduction of the cape effect, and the related drifts (1 km at sea, virtually nil along the coast). The heated average surface areas inferred from the first two calculations are shown in Fig. 3.5.

The full model is shown to lead to surface areas with temperature rises which are 50 to 100 percent larger than the schematic model. The low drift near the coast reproduced in the full model gives an accumulation of heat near the

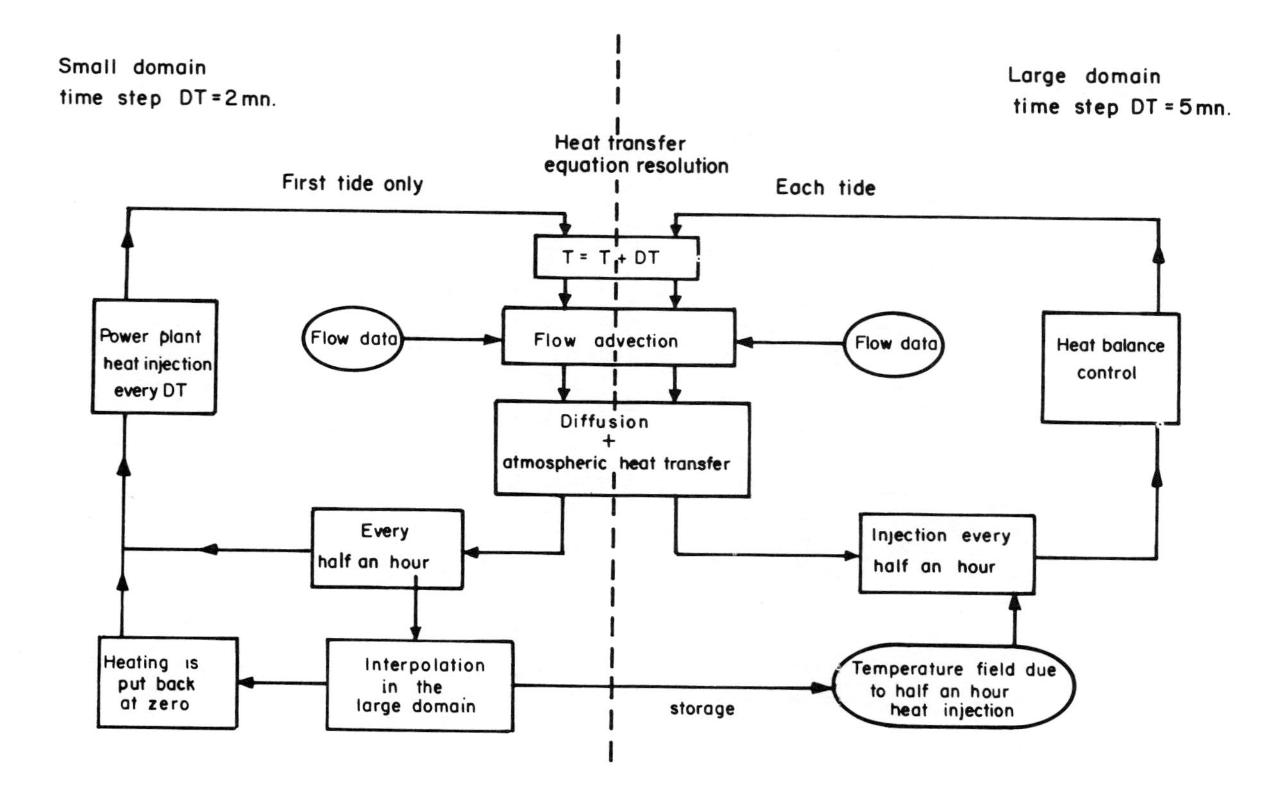

Figure 3.4. Two domain temperature field mathematical model schematic chart.

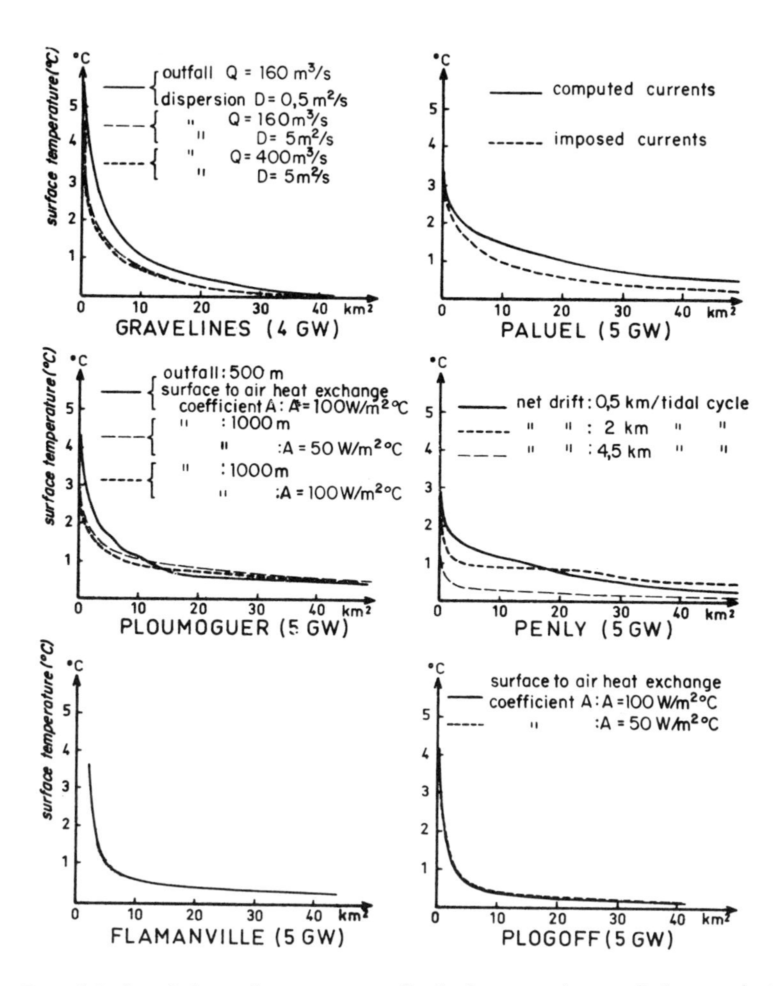

Figure 3.5. Cumulative surface temperature distributions around ocean-sited power plants.

discharge, while the stronger drift of the schematic model leads to a reduction in the surface areas affected by the mean temperature rises.

Flamanville site: The current off Flamanville flows in two directions, towards the south at ebb tide and towards the north at flood tide. Currents are strong (1.2 m/s), and flow relatively parallel to the coast. However, there is a strong water drift flowing south-west, which tends to carry the heated water off to sea (this was ascertained *in situ* and on the reduced English Channel model at the "Institut de Mécanique" of Grenoble).

On the site, only one calculation was performed, using $D = 5\ m^2/s$ parallel to the coast and a $0.5\ m^2/s$ coefficient perpendicular to the coast, and a value for A of $100\ W/m^2.°C$.

The strong currents, on the one hand, (1.2 m/s), and the strong drift towards south-west on the other hand, lead to a very low temperature rise as compared to the other sites (see Fig. 3.3 for the temperature field and see Fig. 3.5 for the curves giving the cumulative surface areas lying within different isotherms).

Penly site: The Penly site is very close to the Paluel site, and the tidal currents are quite similar. However, in regard to residual currents, a completely different behavior can be noted. The cape effect pointed out at Paluel no longer exists here. On the other hand, it would appear that the general drift towards the east which was generally noted in the English Channel, and channeling by the coast in the area (see Fig. 3.3) becomes near Penly a residual current flowing north. By following floats from the site, it has been shown, although under the influence of southerly wind, that a drift up to 4.5 km per tide exists.

Consequently, several calculations were performed with the same $D = 5\ m^2/s$, and $A = 100\ W/m^2.°C$ atmospheric exchange coefficient, but with different drifts: 0.5, 2 and 4.5 km/tide. Figure 3.3 shows the resulting temperature field. Its size is clearly related to the 2 km/tide drift flowing off to sea. Figure 3.5 gives the average surface areas of the heated effluent as function of temperature. Very high sensitivity to the value of the drift can be noted, especially since in this case the heated effluents are carried off to sea.

Ploumoguer site: The currents around this site are comparatively complex because of the topography of the sea-bottom, and the existence of shallows and islands. They generally alternate, and their magnitudes are highly variable. There are certain coastal features, particularly currents ebbing back into the coves. In examining the available measurements, a substantial drift off the coast can hardly be perceived.

Because of the specific geometry of the coast and the possibility of stagnation of the heated water, two parameters were studied in particular; the distance of the outfall from the coast at which dilution could be increased upon reaching a strong current zone, and the value of the atmospheric exchange heat coefficient, an important parameter which determines the dead water temperature. The dispersion coefficient was taken as being constant and equal to $5\ m^2/s$. Figure 3.3 shows the general aspect of the field for the tidal time concerned and Fig. 3.5 shows the curves of the surface areas as function temperature rise. Calculations confirm that the distance of the discharge is of predominant importance in determining the size of the temperature field. The impact of the atmospheric heat exchange is less marked, but is nevertheless felt at the site since the nearby coves constitute areas where heated effluents tend to accumulate.

Plogoff site: At this site, a strong drift towards the west linked to a marked dissymmetry between the flood currents (8 h 30 duration) flowing towards the west, and the ebb currents (3 h 30 duration), flowing towards the east can be noted. The maximum current in a mean spring tide is about 0.7 m/s.

Calculations were performed for $D = 5\ m^2/s$ parallel to the coast, and $D = 1\ m^2/s$ perpendicular to the coast. Calculation of the temperature field was performed for two atmospheric exchange coefficient values: 50 and 100 $W/m^2.°C$. The temperature field is shown in Fig. 3.3, and the curves of cumulative heated surface areas are shown in Fig. 3.5. Atmospheric heat exchange has a very small impact on the size of the temperature field, at this site, since the very strong drift towards the west results in a very small temperature rise.

Conclusion: In the case of the English Channel, and more generally, in the case of tidal seas, the main factors which determine the dimension and magnitude of the temperature field are first of all the advective parameters (drift and dispersion), and then atmospheric heat exchange. The former spread the heated water over a certain surface area through which the latter transfers the heat. By comparing the results obtained for each of the sites, and by comparing several results obtained for the same site for different parameter values, the following conclusions may be drawn:

(1) Drift is the essential parameter for proper dilution. In addition to its magnitude (see Penly), its direction in relation to the coast, and its variation in a horizontal plane are fundamental (see Paluel); a drift carrying the heated effluents off to sea is more favorable than a residual current flowing parallel to the coast, even though it be stronger.

(2) When the flow patterns are complex (see Ploumoguer), the value of the residual drift (average at one point), is felt less. The Lagrangian drift (motion of water mass) then carries the heated effluent towards an area where currents are different, and does not carry it back in front of the plant.

(3) Current magnitude is likewise important, whether generally around the site (see Flamanville), or at the outfall point for initial dilution (see Ploumoguer).

(4) Dispersion does not play a major part in the range of coefficients studies and on the sites along the English Channel, where the existence of strong drifts is more important.

(5) Finally, atmospheric exchanges have a substantial impact in stagnant water areas where heat can stagnate (see Ploumoguer), but are completely inoperative at the temperature rise level with which we are here concerned if proper dilution is obtained by currents (see Plogoff).

4. DILUTION IN THE NEAR FIELD

4.1 Model Used--General

In most of our power plant siting studies along the English Channel, the water intake was through a canal along the coast, and the outfall through underground galleries leading out several hundred meters from the coast. Water was discharged by nozzles placed at the bottom of the sea, of an approximate 4 m diameter, at an approximate 3 m/s velocity, directed horizontally and perpendicular to the direction of the tidal currents. Generally, there is an outfall for each unit, and four units per site. One of the problems posed is therefore how best to position these four structures so as to ensure the occurrence of smallest possible thermal impact between each other, and so that the installation cost, hence the length of the galleries, be as small as possible.

The thermal field in the area very close to each outfall structure and outfall interaction were both studied by means of physical and mathematical models. For the numerical studies we used an aquatic version of the PANACH code developed to describe the visible plumes of the cooling towers. This code can determine the dilution of an outfall in a strong cross current, where the order of magnitude of the current and the order of magnitude of the outfall velocity are the same. The code solves the Navier-Stokes and the continuity equations coupled with the heat transport-diffusion equation three-dimensionally in a steady current. With the coordinate in the direction of the flow being used as an evolution variable, the calculations are successively performed in planes perpendicular to the flow, in a domain schematically depicted as a rectangle. This model has been described elsewhere (Viollet, 1977; Benque, Caudron, and Viollet, 1977), and consequently, only the general characteristics will be described herein.

The model uses the following simplifying assumptions: (1) the density variances have been taken into account only with regard to buoyancy forces (Boussinesq approximation) and to turbulent diffusion reduction; (2) turbulent exchanges have been represented by a turbulent viscosity and diffusivity; and (3) turbulent exchanges and pressure gradients in the direction of the flow are negligible, as compared to the advective transport, thereby allowing for the solution of the equations in successive planes perpendicular to the direction of the ambient current. In the "initial" plane, the outfall is schematically depicted by a temperature field, the size of which is determined so as to properly represent the discharged momentum and thermal power. The equations, discretized in finite differences, are solved by a splitting method. Calibration of the model is performed by means of adjusting turbulent diffusion coefficients, calculated, initially, according to the characteristics of the outfall, by comparison with the experimental results obtained in a flume. The definition of such coefficients limits the validity domain of the model in the jet-entrainment zone. Reduction of the vertical mixture by stratification is taken into account by an empirical formula involving a Richardson Number.

4.2 Comparison with Various Experiments

4.2.1 Experiments in a flume with a single outfall. In connection with the study of the thermal field close to the vicinity of a depth outfall, the results of the PANACH calculation code were compared to the measurements performed in a flume at a 1/50th scale. The conditions represented in the two models were a 1 m/s uniform ambient current, a 45 m^3/s outfall discharge, a temperature rise of 15°C per outfall, and a 3 m/s outfall velocity.

The vertical profiles obtained on the basis of the physical model (Fig. 4.1) and of the mathematical model (Fig. 4.2) show for such a 1 m/s ambient current the same tendencies of hot water upsurge, linked to buoyancy forces, followed by rapid homogenization of the temperature rises at all depths. On the other hand, the two models consistently show that the outfall conditions make for rapid dilution of the heated water. Indeed, the maximum temperature rise drops by a ratio of 2 at 10 m from the outfall, and by a ratio of 6 at 50 m. Such consistency is particularly illustrated by Fig. 4.3, which gives a comparison between the results of the two methods for the normalized decrease of the maximum rise on the jet-axis versus distance from the outfall.

4.2.2 Paluel Site--dilution of four outfalls. The hydraulic studies for the projected Paluel plant, on the coast of Normandy, required the implantation of various reduced models. In particular, a non-distorted 1 to 100 scale maritime model was expected to lead to the definition of the intake structures and to optimization of the positioning of the intake and of the various outfalls in order to reduce the nearby thermal impact and heated effluent recirculation.

The nearby thermal impact depends on the interaction of the plumes from the different outfalls. Because of the variation of the currents, the interaction varies during the tide, although it remains comparatively constant during the period of maximum current. At the scale of the studied domain, the current is then nearly steady and the PANACH model may allow for a preselection of the feasible projects which are then studied on the reduced model.

The calculation code was recalibrated on the reduced model, that is, the turbulent diffusion coefficients were adjusted in such a way that despite schematization of the calculation conditions, the results of the numerical model are approximately consistent with the measurements performed on the reduced model (on which depth and flow are far from being uniform). The temperature field corresponding to all of the four outfalls is obtained by superposition of the plumes calculated for only one outfall (jet-interaction with regard to flow and to buoyancy forces is neglected), thereby allowing for very rapid testing of a large number of outfall patterns.

Figure 4.4 shows the calibration results for a preliminary pattern and for different depths. The heated surface areas are of comparable size although the overall aspects of the plumes differ somewhat. Such differences are accounted for by the fact that the flow is definitely not uniform on the reduced model, and also by the fact that the measurements are insufficiently filtered and still show some fluctuations.

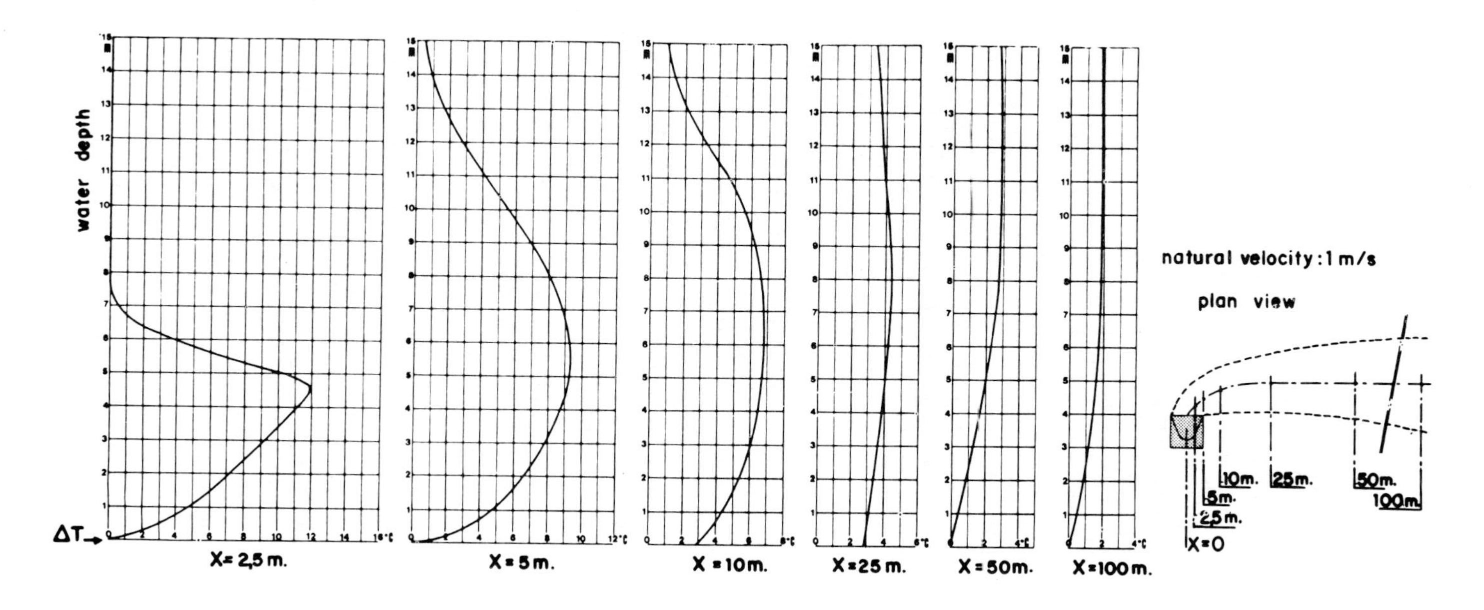

Figure 4.1. Vertical temperature profiles measured on the plume axis ($\Delta t_o = 17°C$).

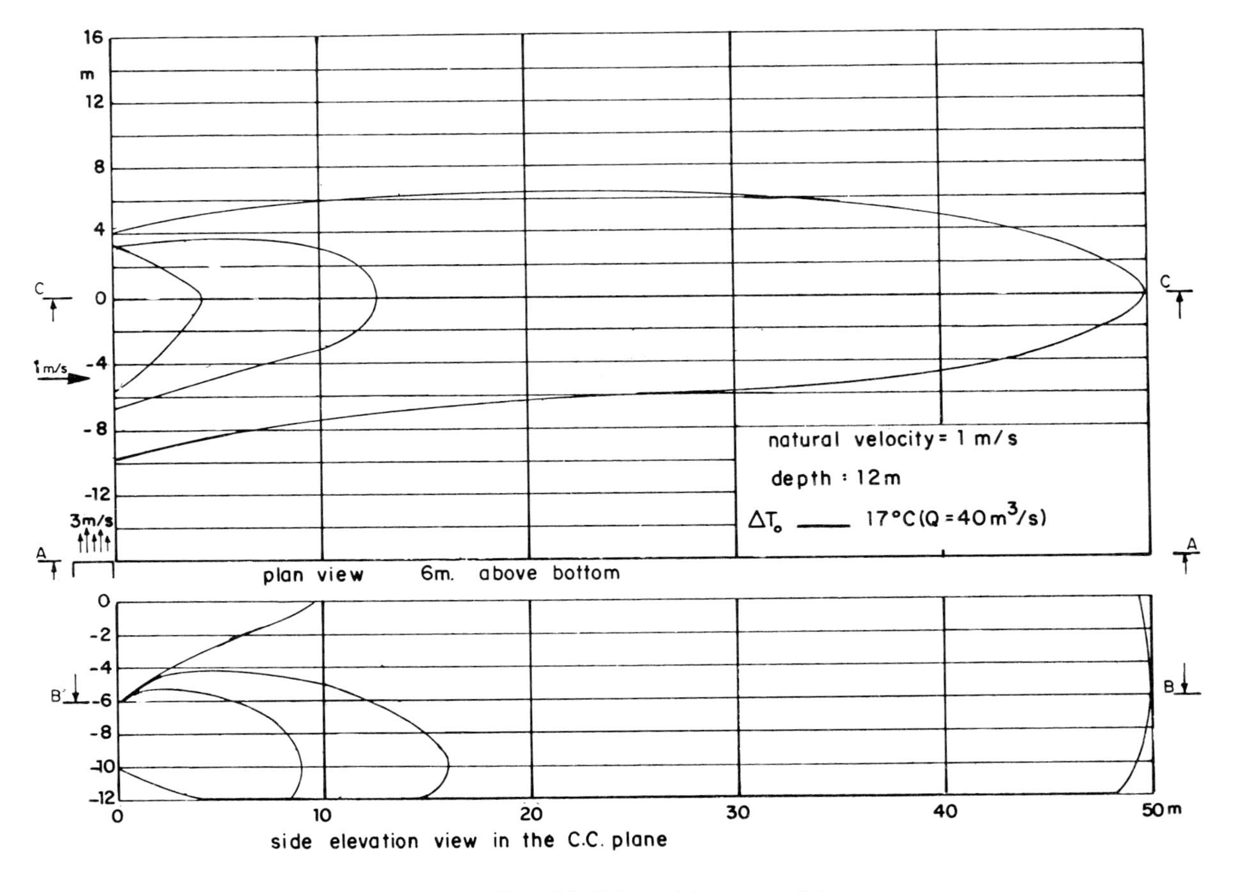

Figure 4.2. Calculated temperature field.

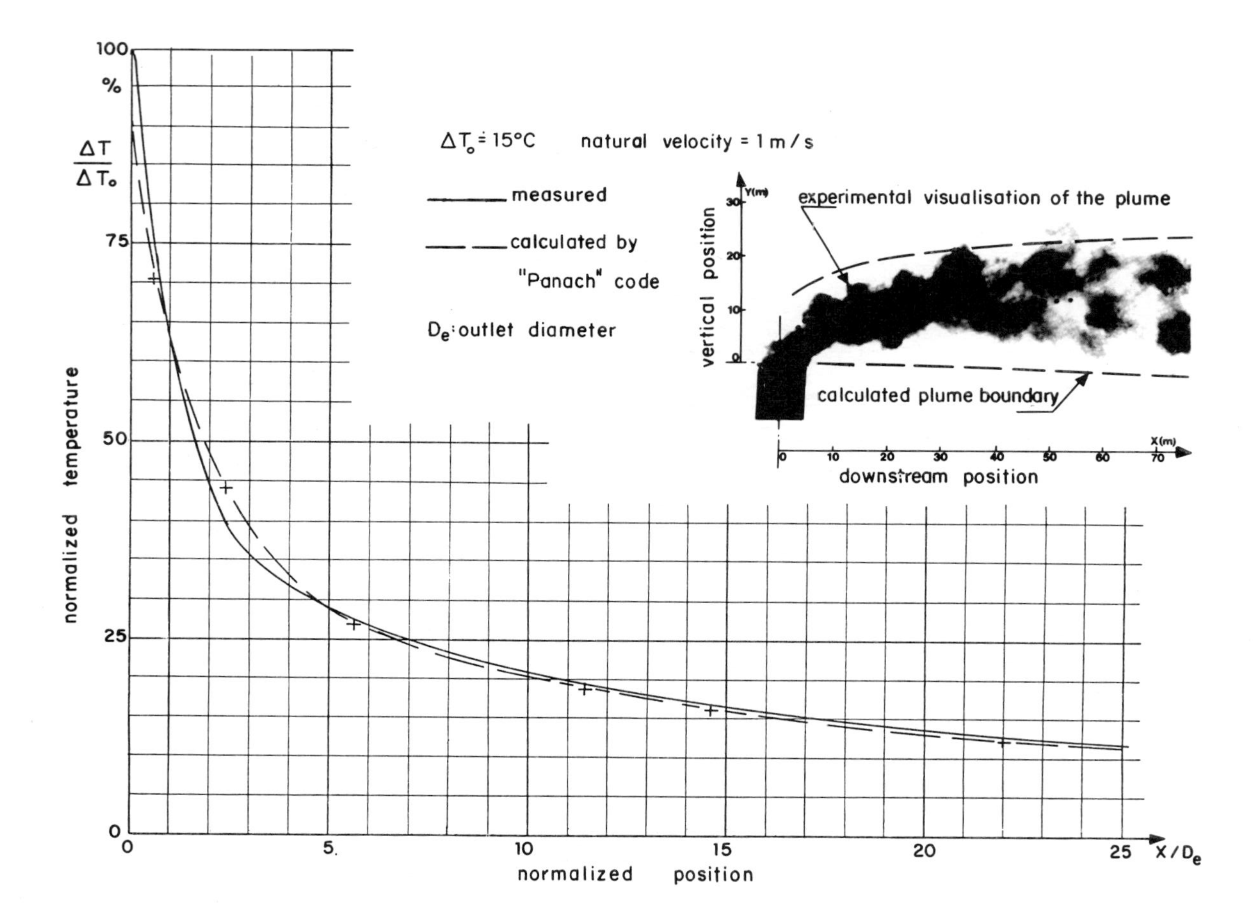

Figure 4.3. Comparison between measured and calculated temperature on the plume axis.

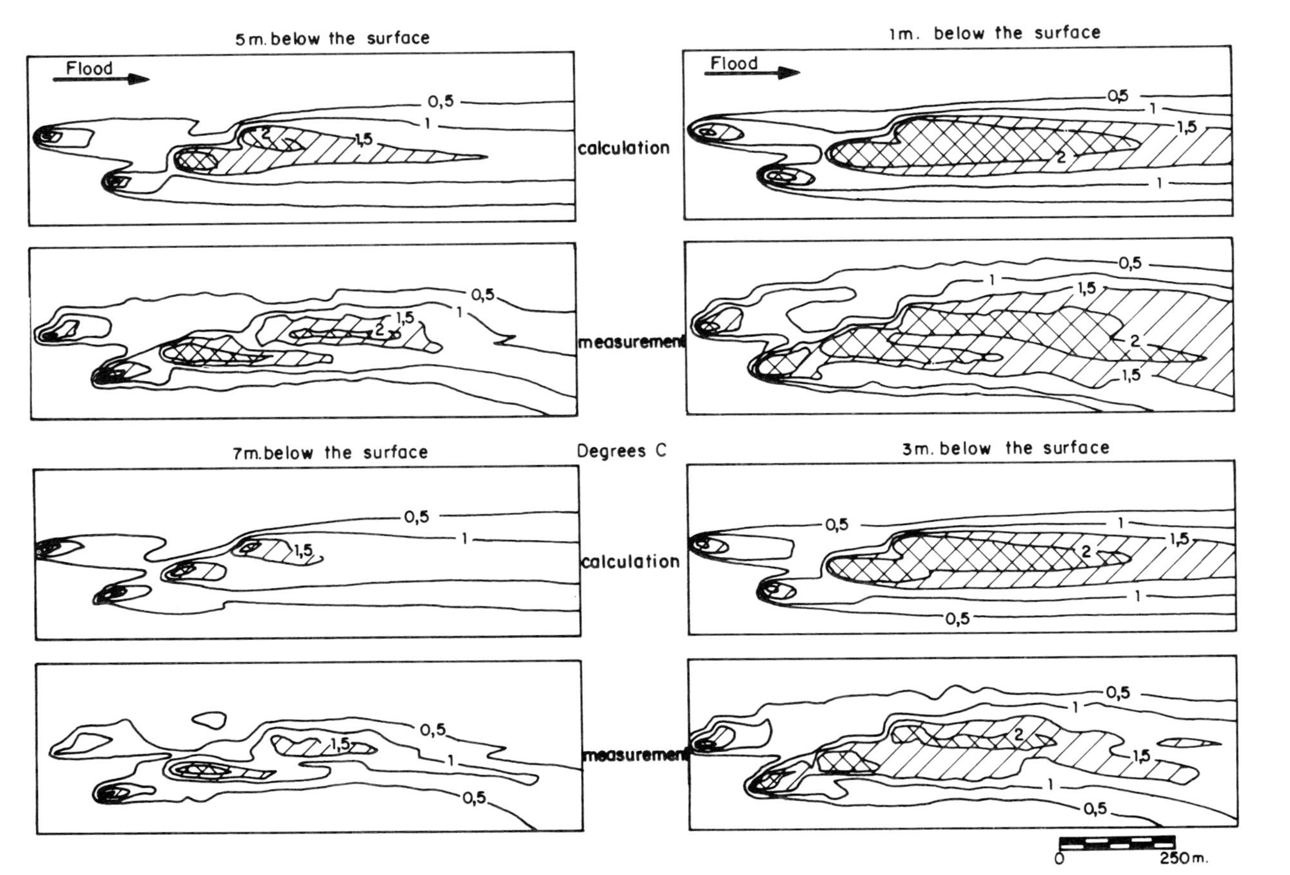

Figure 4.4. Temperature of near field near the outfall-Paluel site. Comparison between measured and calculated results at maximum tidal velocity.

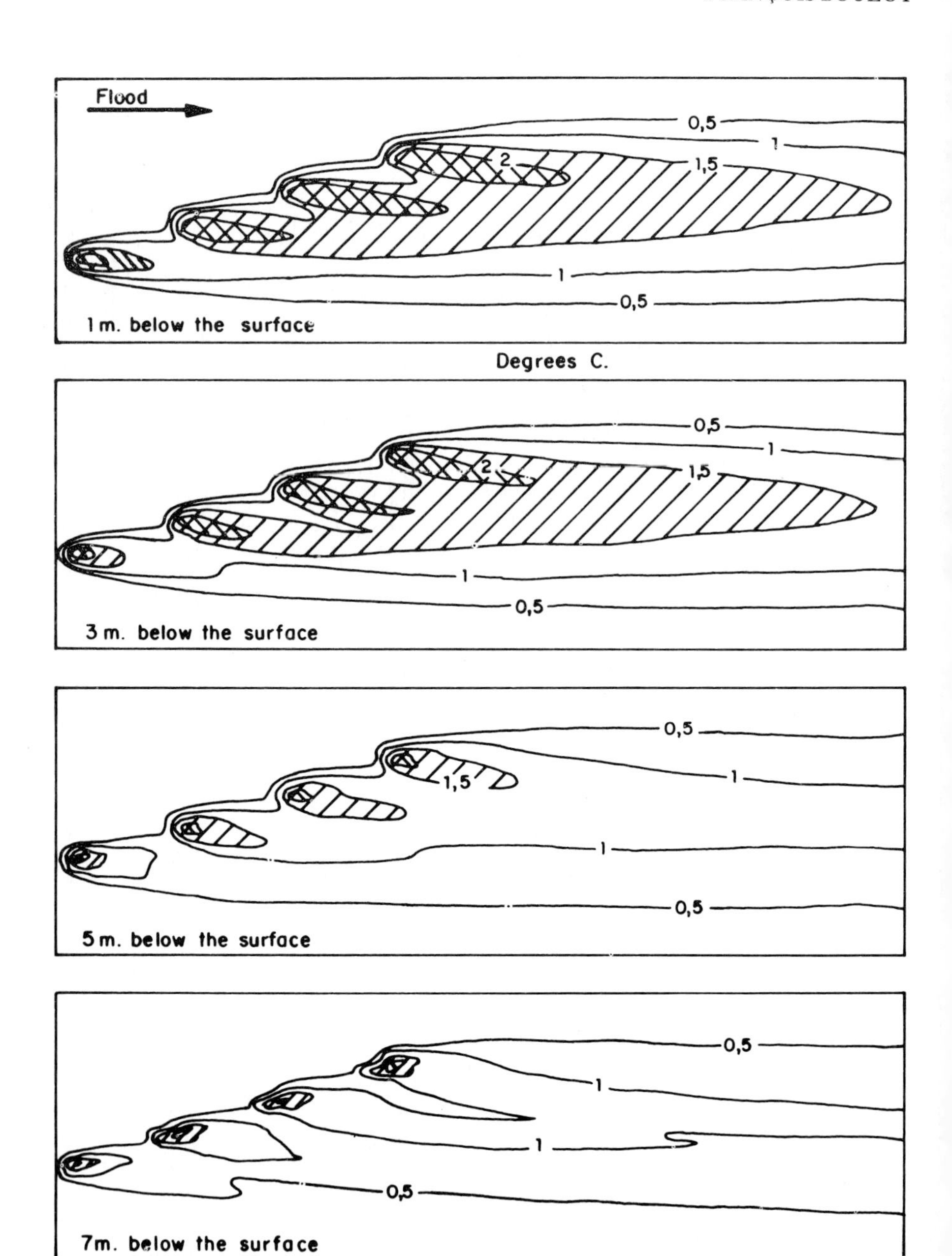

Figure 4.5. Results for the final design of the outfall of the Paluel plant. Calculated temperatures at different depths during flood tide.

In particular, the calculation code has shown that a slight shift of all of the outfalls in relation to the direction of the flow was substantially more effective than the lengthening of the distance between outfalls in avoiding interactions. Figure 4.5 thus shows that a 50 m lag between each outfall reckoned perpendicularly to the flow can limit the interaction to less than 2°C in the heated plumes.

4.3 Model Limitations and Operating Problems

The operating principle of the model and the assumption adopted limit the scope of applications to the study of the outfall dilution in a steady cross current perpendicular to the jet axis in a schematized geometry. The domain of validity is restricted to the nearby thermal field, outside of the immediate vicinity of the outfall nozzle, due to the representation of the initial conditions. Indeed, the outfall is schematically depicted in the initial diagram (perpendicular to the discharge section of the outfall) by a rectangular field, the size of which is calculated so as to ensure accurate reproduction of momentum and discharged thermal power. At the other end, the model ceases to be valid as soon as the turbulent diffusion coefficient ceases being a function of outfall conditions and becomes dependent on ambient medium characteristics. Finally, the ambient current must be sufficiently swift to satisfy the assumptions of the model, in particular, simulation of a warm water wedge upsurge is not possible in the model. The spacing between calculation planes must take account the grid size in the planes, and the magnitude of the different current components, so that the Courant number relating to advection in the calculation planes be not too high; on the other hand, in the case of the study of the thermal field induced by various outfalls, definition of a grid whereby detailed reproduction of initial conditions relating to an outfall may be reconciled with the accounting for the overall thermal field generated by upstream outfalls is difficult or costly.

The PANACH three-dimensional calculation code for the dilution of an outfall in a cross current was validated through comparison with various reduced model experiments. Schematization of calculation conditions allowing for prompt implementation of the model makes it a very handy tool specially adapted for preliminary studies, or even for project studies in simple geometries.

Other models which are completely three-dimensional and solve the full Navier-Stokes equations are being developed, and have already yielded encouraging results. They will be used to deal with a larger variety of outfall conditions but at a much higher cost.

5. CONCLUSIONS

We have described the modeling patterns used in the "Laboratoire National d'Hydraulique" at Chatou, to study the dilution of power plants planned by Electricité de France on the French coast of the English Channel, some of which are currently being built. The purely numerical aspects of equation solutions have not been discussed, since this was not the subject of this paper. However, we have attempted to show that, after determination of a proper numerical scheme, there still remained a certain number of difficulties to be solved, particularly in regard to taking into account of actual geometry of the area of the effluent discharge. In the future, it may be expected that the finite element method, thanks to improved description of the boundaries, and to the possibility of using very different size elements, will allow for the solution of the remaining problems.

In regard to the quality of these results in actually reproducing natural reality, a clear answer cannot as yet be given, since the first plant which was studied will only be starting operation during 1980. However, some evidence does exist. First, in regard to the near field, the PANACH code for dilution in a cross current yielded results consistent with those observed *in situ* (the cooling tower visible plume problem, Caudron and Viollet, 1979). For the far field, the most important phenomena is tidal current advection; the agreement of the results of our calculations with *in situ* measurements has been shown. Obviously some uncertainties still remain in regard to the dispersion tensor. In our opinion, only measurements *in situ* of a quite fundamental nature would allow for confirmation of what we have stated, and for model improvements if necessary.

The most important lack of information bears on the residual current field. The consequences of this ignorance are not too serious for a short-lived pollutant such as heat, which passes off into the atmosphere. For a long-lived pollutant, however, the residual current field is much more important because it can, through water renewal in the discharge area alone, prevent pollutant concentration from continually increasing. Knowledge of the residual velocity field is a difficult problem, but of such importance that thorough research combining *in situ* measurements and modeling attempts must be carried out.

REFERENCES

Benque, J. P. and Boulot, F. (1975). Prises et rejets des centrales thermiques en bord de mer. *Proc. Congr. Int. Assoc. Hydraul. Res. 16th, Sao Paolo, Brazil.*

Benque, J P. and Caudron, L. (1977). Calcul tridimensionnel des rejets d'eau chaude dans un courant traversier. *Proc. Congr. Int. Assoc. Hydraul. Res. 17th, Baden Baden, West Germany.*

Benque, J. P. and Warluzel, A. (1979). Dispersion dans une mer à marée. *Proc. Congr. Int. Assoc. Hydraul. Res. 18th, Cagliari, Italy.*

Benque, J. P., Ibler, B., Keramsi, A., and Labadie, G. (1980). A finite element method for Navier Stokes equations. *Third International Conference on Finite Elements in Flow Problems, Banff.*

Benque, J. P., Ibler, B., and Labadie, G. (1980). A finite element method for the calculation of compressible viscous flows. *First International Conference on Numerical Methods for Non-linear Equations, Swansea.*
Boulot, F. (1977). Echauffement du milieu naturel induit par le rejet de centrales. *J. Thermoécologie.*
Caudron, L., and Viollet, P. L. (1979). Calcul tridimensionnel permanent des panaches de réfrigérants atmosphériques. *Proc. Congr. Int. Assoc. Hydraul. Res. 18th, Cagliari, Italy.*
Caudron, L. and Viollet, P. L. (1979). Les panaches d'aéroréfrigérants au voisinage des centrales thermiques. *La Météorologie* **8**, Ser. 18.
Cunge, J. A. and Preissmann, A. (1961). Calcul des intumescences. *Proc. Congs. Int. Assoc. Hydraul. Res. 9th, Dubrovnik.*
Daubert, A. and Graffe, O. (1967). Quelques aspects des écoulements presque horizontaux à deux dimensions en plan et non permanents. *La Houille Blanche* **8**.
Daubert, A. (1974). La dispersion dans les écoulements filaires. *La Houille Blanche* **1**.
Fischer, H. B. (1979). "Mixing in Inland and Coastal Waters." Academic Press.
Lomer, J. F. (1978). La dérive en mer à marées. Doctoral thesis, University of Paris, VI.
Nihoul, J. and Ronday, F. Effect of the tidal stress on the residual circulation of the North Sea. Université de Liège.
Viollet, P. L. (1977). Etude des jets dans des courants traversiers et dans des milieux stratifiés. Thesis, University of Paris.

DISCUSSION

Peter Mangarella

Woodward-Clyde Consultants,
San Francisco, California

How did you specify the boundary condition in transport simulation (i.e., the temperature boundary condition) along the boundary between the coarse and fine grid? Specifically, did you take into account the possibility of heat exchange from the coarse grid boundary condition into the fine grid and, secondly, was the boundary condition flow direction dependent?

REPLY

In the procedure we use now, there is no boundary condition between the coarse and fine grid. The fine grid we use is only a means of spreading the temperature before injection in the coarse one.

Thanks to the linearity of the solved equation, only the supplement of temperature due to about half an hour of injection is computed in the fine grid (before it reaches the boundaries). Then, this result is added to the temperature field in the coarse grid (which covers also the region of the fine one). In fact, we are interested only in the field obtained in the coarse grid. The fine grid is simply used to add realistic puffs of temperatures to the computed field in the coarse grid.

TWO-DIMENSIONAL TIDAL MODELS FOR THE DELTA WORKS

J. J. Leendertse

The Rand Corporation
Santa Monica, California

A. Langerak

Rijkswaterstaat, Delta Service
The Netherlands

M. A. M. de Ras

Rijkswaterstaat, Data Processing Division
The Netherlands

1. INTRODUCTION

In 1953 a disastrous storm surge occurred in the North Sea. In the southwestern part of the Netherlands dikes were overtopped by waves and tides, causing them to fail. Large sections of the country flooded and many lives were lost. As a result of this disaster, plans were made to improve the protection against floods. In this region many inlets and islands exist, and it was quite logical that the plan for improvement would include shortening of the coastline by closing the inlets as otherwise the dikes over many hundreds of miles would have to be heightened. According to the so-called "Delta Plan," only the New Waterway to Rotterdam and the Westerscheldt giving access to Antwerp would not be closed (Fig. 1.1).

The closure of tidal inlets is technically difficult. Reduction of the cross section initially increases the velocities in the closure opening as the work progresses. In the plan, the smaller openings were closed first so that experience could be obtained for the closure of the Eastern Scheldt (Oosterschelde), for which a five-mile-long dam was designed.

During its initial construction stage, opposition against its execution became very evident. This opposition was to a large extent due to environmental reasons--the closure would change the Eastern Scheldt from a tidal estuary to a fresh or brackish lake. On the other hand, if the closure was not accomplished, the long dikes around the Oosterschelde would have to be heightened, which

ISBN: 0-12-25852-0

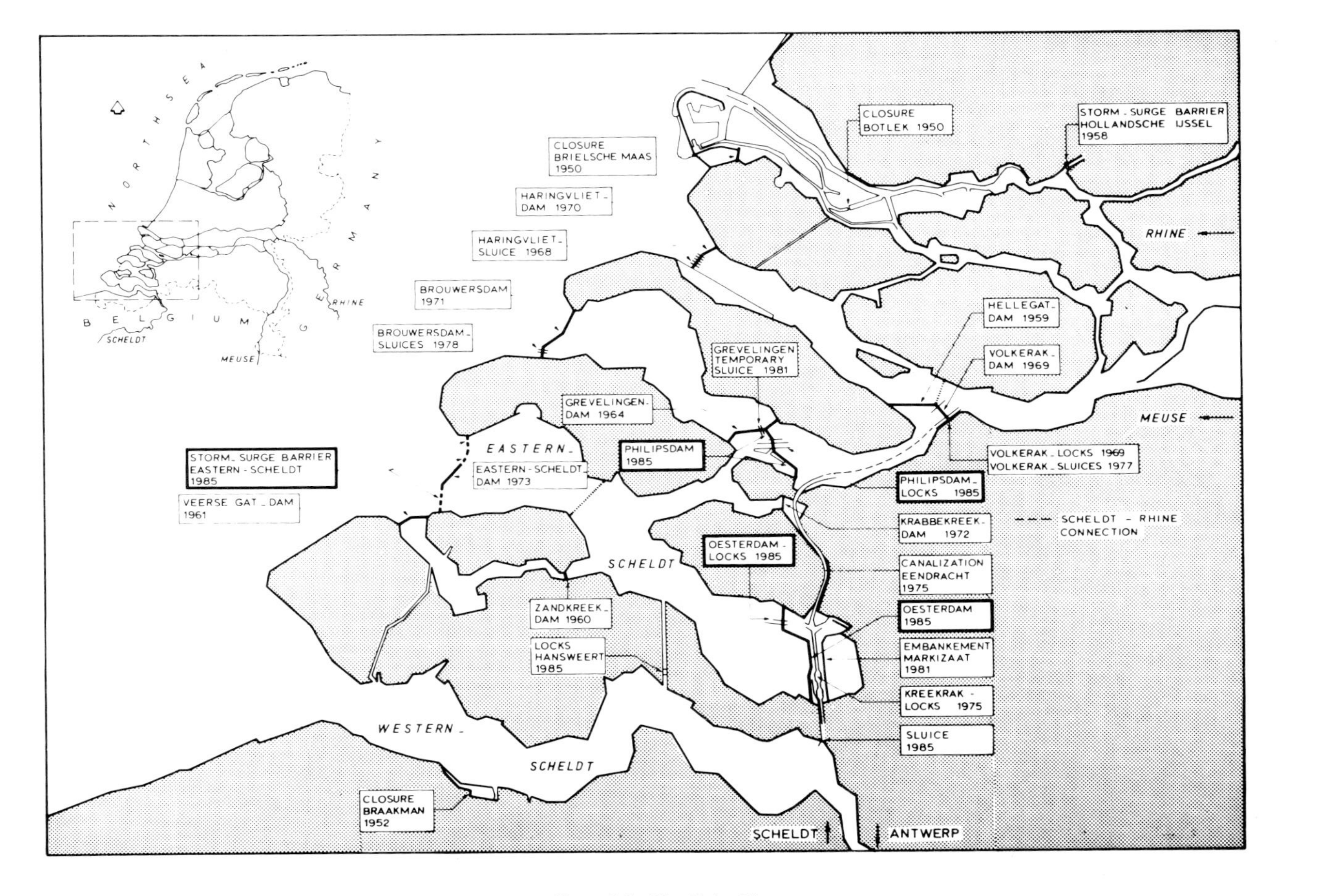

Figure 1.1. The Delta Plan.

would be very costly and would naturally also have environmental impacts. A third possibility was to build a storm surge barrier by which the Oosterschelde would maintain its character as a tidal estuary, even though the tidal range would be reduced. During a storm surge the barrier would be closed.

After a study of the impacts of these three alternatives on such varied subjects as the environment, the safety of the area, transportation costs, the local and national economy, and employment, the Netherlands Cabinet decided to build the storm surge barrier. This barrier is being built across the three main channels of the inlet, having a total length of nearly three kilometers. The barrier will reduce the present 80,000 m^2 cross sections of the three channels to an effective aperture of 14,000 m^2, by which the discharges are reduced and also the tidal range in the Oosterschelde, as mentioned above. The distribution of the flow through the barriers is chosen in proportion to the maximum values of the discharges through the present channels at mean tide in order to maintain the present morphological pattern as much as possible.

In addition to the construction of this barrier, the construction of secondary dams, sluices, and locks associated with these dams is underway, as shown in Fig. 1.1. The Oesterdam and the Philipsdam reduce the tidal area of the Eastern Scheldt by about 20% and make the waterway between Antwerp and the Rhine free of any tidal influence.

The model studies with the two-dimensional model are directed toward investigation of the change in offshore tidal currents and its effect upon the sediment transport in this offshore area. Furthermore, information is supplied to wave diffraction/refraction models, and with the two-dimensional model the barrier closing strategies are being verified. These closing strategies were determined with a much simpler mathematical model.

The model is now also being used extensively in the prediction of salinity and water quality in the Oosterschelde as a result of barrier construction and the construction of the secondary dams. In particular, the model will be used in support of studies for the closure of the secondary dams. It is expected that in the latter phases of the execution the models will be used extensively in planning the different phases of construction.

2. THE MODEL AREA

After construction of the barrier, the flow in the three tidal channels will be reduced considerably and it can be expected that the influence could be noticed over a section of the offshore area of the Delta region. In case of a closure during a storm surge, this effect would be even larger. To be effective in model investigations of this type, the boundaries of the model have to be chosen at such a distance from the site of construction that the influence of the construction is insignificant. In 1966, during studies for the closure of the Haringvliet, model experiments were made with a model of the Southern North Sea, as described in Leendertse (1967). Comparisons were made of currents and water levels for the conditions existing in 1960 and for the conditions after the Delta Works were completed for the plan with the Oosterschelde completely closed.

It appeared that the effect of construction on water levels and currents is insignificant at a distance of about 25 km out of the coast. Consequently, the seaward boundary parallel to the shore was taken at this distance. In the north the model area was bounded near Scheveningen.

Within the model area numerous discharges are present. In addition to the New Waterway, the discharges through the sluices in the Haringvliet and the discharge of the river Scheldt are important. Particularly relevant in this study are the discharges through the sluices and locks on the Volkerak and the Kreekrak sluices, as they influence considerably the salinity in the Oosterschelde.

Not only do fresh water discharges exist in the Oosterschelde, but also salt sinks. Sea water enters through the locks into the fresh water channel system in the interior region. The salinity in the Oosterschelde is not only dependent on the local discharges, but it is influenced also by the Rhine discharge through the New Waterway. With north and northwesterly winds, water masses from offshore of the New Waterway can be transported along the coast and then enter the Oosterschelde.

To study the influence of the New Waterway discharge on the salinity of the Oosterschelde this waterway to Rotterdam with a major discharge of the Rhine was included in the model. The southwestern boundary of the model was chosen near Blankenberghe in Belgium. The model area includes the Eastern Scheldt as well as the Western Scheldt.

The tides are predominantly semidiurnal, with large variations in range over the tidal area. In the southwestern part of the model area the amplitude of the M_2 tidal component exceeds 2 meters, and in the northern part this component is about .75 meters.

A number of overtides are present, particularly in the northern part of the model area. Figure 2.1 shows the distribution of the harmonic components at four different locations in the model area.

The amplitudes of the overtides increase and decrease with the semidiurnal tidal heights, as shown in Fig. 2.2. In this figure the observed tide is plotted, together with the tidal variation in a frequency band of the quarterdiurnal tide. From this behavior one can draw the conclusion that the semidiurnal tidal wave propagation through the system is far from linear.

The amplitudes of the diurnal tide are small and about the same over the whole system. The mean tide levels vary considerably. The North Sea and the continental shelf of the Atlantic Ocean near England and Norway are relatively shallow. Consequently, meteorological disturbances in this area will generate long-period water level fluctuation and storms in the North Sea may generate storm surges of considerable height. These long-period water level fluctuations can have considerable influence on the propagation of the semidiurnal tide in the shallow North Sea, and the time as well as the magnitude of High Water may differ considerably from the predicted tide.

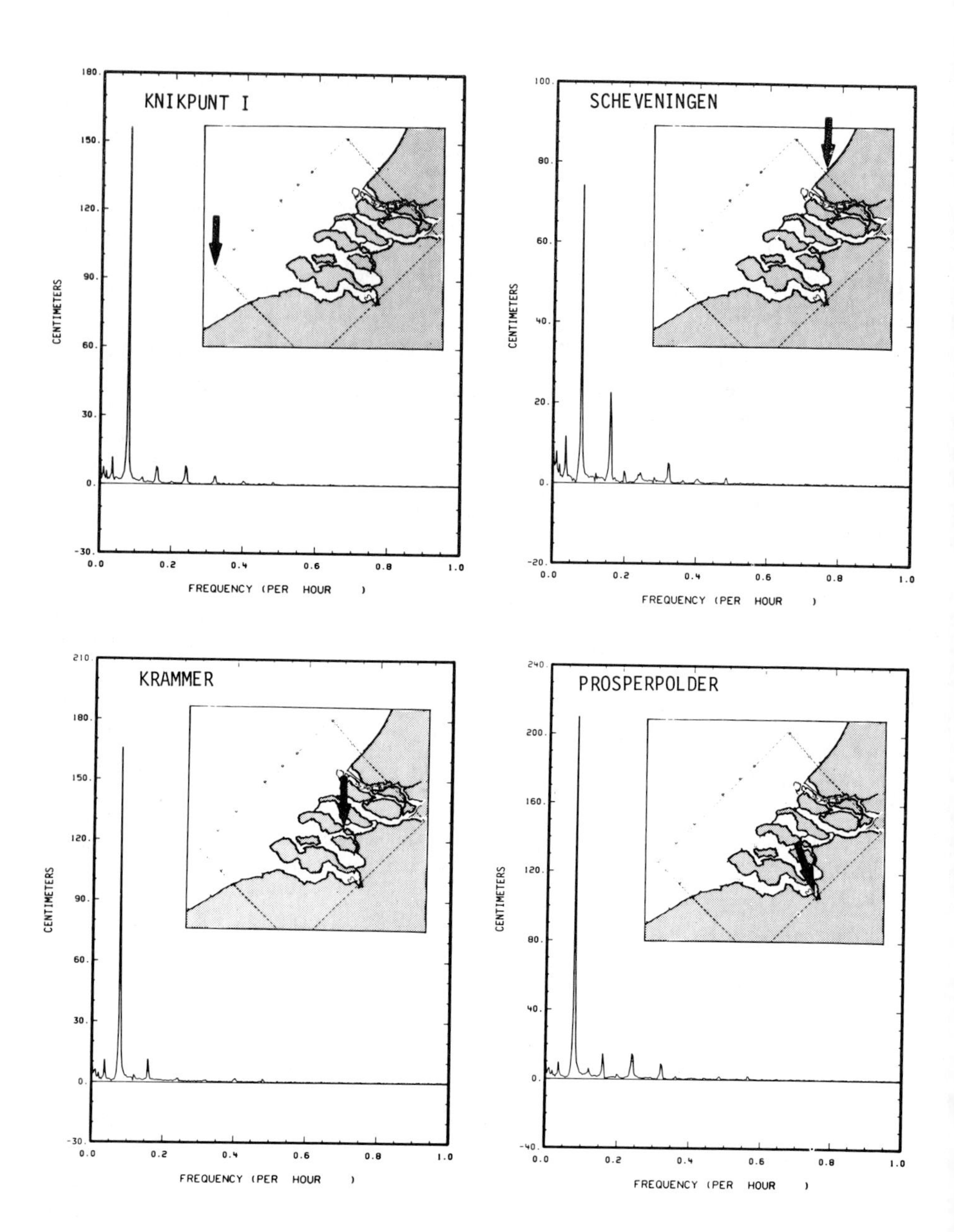

Figure 2.1. Distribution of tidal components at four locations in the model area.

The variability in the phase and amplitude of the tidal wave has severe consequences in the modeling effort, as the predicted tides cannot be used effectively for adjustment and verification of the model and its submodel. An actual simulation of a period in which sufficient boundary and verification data have been collected had to be used.

In the model area we found a rather diverse flow condition to be modeled. In the offshore area the variation in bottom slope is relatively small if we discount the large sand waves which are present at many locations. In the estuaries the tidal flow is more or less one-dimensional in the deep and shallow channels. On the tidal flats, when the water depth is sufficient, this flow may be more two-dimensional. The channels pose particular problems in finite difference modeling, as sufficient resolution should be available to represent these channels in the finite difference bathymetry of the model. In view of the large area to be modeled, the grid size was chosen at 800 m, which is half the grid size of an earlier Randdelta model which was operational in the period of 1967-1977.

With this grid size many of the tidal channels were difficult to represent, and for this reason model investigations with a grid size of 400 m were also planned. It appeared, however, that with a judicious depth approximation on an 800 m grid models could be obtained which, for practical applications, represented very well the complicated physical processes.

The water in the model region is generally quite well mixed, with the exception of the areas near major discharges such as the Keeten-Volkerak. At such locations the salinity decreases rapidly toward the discharge and only partial mixing may be present. In those areas considerable variations in the horizontal velocities and the salinity may be present in a vertical at a given time. This complicates modeling, as it is assumed in the formulation of the model that the velocities as well as the salinities are more or less constant in any given vertical.

3. MODELING INSTRUMENTS

Field data used in mathematical model studies are generally taken at equally spaced time intervals. The data often contain variations in which we are not interested, and which are eliminated by digital filters. Since it is not possible to determine beforehand what filter is needed, a program was made to vary the filter characteristics. The filters which could be handled are all nonrecursive and time-invariant. In the modeling studies we typically used low-pass filters, high-pass and bandwidth filters. Low-pass filters were used for the elimination of noise out of records and for the elimination of the tides to obtain the effects of meteorological disturbances. High-pass filters were used to eliminate the meteorological disturbances from the records. Bandwidth filters were used to eliminate certain tidal components from the records. For example, Fig. 2.1 shows the overtides in a tidal record taken at the northern corner of the model area. In the inset of the graph the amplification factor of the filter is shown, with its peaks at 0.166, 0.5 and 1.0 cycles per hour.

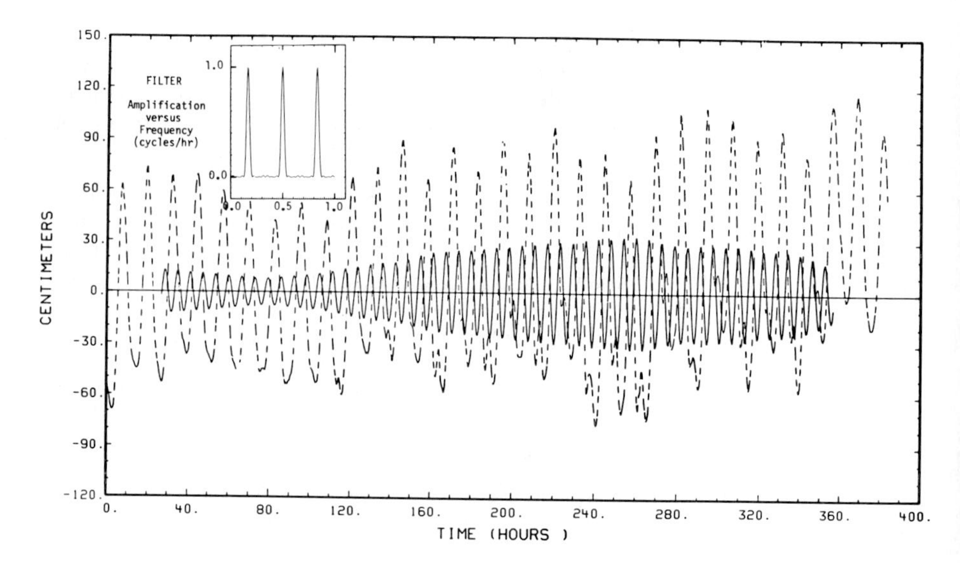

Figure 2.2. Observed tide (---) at a station in the most northerly corner of the model area, together with the quarterdiurnal overtide (—) obtained from the record by a bandpass filter.

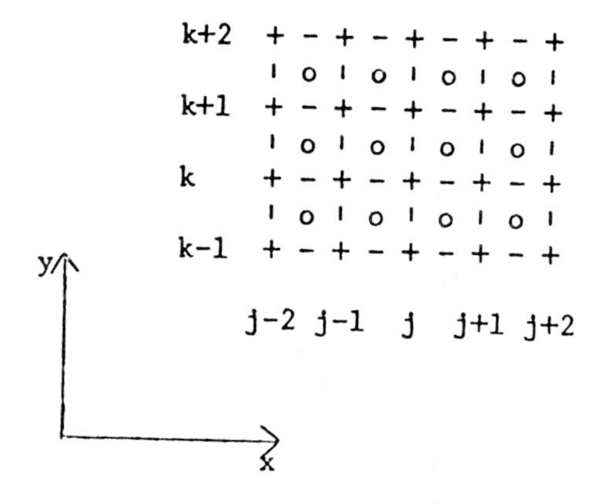

Figure 3.1. Space staggered grid. + is the water level (ζ) and mass density (ρ_A); O is the water depth (h); – is the U velocity (u); | is the V velocity (v).

In addition to these analyses in the time domain, Fourier analyses of the observation series were made in the frequency domain. The distributions of tidal components shown in Fig. 2.1 were made by Fourier analysis. Fourier series were extensively used as inputs in models as the data management is much easier than with time series.

Use was also made of cross-spectral analysis. With cross-spectral analysis optimal linear relations between time series can be determined which then in turn can be used for prediction. Cross-spectral analysis provides the data for building simple models for the relation between two or more variables. A group of programs particularly for this purpose was built. Up to now cross-spectral analysis has hardly been used by finite difference modelers, even though this analysis method is commonly used by other investigators of time-varying phenomena.

The actual simulations for the investigations were made with the WAQUA or SIMSYS2D system. This system consists of an interlocking system of programs for data processing, two-dimensional simulation and graphical representation, designed particularly for engineering investigations. The input data processor is the first program used for each simulation. Datasets which are easy to assemble by the model engineer are reordered in the form suitable for simulation. Simultaneously checks are made as to consistency of the data, and an annotated report is written for documentation purposes. The data display systems permit nearly all time-varying data and all spatially-varying data to be graphed or charted, allowing a visual check on those datasets. In routine investigations these graphs are also put on microfiche and archived.

The actual simulations are made with a program which computes the vertically integrated equation of motion, the equation of continuity, and transport equations by use of finite differences. The relation between density and salinity is expressed by an equation of state. Two sets of finite difference equations are used alternately, and the sets are solved implicitly for each direction.

The computations are made on a rectangular grid system with variables for the water level (ζ) and the concentration (P) on points of the grid at integer values. The velocity variables are located between the water level points so that for the computation of the new velocity value the pressure gradient is directly available without averaging (Fig. 3.1). A more detailed description of the basic method used in the system can be found in Leendertse and Gritton (1971), but modifications have been made to obtain a higher order of approximation.

The first set of equations refer to computations centered on time level $n\Delta t$, and can be written as follows

Momentum Equation

$$\delta_t u - f\bar{\bar{v}} + A(x) + g\overline{\delta_x\zeta}^t + \tfrac{1}{2}\,\frac{g}{\rho}(\bar{h}^y + \bar{\zeta}^x)\delta_x\rho + R(x)\bar{u}^t$$

$$- \frac{\theta\rho_a W^2 \sin\psi}{\rho(\bar{h}^y + \bar{\zeta}^x)} - \delta_x \boldsymbol{p} - k\nabla^2 u_- = 0 \qquad \text{at } j+\tfrac{1}{2}, k, n \tag{1}$$

Continuity Equation

$$\delta_{+\frac{1}{2}t}\zeta + \delta_x\left[(\bar{h}^y + \bar{\zeta}_+^x)\, u_+\right] + \delta_y\left[(\bar{h}^x + \bar{\zeta}^y)\, v\right] = 0 \tag{2}$$

Transport Equation

$$\begin{aligned}\delta_{+\frac{1}{2}t}[P_i(\bar{\bar{h}}+\zeta)] &+ \delta_x\left[(\bar{h}^y + \bar{\zeta}_+^x)\, u_+\bar{P}_{i_+}^{x}\right] + \delta_y\left[(\bar{h}^x + \bar{\zeta}^y)\, v\bar{P}_i^y\right] \\ &- \delta_x\left[(\bar{h}^y + \bar{\zeta}_+^x)\, D_{x_+}\delta_x P_{i_+}\right] - \delta_y\left[(\bar{h}^x + \bar{\zeta}^y)\, D_y\delta_y P_i\right] \\ &+ \sum_{l-1}^{i-1} (\bar{\bar{h}} + \zeta_+)\, K_{il} P_{l_+}\alpha_i + \overline{(\bar{\bar{h}} + \zeta)\, K_{ii} P_i}^{t/2} \\ &+ \sum_{l=i+1}^{l_{\max}} (\bar{\bar{h}} + \zeta)\, K_{il} P_l \beta_i + S_i = 0 \quad at\ j,\ k,\ n\end{aligned} \tag{3}$$

where $A(x)$ is the advection term, f is the Coriolis parameter, g is the acceleration of gravity, h is the distance between bottom and reference plane, k is the horizontal velocity diffusion coefficient, k_{il} is the reaction coefficient between constituent i and l, $l_{\max}$ is the maximum number of constituents, P_i is the constituent i, p is atmospheric pressure, $R(x)$ is the bottom stress coefficient, S is the source of sink of constituent P_i per time unit, u is the vertically averaged velocity component in x direction, v is the vertically averaged velocity component in y direction, w is the wind speed,

$$\alpha_i = \begin{cases} 0 & i = 1 \\ 1 & 1 < i \leqslant l_{\max} \end{cases}$$

$$\beta_i = \begin{cases} 0 & i = l_{\max} \\ 1 & 1 \leqslant i < l_{\max} \end{cases}$$

ζ is the water level elevation relative to a horizontal reference plane, Θ is the wind stress component, ρ is the density of water, ρ_a is the density of air, and ψ is the angle between wind direction and the positive y direction.

Averages and differences are symbolically represented by

$$\bar{F}^x = \tfrac{1}{2}\{F[(j+\tfrac{1}{2})\Delta x,\ k\Delta y,\ n\Delta t] + F[(j-\tfrac{1}{2})\Delta x,\ k\Delta y,\ n\Delta t]\}$$

$$\delta_x F = \frac{1}{\Delta x}\{F[(j+\tfrac{1}{2})\Delta x,\ k\Delta y,\ n\Delta t] - F[(j-\tfrac{1}{2})\Delta x,\ k\Delta y,\ n\Delta t]\}$$

$$\begin{aligned}\bar{\bar{F}} = \frac{1}{4}\{&F[(j+\tfrac{1}{2})\Delta x,\ (k+\tfrac{1}{2})\Delta y,\ n\Delta t] + F[(j+\tfrac{1}{2})\Delta x,\ (k-\tfrac{1}{2})\Delta y,\ n\Delta t] \\ &+ F[(j-\tfrac{1}{2})\Delta x,\ (k+\tfrac{1}{2})\Delta y,\ n\Delta t] + F[(j-\tfrac{1}{2})\Delta x,\ (k-\tfrac{1}{2})\Delta y,\ n\Delta t]\}\end{aligned}$$

These are shown only for x, but also used for y and t. A special notation is used to indicate shifted time levels

$$\begin{aligned}
\delta_{1/2t} F &= \frac{2}{\Delta t}\{F[(j\Delta x, k\Delta y, (n+1/2)\Delta t] - F(j\Delta x, k\Delta y, n\Delta t)\} \\
F_+ &= F[j\Delta s, k\Delta y, (n+1/2)\Delta t] \\
F_- &= F[j\Delta x, k\Delta y, (n-1/2)\Delta t] \\
\overline{F}^{t/2} &= 1/2\{F[j\Delta x, k\Delta y, (n+1/2)\Delta t] + F[j\Delta x, k\Delta y, n\Delta t]\}
\end{aligned}$$

F_*, $F_\dagger$ are functions at time levels in the range of $(n-1/2)\Delta t$ to $(n+1/2)\Delta t$; its value is generally obtained by iteration.

The second set of difference equations are centered around time level $n+1/2$, and are written

$$\begin{aligned}
&\delta_t + f\overline{\overline{u}} + A(x) + g\delta_y\delta^t + 1/2\,\frac{g}{\rho}\,(\overline{h}^x + \overline{\zeta}^y)\,\delta_y \boldsymbol{p} + R(y)\overline{v}^t \\
&\qquad - \frac{\Theta\rho_a W^2\cos\psi}{\rho(\overline{h}^x + \overline{\zeta}^y)} - \delta_x \boldsymbol{p} - k\nabla^2 v_- = 0 \ at\ j,\ k+1/2,\ n+1/2 \qquad (4)
\end{aligned}$$

$$\delta_{+1/2t}\zeta + \delta_x\left[(\overline{h}^y + \overline{\zeta}^x)\,u\right] + \delta_y\left[(\overline{h}^x + \overline{\zeta}^y_\dagger)\,v_+\right] = 0\ at\ j,\ k+1/2,\ n+1/2 \qquad (5)$$

$$\begin{aligned}
&\delta_{+1/2t}\left[P_i(\overline{\overline{h}}+\zeta)\right] + \delta_x\left[(\overline{h}^y + \overline{\zeta}^x)\,u\overline{P_i}^x\right] + \delta_y\left[(\overline{h}^x + \overline{\zeta_\dagger}^y)\,v_+\overline{P_{i_+}}^y\right] \\
&\qquad - \delta_x\left[(\overline{h}^y + \overline{\zeta}^x)\,D_x\delta_x P_i\right] - \delta_y\left[(\overline{h}^x + \overline{\delta_+}^y)\,D_{y_+}\delta_y P_{i_+}\right] \\
&\qquad + \sum_{l=1}^{i-1}(\overline{\overline{h}}+\zeta)\,K_{il}P_l\alpha_i + \overline{(\overline{\overline{h}}+\zeta)\,K_{ii}P_i}^{t/2} \\
&\qquad + \sum_{l=i+1}^{l_{max}}(\overline{\overline{h}}+\zeta_+)\,K_{il}P_{l_+}\beta_i + S_i = 0\ at\ j,\ k,\ n+1/2 \qquad (6)
\end{aligned}$$

The investigator can use different options for the advection terms, for example

$$A_1(x) = 0 \qquad (7)$$

$$\begin{aligned}
A_2(x) &= 1/3\left\{\left(\overline{u_*}^x\,\overline{u_*}^x\right)_x + 1/2\overline{\left[u_*(u_* + \Delta^2\nabla^2 u_*)\right]_x}^x \right. \\
&\qquad \left. + 2\overline{\overline{v}}\overline{(u_*)_y}^y + \overline{\overline{v}^x(u_*)_y}^y\right\} \qquad (8)
\end{aligned}$$

$$A_3(x) = u_*\overline{(u_*)_x}^x + \overline{\overline{v}}\overline{(u_*)_y}^y \qquad (9)$$

Use of Eq. (7) omits the advection, while use of Eq. (8) introduces conservation of vorticity and vorticity squared in the computation. This is exactly the case if the water depth is not variable, otherwise only by approximation.

Equation (9) is a simple representation of the advection term which has been widely used by many investigators. It is particularly suitable for flow with weak contributions of the advection term. $A_1(y)$, $A_2(y)$ and $A_3(y)$ are computed in a similar manner.

For the bottom stress terms $R(x)$ and $R(y)$, the investigator has two options, here written only for $R(x)$

$$R(x) = g \frac{[(u_*)^2 + (\bar{\bar{v}})^2]^{1/2}}{(\bar{h}^y + \bar{\zeta}^x)(\bar{C}^x)^2} \tag{10}$$

$$R(x) = \frac{\overline{(a_3)}^x \, (\bar{e}^x)^{1/2}}{(\bar{h}^y + \bar{\zeta}^x)} \tag{11}$$

where e is the subgridscale energy intensity.

The subgridscale energy is computed as a constituent. The source term now contains a source and a sink. The source is the energy loss for the main flow and the sink is decay of the turbulent energy.

The energy generation in the water column is a function of the square of the velocity and the square root of the energy intensity

$$S' = a_3\sqrt{e}\ U^2 \tag{12}$$

where U is the the magnitude of the velocity; and a_3 is the constant. This expression is in agreement with the finite difference representation [Eq. (12)] which can be used for the bottom stress computation.

The decay in the water column is

$$DH = C_2He^{3/2}/L \tag{13}$$

where C_2 is the constant; L is the length scale; and H is the temporal depth equal to $(h+\zeta)$.

The length scale of the turbulence is unknown and varying in time and place. To come to a workable model it is assumed that it is a linear function with depth. The source term can then be expressed

$$S = -S' + DH\sqrt{e}\ (-a_3u^2 + a_2e) \tag{14}$$

where a_2 is equal to C_2H/L.

The bottom stress coefficients [Eqs. (10) and (14)] are expected to be the same, thus,

$$a_3\sqrt{e}\ \frac{U}{H} = \frac{g}{C^2 H}\ U^2 \tag{15}$$

and, as in steady flow in a channel with equal depth, energy generation and decay are in equilibrium, we find

$$e = \frac{a_3}{a_2}\ U^2 \tag{16}$$

from which we can derive

$$a_3 = \left(\frac{a_2 g^2}{C^4}\right)^{1/3} \tag{17}$$

With this model of the turbulent energy we find that the average energy intensity in a vertical is dependent in part on the memory of previous events and is governed by the square of the velocities and inversely related to the Chezy value. Thus, increasing roughness increases the subgridscale energy intensity. The decay coefficient a_2 determines the general level of the subgridscale energy in the system and is the only coefficient required for this energy computation in addition to the other variables used for the solution of the hydrodynamic equations.

From experiments it is found that the Chezy value is not truly a constant, but that it is weakly dependent on the depth. We introduce the bottom roughness as a Manning's coefficient and compute periodically the Chezy value by use of

$$C = \frac{l}{n}\ \overline{\overline{H}}^{1/6} \quad \text{(metric system)} \tag{18}$$

Our extensive experimentation with models of estuaries with a considerable horizontal salinity gradient and a sandy bottom indicate that the Chezy value is dependent on the direction of the current as well as the steepness of the gradient.

The following approximation is used optionally for computation

$$C = \frac{\overline{\overline{H}}^{1/6}}{n}\left\{1 + a_1\ \frac{(\overline{u\delta_x s}^{x} + \overline{v\delta_y s}^{y})}{((\overline{u}^x)^2 + (\overline{v}^y)^2)^{1/2}}\right\} \tag{19}$$

Use of Eq. (19) increases the effects of bottom stress during flood and gives a decrease during ebb.

In addition to the advection of momentum in Eqs. (1) and (4), a term is used to compute the dispersion of momentum. The horizontal momentum diffusion is generally small except when density differences over the vertical are present.

The dispersion coefficient is optionally made a spatial variable to be determined during the adjustment phase of a model investigation. Consequently, it is, for example, possible to represent an offshore area where considerable two-dimensional turbulence is present with a different coefficient than the adjacent estuary. In addition to this time-invariant coefficient, another coefficient is introduced optionally which makes the momentum dispersion a function of the local vorticity gradient.

When a sharp spatial gradient of the velocity is present it can be assumed that larger momentum transfers are occurring than we would expect with a weak gradient. Consequently the momentum diffusion coefficient we have introduced in the semi-momentum equation is

$$K = K_o + K' \mid \delta_y w \mid (\Delta S)^2 \tag{20}$$

where

$$w = \delta_y u - \delta_x v \tag{21}$$

In Eq. (4) we use a similar coefficient with the derivative of the velocity in x direction.

Strictly speaking, the dispersion term in the semi-momentum equations should have contained the dispersion coefficient in the first derivative, as it is varying in space. Because of the complexity which would result, the more simple expressions are used. This seems justified, as the spatial variability is generally small and the contribution of the term relatively insensitive to the computation results.

Tidal flats and marshes are approximated in the simulation by taking computational points which represent a certain area into the computation when the water level is flooding. When the area is dry the point does not participate in the flow and transport simulation.

The simulation of these areas presents a number of unusual computational problems which resulted in the design of rather extensive computational procedures. A major problem is that the changes are discrete. When a certain area is taken out of the simulation or when the area floods, the sudden change in the flow generates a small wave, which propagates from its point of origin. Such a wave can cause flooding or drying of adjacent areas, which in turn generates waves. Stability problems can arise in large simulations with extensive tidal flats. In practice, however, this problem can be alleviated by assessing flooding and drying at larger intervals than the computation step. Waves generated then have time to decay. Nevertheless, at each time step a check has to be made for consistency in the equation of continuity.

The design of the boundary change made us aware that this flooding and drying of tidal flats is a process by which energy in the tidal frequencies is transferred to higher frequencies.

In reviewing accurate tide level observations made over a number of ranges across a tidal flat at 500 m intervals, evidence was found that water level variations of considerable magnitude are present during flooding. These water level fluctuations would normally not be recorded, as tide gauges have considerable damping in that frequency range.

With computation procedures in the model, the energy in these high frequency ranges is now simulated by waves in a band which still can be represented by the finite difference equations.

Another major modeling problem was found in the representation of the storm surge barrier. For steady flow it is not difficult to compute the flow through a barrier if the sill height, the width of the barrier opening, the shape coefficient and the water levels on each side of a weir are given. The computation becomes considerably more difficult for time-varying flow. If the formulas for the steady state condition are used in a model with time-varying water levels and subcritical flows, the system can become unstable. As a consequence, the model system contains a barrier computation procedure in which the dynamic effects are accounted for.

Since the model system would be used extensively for the solution of engineering problems concerning a storm surge barrier, the investigator should be able to control, in a time-varying sense, the width of the opening and be able to lower a gate at a certain time with a varying speed. It was also anticipated that the model would be used for simulating the closure of tidal inlets by heightening a submerged dam, so the sill height in the model had to be time-varying too.

To account for the different flow conditions the applicability of certain flow equations was determined from the ratio of the upstream and downstream water level to the height of the underside of the gate in relation to the height of the sill, as shown in Fig. 3.2.

The transition of one flow condition to another is generally not clearly defined, and we introduced gradual transitions for applicability from one formula to another in about the same manner as is being used for one-dimensional computations for the Delta Works.

In the subcritical flow condition [4] with a free surface the downstream water level influences the flow rate. In steady flow, according to the Dutch practice, the discharged can be expressed by

$$Q = \mu_1 B H_2 \sqrt{2g(H_1 - H_2)} \tag{22}$$

where μ_1 is the coefficient; B is the width of the structure; H_1 is the vertical distance between water level upstream and the height of the sill; and, H_2 is the vertical distance between water level downstream and sill.

In the computation algorithm the barrier is designated at a water level point with flow in either u or v direction or with flow in both directions. Downstream of this point the flow is computed by use of the following semi-momentum equation

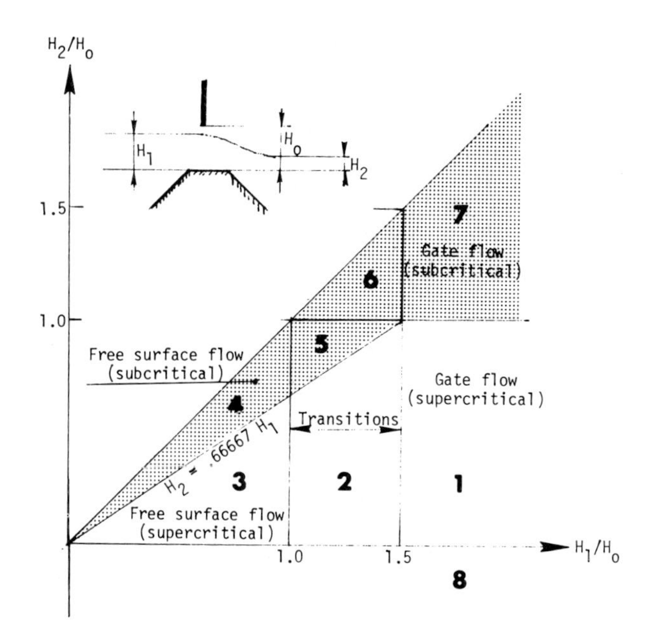

Figure 3.2. Basic flow conditions of the model barrier.

$$\delta_t u + g\overline{\delta_x \zeta}^t + \frac{g\bar{u}^t |u_*|}{(\bar{h}^y + \bar{\zeta})(\vec{C})^2} + \frac{1}{2\Delta x}\left\{\frac{(\bar{h}^y + \bar{\zeta})}{\mu B H_2}\right\} \bar{u}^t |u_*| = 0 \quad at\ j+½,\ k,\ t \tag{23}$$

where B is the effective width; H_2 is the vertical distance between sill and downstream water level; and, $\vec{C}$ is the downstream Chezy value.

This momentum equation has much in common with the normal momentum equation used in the system.

The first term is the acceleration, the second the pressure term, the third the friction loss, and the fourth term represents the gradient due to energy loss over the barrier. We omit coriolis force and the pressure gradient due to density differences; also, atmospheric influences are not included.

In a steady state condition with omission of the bottom stress we would have obtained

$$g\overline{\delta_x \zeta}^t + \frac{1}{2\Delta x}\left\{\frac{(\bar{h}^y + \bar{\zeta})}{\mu_1 B H_2}\right\}^2 \bar{u}^t |u_*| = 0 \tag{24}$$

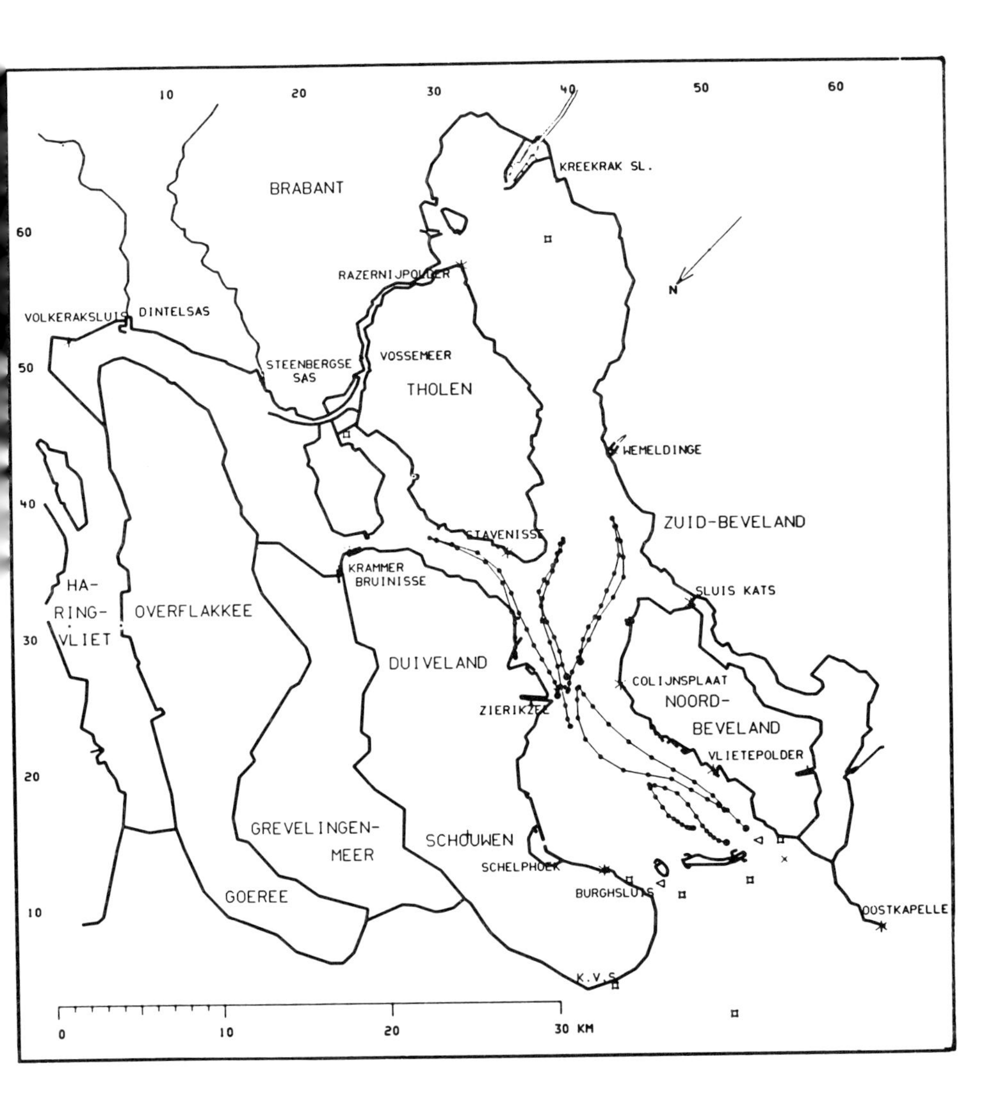

Figure 3.3. Trajectories during one tidal cycle from five particles released simultaneously.

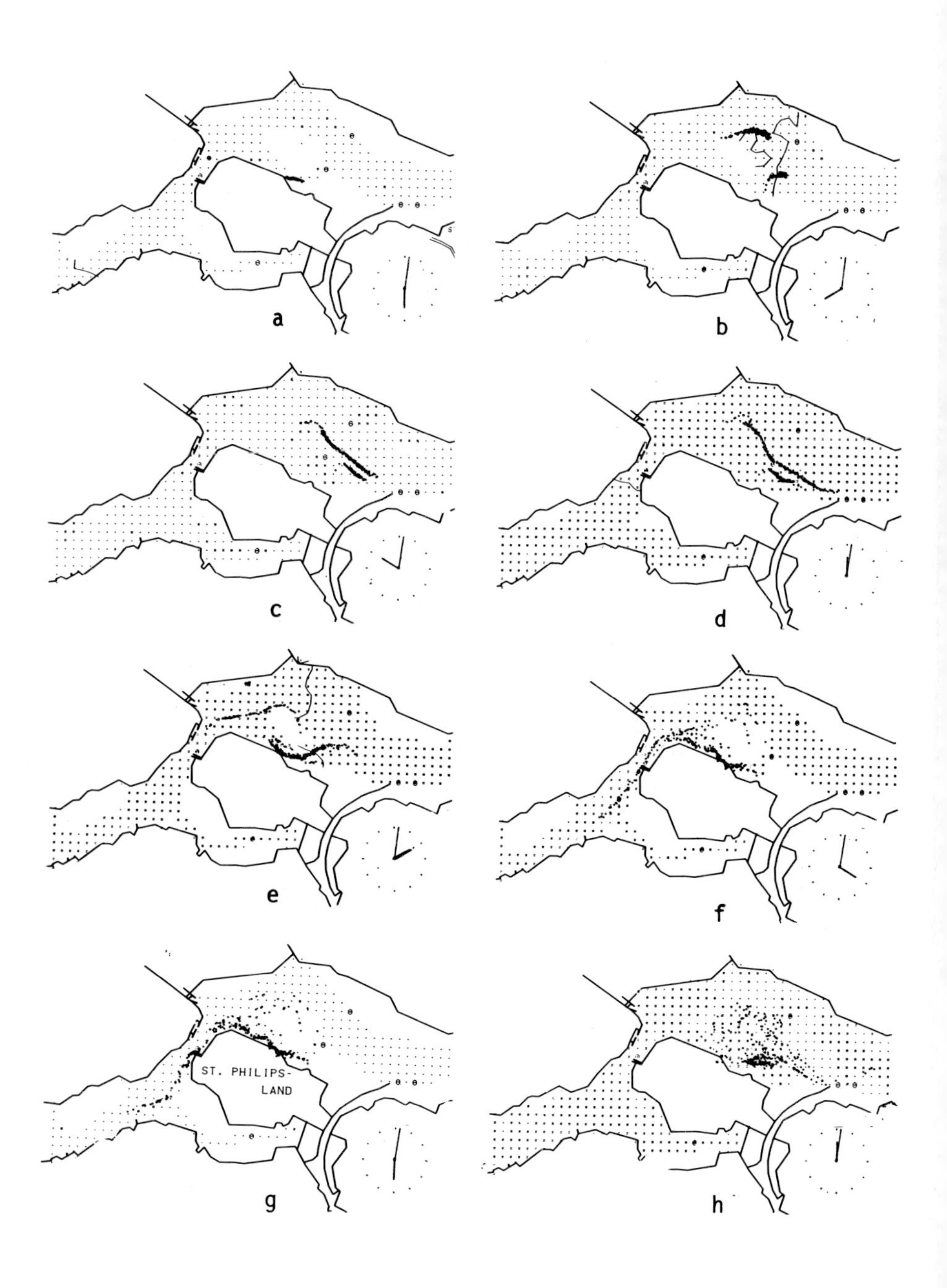

Figure 3.4. Movement and dispersion of a cloud of particles. (Particle distributions are shown at two-hour intervals except that the last distribution is at eighteen hours after the initial release.)

which is the equivalent of the subcritical flow equation [Eq. (22)].

The system has many diagnostic tools. For example, particles can be released at arbitrary times at arbitrary locations and followed over a period of time, and their track can be recorded (Fig. 3.3). It is also possible to release a cloud of particles and follow the movements of each of the particles in the cloud. For each particle, the location is determined from the current field and a stochastic displacement of which the magnitude is determined from the intensity of the subgridscale energy. A random number generator is used to determine the stochastic displacement, and the displacement in the direction of the current and perpendicular to the current direction are uncorrelated and can have different magnitudes. This approach appears particularly attractive for the simulation of dye releases, as no resolution problems occur as with finite difference representations (Fig. 3.4).

Other diagnostic tools which are available are integrations of mass or dissolved substances through particular cross sections in the system or through all sections. The latter integration provides information about net transports if periodic tidal inputs are used. Furthermore, the capability exists to abstract and plot the contributions of all terms of the hydrodynamic equations.

Most important are, however, time histories of all variables and charts of all variables which can be displayed.

4. THE MODELS

Models with different grid sizes are used in the studies. The grid sizes are 1600 m, 800 m and 400 m. Presently the most commonly used models have a grid size of 800 m. These models are:

RDII: This Randdelta model covers the complete model area as shown in Fig. 4.1. The computations are made on a grid of 108 by 147 points. This model has a good resolution for the offshore area. The resolution of the estuaries is poor, as channels are present which are smaller than the grid size. These areas are difficult to adjust, and to make this process less costly, separate models were made of three subregions. The construction of these submodels is relatively simple, as a program is available which constructs the model input of the submodel from the original model. Only boundary conditions have to be inserted separately. The submodels are extensively used for engineering investigations in those experiments which do not influence the boundary conditions.

KTVL II: This submodel covers only the waters of Keeten-Volkerak, Mastgat and Zijpe. This section of the RDII model appeared the most difficult to represent, as extensive tidal flats exist and the system is poorly represented on an 800 m grid. In addition, only partial mixing is present, and the tidal range is large. The dimensions are 35 by 20 points. Figure 4.2 presents computed chlorosity distributions.

OOST: This mode of the Eastern Scheldt includes the KTLV II model. It is the most commonly used model, with dimensions of 62 by 69 points (Fig. 4.3).

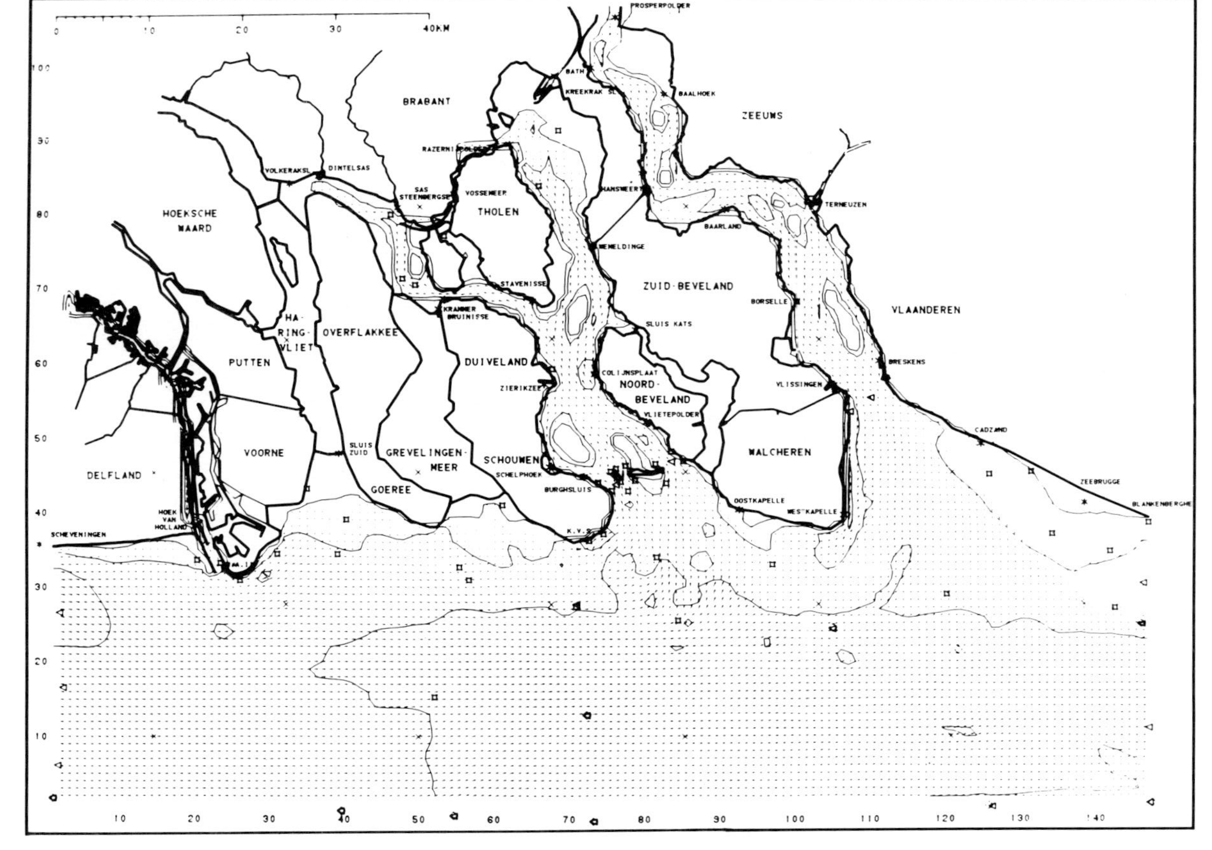

Figure 4.1. Randdelta II Model. (Shown are velocity vectors and the 10, 25, 50, and 75 cm/sec isovelocity contours. The situation on 2 September 1975 at 7:00 hours is shown.)

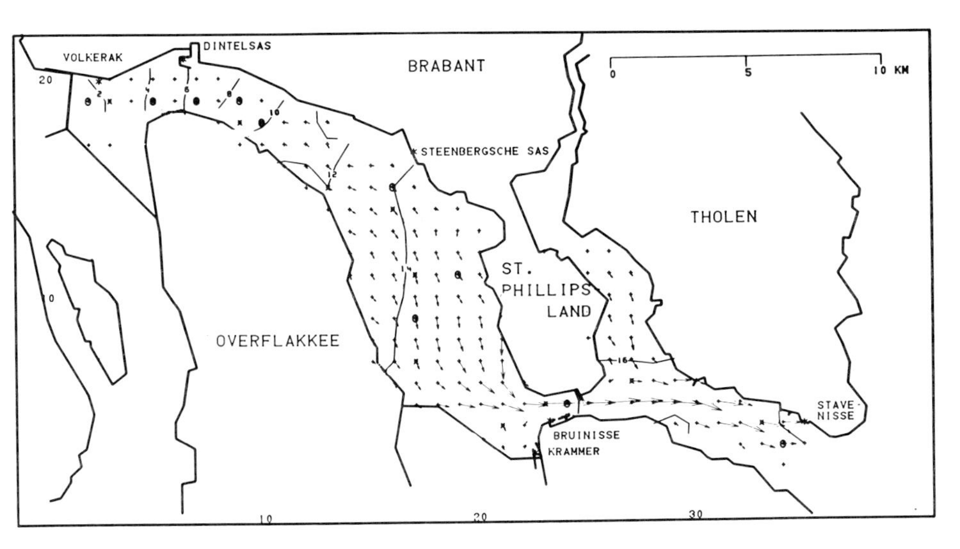

Figure 4.2. Keeten-Volkerak Model. (Shown are velocity vectors and contours of equal salinity.)

WEST: This model is of the Western Scheldt. It has been used mainly for adjustment purposes. Several boundary conditions are used at the upstream end of the model area. Near the border between the Netherlands and Belgium we can use a water level boundary, a velocity boundary, or a channel representation of the Scheldt River on the two-dimensional grid system (Fig. 4.4).

Since experiments with the RDII model are rather expensive in computation and data management, testing of boundary conditions for this model was initially done on a model with a grid size of 1600 m.

To obtain more detailed information, a group of models is being developed which are based upon a 400 m grid. The basic model is the SCHELDES model. This model covers both the Eastern Scheldt and the Western Scheldt, and has grid dimensions of 212 × 112 points. Submodels are being made; one submodel (KTVL I) is already adjusted and verified (Fig. 4.5).

5. RDII MODEL SETUP

The tide boundaries were obtained from a 16-day survey in which water pressures and water levels in the offshore area were measured. The survey group computed the water levels from the observed pressures, taking into account the barometric pressure variations, sea water temperatures, salinity, and bottom settling. The reference levels of these time series were computed from water level

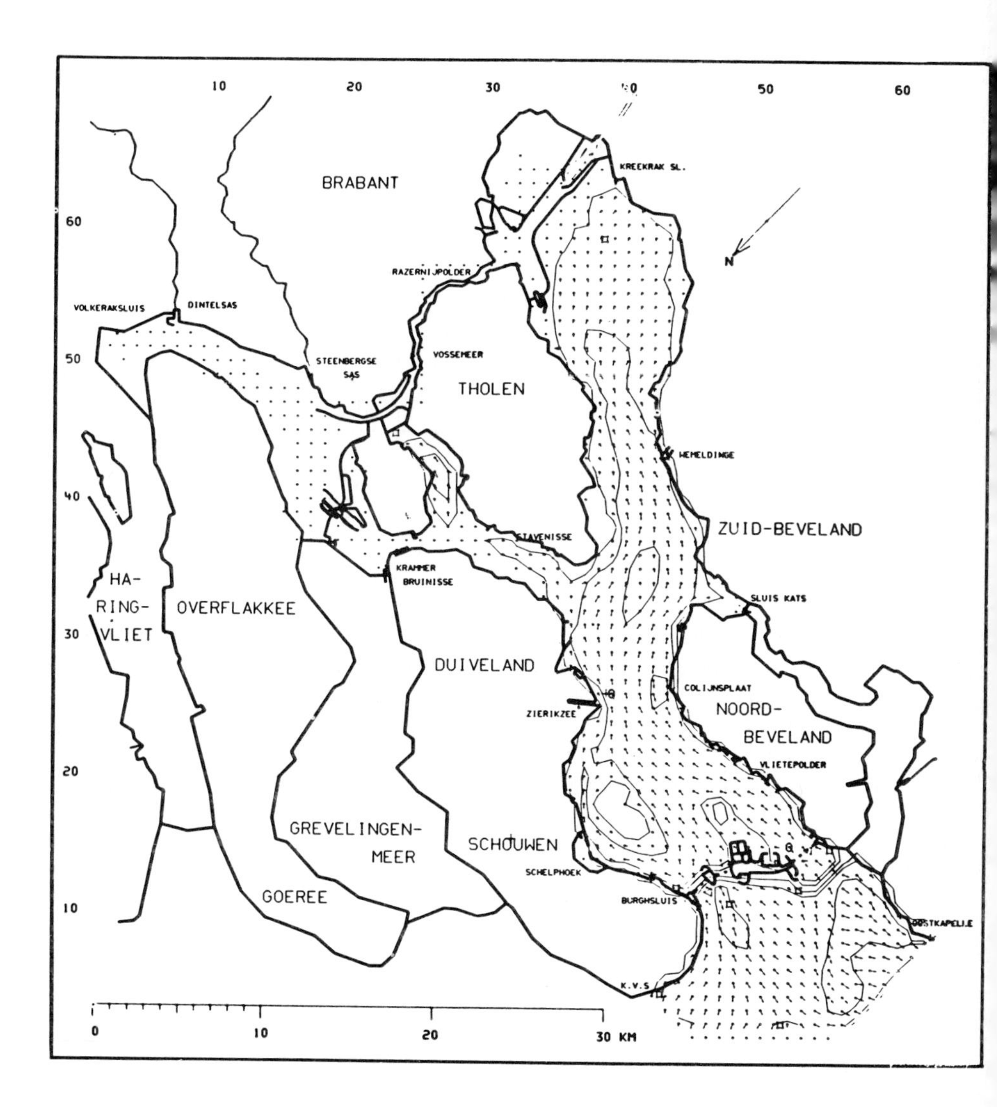

Figure 4.3. Oosterschelde Model. (Shown are velocity vectors, locations of points which are under water, and the 10 and 25 cm isovelocity contours. Simulation with barrier in operation and the Oesterdam and Philipsdam closed.)

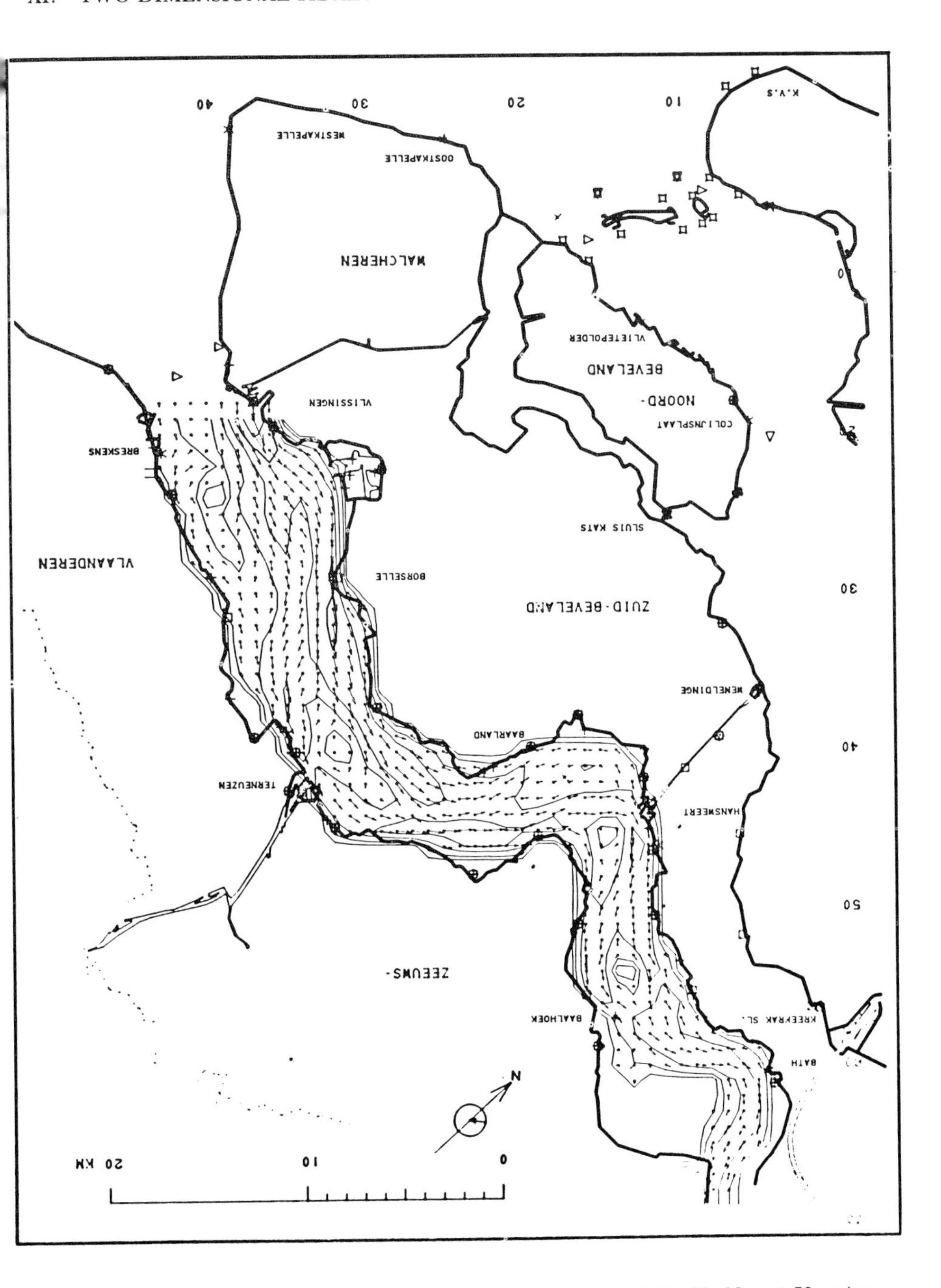

Figure 4.4. Westernschelt Model. (Shown are velocity vectors and the 10, 25, and 75 cm/sec isovelocity contours.)

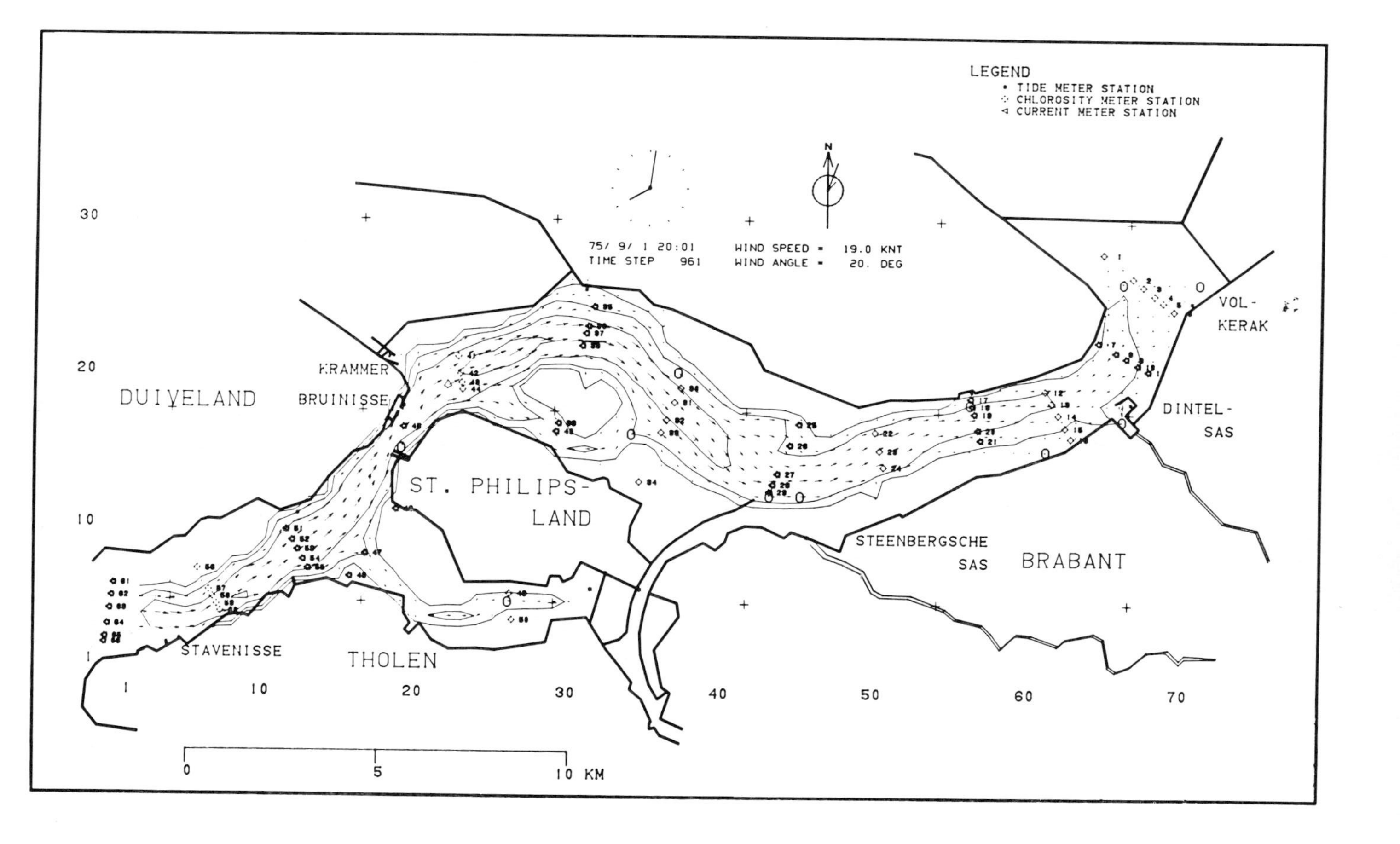

Figure 4.5. Keeten-Volkerak Model (400 m grid). (Shown are the isovelocity contours of 10, 25, 50, and 75 cm/sec and the velocity vectors.)

recording of shore stations by use of observed velocities in ranges perpendicular to the shore which contained the pressure recorders (Van de Ree, *et al.* 1978).

Even though the field data as received by the modeling group was of a high quality, particularly if one considers the wave action in the relatively shallow area and the extensive shipping traffic, a considerable amount of noise was present in the water levels to be used as boundary conditions of the model.

In the first experiments we used the data as processed by the survey group directly and made linear interpolations of the water levels between the boundary stations. This appeared completely unacceptable. Random in- and outgoing currents would be generated across the boundary, caused by the estimated errors of only a few centimeters in the data (Fig. 5.1).

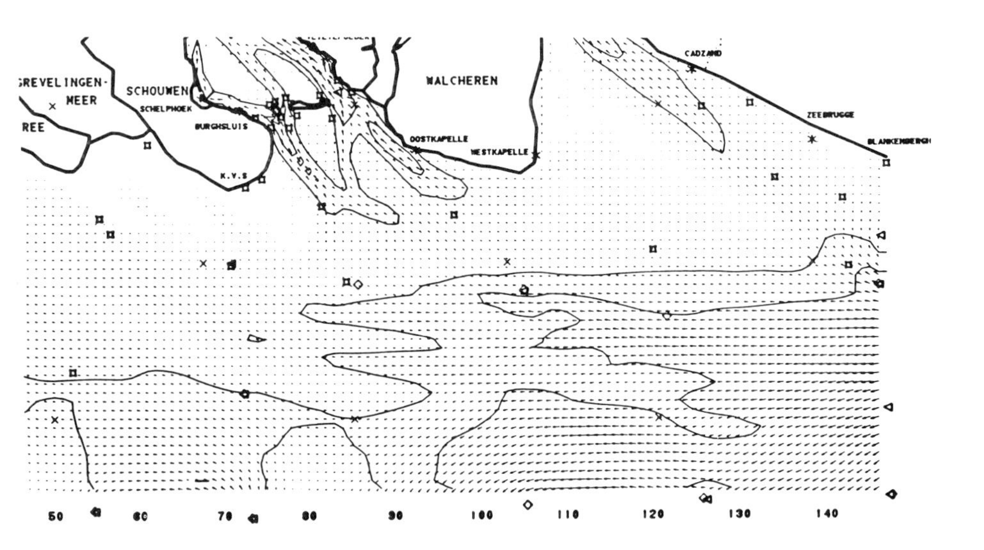

Figure 5.1. Computed flow field near boundary obtained directly from measurement.

Subsequently we made a cross spectral analysis of each boundary station to three adjacent fixed stations which were either a shore station or a water level station on an offshore instrument tower. From this cross-spectral analysis response functions were obtained and the water level at the boundary station could be predicted with a linear model from the fixed station. Since the fixed stations were more accurate and contained less noise, the hindcasted water levels contained less noise than the original records. In the model we used the weighted average of the three predictions with a weighting factor inversely related to the distance of the cross-spectral analysis. Since adjacent boundary stations used predictions obtained from at least two of the same fixed stations, a smooth boundary relatively free of errors could be computed (Fig. 5.2). The reference levels were checked by filtering all boundary records with a low pass filter and comparing the mean levels.

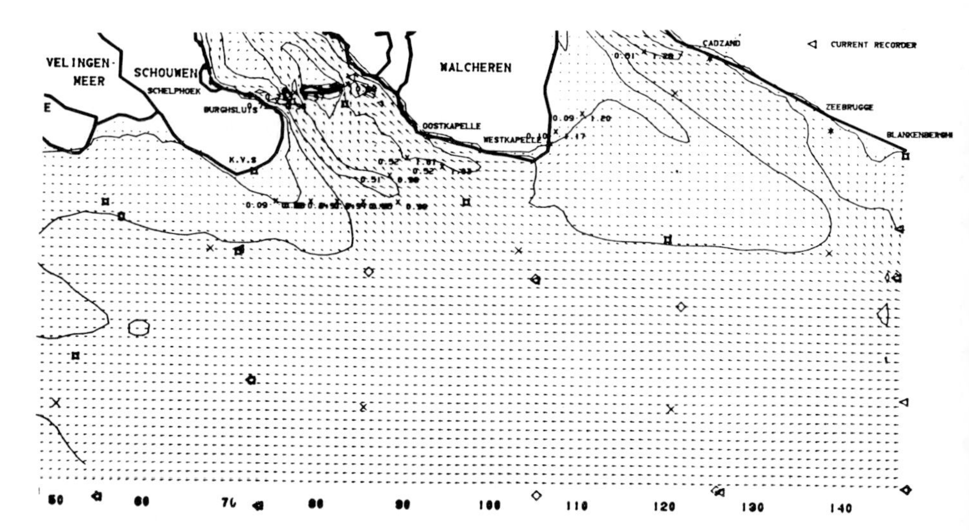

Figure 5.2. Computed flow field near boundary obtained by hindcasts.

The major difficulty we experienced was at the northeastern boundary of the model. The discharge of the Rhine River causes large variations in density at this boundary, and the water level station at Scheveningen was influenced to a considerable extent by a local fresh water discharge which influenced the density. A more complete description of this boundary determination is found in Van de Ree, *et al*, (1978).

From the available processed time series of the boundary a 125-hour period was selected as the period of simulation. In this period in the beginning of September 1975, measurements were made of discharges in several ranges across the Eastern and Western Scheldt which could be used for adjustment and verification of the model.

At the open boundaries a linear interpolation of amplitudes and phases of the Fourier components at the boundary station was used as input. This procedure generated a smooth boundary at every time step of the simulation.

For the adjustment experiments a velocity boundary was used at the Scheldt boundary, as such a condition would induce large water level errors in the Western Scheldt if the Manning's coefficients were improper. After this adjustment we used initially the water level as a boundary and subsequently an approximation of the flow and impedance characteristic of the river was made with a channel representation on the two-dimensional grid. Similarly a channel representation was made for New Waterway to Rotterdam and beyond. Presently one-dimensional river representations are being prepared.

The initial Manning's coefficients were estimated in part from the height of sand waves evident in the hydrographic surveys and in part on the basis of other one- and two-dimensional model results. The propagation of the tide in the estuaries appeared very sensitive to the approximation of the depth. Cross sections obtained from the hydrographic surveys were drawn on all ranges of the two-dimensional model, and subsequently the model cross sections were estimated in such a manner that cross sections in the model and prototype were essentially the same (Fig. 5.3). This is the basis of the finite difference approximation [Eqs. (1) through (19)] and is in contrast with approximations by others which emphasize the mean depth.

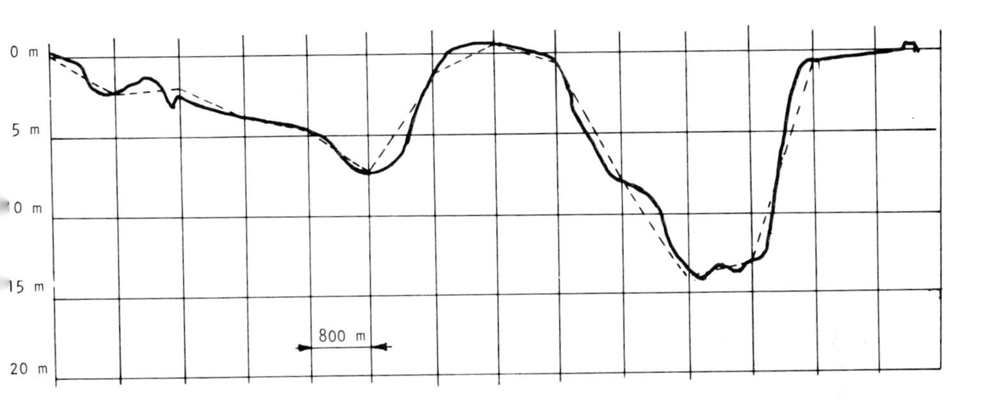

Figure 5.3. Sample cross-section prototype and model taken from working drawings.

To check the data input, all time-dependent inputs, all spatial variables, and all locations where model abstractions are made were plotted by the graphic display system for visual inspection.

Of particular importance is the check on the depth representation on the progress of flooding and drying of tidal flats and marshes. The display system permits mapping of areas which are under water at given levels. By drawing flooding charts of a mean water level at 0.5 m intervals, it is possible to check if all main channels can transport during ebb.

Large datasets are required for producing the spatial displays of velocity, transport, constituent distributions, and the water levels. Map outlines have to be digitized and titles and labels have to be located on their proper spots. The instructions are simple, but, due to large amounts of data, quite extensive. It appears that the preparer of such data should have an engineering background to produce maps and charts acceptable in the engineering profession.

6. ADJUSTMENT

Adjustments initially were made based upon curve fitting and determination of the standard deviation of computed and observed data. In the later phases of the investigation cross-spectral analysis was used, using the same time spans in model and prototype.

The parameters used in the adjustment are the depth, the Manning's number coefficient, the coefficient which determines the change in the Chezy coefficient as a function of the salinity gradient, and the viscosity coefficients. The wind stress coefficient was not adjusted. For the salinity we used summer average conditions with adjustments in those areas where data was available during the survey period.

The first step in the process was to see if the flow patterns were proper. We verified if the current variations in the channels were not large. This was checked by computing and printing velocity magnitudes at the water level points, at the depth points, and at locations midway between the u and v points. If large variations did occur, the local approximation of the cross-section was reviewed in the cross-sectional drawing and corrections made. This procedure is quite effective and consistent with the physical processes, which in natural channels would not maintain such variations.

A large number of sensitivity tests were made. For example, we determined how long the memory of the system is, and the sensitivity of depth changes on amplitudes and phases of the tide propagation.

Sensitivity tests were also made of the influence of the time step on amplitudes and phases of the tide. Increasing the time step increases the amplitude gradually and causes phase lag in the interior of the estuaries, as could be expected from an analysis of the propagation factor of the finite difference scheme. Time steps of 1.25 min were used for the more important simulations. This represents a Courant speed of 2 to 4 for the main part of the system (Figs. 6.1 and 6.2). We also made sensitivity tests of the viscosity coefficients. Increasing the viscosity reduced the amplitudes in the Eastern Scheldt. In the Western Scheldt increasing the viscosity causes the amplitudes to decrease (Fig. 6.3). In the figure the water level of one of the computations is also shown to indicate the phase relation. The water level records were low pass filtered to eliminate oscillations with periods smaller than one hour to make amplitude and phase relations clearer. Note that M_4 is more suppressed than M_2 by the viscosity terms. With the viscosity coefficients only minor adjustments in the amplitudes can be made. The momentum diffusion acts mainly as a weak low pass filter during computation, suppressing the higher frequency oscillations caused by the submergence of the tidal flats.

The choice of the advection term [Eq. (8) and (9)] did not significantly influence the generation of the overtides; however, computation by use of Eq. (8) appeared to have long-duration stability. One of the submodels was stable when a period of about two months was simulated. The use of Eq. (9) was stable for only a few tidal cycles when a first order approximation was used for the time expression in u_*. After a few tidal cycles nonlinear instabilities were generated by generation of eddies in the flow field. The Arakawa expression [Eq. (8)], which

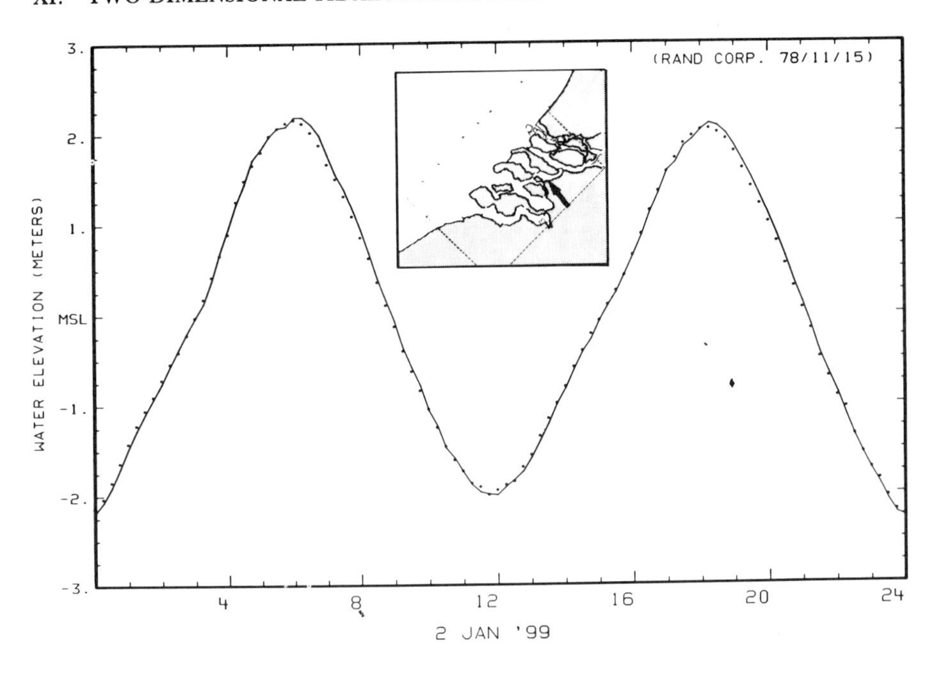

Figure 6.1. Computer water levels with a timestep of 1.25 min (···) and 2.5 min (—) .

conserves vorticity and the squared vorticity in constant depth, prevented this occurrence.

The main adjustment of the tidal amplitudes and phases was accomplished by use of the Manning's value (Fig. 6.4). Extensive use was made of the submodels, not only for the Manning's coefficients, but also for the adjustment of the bathymetry and for the dispersion coefficients in the water quality part. The adjustment of the hydrodynamics and the water quality has to be made simultaneously, as the density and salinity distributions and the diffusion coefficient have influence on the tidal propagation.

The KTVL II model was operated for a several-month (real time) simulation to adjust the diffusion coefficient in the partially mixed sections (Fig 6.5). Comparisons were made of quasi-equilibrium salinity distributions measured in a field experiment and in the model. The quasi-equilibrium condition was obtained by maintaining an approximate constant flow of 50 m^3/sec in the system. These tests indicated that the computations are conservative. The total inflow over one tidal cycle matched completely the integrated time varying flow at the new boundary of the model, as we expected from the conservative properties of the finite difference scheme. To assess the influence of machine truncation, simulations were made with the same programs and same inputs on a 32 and 36 bit machine. The

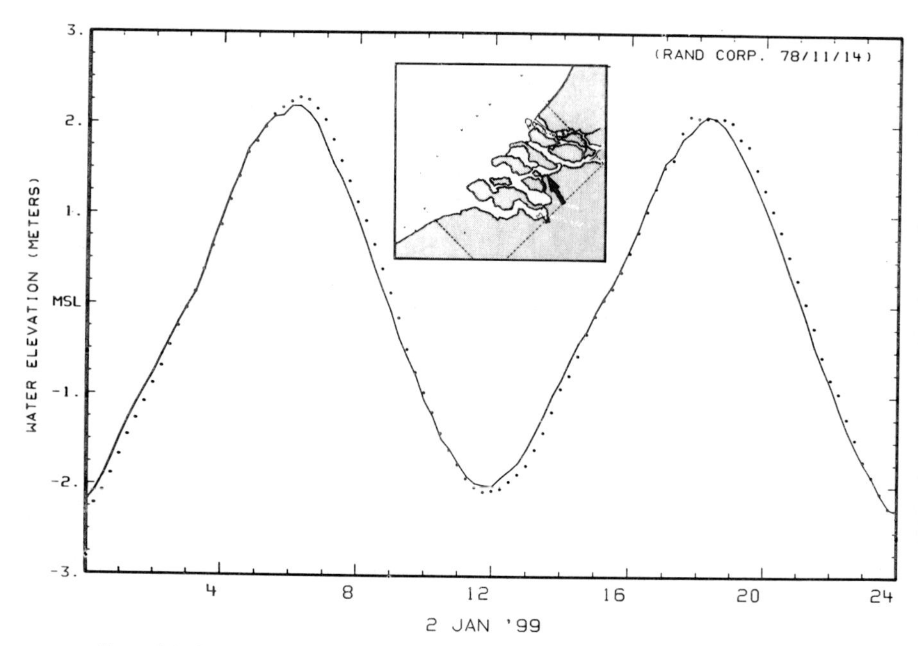

Figure 6.2. Computed water levels with a timestep of 2.5 min (—) and 5 min (···).

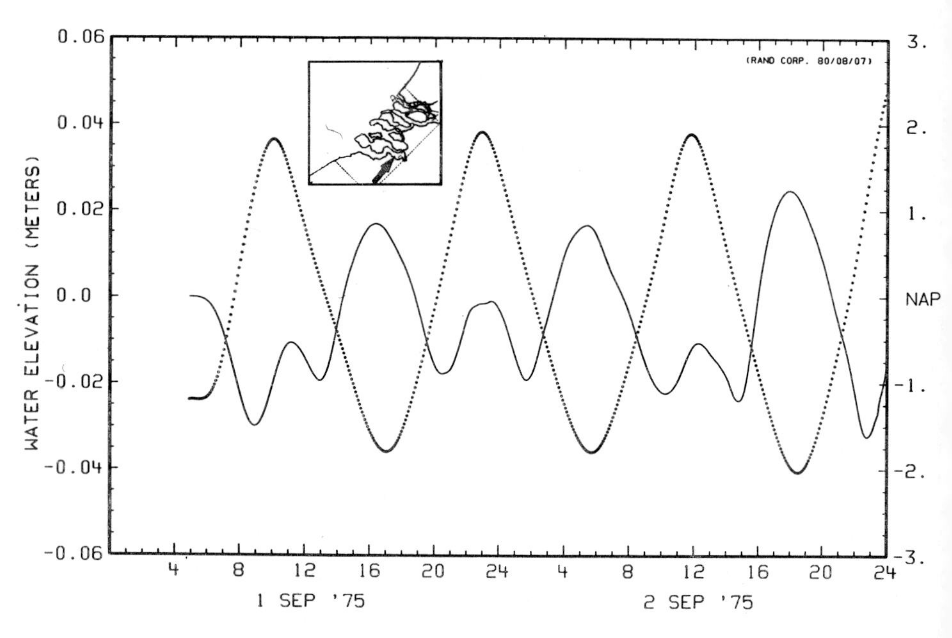

Figure 6.3. Difference (—) between the water levels computed by use of a viscosity of 10 m^2/sec and no viscosity (left scale). The water level (···) of one of the computations is also shown in the graph (right scale).

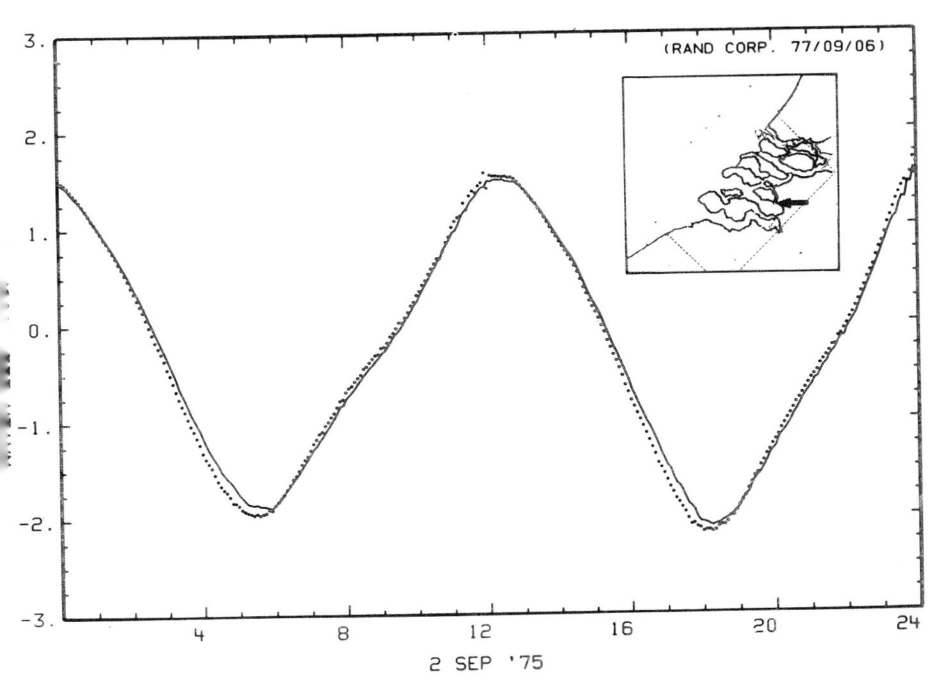

Figure 6.4. Comparison of water levels computed with a 10% difference in Manning's n value.

simulation results of the water levels were identical in the practical significant decimal. Concentration differences were somewhat larger being of an order of 10^{-4} after a day's simulation. It was initially not necessary to make an adjustment of the diffusion coefficients from our original estimate of 10 m^2/sec in those parts of the system where the waters are well mixed. Later, when dye release experiments were made, this value appeared too high, probably due to numerical dispersion by steep gradients [see Chap. 12].

7. MODEL VALIDATION

For the validation of the model data at the end of the five-day simulation period was used. The agreement between observed and computed water levels was good. The standard deviations for nearly all stations were only 1 to 2 cm larger than during adjustment. Typically standard deviations were 5 to 6 cm (Fig. 7.1). This is particularly favorable taking into account that the accuracy of the boundary inputs is about 4 cm and the accuracy of the shore-based water level gauges is 2 to 4 cm.

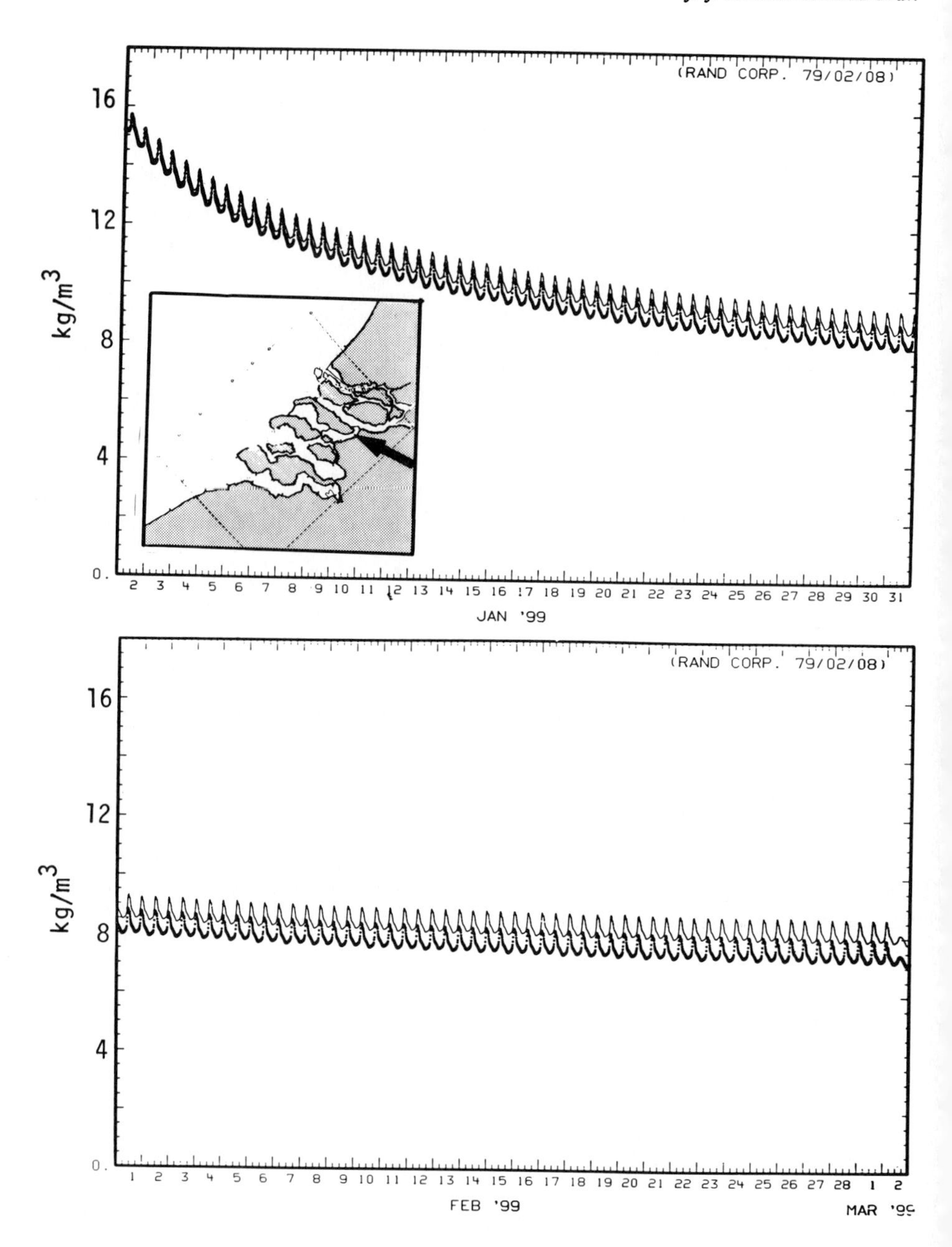

Figure 6.5. Transient salinity after initiation of fresh water flow into the KTVL II Model with homogeneous salinity at start for two slightly different distributions of the dispersion coefficient.

Of particular importance were the comparisons of the observed and computed currents and mass transports through three sections. Fig. 7.2 shows the observed and mass transports through three sections in Eastern Scheldt. The accuracy of the observed transports is estimated to be 10 to 15 percent.

Not only the flow in the Eastern Scheldt is well represented by the model but also the flow field offshore. Figures 7.3 and 7.4 present computed and observed velocity components parallel to the shore. The observed currents were recorded approximately at mid-depth and do not necessarily represent completely the vertically averaged current which is computed in the model.

Since the model is being used for the study of phenomena with frequencies lower than the tide such as for water quality and storm surges, comparisons were also made of low pass filtered water level and current data. Figure 7.5 represents the variations of the mean water levels at stations in the Eastern Scheldt. The relatively large variations are partly caused by external influences, mostly by local wind and density effects. A filter is used in these graphs, which does not completely eliminate the semidiurnal tide, as otherwise too much data has to be eliminated at the end of the time series.

The comparison of the observed and computed mean currents is also very favorable as shown in Figs. 7.6 through 7.9. The mismatch at the beginning of all these time series is caused by the initiation of the model from a condition at rest. The low frequency current variations are in part generated outside the model and are generated by the model by the water level boundary inputs. In part these variations are caused by wind in the model area and also by the density differences. We used a time varying but spatially invariant wind field for the adjustment phase and the validation computation. Since an offshore station is used for the input wind data, it can be expected that the wind stresses in the estuaries are overestimated.

The validation of the proper representation of the salinity distributions in the Eastern Scheldt is based upon experiments described in another paper at this symposium [Chap. 12]. The partially mixed section was validated on basis of a simulation with a duration of two months real time during which 50 m^3/sec was discharged in the Keeten-Volkerak The results were compared with field data from a period in which the same prototype discharge was obtained.

For the offshore area, we found that after five days water of the New Waterway approached the Eastern Scheldt (Fig. 7.10). It was fortunate that a sequence of wind data similar to the one used in the five-day simulation could be found for which salinity distributions offshore the Eastern Scheldt were measured (Fig. 7.11). From the insights gained in the physical processes by this model investigation and from simulation results it is concluded that simulations can be made with an accuracy of about 5 to 6 cm for the water heights. We concluded also that the tidal heights and water movements in the estuary are influenced to a considerable extent by the salinity distributions. The largest uncertainty in making predictions is the roughness of the bottom, which will change when the tidal flows in channels are reduced. This uncertainty is of much more concern to the investigators than the computational properties of the finite difference model used.

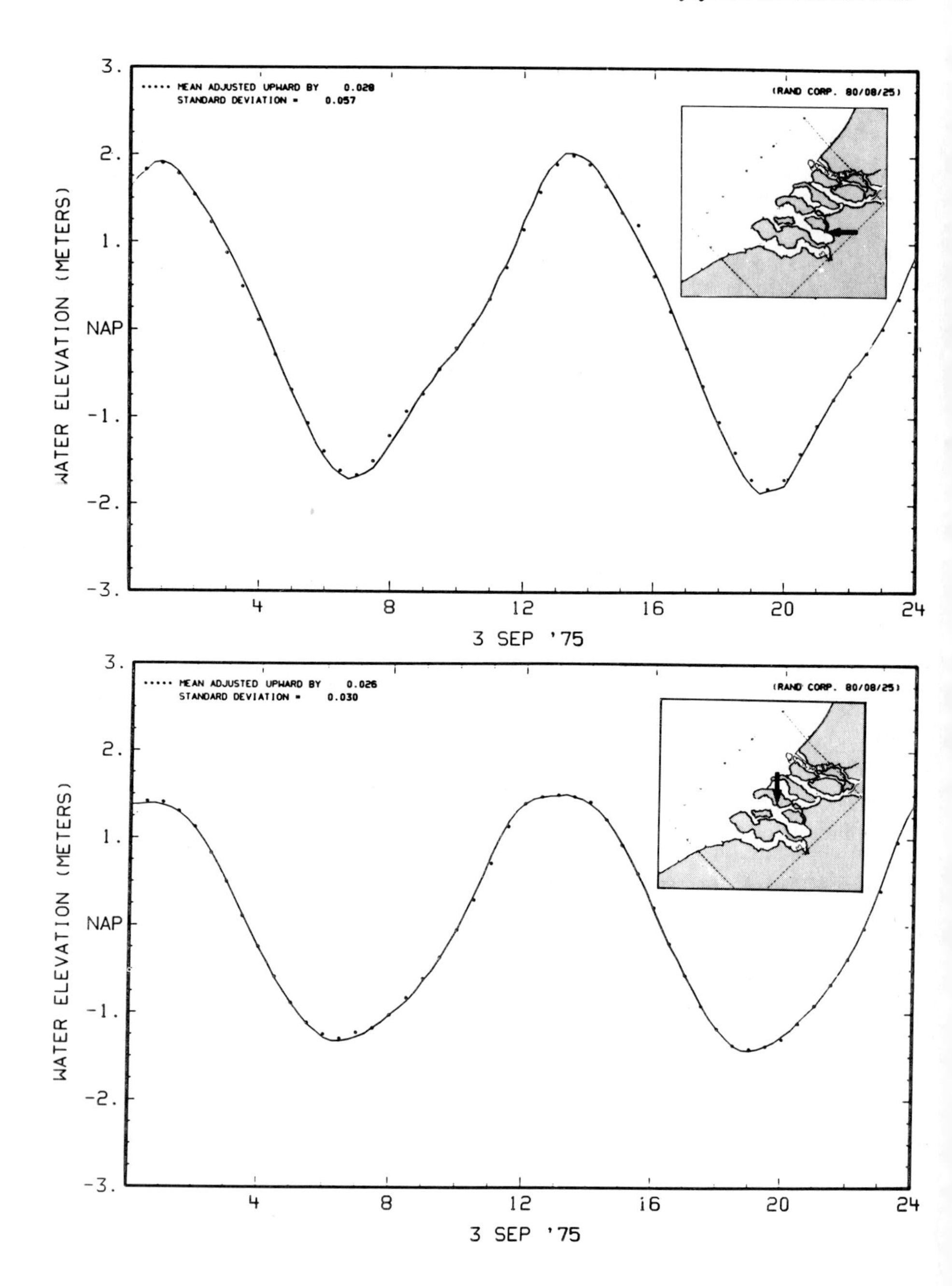

Figure 7.1. Validation results at two water level stations in the Oosterschelde [observed (···), computed (—)].

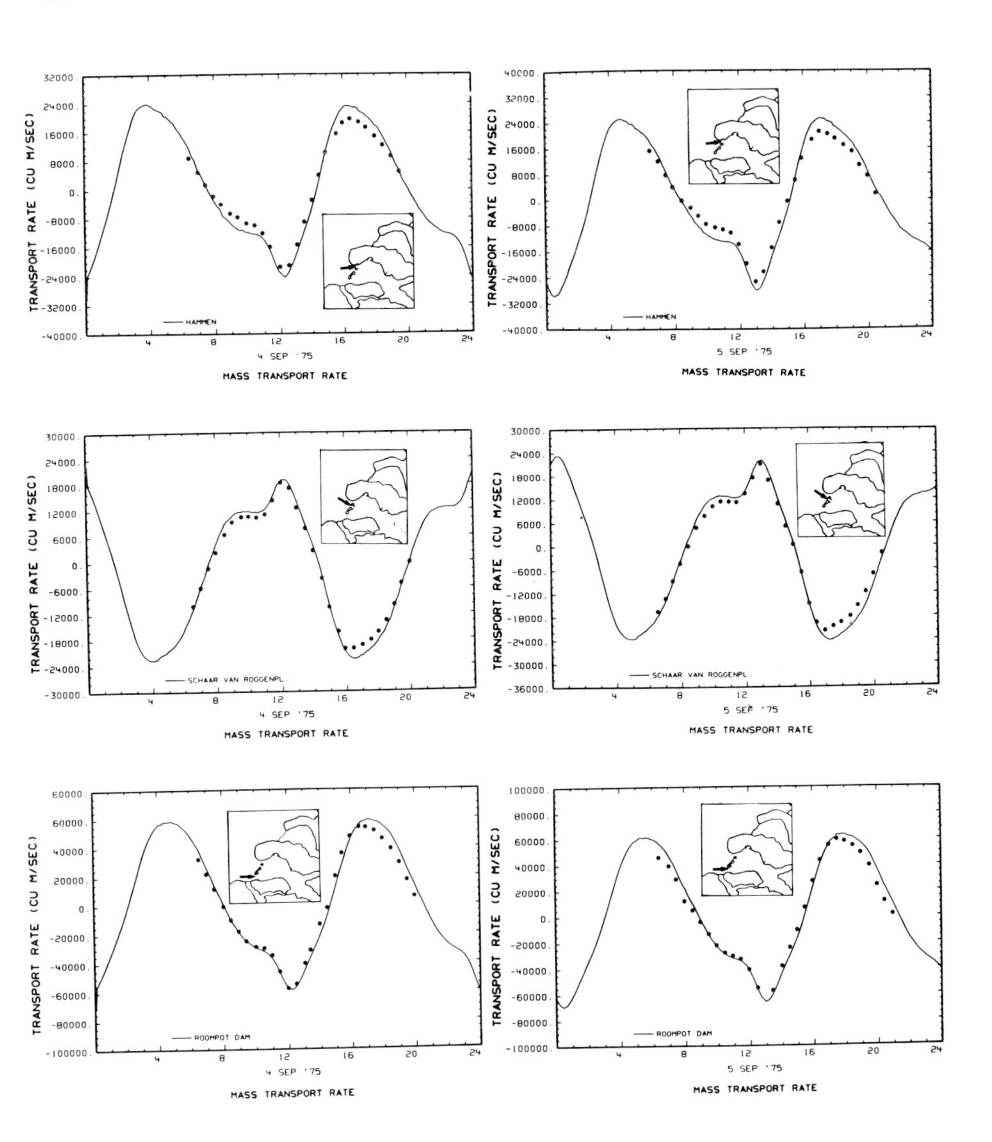

Figure 7.2. Comparison of observed (···) and computed (—) flow through tidal channel cross-sections used for validation of the Oosterschelde Model.

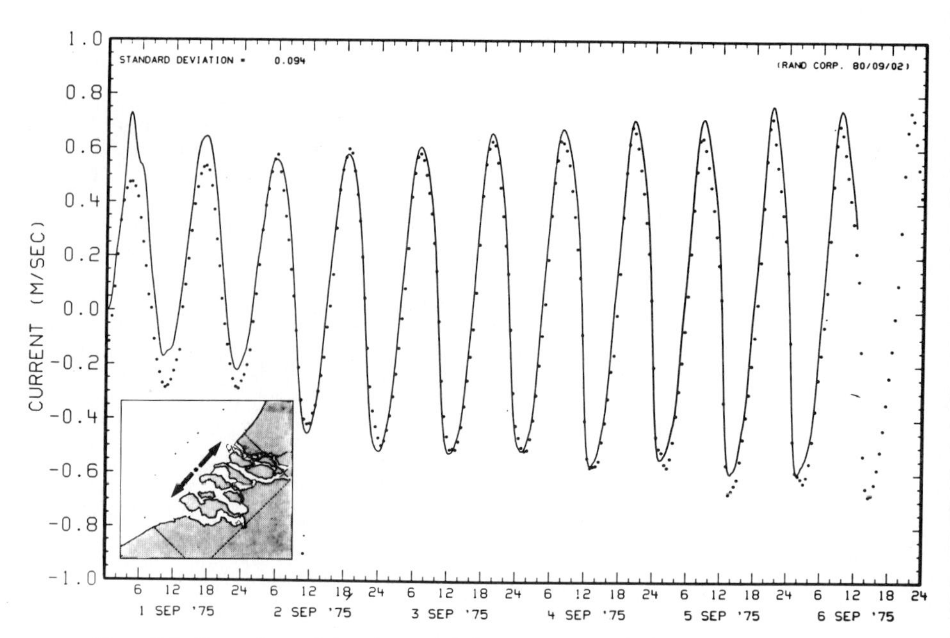

Figure 7.3. Comparison of the observed (···) and computed (—) velocity component which is parallel with the shore at a station in the middle of the model area.

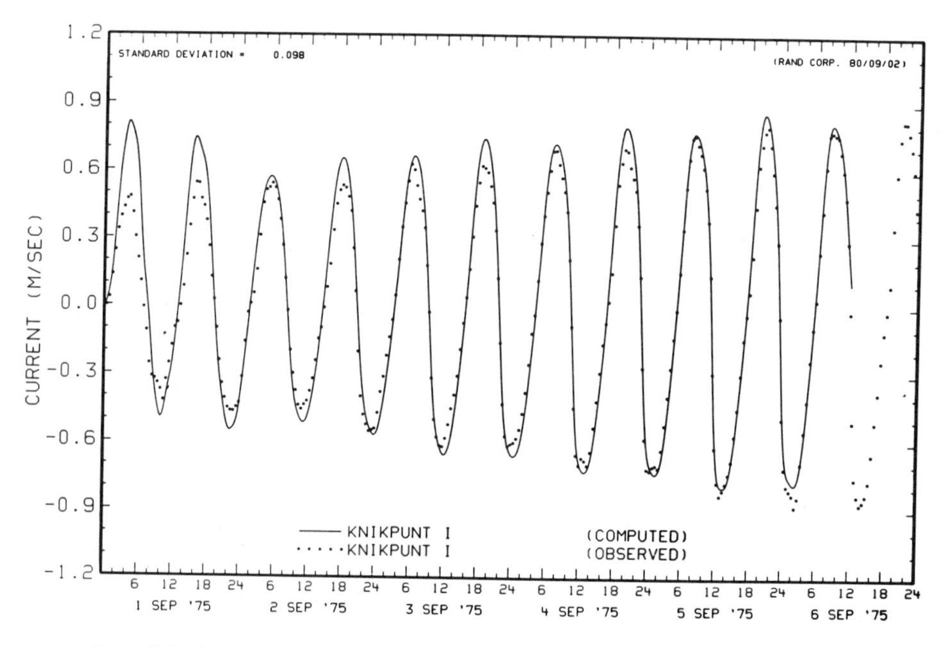

Figure 7.4. Comparison of the observed (···) and computed (—) velocity which is parallel with the shore at the most westerly point of the model area.

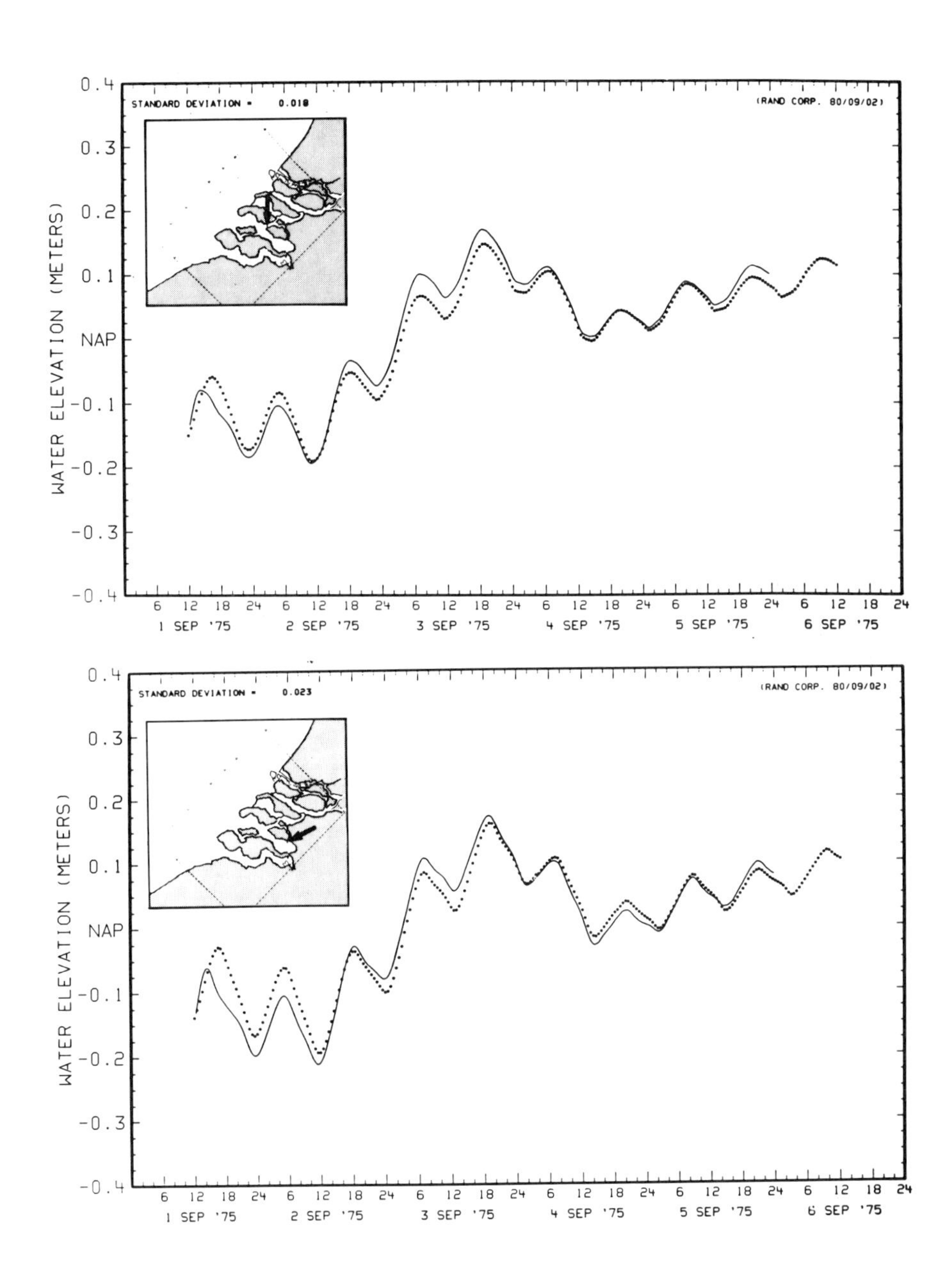

Figure 7.5. Variations of the computed (—) and observed (···) mean water levels at two stations in the Eastern Scheldt.

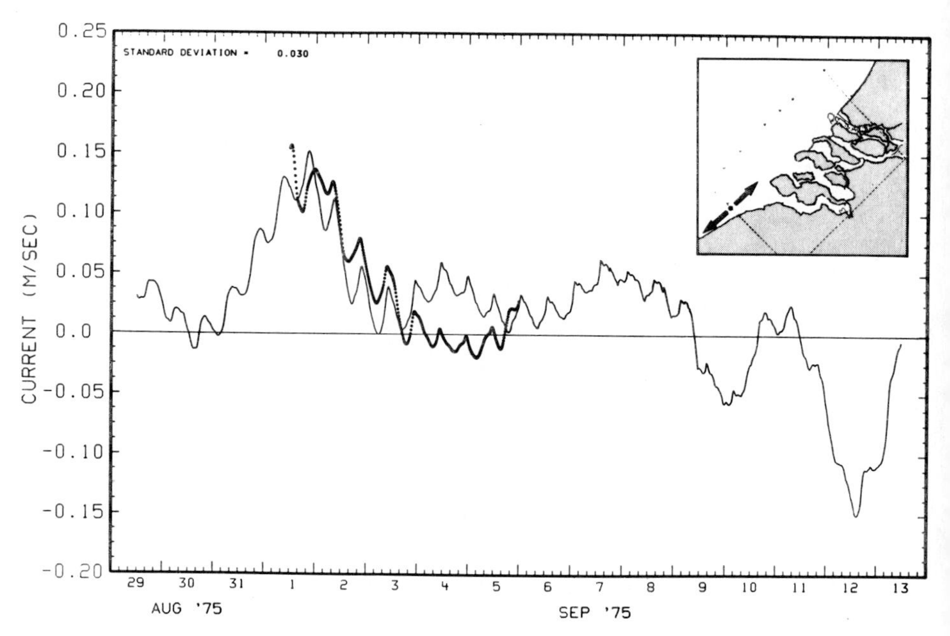

Figure 7.6. Low frequency variations of the observed (—) and computed (···) current component which is parallel to the shore at a station near the southerly boundary in the model area.

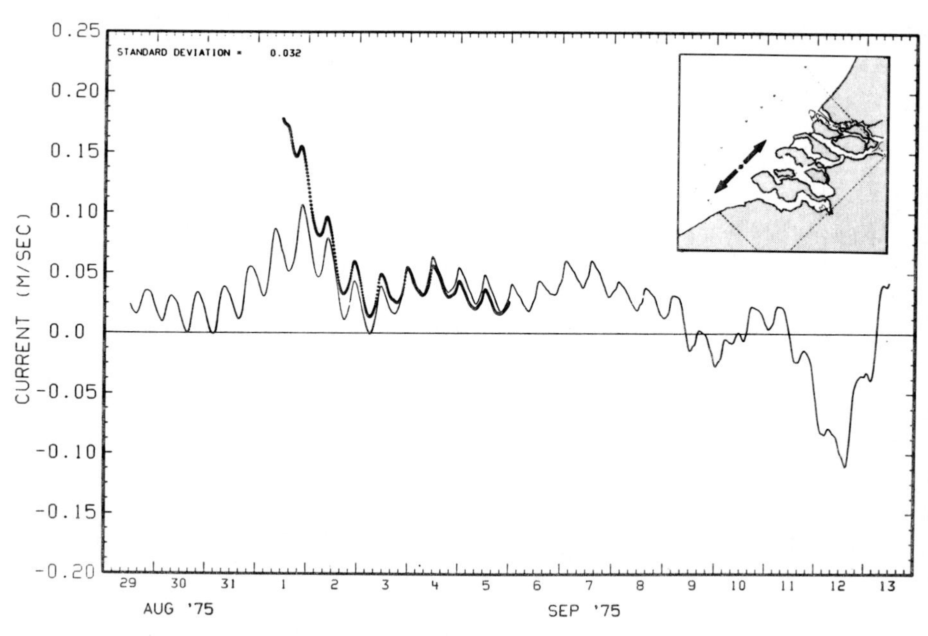

Figure 7.7. Low frequency variations of the observed (—) and computed (···) current component which is parallel to the shore at an offshore station.

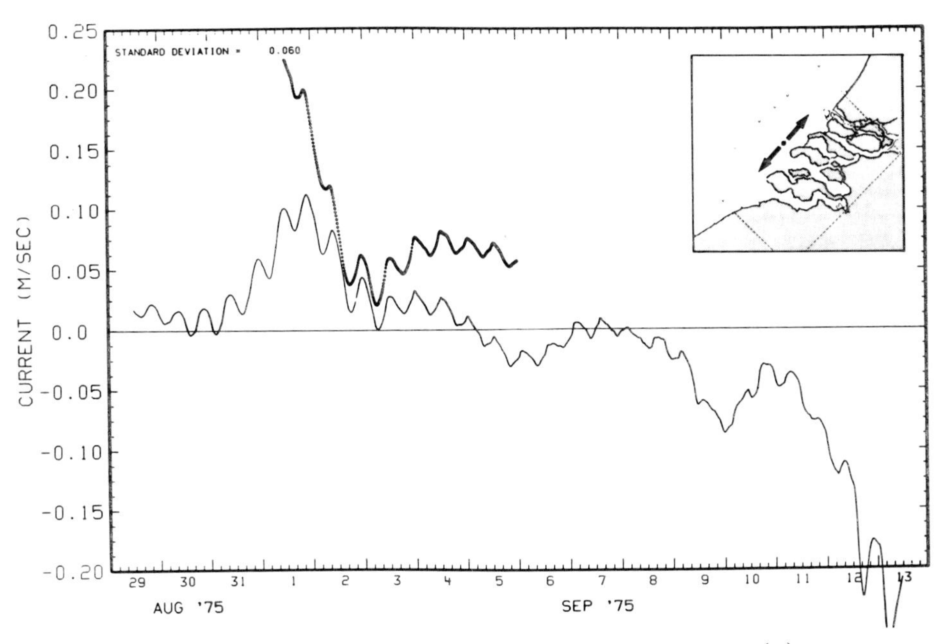

Figure 7.8. Low frequency variations of the observed (—) and computed (···) current component which is parallel with the shore at a relatively shallow station in the offshore area.

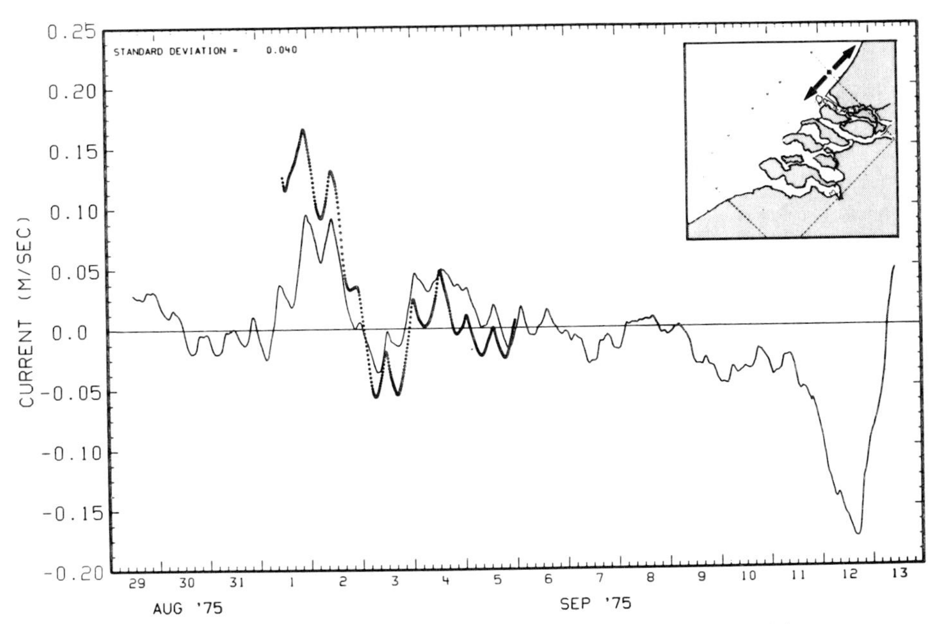

Figure 7.9. Low frequency variations of the observed (—) and computed (···) current component which is parallel with the shore at a station at the northerly boundary of the model area.

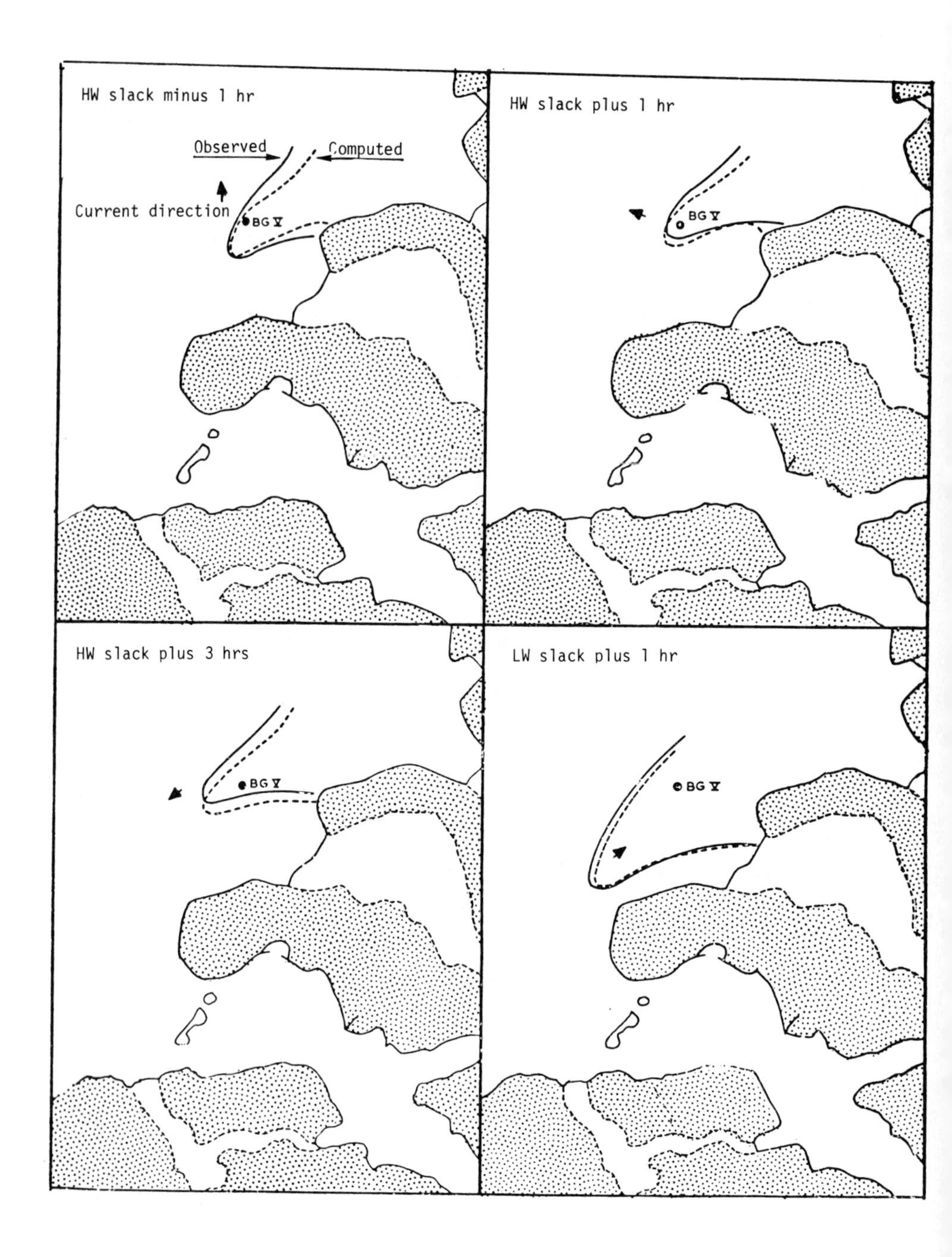

Figure 7.10. Comparison between observed and computed chlorosity distributions.

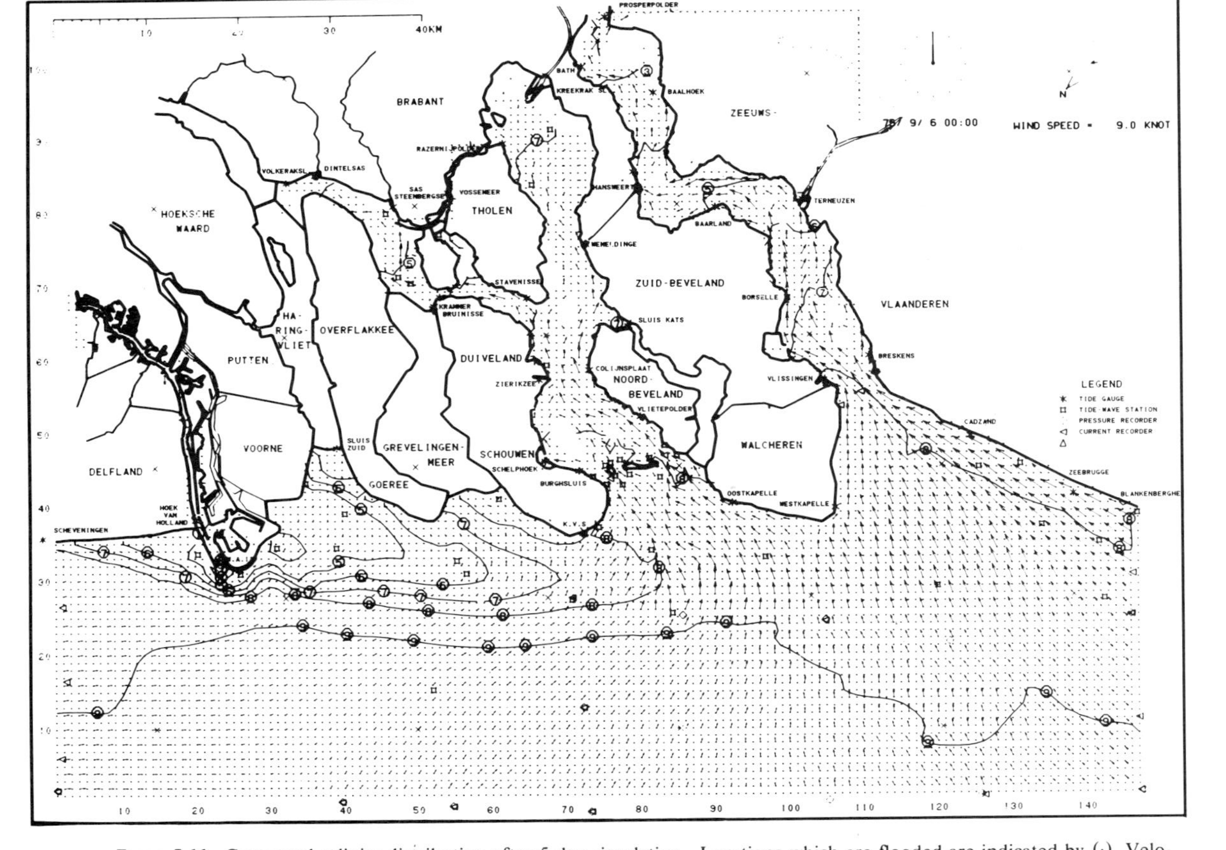

Figure 7.11. Computed salinity distribution after 5-day simulation. Locations which are flooded are indicated by (·). Velocity vectors plotted every other grid point in each direction. The velocity vector scale is one grid unit for a velocity of 0.5 m/sec. The salinity isocontours are: (1) 8 kg/m^3, (2) 12 kg/m^3, (3) 16 kg/m^3, (4) 20 kg/m^3, (5) 24 kg/m^3, (6) 26 kg/m^3, (7) 28 kg/m^3, (8) 30 kg/m^3, (9) 31.5 kg/m^3.

8. EXPERIMENTS

The model and its submodels are being used for a number of different purposes, all related to the execution of the storm surge barrier and the secondary dams in the system. Many experiments were made during the adjustment phase when model results were required and it was considered that the model was already validated with relation to the problem to be solved. In many instances these experiments referred to the study of alternatives by making comparisons between those cases.

Of particular interest is that the model was used to gain insight or confirmation of assumptions made in studies with other models. It was also used for validation of other models. For example, the one-dimensional model of the Oosterschelde uses a boundary relatively close to the mouth as does the two-dimensional model and the hydraulic model. The construction of the barrier would change the boundary condition of these models and by comparing two simulations for present and future conditions an assessment of change of the boundaries of those models could be made. Figure 8.1 represents the water level at one of the boundaries of the one-dimensional model for the present condition and a future condition.

As an example of the many experiments which are being made we refer to the next paper of this symposium.

9. CONCLUDING REMARKS

The major part of the work was done by two task groups, one group for building the program system for data management, simulation and representation. The other group was mainly involved in assembly of data, making approximations for inputs--in other words, building the actual model--and the strategic design of the experiments. The authors were involved with the simulations and the initial interpretation of results.

The efforts of both groups were extensive, their jobs demanding and results often disappointing. Tenacity and endurance appears to be a prerequisite for modeling. Considerable time was spent in acquiring the experience and technology to make the model studies.

The computational principles used in this study were already set at its start. During the investigations many additions and modifications were made to obtain an efficient general purpose system. The time spent on analytical studies as to the behavior of the modeling system was a minimum as such information could be readily obtained from model results and literature.

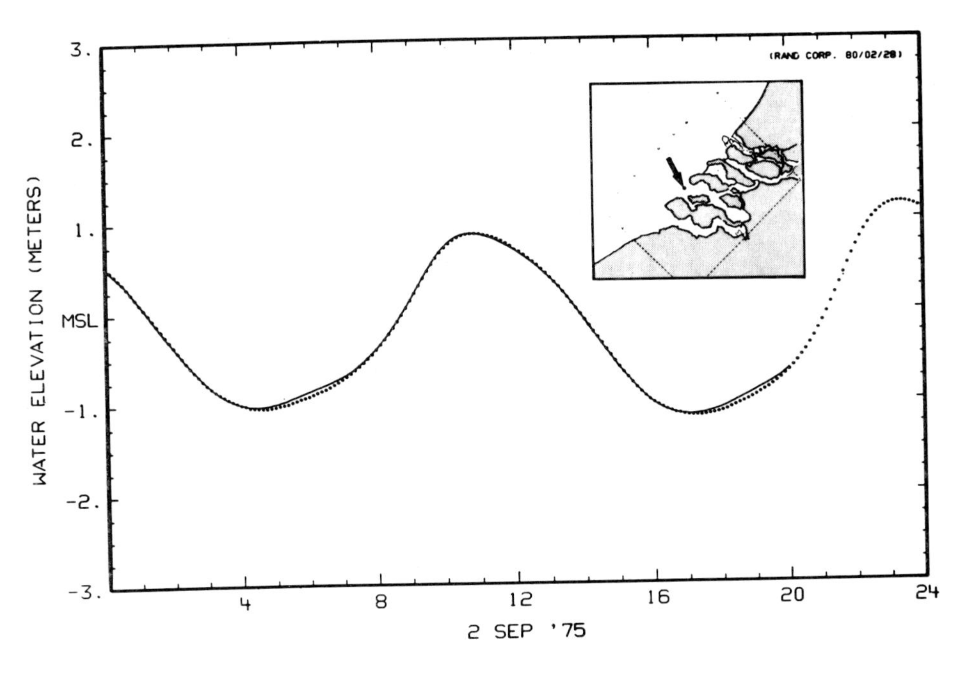

Figure 8.1. Comparison of water levels at a boundary of the one-dimensional model for a simulation with (—) and without (···) the storm surge barrier.

Acknowledgements. Many have assisted us directly or indirectly in the model studies. Mr. A. B. Nelson, Ms. C. N. Johnson and Ms. M. Fujisaki wrote the many programs which made mathematical simulation possible. Mr. J. Vincent and Mr. E. Fasel prepared much of the input data for the simulations. Ms. K. Sparr and Ms. S. Ritter made the general routine simulations. Ms. J. Douglas kept our records in good order and prepared the manuscript. Mr. J. Voogt supported us through the many pitfalls in preparing a model of this size and complexity. He is particularly acknowledged for the guidance and moral support he has given us in the many moments of despair.

REFERENCES

Leendertse, Jan J. (1967). Aspects of a Computational Model for Long-Period Water-Wave Propagation. The Rand Corporation, RM-5294-PR.

Leendertse, J. J. and Gritton, E. C. (1971). A Water-Quality Simulation Model for Well Mixed Estuaries and Coastal Seas: Vol. II, Computation Procedures. The Rand Corporation, R-708-NYC.

Van de Ree, W. J., Voogt, J. and Leendertse, J. J. (1978). A Tidal Survey for a Model of an Offshore Area. The Rand Corporation, P-6248.

DISCUSSION

Timothy J. Smith

Insistute of Oceanographic Sciences,
Crossway, Taunton, Somerset, England

The author made a brief mention of the use of the depth mean sub-grid scale energy to parameterize the bed shear stress. I would like to ask the author to expand on this point, in particular concentrating on the advantages of this approach over a quadratic friction law. Noting that the depth mean sub-grid scale energy equation reduced to the quadratic friction law under equilibrium conditions, under what conditions are the advective and diffusive transport of sub-grid scale energy important? Finally, I would like to ask the author whether a depth-mean dissipation length scaled on the local flow depth is sufficient, or whether the spatial variation of the length scale must be explicitly specified.

REPLY

Simulations using the depth averaged subgridscale energy to compute the bed shear stress give nearly the same simulation results as use of the quadratic friction law, as could be expected. This approach becomes important, however, when other inputs exist of small scale energy such as the turbulence generated by a barrier, turbulence generated by flow divergence and wind.

The subgridscale energy is used for dispersion simulation of a cloud of particles in combination with a Monte Carlo method. A whole new field of application will likely develop, at which time we will see the relative importance of the length scale.

PREDICTIVE SALINITY MODELING OF THE OOSTERSCHELDE WITH HYDRAULIC AND MATHEMATICAL MODELS

J. Dronkers

Rijkswaterstaat, The Hague, The Netherlands

A. G. van Os

Delft Hydraulics Laboratory, The Netherlands

J. J. Leendertse

The Rand Corporation, Santa Monica, California

1. INTRODUCTION

Construction of the Delta Works in the southwestern part of the Netherlands has formed a complicated system of basins, rivers, canals, locks and sluices (see Fig. 1), some fresh, some brackish, and some salt. The Oosterschelde, which will be partially closed by a storm surge barrier, has an important fish nursery as well as oyster and mussel cultures. To maintain these functions, the salinity in the estuary should as much as possible approximate the present high values. Since new secondary dams will be built and, in addition, the tidal range (thus, the mixing) will be reduced, the need exists to make predictions of salinity distributions after completion of the works under different alternative conditions for discharge sluices and navigation locks. Thus, research is needed to understand the dominant transport mechanisms in the estuary and to develop reliable tools for the prediction of salinities and concentration of other substances in the future.

Research using available prototype information, analytical models, physical scale models, and numerical models has been conducted, with special attention to the following items:

- what are the dominant mechanisms?

- is there any influence of the distortion of a physical model or the grid size of a numerical model on modeling of transport processes?.

- can the influence of density differences be neglected?

ISBN: 0-12-25852-0

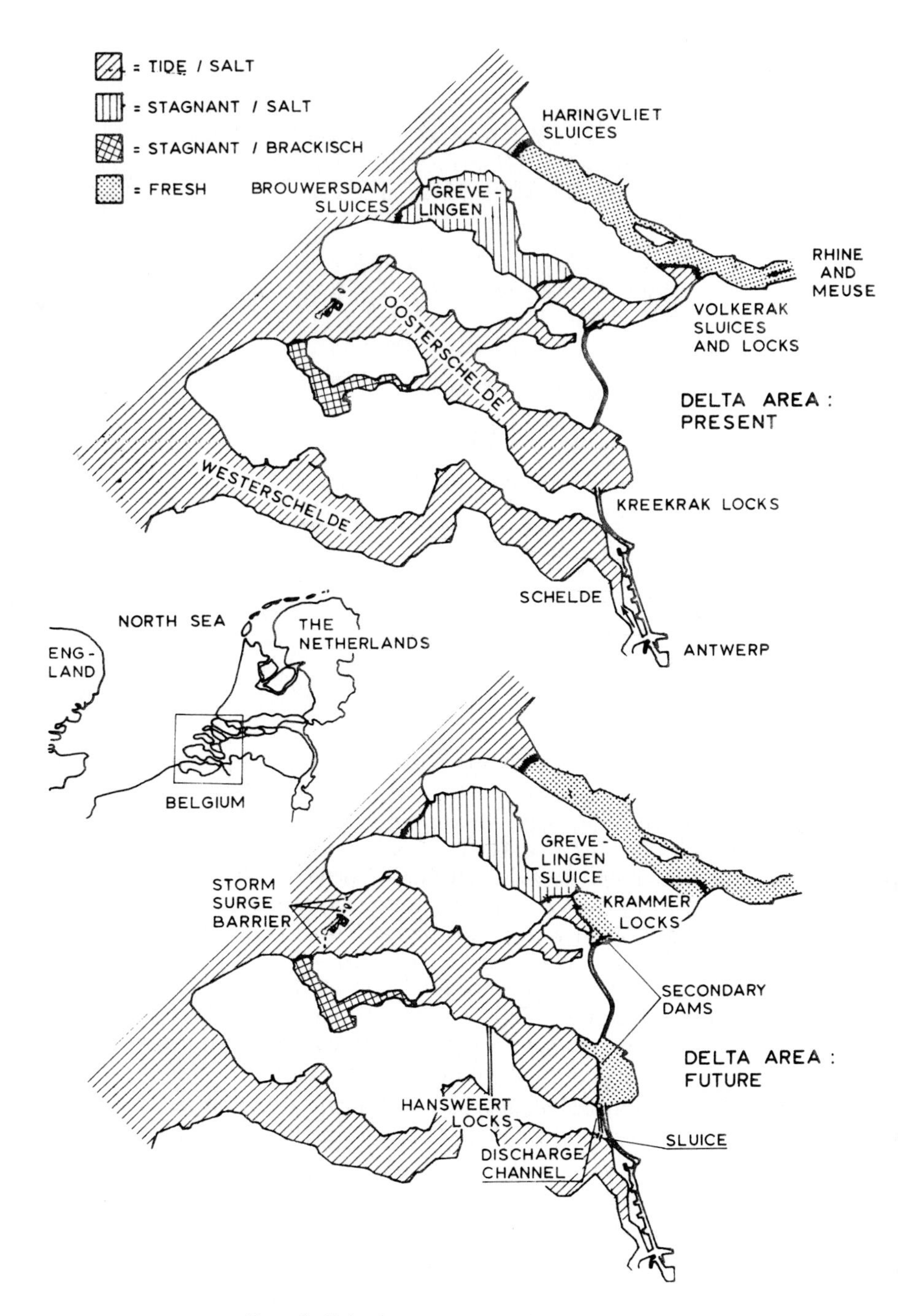

Figure 1. Delta Area, present and future situation.

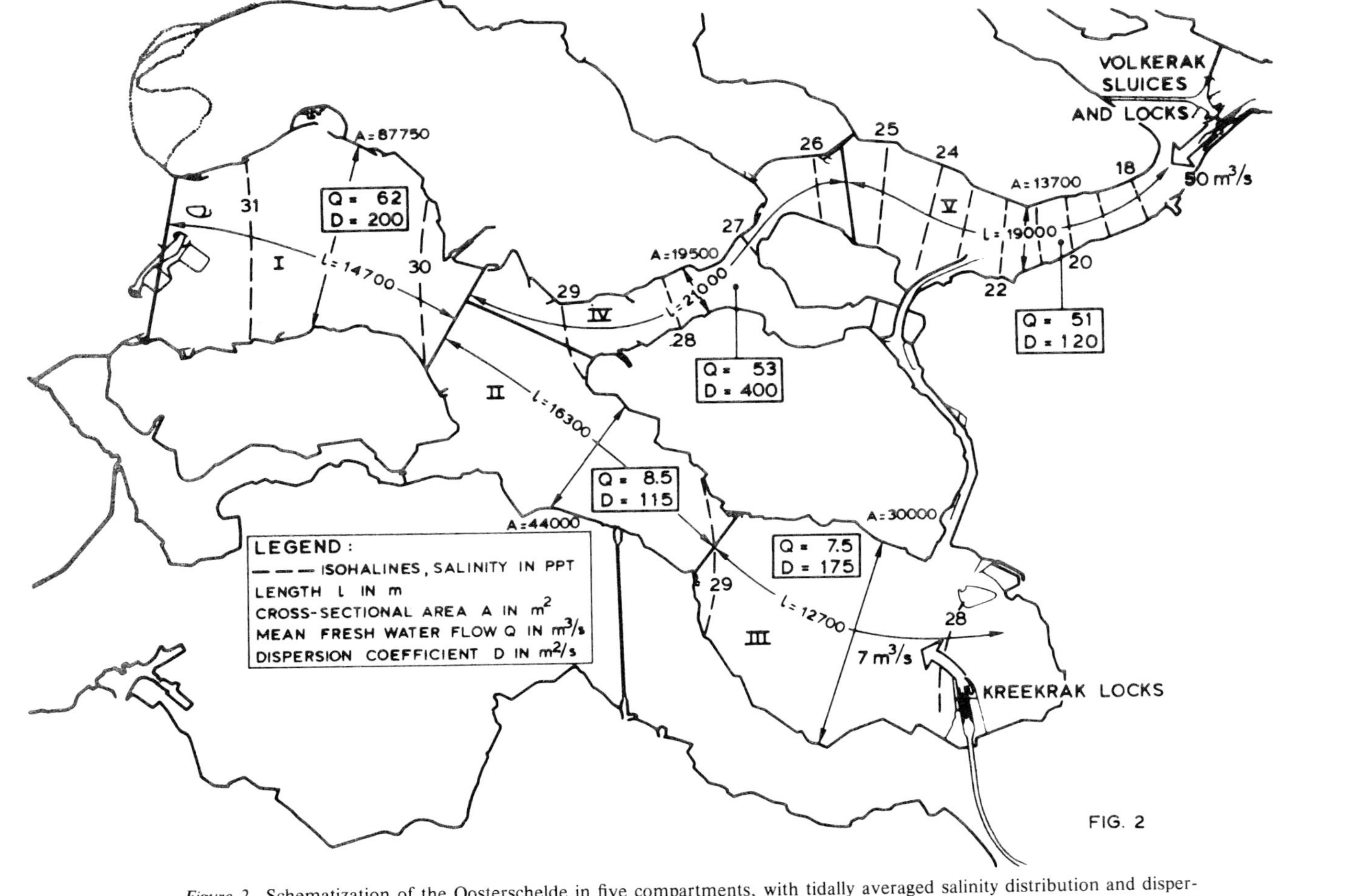

Figure 2. Schematization of the Oosterschelde in five compartments, with tidally averaged salinity distribution and dispersion coefficients; discharge at Volkeraksluices = 50 m^3/s.

With respect to the latter, the Oosterschelde is a well-mixed estuary (Dronkers, 1981), except in the northeastern branch, where stratification occurs because of the fresh water discharge. The model studies presented here are concerned only with the major part of the Oosterschelde, in which density differences play a minor role in the mixing processes.

2. DESCRIPTION OF THE ESTUARY

The Oosterschelde can be characterized as a well-mixed estuary. Typical geometrical scales are : length, 50 km; width, 4 km; depth, 10 - 20 m. The tidal range varies between 3 m and 4 m and the maximum tidal velocity between 1 m/s and 2 m/s. The tide is predominantly semidiurnal. The major portion of the fresh water enters the northern branch of the estuary via the Volkeraksluices at a constant rate of 50 m^3/s. A smaller amount of the fresh water (less than 10 m^3/s) enters the southern branch via the Kreekrak locks (see Fig. 2).

The bottom of the Oosterschelde is sandy; some siltation takes place in the marsh areas near the boundaries. The channels (typical width, 1000 m; typical depth, 20 m) form a complicated network (see Fig. 3), especially in the seaward part. In between and at the sides of the channels there are large sandy flats (nearly 50 percent of the surface of the estuary), which are flooded by the tide.

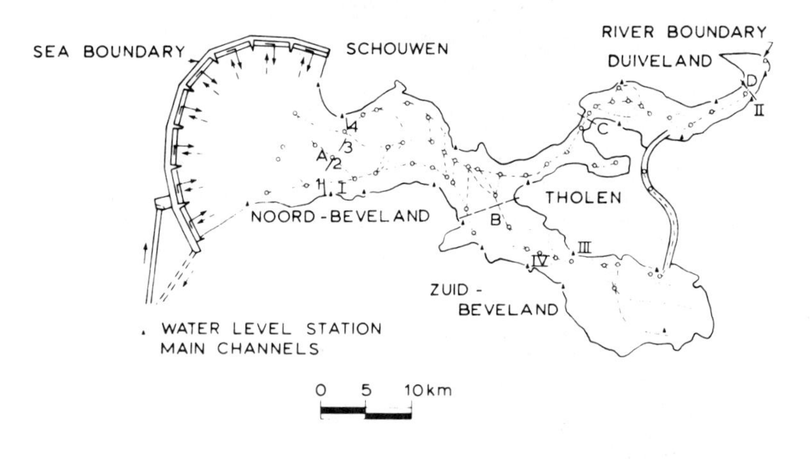

Figure 3. View of the hydraulic model, and channel system in the Oosterschelde.

Due to the nearly constant fresh water inflow, the tidally averaged salinity distribution changes only slightly during a year (a few ppt). Therefore the longitudinal dispersion coefficient D can be easily determined from the one-dimensional tidally-averaged advection-dispersion equation

$$Qc = DA \frac{dc}{dx}$$

where c is the tidally and cross-sectionally averaged salinity; Q is the fresh water discharge; A is the tidally averaged cross section area. The average salinity distribution in the Oosterschelde for the year 1978 is shown in Figure 2, together with a schematization of the estuary in five compartments. For each compartment the length, the average cross section area, the yearly average of fresh water discharge and the computed value of the longitudinal dispersion coefficient are indicated. It has been shown that the influence of the channel connecting Compartments III and V is very small when the fresh water discharge is 50 m^3/s (Dronkers, 1981).

In Fig. 4 the vertical salinity distribution in the two northern Compartments IV and V is shown at low water slack and high water slack. The influence of density differences appears to be important only in the most landward Compartment V. The high value of the dispersion coefficient in Compartment IV probably results from the fact that the tidal velocities are approximately 50 percent higher in this compartment than in other parts of the estuary.

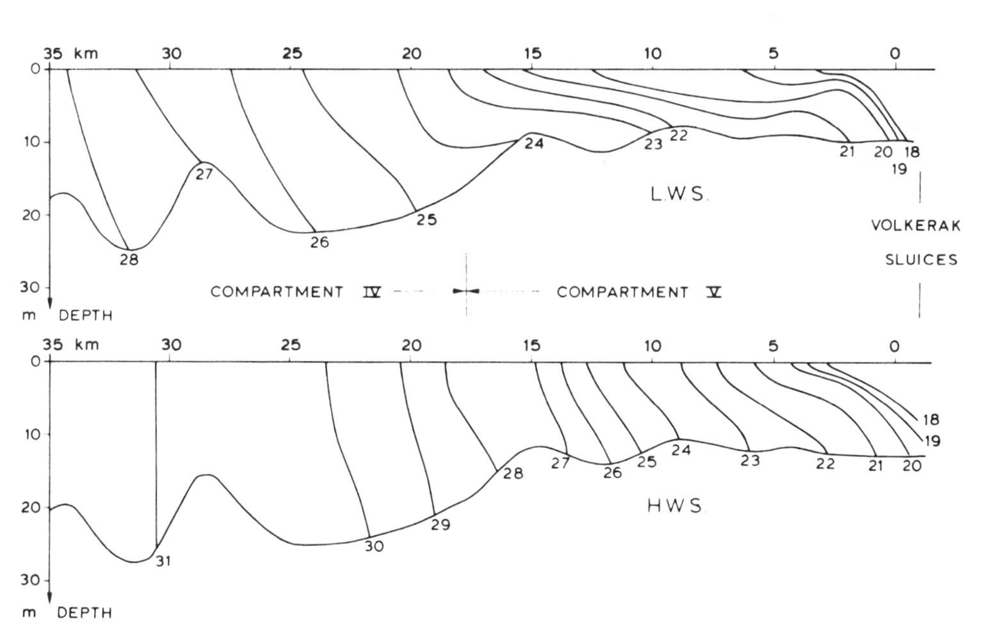

Figure 4. Isohalines in a longitudinal cross section of the Compartments IV and V at Low Water Slack (LWS) and High Water Slack (HWS).

The residence time of the water in the estuary is large. These times are indicated in Figure 5 as computed with the dispersion coefficients which are determined from the tidally averaged salinity distribution (see Figure 2). At many

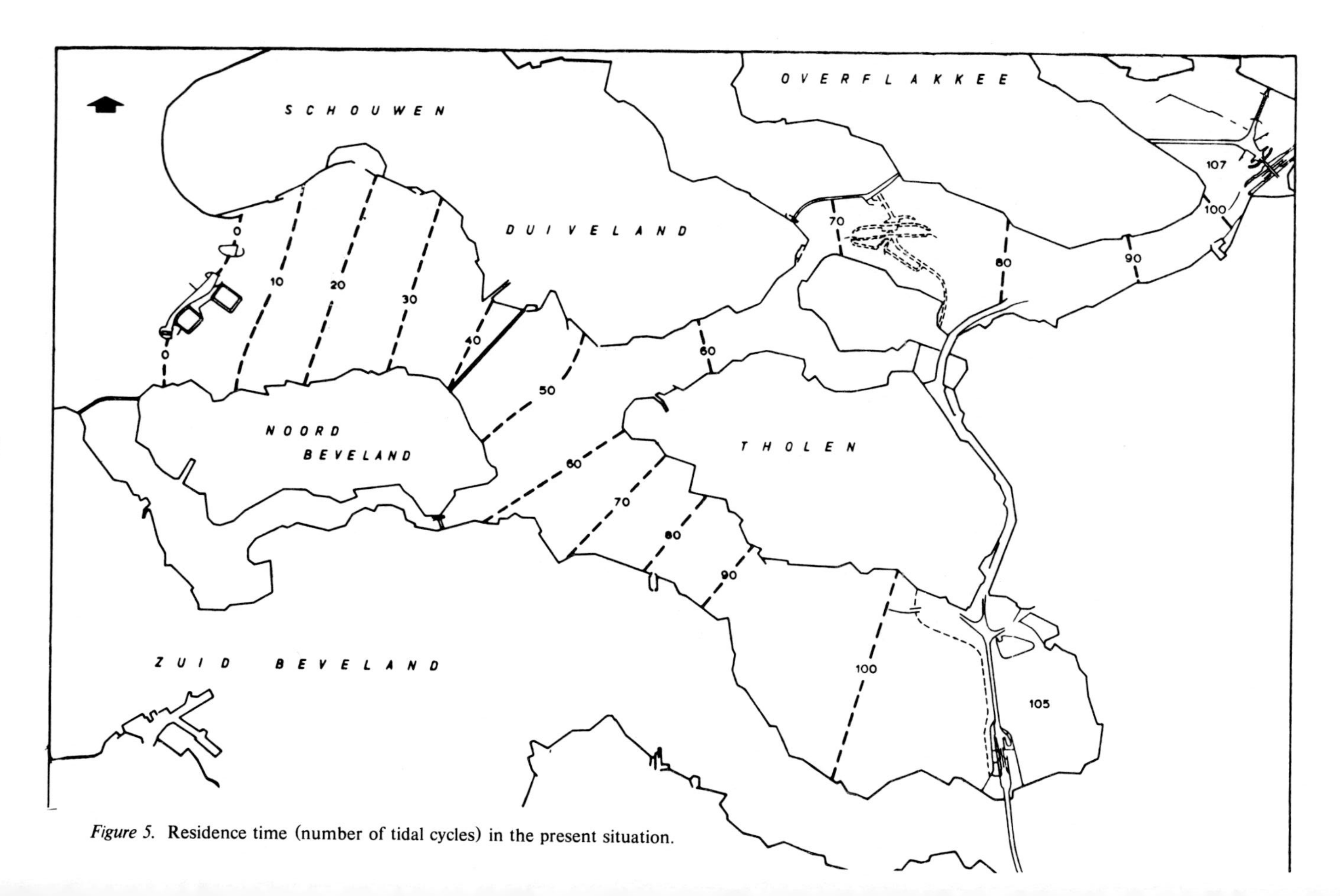

Figure 5. Residence time (number of tidal cycles) in the present situation.

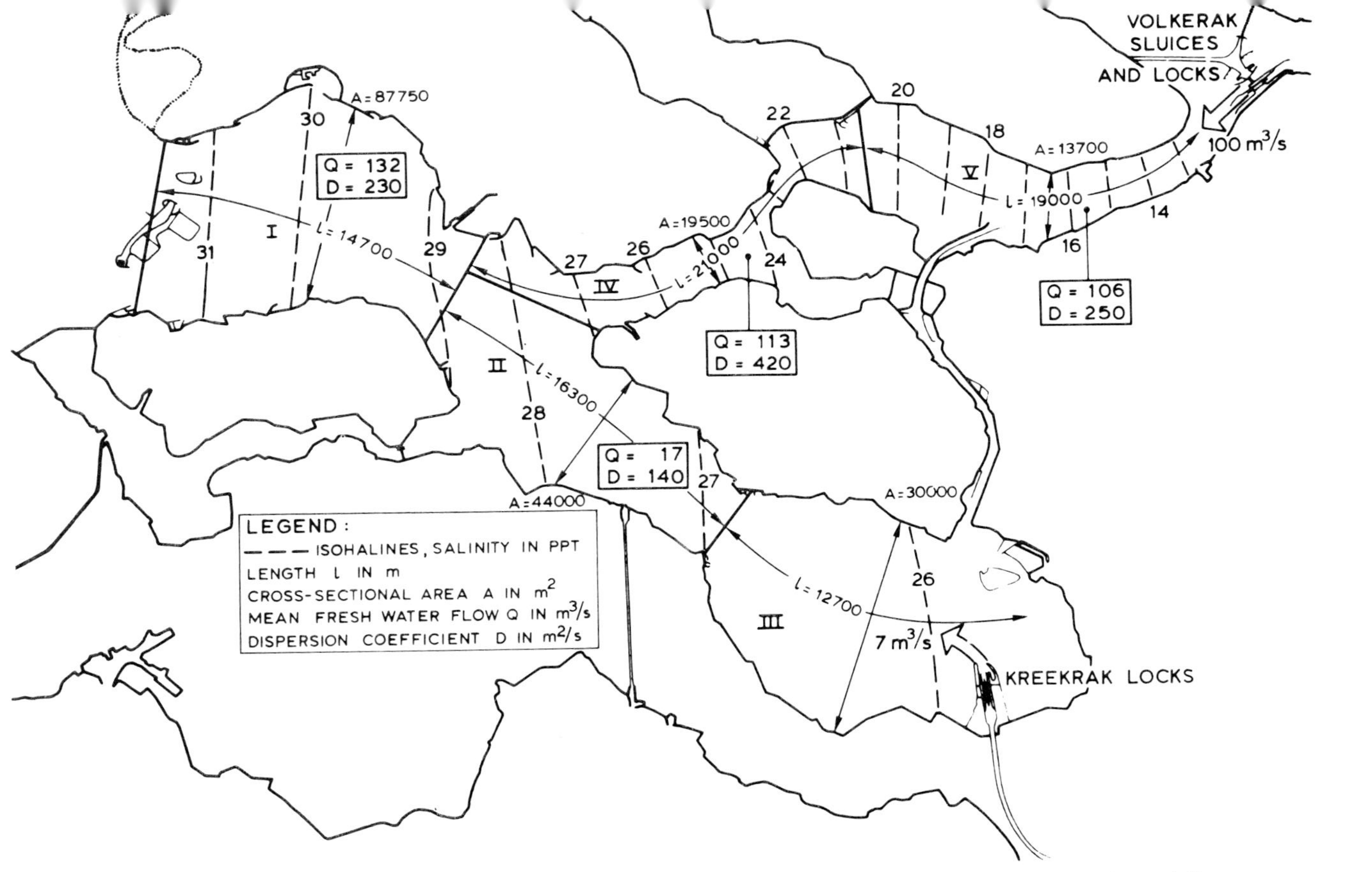

Figure 6. Schematization of the Oosterschelde in five compartments, with tidally averaged salinity distribution and dispersion coefficients; discharge Volkeraksluices = 100 m^3/s.

locations in the estuary a steady fresh water discharge is simulated using the tidally averaged one-dimensional model. For each simulation the resulting fresh water content of the estuary is computed and divided by the discharge, yielding a value for the residence time at the location of the discharge. The discharges are taken on the same order as the discharges in nature; these are so small that the computed residence times do not depend explicitly on the actual values of these discharges. The time scale for renewal of the waterbody in the estuary amounts to approximately 100 tidal cycles. Water quality models should be able to run such periods.

The influence of the fresh water inflow on the dispersion coefficients is shown in Fig. 6. During a period of 5 months the fresh water inflow at the Volkeraksluices was set at 100 m^3/s. In this period a new stationary salinity distribution was established. The dispersion coefficients derived from this salinity distribution do not differ much from the coefficients associated with the 50 m^3/s fresh water inflow, except in Compartment V. This confirms that the influence of the density differences is restricted to this part of the estuary. Because of uncertainties with respect to the influence of the channel connecting the Compartments III and V for a discharge of 100 m^3/s, it is not possible at this time to arrive at a reliable estimate of the dispersion coefficient in Compartment III.

3. THE HYDRAULIC SCALE MODEL

The hydraulic scale model represents an area of about 180.000 acres (see Fig. 3). The vertical scale could not be larger than $n_d = 100$, in order to maintain turbulent flow during nearly the whole tidal cycle. To keep the horizontal dimensions practical the model had to be distorted. A distortion of a maximum of 4 has been found not to influence the horizontal flow pattern even near a local narrowing (Delft Hydraulics Laboratory, 1972), so this distortion was adopted. The scales used have the following values: n_l = length scale = 400; n_d = depth scale = 100; n_u = velocity scale = 10; n_t = time scale = 40; n_C = roughness scale = $(n_l/n_d)^{1/2}$ = 2; n_g = scale of acceleration of gravity = 1; and $n_I = n_d/n_l = .25$. The model has a fixed bed and contains the whole Oosterschelde Estuary and an important part of the sea. It is run with homogeneous fresh water. The location of the sea boundary has been chosen such that an important narrowing of the mouth of the estuary will hardly influence the tidal motion at this boundary. The relatively high bed roughness required in the model because of the distortion of the scales in the horizontal and vertical sense was obtained by small concrete cubes or selected pebbles.

After construction, the model was calibrated first to be able to reproduce a prototype tide as accurately as possible. The vertical tidal motions (water levels as a function of time) for 25 stations over the region, and the horizontal tidal motions (discharge as a function of time) for four cross sections, were compared for model and prototype. The calibration, carried out with the tide of September 11, 1968, was followed by a verification with the tide of October 11, 1971 (see

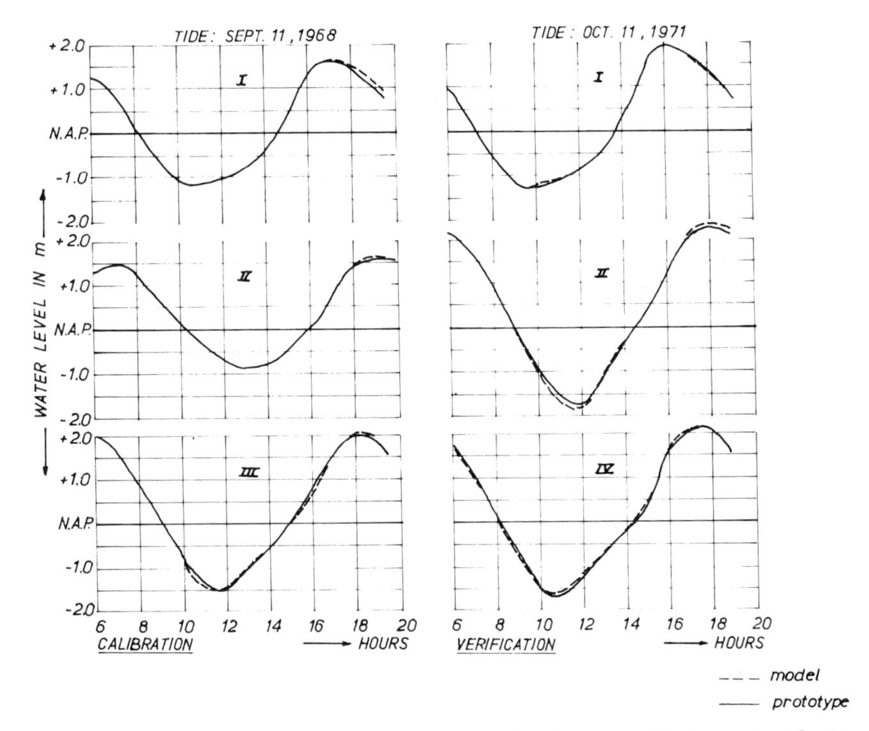

Figure 7. Comparison of water levels in hydraulic model and nature: calibration and verification.

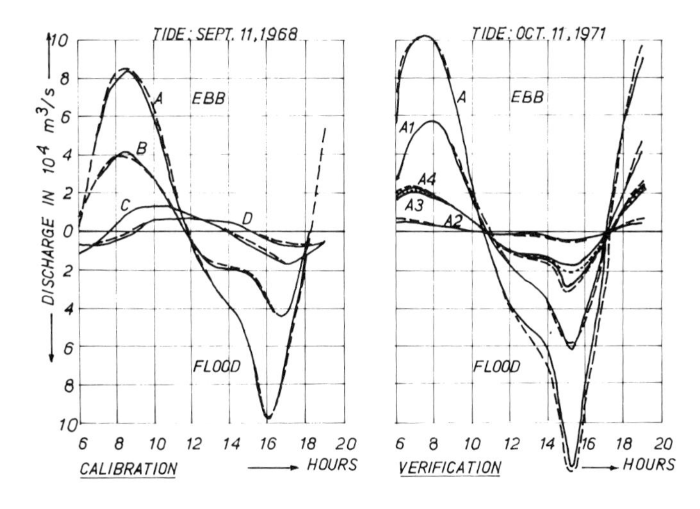

Figure 8. Comparison of discharges in hydraulic model and nature: calibration and verification.

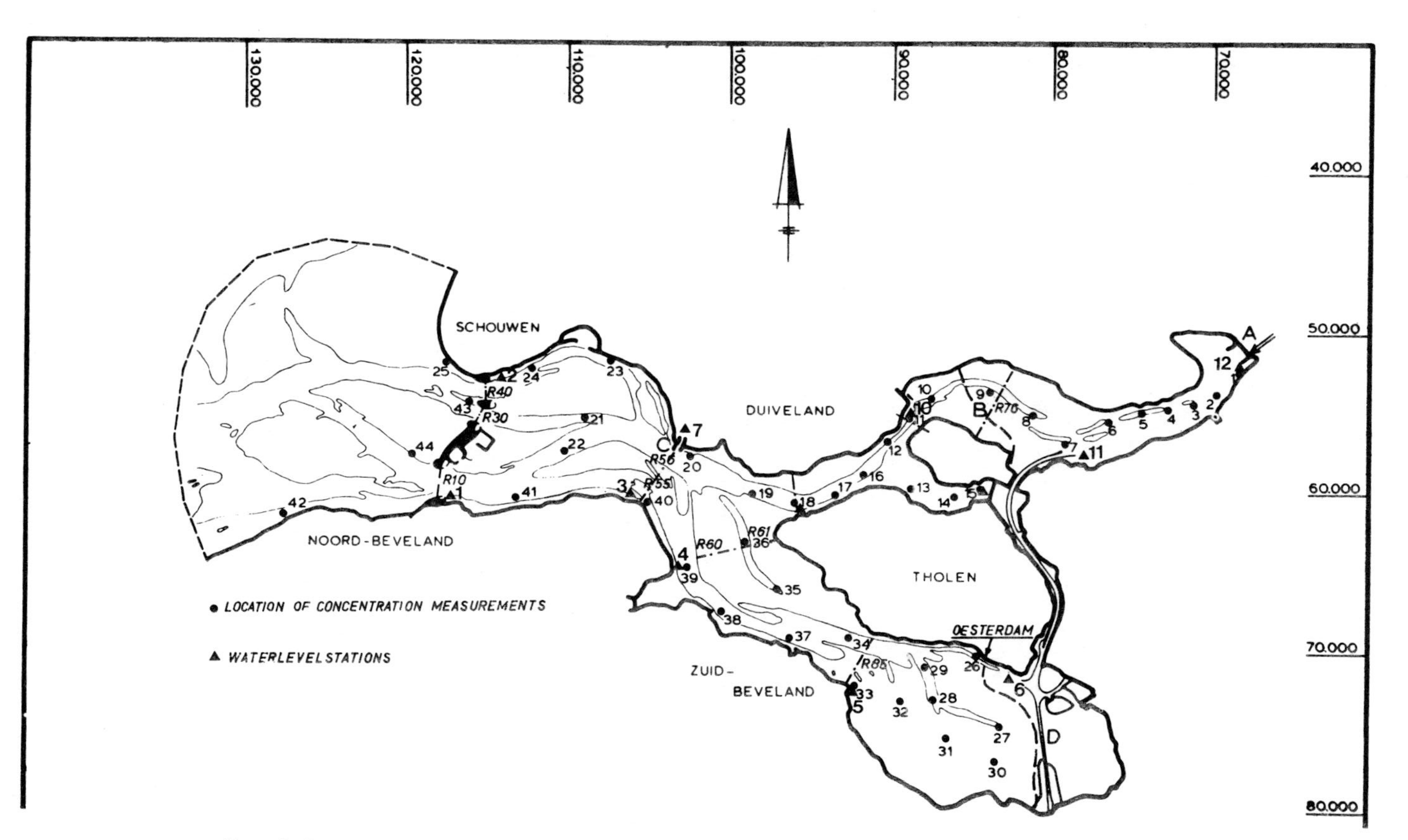

Figure 9. Location of concentration measurements, water levels and discharges in the hydraulic model; location of dye injections.

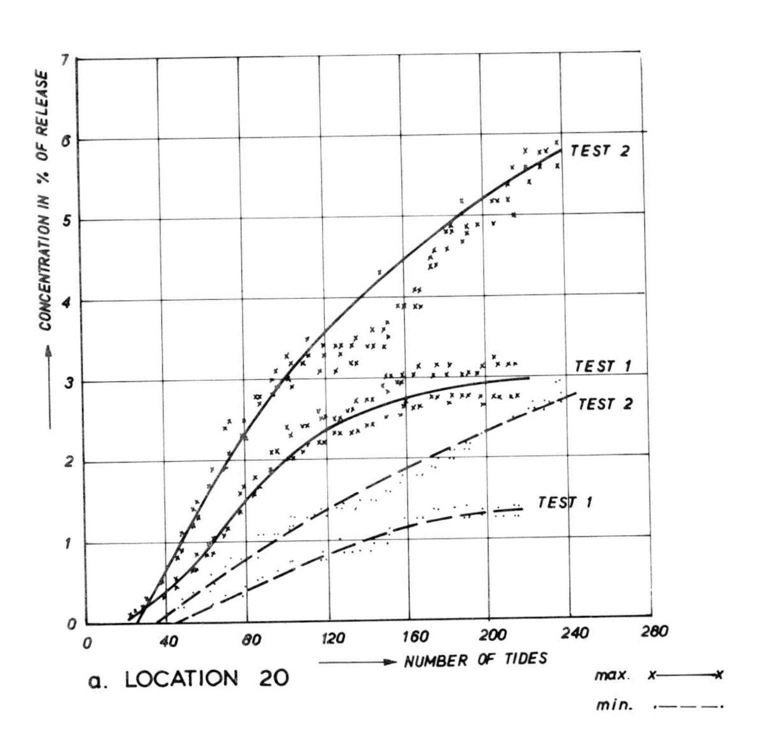

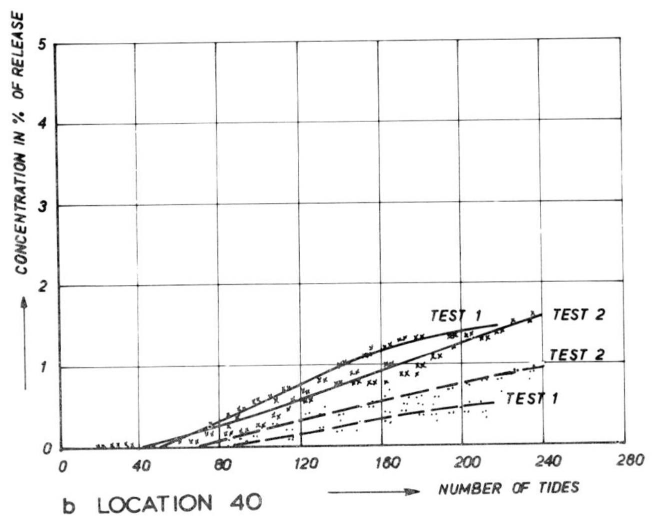

Figure 10. Development of dye concentration in the hydraulic model as a function of time.

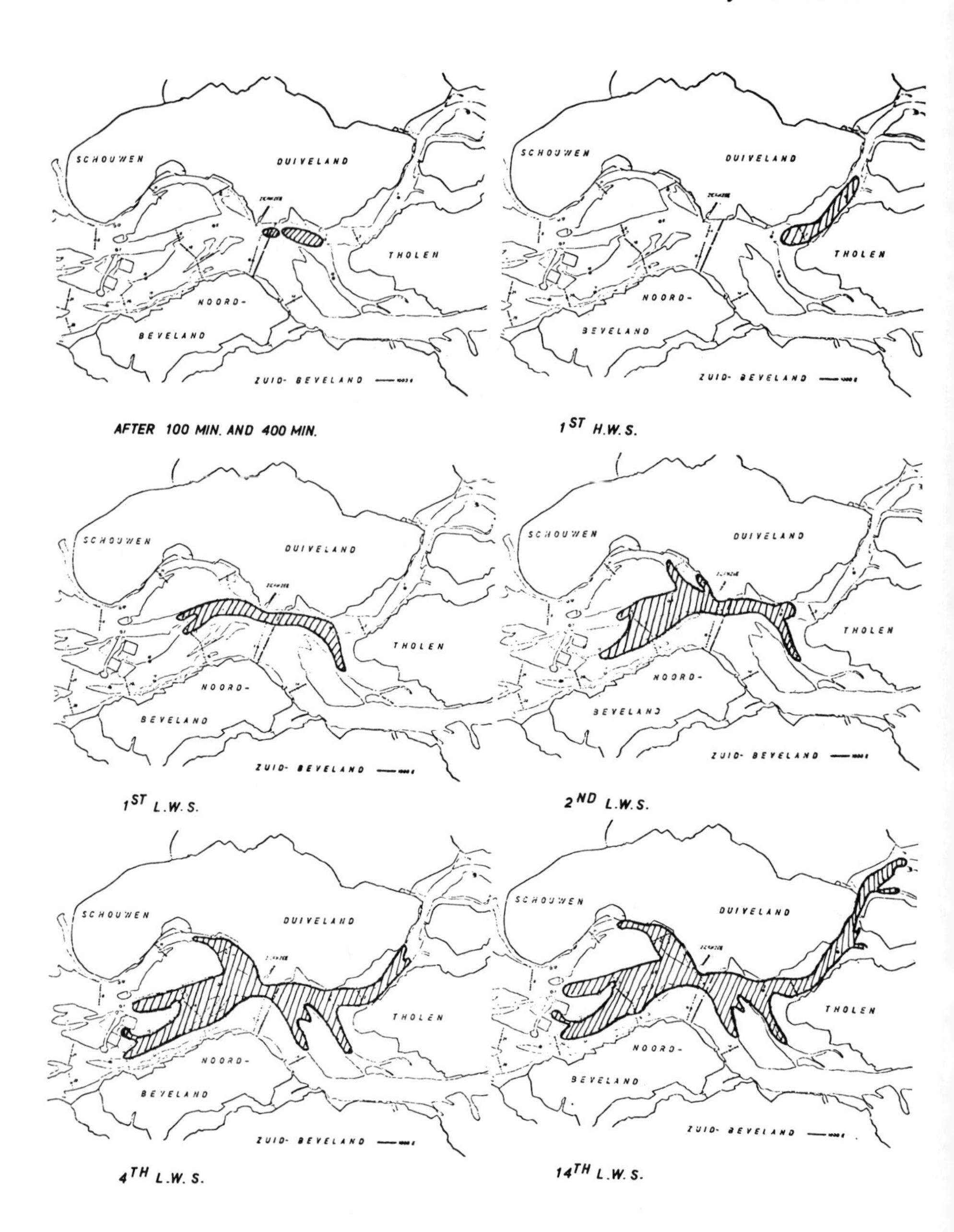

Figure 11. Development of dye patch released near Zierikzee: 10^{-5} kg/m^3 concentration contour.

Figs. 7 and 8). It was concluded that for large scale water movement (water levels and discharges) the model simulates nature very well. Since the effects of local narrowing can be expected to be reproduced too (Delft Hydraulics Laboratory, 1972), the future water movement can be simulated satisfactorily in this model if the storm surge barrier is modeled correctly (Stroband and Van Wijngaarden, 1977).

To study the transport mechanisms in the Oosterschelde model the dye release tests given in Table 1 were performed (Delft Hydraulics Laboratory, 1980a and 1980b). For the continuous release experiments the dye (Rhodamine WT) was discharged into the model with an equivalent prototype discharge of 50 m^3/s for Test 1 and 25 m^3/s for Test 2 and concentration of 50×10^6 kg Rho/ m^3. The tests lasted 240 tides (approximately 4 months). Figures 10a and 10b give the Rho-concentration at positions 20 and 40 (see Fig. 9) as a function of time for Test 1 and Test 2. The change in time scale is clearly visible. The instantaneous release experiments were performed by releasing 5 l (prototype equivalent 800.000 m^3) of water with a Rhodamine concentration of 0.4 kg Rho/m^3 at low water slack tide. The dispersion of these dye patches was measured at low water slack tide and at high water slack tide during 22 tidal cycles. Figure 11 gives some results of these measurements for Test 3. Extended information of the model dye release studies is given in Delft Hydraulics Laboratory (1980a, 1980b).

Table 1.
Dye Release Tests in Hydraulic Model

No.	Situation	Dye Release	Location (see Figure 9)
1	present	continuous	Volkeraklocks (A)
2	future	continuous	Krammerlocks (B)
3	present	instantaneous	Zierikzee (C)
4	future	instantaneous	Zierikzee (C)
5	present	continuous	Kreekraklocks (D)

4. THE MATHEMATICAL MODEL

The mathematical model used in this study is the submodel Oost of the Randdelta II described in Chapter 11. The model was adjusted and validated on boundary conditions obtained from a Randdelta II simulation. The comparisons for the adjustment and verification were made with the effects of density, earth rotation, and the wind included in the model. Bottom roughness was adjusted in blocks to obtain agreement on water levels. Comparisons were also made of the mean levels of the observed and computed data by low pass filtering. The comparisons of the observed and computed transports through the channels in the inlet are also good.

Hydrographic surveys indicated large spatial variations in bottom roughness which were not reflected in the model. Field surveys and model experiments indicated further that in certain areas the height of the sandwaves (thus roughness) increases with larger amplitudes (see Chapter 11). This was not simulated in the model, and data from September 2 and 3, 1975 were used only in the adjustment process.

To obtain horizontal tidal movements comparable with the hydraulic model, new boundary conditions at the seaward end of the model had to be prepared. For the experiments in the hydraulic model a tide with smaller amplitude than the average tide was used. From the tide records at two shore stations at the open boundary the average tide of 1961 was determined and a comparable actual curve of this amplitude was found in the records of the stations. The curves found closed quite well in a 12-½ hour period. A Fourier composition of each curve was prepared, and 86 percent of their amplitudes were comparable with the tide used in the hydraulic model.

Subsequently, estimates of the response functions of the six input boundary stations to each of the two shore stations from computed data in the Randdelta model were made by use of cross-spectral analysis. From the tide curves at the shore stations estimates could be made of the boundary tides by use of these response functions. Since two estimates were available--one from each shore station--a weighted average was used with a weighting factor inversely related to the distance between boundary station and shore station. Linear interpolation of water levels was used for the model boundary between the model boundary stations.

In the hydraulic model experiments the rotators for simulating the effect of the earth's rotation were turned off, as they could influence the dispersion. Consequently, in the mathematical model experiments, the Coriolis-force had to be set to zero. Since the input tide boundaries were based upon an actual simulation, a sensitivity experiment had to be made to study the effect of the earth's rotation. Figure 12 presents the integrated water transport with and without the effect of the earth's rotation in the most northern channel of the inlet.

For the actual experiments the wind force was set to zero and the coupling between salinity and the hydrodynamics of the system was removed.

To facilitate the analysis of model results, at half-hourly intervals during the computations, graphs with concentration contours were prepared. A large number of contour intervals were chosen to facilitate comparisons with measurements and estimates in the hydraulic model (Fig. 13).

In order to compare the results with the hydraulic model, scale factors were used for the longitudinal and transverse dispersion coefficients in the mathematical model. The longitudinal dispersion coefficient was computed from the Elder-type equation

$$\epsilon_x = 6\, du_*$$

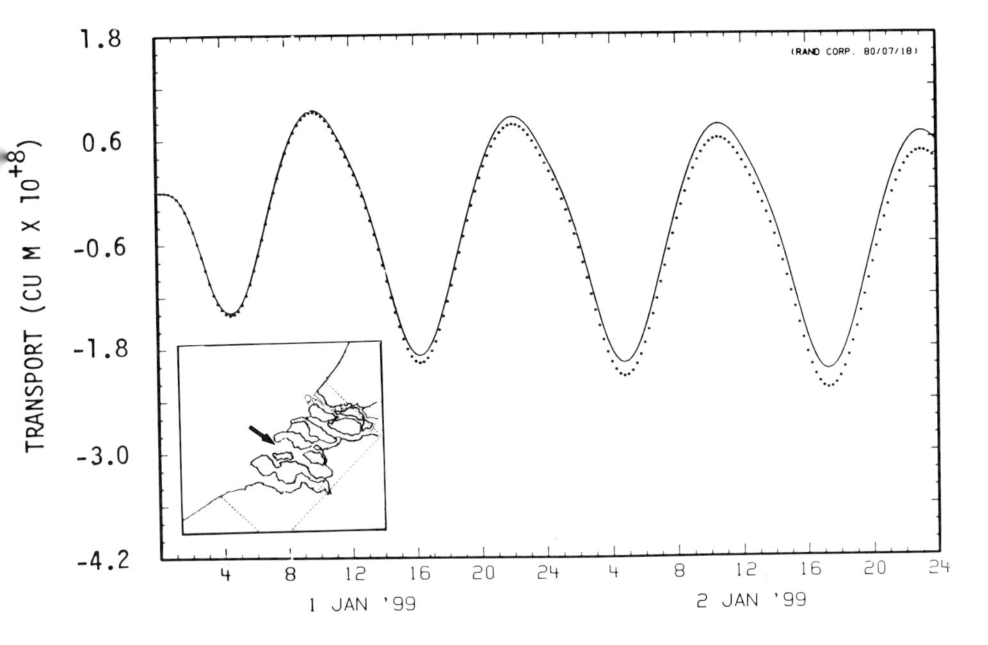

Figure 12. Integrated mass transport through the most northerly opening (Hammen) with and without Coriolis force.

where d is the temporal depth, $u_* = \sqrt{g(u^2 + v^2)}/C$ = shear stress velocity, C = Chézy-coefficient, u, v = velocity components. The transverse dispersion was represented by an average time independent isotropic dispersion coefficient for each point in the grid. This dispersion coefficient was computed from the local velocities from a previous simulation and based on the equation

$$\epsilon_y = du_*$$

The scaled dispersion coefficients ϵ_x^x and ϵ_y^s, used in the model, are found by multiplying ϵ_x and ϵ_y by the factor

$$\frac{n_l \cdot n_C}{n_d} = 8$$

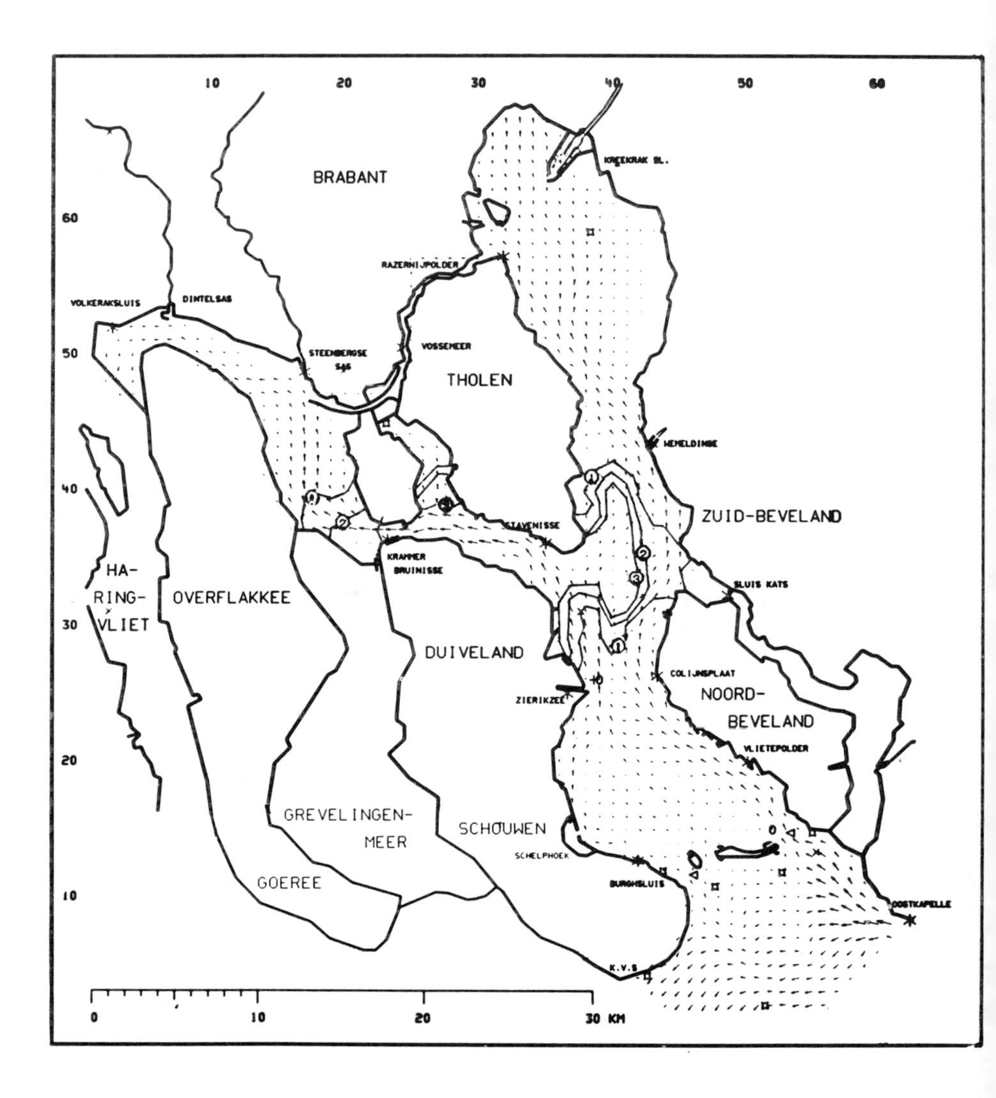

Figure 13. Dye distribution chart obtained from the numerical simulation representing distributions near the 4th high water slack after injection. Isolines at (1) 10^{-5}kg/m^3, (2) 2.10^{-5}kg/m^3, (3) 3×10^{-5}kg/m^3. Dye release point (0) in mathematical model is a few hundred meters from the release point (+) in the hydraulic model toward the center of the estuary.

5. CONTINUOUS DYE RELEASES, MODEL-PROTOTYPE COMPARISONS, AND PREDICTIVE USE

In the hydraulic scale model, the fresh water inflow in the Oosterschelde is simulated by continuous injections of a neutrally buoyant dye at the Volkerak locks and the Kreekrak locks (in separate experiments) until a stationary concentration distribution is reached. The dispersion coefficients computed from these concentration distributions are given in Table 2, together with the values obtained from field measurements. The overall agreement is satisfactory. The largest differences occur in Compartment I. A possible explanation for this difference between model and nature may be found in the distortion of the scale model: with respect to nature the vertical scale is four times larger than the horizontal scale. The difference observed in Compartment V may be traced to the density differences occurring in nature.

Table 2.
Model Comparisons and Predictions

Compartment (see Fig. 2)	Dispersion Coefficients in m^2/s		Predicted Changes Due to Barrier and Dam Construction*	
	D_{EXP} From field measurements	$D_{H.M.}$ From hydraulic model	D/D'	u/u'
I	200 ± 40	700 ± 250	6.4 ± 2.5	1.75
II	115 ± 25	180 ± 30	2.8 ± .8	1.7
III	175 ± 50	130 ± 35	3.25 ± 1.2	2
IV	400 ± 50	380 ± 60	16.5 ± 5	6
V	120 ± 20	75 ± 15	–	–

**Note: D and u are values for present conditions. D' and u' are predicted future values.*

In the hydraulic model, the influence of the tidal motion in the Oosterschelde on the dispersion coefficients is investigated by repeating the previous experiments with continuous injection of dye for a situation where secondary dams and the storm surge barrier are present (future situation, see Fig. 1). Table 2 also gives the change of the dispersion coefficients from the present to future situation, and the corresponding reduction of mean tidal velocities. In Compartments II, III and IV the dispersion coefficient changes approximately proportionally to the square of the mean tidal velocity. A possible explanation is based on the following qualitative arguments. The one-dimensional dispersive transport

$$\dot{M}_D = -DA\frac{dc}{dx} \tag{1}$$

where D = dispersion coefficient; A = tidally averaged cross section; and, c = tidally and cross-sectionally averaged constituent concentration, may also be written as

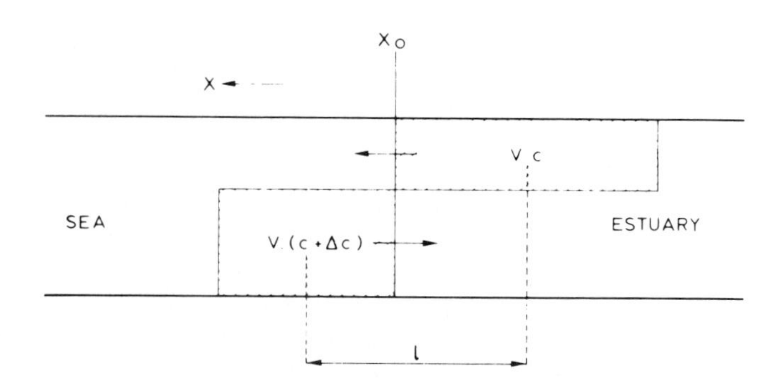

Figure 14. Schematic representation of the renewal of estuary waters by dispersion.

$$\dot{M}_D = -\frac{V}{T}\Delta c \qquad (2)$$

where V = volume of water exchanged in a tidal period through the considered cross section by dispersion; T = tidal period; and, Δc = concentration difference of the exchanged water volumes. A schematic representation is given in Fig. 14. In this simplified model Δc may be written as

$$\Delta c = l\frac{dc}{dx} \qquad (3)$$

where l = averaged distance over which exchange of water takes place in a tidal period as a consequence of the mixing processes. Substituting (2) and (3) in (1) yields for the dispersion coefficient the expression

$$D = \frac{l \cdot V}{A \cdot T} \qquad (4)$$

If the horizontal large scale water motion is mainly responsible for the renewal of estuary waters, it may be assumed that l and V are both proportional to the tidal excursion of fluid particles. In that case the dispersion coefficient depends on tidal parameters as the square of the tidal velocity. The foregoing implies that the dispersion coefficient does not depend on the water depth or on the turbulent mixing. Consequently the vertical scale of the distorted hydraulic model has no influence on the dispersion coefficient. Comparison of instantaneous dye release measurements in nature, in a hydraulic scale model with a distortion of 3 and a hydraulic scale model with a distortion of 37.5 has shown this to be true in another Delta Estuary, The Brouwershavense Gat, (Delft Hydraulics Laboratory and Rijkswaterstaat, 1974).

From Table 2 it follows that both the conditions $D \propto u^2 T$ and $D_{HM} = D_{EXP}$ (in scaled units) are approximately satisfied in the Compartments II, III, and IV of the Oosterschelde. In Compartment I the model described above apparently fails. This points to a possible influence of the distortion on the longitudinal dispersion in this region of the estuary. In this context one should focus on mixing processes which depend on the width-to-depth ratio of the channels.

In the Oosterschelde the width-to-depth ratio is on the order of 100. The time scale for lateral mixing τ_y can be computed from the expression

$$\tau_y = \frac{b^2}{2\epsilon_y} \tag{5}$$

where b is the characteristic width, 1000 m. It is assumed that the lateral dispersion coefficient ϵ_y is given by an expression of the type (Fischer, *et al.*, 1979)

$$\epsilon_y = \gamma d u_* \tag{6}$$

where d is the characteristic depth, 20 m, u_* is the average shear velocity = $\sqrt{g/C^2} = 0.05$ m/s, $\gamma \approx 1$. The time scale for lateral mixing is found to be on the order of 10 tidal periods. In the hydraulic model the width-to-depth ratio is, by a factor of 4, smaller than in nature. The roughness of the hydraulic model is larger (Chézy-coefficient C smaller) and therefore the relative shear velocity u_*/u is larger than in nature. The coefficient γ decreases only weakly with decreasing width-to-depth ratio (Okoye, 1970) and therefore Eqs. (5) and (6) show that the time scale for lateral mixing of the hydraulic model is at most a few tidal periods--much less than in the estuary. The difference in lateral mixing between model and nature affects mixing mechanisms, which may contribute substantially to the longitudinal dispersion in wide estuaries like the Oosterschelde. A mathematical description of these mixing mechanisms has been given by Okubo (1967) for a schematic estuary.

One mechanism affected by lateral dispersion is related to the horizontal residual circulations which are present in most of the channels of the estuary (see also Section 6). A patch of dye which has initially a homogeneous distribution in the cross-section will be stretched in a longitudinal direction when a residual current with opposite directions at both boundaries is present. The longitudinal size of the patch (the second moment of the cross-sectionally averaged concentration) increases faster when the lateral mixing is small. Thus, the longitudinal dispersion coefficient D increases with increasing lateral mixing time τ_y. This implies that in the hydraulic scale model the influence on the longitudinal dispersion of horizontal residual circulations within the channels is smaller than in nature. Another important mixing mechanism involving lateral dispersion is related to the non-homogeneous distribution in the cross-section of the oscillating velocity component. Consider a patch of dye which at a given phase of the tidal cycle is homogeneously distributed in lateral direction. Suppose further that the tidal velocity is significantly larger in the center of the channel than near the boundaries. Then the patch of dye will be stretched in longitudinal direction in a first stage of the tidal cycle, and next be contracted in a second stage. In the absence of lateral mixing the patch of dye would return to its initial state and no longitudinal dispersion would occur. In the opposite case of instantaneous lateral mixing the patch of dye would not be stretched and longitudinal dispersion would not occur either. But if the time scale for lateral mixing is on the order of a tidal period, the patch of dye is stretched to a certain extent in the first stage of the tidal period, but at the same time it is mixed to a certain extent in lateral direction. Therefore the patch of dye is not contracted to its initial state in the second stage of the tidal period, and longitudinal dispersion results. The longitudinal dispersion decreases with decreasing lateral mixing time τ_y, when $\tau_y << T$, but increases with decreasing τ_y when $\tau_y >> T$. The last case applies to the Oosterschelde estuary. This implies that the longitudinal dispersion in the hydraulic model is larger than in the estuary if the above described mechanism yields an important contribution to the overall dispersion.

The conclusion is that the distorted width-to-depth ratio in the hydraulic model may in certain regions of the estuary influence significantly the mixing processes, yielding a larger or a smaller longitudinal dispersion coefficient depending on which of the aforementioned mixing mechanisms is dominant. Also the difference between the dispersion coefficients in the hydraulic model and nature in Compartment I might be traced to the distortion of the model. The higher dispersion coefficient found in the hydraulic model points to a mixing mechanism which is enhanced by a decreasing lateral mixing time τ_y (when $\tau_y > T$), or by an increasing lateral dispersion ϵ_y. Thus, one may also expect that the dispersion coefficient in Compartment I of the hydraulic model varies with tidal velocity faster than the square. This conforms to the results presented in Table 2. An alternative interpretation in terms of mixing mechanisms is presented in Section 7.

6. INSTANTANEOUS RELEASES OF DYE: COMPARISON OF HYDRAULIC AND MATHEMATICAL MODELS

In both hydraulic and mathematical models a non-buoyant dye was injected at low water slack tide at the location designated by a dot in Figs. 15a-15f. Both models were driven by similar periodic tides at the seaward boundary and no fresh water was discharged into the system. In the hydraulic model the magnus effect rotators (to simulate the Coriolis-effect) were not used and in the mathematical model the effect of the earth rotation was set at zero for the experiment described in this section. The influence of the earth rotation was studied by comparing the results of this experiment with another experiment which only differed by the presence of a Coriolis-force. Hardly any difference could be noticed between the concentration distributions, and therefore it is plausible that the earth rotation has no significant influence on the mixing processes in the Oosterschelde estuary. In

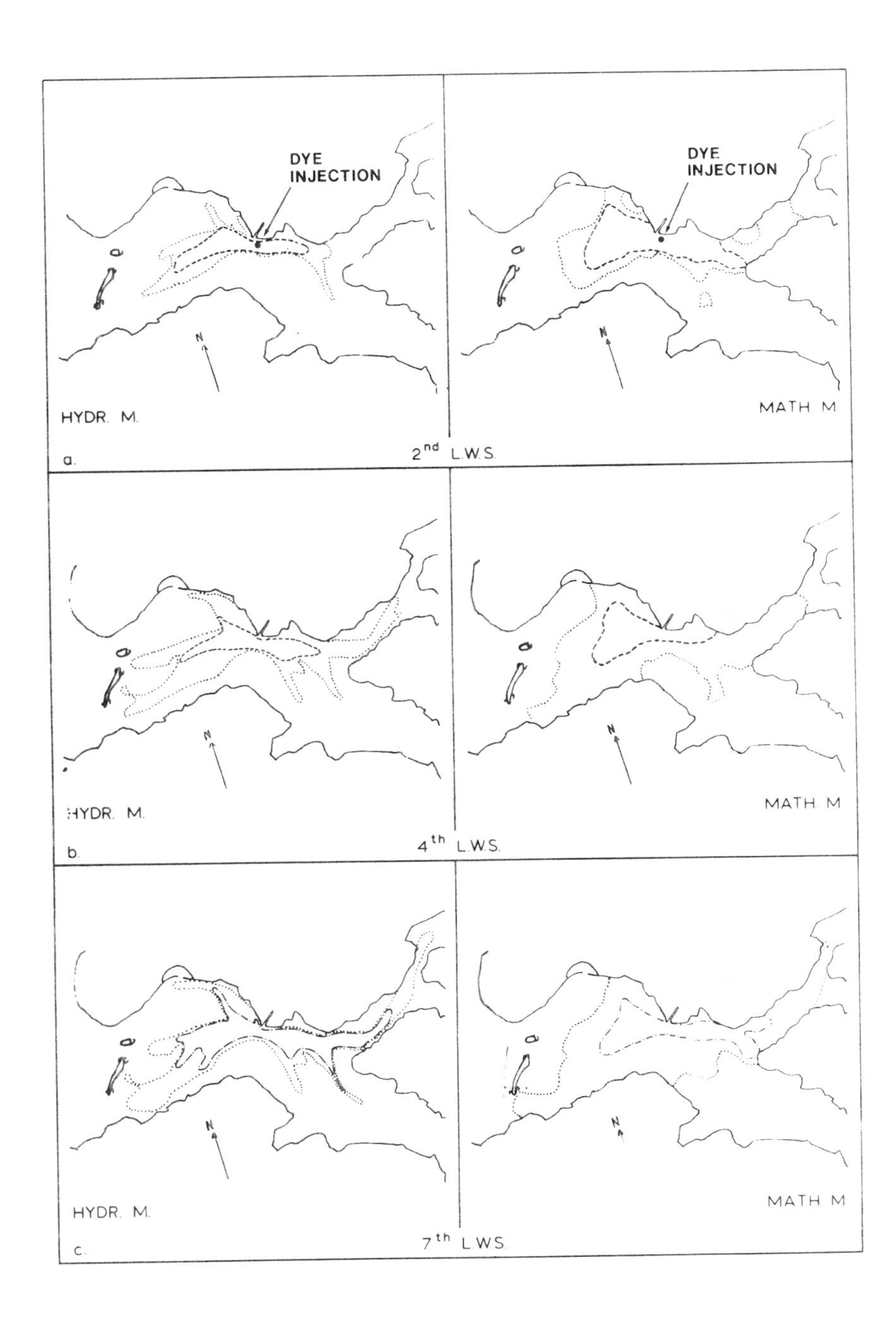

Figure 15. Isoconcentration lines in the hydraulic and mathematical models at (a) second, (b) fourth, and (c) seventh Low Water Slack (LWS) after injection for an instantaneous injection of 2 g dye at LWS at the location indicated by the dot. Isolines: --- 3×10^{-5} kg/m^3; −·− 2×10^{-5} kg/m^3.

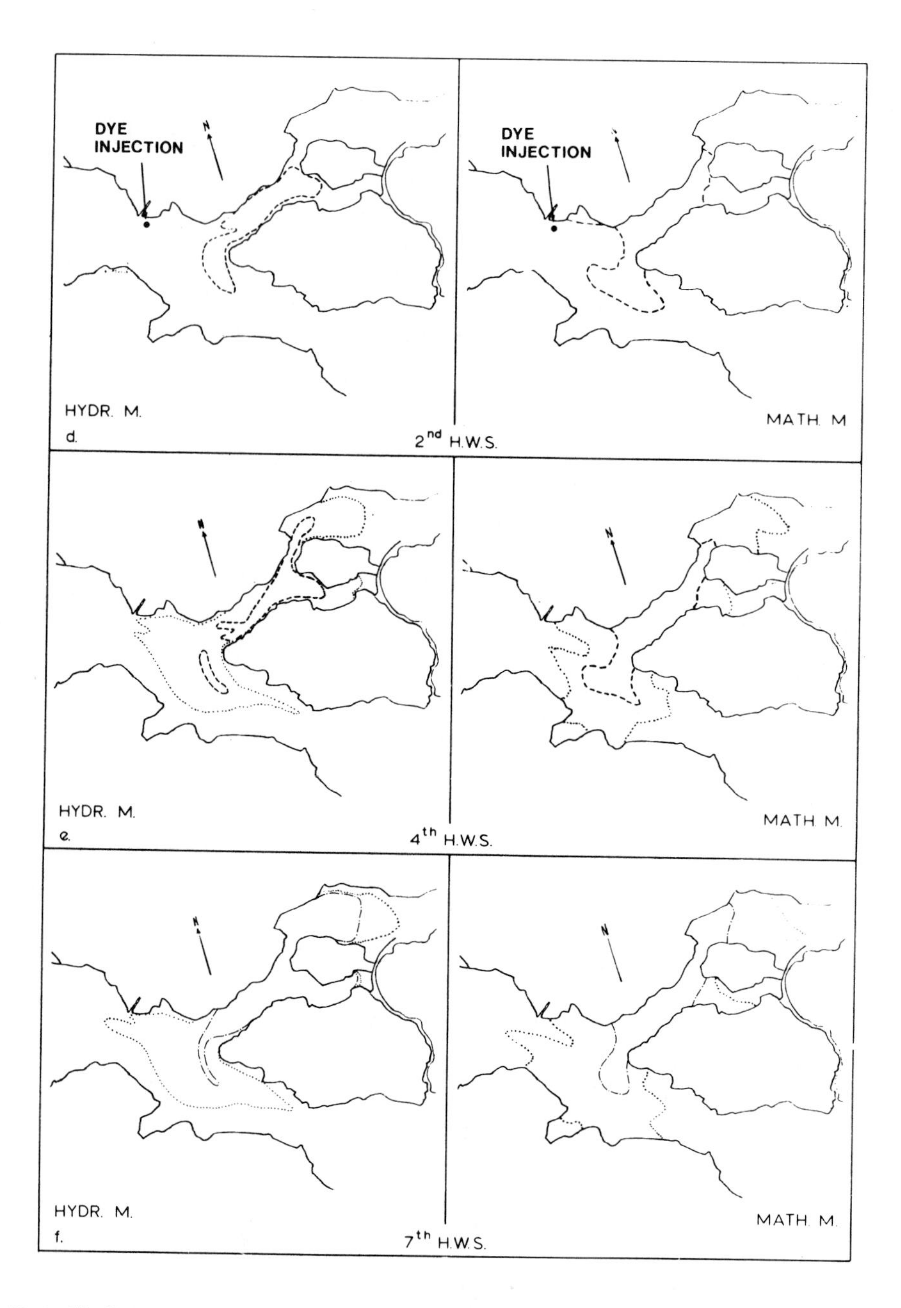

Figure 15, Cont. Isoconcentration lines in the hydraulic and mathematical models at (d) second, (e) fourth, and (f) seventh High Water Slack (HWS) after injection for an instantaneous injection of 2 g dye at LWS at the location indicated by the dot.

Figs. 15a-15f isoconcentration lines in the hydraulic and mathematical models are shown at high water slack and low water slack 2, 4, and 7 tidal periods after the moment of injection. The figures show that the longitudinal spreading of the cloud of dye is very similar in both models, but that the lateral spreading in the mathematical model is larger than in the hydraulic model. A possible explanation is that the lateral dispersion coefficient ϵ_y used in the mathematical model is too large; numerical dispersion might also be important. A rough estimate for the lateral dispersion coefficient has been obtained using the measured concentration in the hydraulic model. The lateral concentration distribution in the cross section at the location of the highest concentration is compared to a Gauss distribution. It is assumed that the variance σ_y^2 of the Gauss distribution increases linearly with with a slope equal to 2 ϵ_y. This yields

$$0.25 \leqslant \frac{\epsilon_y}{d\ u_*} \leqslant 1$$

where u_* = mean shear velocity. In the hydraulic model in addition to turbulence and transverse circulations the lateral dispersion is caused by large-scale advective transport in the channel network, as shown in Fig. 16. In the mathematical model the value $\epsilon_y/u_*d = 1$ is used for the contributions of turbulence and transverse circulation only. The resulting lateral spreading in the mathematical model is too large, because the advective transport in the channel network is to some extent accounted for twice.

Unfortunately, in computations with a smaller value for the lateral dispersion coefficient (which would also provide a test for the influence of the vertical scale in a distorted model) numerical problems arose, related to the occurrence of very steep concentration gradients. In that case there is probably also important numerical dispersion. These problems could be partially solved by using a smaller grid size than the coarse 800 m grid of the present model. The numerical problems tend to disappear after a period of initial mixing.

Another difference between the hydraulic and the mathematical model can be seen in the dye intrusion into the southeastern branch of the Oosterschelde. Figures 15d-15f show that in the hydraulic model the dye intrusion is strongest at the north side of this branch, while in the mathematical model the dye enters this branch mainly via the south side.

As do most wide estuaries, the Oosterschelde has channels with a dominating ebb current and channels with a dominating flood current. Often the morphology of an estuary will reveal whether a channel is ebb or flood dominated. According to Van Veen (1950) an ebb channel is generally shallow at its seaward end while a flood channel is shallow at its landward end. In the hydraulic model residual circulations were measured at a few cross sections, with the results shown in Fig. 17a. These circulations were measured by integrating the measured velocities over the tidal cycle and over the depth to obtain a residual flux in m^2/s, and integrating the result over that part of the width containing a residual flow in the given direction. Thus the residual circulation within the channels as given in Figure 17a was

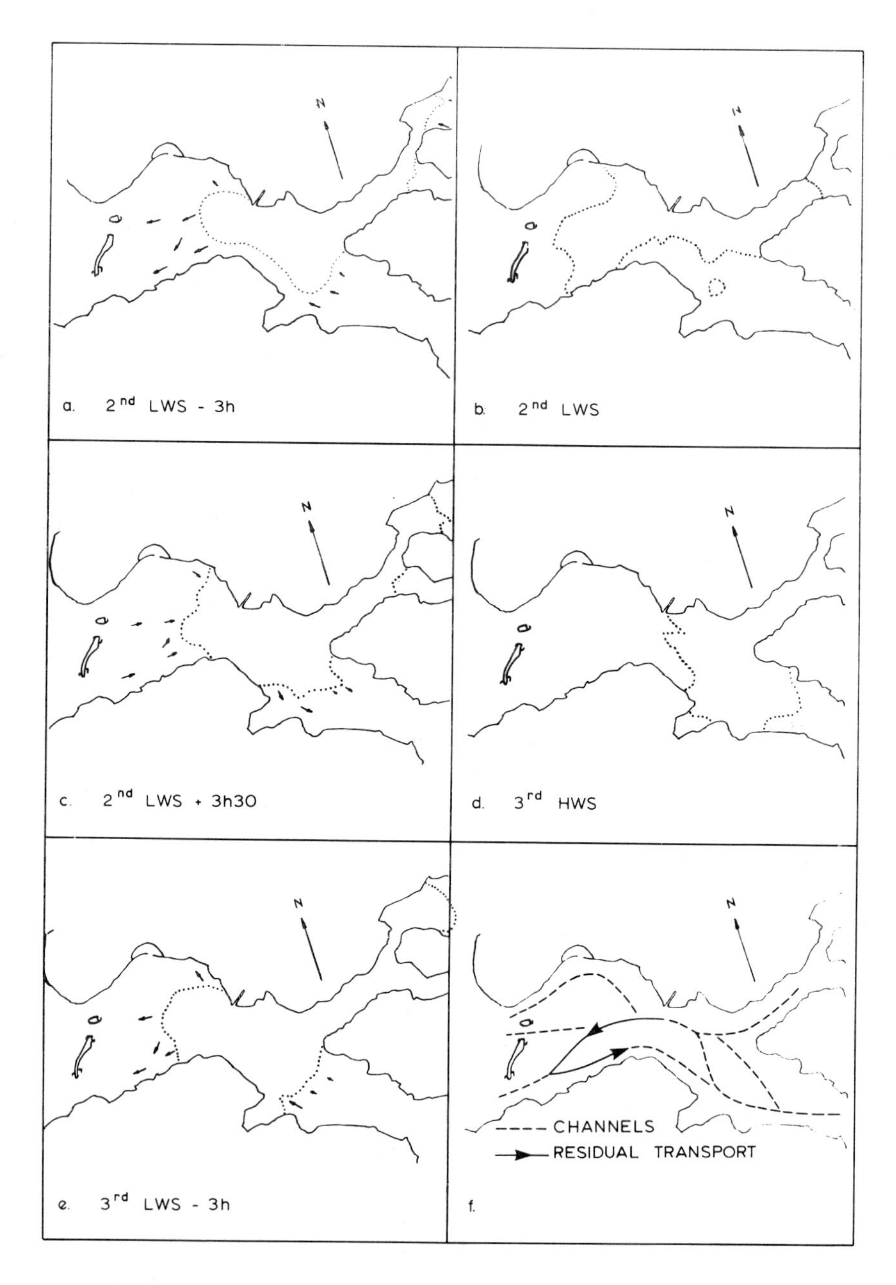

Figure 16. Evolution of the contour line for concentrations greater than $5 \times {}^{-6}$ kg/m^3 in the mathematical model during the third cycle after injection: (a) three hours before the second LWS after injection; (b) second LWS; (c) three hours, thirty minutes after second LWS; (d) third HWS after injection; (e) three hours before the third LWS after injection; (f) spreading by longitudinal advection in the channel network.

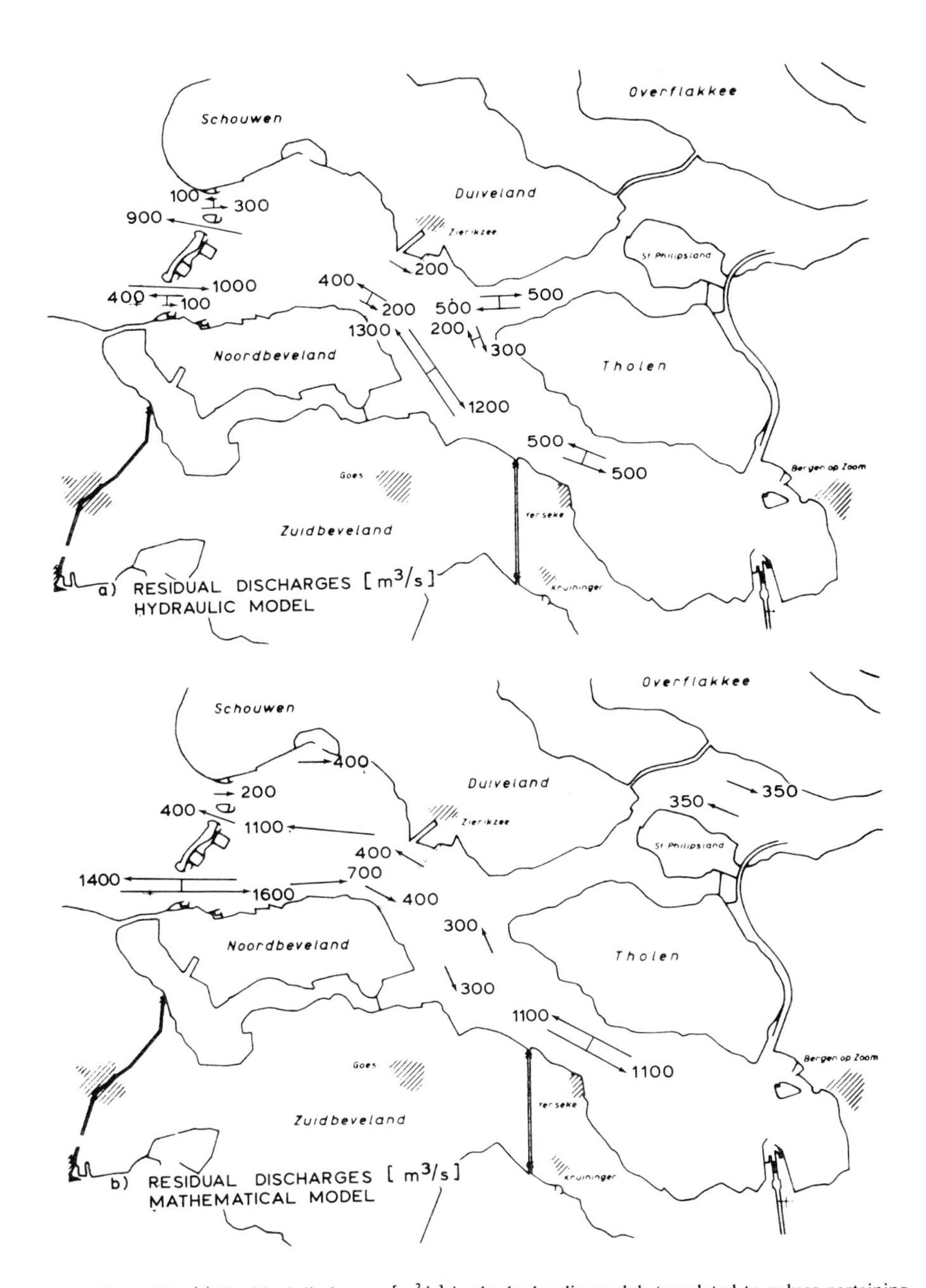

Figure 17. (a) Residual discharges [m^3/s] in the hydraulic model, translated to values pertaining to the actual estuary. Connected arrows indicate circulations within the same main channel. (b) Residual circulations in the mathematical model. The arrows indicate the direction of the residual discharge per unit width.

obtained. A much more complete view of residual circulations can be obtained from the mathematical model, as sketched in Fig. 17b. The circulation cells computed by the mathematical model are small, implying that the residual discharges may be very different in cross sections which are close to each other. This may explain in part the differences shown in the two figures. It is most probably a local difference in the residual circulation pattern which causes the dye intrusion into the south-eastern branch to be different in the hydraulic and the mathematical model. In the seaward part of the estuary the dye intrusion seems to be about the same in all the channels, independent of whether there is a residual ebb or a residual flood flow (see Fig. 15a-c). Also the average residual discharge in the main channels is very small as compared to the tidal discharges. From this it may be concluded that the mixing processes are not dominated by residual circulations in this part of the estuary.

7. QUALITATIVE DESCRIPTION OF LARGE-SCALE MIXING PROCESSES IN THE OOSTERSCHELDE

For practical reasons the patch of dye in the hydraulic model could only be sampled at high water slack and low water slack, whereas in the mathematical model the position and the spreading of dye can be analyzed at any moment. From this analysis insight can be gained into the processes which are responsible for the mixing in the estuary. The different mixing processes which appear in the mathematical model will be illustrated by the evolution of the dye mass in time. In Figs. 18a-18e the evolution of the patch of dye during the first tidal cycle after injection is shown. During the flood period (see Figs. 18a, 18b and 18c) the patch of dye enters the three channels indicated by Fig. 18f; a large amount of dye remains in the dead zone along the northern shore. In Channel 2 the tide precedes the tide in Channel 1: ebb starts in Channel 2, while Channel 1 is still in flood. In Channel 1 slack tide at the north side precedes slack tide at the south side. Because of the larger depth at the south of Channel 1 the water masses experience less friction and therefore decelerate less when the water surface slope changes sign. In this way a large-scale circulation is created causing the exchange of estuary waters. This conforms to the evolution of the patch of dye shown in Figs. 15c and 15d; a schematic representation is given in Figure 19. The same mechanism also explains the relative maximum in salinity observed in nature after each high water slack (see Fig. 20) at the salinity station indicated in Fig. 18f. The salinity record suggests that high salinity seawater is trapped between relatively low salinity estuary water.

Figure 18 also shows that, during the ebb, part of the dye in Channel 3 is separated from the main body of the dye mass. This is a consequence of a local residual circulation pattern in which the flood current dominates at the northern side of Channel 3.

Comparison of Figs. 18a and 18e shows the enormous dispersion of the dye in one tidal period. The large-scale processes playing a role in the dispersion are

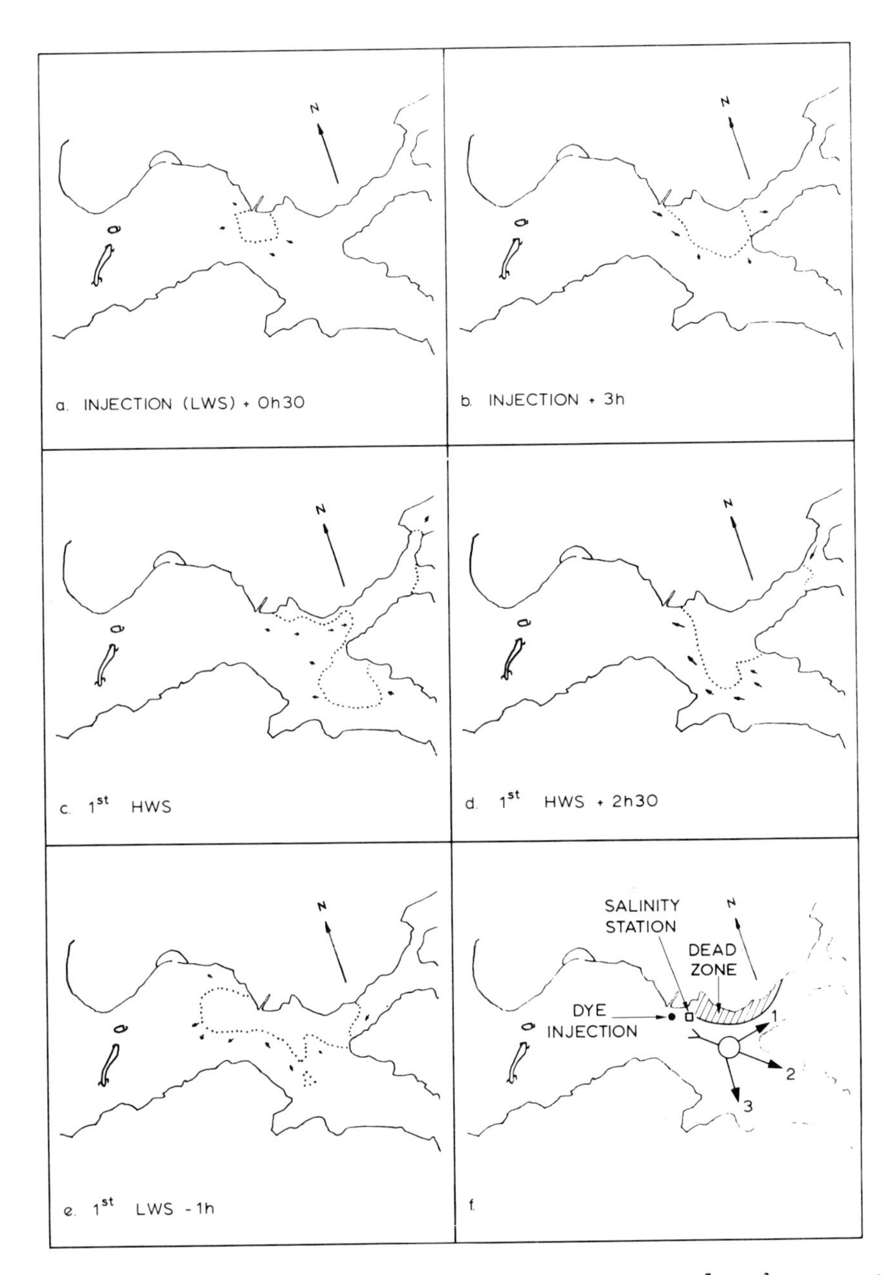

Figure 18. Evolution of the contour for concentrations greater than 10^{-7} kg/m^3 during the first tidal cycle after injection in the mathematical model. (a) thirty minutes after injection; (b) three hours after injection; (c) first HWS after injection; (d) two hours thirty minutes after injection; (e) one hour before first LWS after injection; (f) indication of dead zone; channel junction and salinity station.

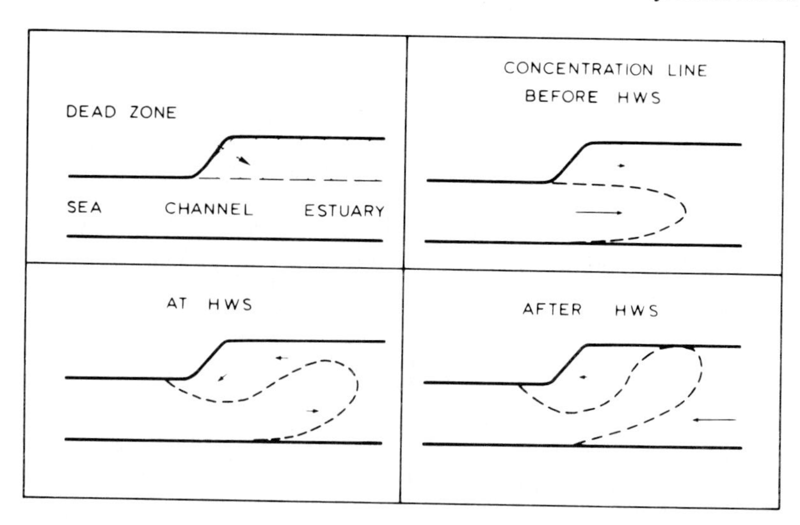

Figure 19. Schematical representation of renewal of estuary waters by return currents in shallow regions near the boundaries.

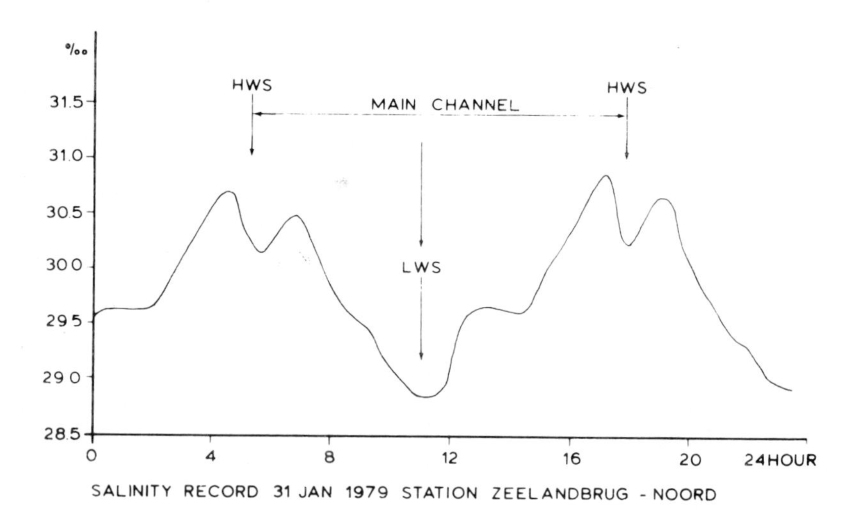

Figure 20. Record of measured salinity values in the Oosterschelde Estuary at the station indicated in Figure 15f.

the retention of water masses in dead zones along the boundary, together with return currents in the shallow parts of the cross section about slack tide, the division of water masses at channel junctions, together with tidal phase difference between connecting channels, and residual circulations in the channel network.

In the seaward part of the Oosterschelde estuary a fluid particle passes several channel junctions during each tidal excursion. Therefore, by choosing different channel branches during flood and ebb, a fluid particle may experience in one tidal period an important residual displacement in longitudinal direction. This implies that the longitudinal dispersion increases when the exchange of water masses between different channel branches increases. The exchange of water masses between the channel branches is influenced by residual circulations and tidal phase differences, but also by lateral mixing. Especially when the time scale for lateral mixing in the channel decreases and becomes comparable to a tidal period, the probability that a fluid particle will choose different channel branches during ebb and flood increases, yielding an increase in longitudinal dispersion. This is another mechanism which may explain the fact that in the seaward part of the hydraulic model a larger longitudinal dispersion coefficient was found than in the Oosterschelde estuary.

In the hydraulic model a second experiment with instantaneous injection of dye was carried out for the future Oosterschelde with storm surge barrier and secondary dams. In Figure 21 the dye patches in the present and future situation are compared at the 4th LWS after injection. It appears that in the future situation the reduction of the dispersion is larger in compartments I and IV than in compartment II. This agrees with the findings of the one-dimensional analysis for the continuous injection as given in Table 2.

8. SUMMARY AND CONCLUSIONS WITH RESPECT TO PREDICTIVE ABILITIES

The construction of a storm surge barrier and two secondary dams in the Oosterschelde will reduce the movement of water and salinity in this estuary. The goal is to secure the surrounding land from flooding and to keep the salinity in the estuary in the same range as in the present situation. A research program is being carried out to predict the results of the construction. The required accuracy is set by the environmental standards for the estuary (salinity range not detrimental for the ecology of the estuary) and the accuracy that can be obtained by measurements in nature. On the basis of these criteria an accuracy of the salinity predictions of 2 ppt is asked for on a total salinity range of approximately 8 ppt. Since the research program is still in progress, the following conclusions are preliminary; a firm statement regarding the predictive abilities of the various models cannot be given as yet. The first objective of the research is to obtain insight into the various mechanisms that dominate the water and mass transport in the estuary. This is done by evaluating and comparing measurements in nature, in a hydraulic scale

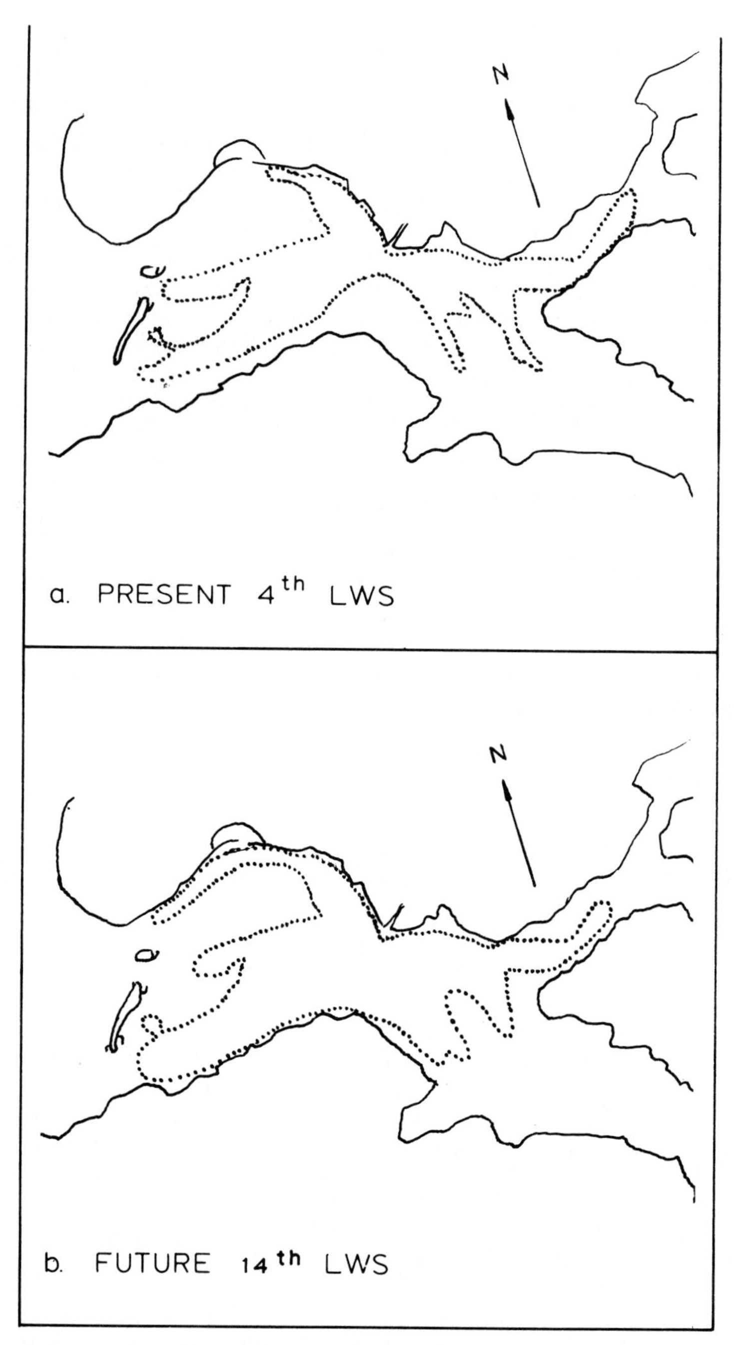

Figure 21. Isoconcentration lines (5×10^{-6} kg/m^3) in the hydraulic model at the fourth LSW after injection. (a) Present situation; (b) future situation.

model and a two-dimensional depth integrated numerical model of the Oosterschelde. The results of this research showed that the major part of the estuary is homogeneous in the vertical direction with a rather small longitudinal salinity gradient. As density currents play a secondary role in this part, depth integrated models with a gradient-type transport formulation can be used. The northeastern branch of the estuary is partly stratified with a dominating role of density currents. Here three-dimensional or two-dimensional width integrated models should be used.

The research reported here focused on the homogeneous part of the estuary. In this part of the Oosterschelde the mixing is dominated by trapping and returning currents in the shallow parts of the cross section about slack tide, and by division and confluence of water masses at channel junctions together with phase differences between the channels. Residual circulations in the channel network seem to be only locally important. Currents caused by the Coriolis-force do not influence the mixing processes in the estuary significantly.

The analysis of the transport processes leads to the conclusion that the time scale of lateral mixing is an important parameter in this estuary especially in Compartment I. This means that with respect to the mixing processes in this compartment scale effects may arise in the hydraulic scale model. In the mathematical model attention must be paid to a correct choice of the lateral dispersion coefficients.

The 800 m grid used in this model research for economical reasons is rather coarse for dye patch dispersion studies in this estuary with tidal channels of a typical width of 500-1000 m and large tidal flats. Therefore, it is surprising that, nevertheless, rather small circulation cells are reproduced in the mathematical model. The circulations computed in the mathematical model are generally in good correspondence with the residual discharges measured in the hydraulic model.

The time scale for renewal in the estuary is approximately 80-100 tidal cycles. The construction of the storm surge barrier and the secondary dams will increase this by a factor of 3 to 4. Extra emphasis should therefore be given to the stability of the models. The WAQUA-system proved to be mass-conservant. Long-term calculations (a two month period) showed the scheme to remain stable.

The longitudinal dispersion of the dye patches in the hydraulic model is, generally speaking, well simulated by the mathematical model for the present situation.

The following preliminary conclusions can be drawn with respect to the predictive abilities of the two-dimensional depth integrated mathematical model for water movement and water quality in the Oosterschelde:

- Once the model is calibrated and verified on the available prototype and physical model data the tidal water movement in the estuary is simulated very accurately (within the accuracy of flow measurements, estimated at about 10%-15% of the total maximum flow rate).

- As far as the velocity distribution is concerned extrapolation to the future situation can be performed since the problem is of the wall shear type. Extra verification is possible by comparison of mathematical and physical model research.

- In the homogeneous part of the Oosterschelde the reproduction of the longitudinal dispersion in the hydraulic model is accurate enough to insure an accuracy within the required standard of 2 ppt. The results of the mathematical model are similar to those of the hydraulic model. Further calibration on the results of transport studies in the hydraulic model for the future situation is possible. Therefore it is likely that salinity intrusion research in the vertically homogeneous part of the estuary can be performed with the mathematical model with the required standard of 2 ppt. Salinity intrusion research in the partly stratified part of the estuary requires a large calibration effort for the dispersion coefficient on the basis of prototype measurements, which amounts to data fitting. The homogeneous hydraulic model cannot be of any help here. Extrapolation of these dispersion coefficients to the future situation remains uncertain. Research with other models should be preferred.

- When using the model for research of calamities special attention should be paid to the proper value of the lateral dispersion coefficient. Calibration against prototype measurements should be carried out, and a smaller grid size is needed. It should be recognized that in problems involving a limited scale in location and time the residual currents in the channel network may play an important role.

REFERENCES

Delft Hydraulics Laboratory (1972). Discharge coefficients of closure gaps in distorted models. DHL Rep. M 731-XIV (Dutch text).

Delft Hydraulics Laboratory (1980a). A study of the dispersion mechanisms in the Oosterschelde; Part I, Continuous dye releases. DHL Rep. M 1603-I (Dutch text).

Delft Hydraulics Laboratory (1980b). A study of the dispersion mechanisms in the Oosterschelde; Part II, Instantaneous dye releases. DHL Rep. M 1063-II (Dutch text).

Delft Hydraulics Laboratory and Rijkswaterstaat (1974). Scale research in nature and hydraulic scale models of the Brouwershavense Gat-Grevelingen estuary. DHL Rep. M 1010 (Dutch text).

Dronkers, J. (1981). Salinity fluctuations in the Oosterschelde estuary: a qualitative interpretation of field measurements. Submitted to *Neth. Jour. of Sea Res.*

Fischer, H. B., Imberger, J., List, E. J., Koh, R.C.Y., and Brooks, N. H. (1979). "Mixing in Inland and Coastal Waters." Academic Press, pp. 151.

Okoye, J. K. (1970). Characteristics of Transverse Mixing in Open-channel Flows. Rep. KH-R-23, California Institute of Technology.

Okubo, A. (1967). The effect of shear in an oscillatory current on horizontal diffusion from an instantaneous source. *Int. Jour. Oceanol. Limnol.* **1**, 194.

Stroband, H. J. and Van Wijngaarden, N. J. (1977). Modeling of the Oosterschelde Estuary by a hydraulic model and a mathematical model. *Proc. 17th IAHR Cong., Baden-Baden,* Paper A 109, DHL Pub. No. 192.

Van Veen, (1950). Ebb- and flood-channel systems in the Dutch tidal waters. *Koninklijk Ned. Aardrijkskundig Genootschap* **67**, 303-335 (Dutch text).

A TWO-DIMENSIONAL, LATERALLY AVERAGED MODEL FOR SALT INTRUSION IN ESTUARIES

P.A.J. Perrels and M. Karelse

Delft Hydraulics Laboratory, The Netherlands

1. INTRODUCTION

Several years ago the Delft Hydraulics Laboratory (DHL) began the development of a predictive numerical model for the calculation of density currents in estuaries. The model is intended for three kinds of applications: as a tool for fundamental research with respect to mathematical and numerical modeling of salt intrusion in estuaries; as a tool for the prediction of the consequences of future alterations in estuaries with respect to tidal motion and salt concentration; and as a tool for the interpretation of phenomena and measurements from the Delft Tidal Flume (the flume is shown in Figure 1 and described in Appendix IV). The development began with the inventorying of the relevant physical phenomena (Section 2). Then the mathematical model could be formulated (Section 3) together with the approximations and schematizations made. Next, a numerical approach could be selected (Section 4). A review of existing two-dimensional, laterally integrated models was also conducted (Section 5). Finally, the model is being calibrated and verified to test its (predictive) abilities (Sections 6 and 7). For the calibration and verification the following series of successive phases are distinguished:

(1) Homogeneous calibration in a single flume, to test the tidal motion in strict two-dimensional (2-D) circumstances;

(2) Inhomogeneous calibration in a single flume, to test the tidal motion and the salt distribution in strict 2-D situations;

(3) Homogeneous and inhomogeneous calibration in a branched flume to test the modeling of the phenomena around a node;

(4) Verification in a flume, to test predictive abilities for strict 2-D situations;

(5) Calibration and verification with field data to test the abilities to represent 3-D features with a 2-D model.

Phases (1) and (2) have been completed and the results can be found in this report. From phase (4) only preliminary results are given. Phases (3) and (5) have not yet been started.

ISBN: 0-12-25852-0

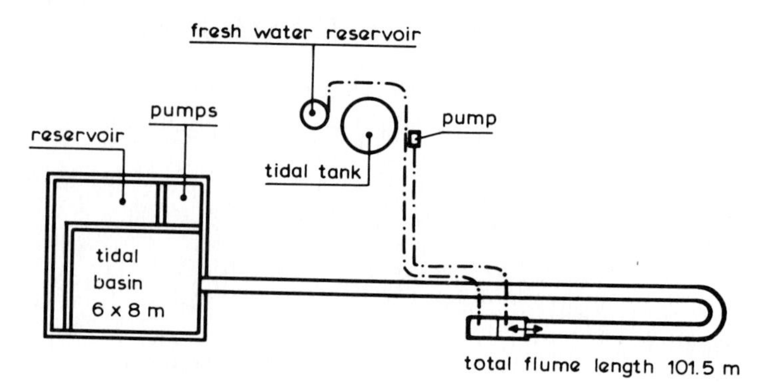

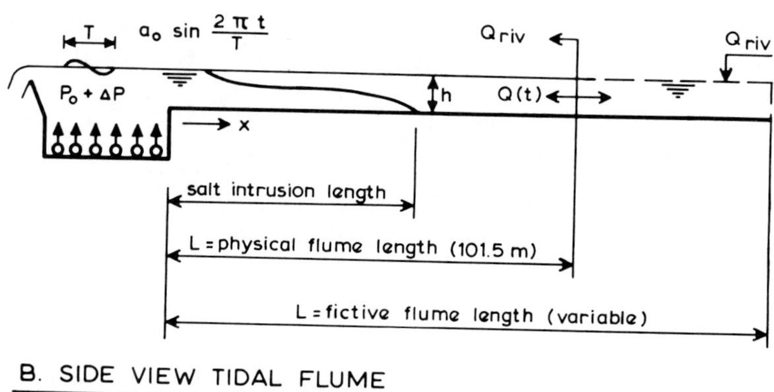

Figure 1. Set up of the tidal salinity flume with notations.

2. CHOICE OF THE MODEL

As indicated in the introduction the aim of the DHL is to build a predictive salt intrusion model for estuaries. A crucial point is that the influence of the density differences on the transport phenomena is correctly modeled. Unless restrictions are made for the possible applications, a three-dimensional model is required consisting of the equations for the conservation of momentum, the conservation of mass (water), and the conservation of salt content. This would call for a large computer capacity; in many interesting situations, however, such as a tidal flume and to a certain degree also an estuary like the Rotterdam Waterway, several restrictions can be made. First, the application was restricted to estuaries where the phenomena are dominated by tidal influences, with a negligible influence of the wind. The estuaries are supposed to have a significant fresh water discharge, which in combination with the tide results in a vertical gravitational circulation. Moreover the application is restricted to laterally uniform estuaries in which the transverse gravitational circulation and the tidal pumping mechanism are negligible. Dispersion is supposed to be mainly due to tidal trapping by the groynes and the small harbors (Fischer, *et al.*, 1979). The main branches and the large harbors influence the mixing process by phase effects. A further restriction is that only long wave phenomena are considered. The shallow water approximation can then be made and only gradual variations in time and the longitudinal direction are permitted. As a result the flow is of a boundary layer type.

An important requirement is that the model should represent a wide range of possible stratified situations. The model should further provide detailed information at a time scale which is much smaller than the tidal period. Hinwood and Wallis (1975a, b) review models for tidal waters up to 1975 and give a classification of tidal models. Here we may distinguish seven groups: zero-dimensional models; one-dimensional models; one-dimensional two-layer models; two-dimensional models in plan; two-dimensional models in side view; two-dimensional two-layer models; and three-dimensional models. Given the restrictions and requirements just mentioned, one-dimensional and two-layer models are insufficient. At least a two-dimensional, laterally-averaged model is required; a three-dimensional one would be better, but from an economic point of view is hardly feasible. Moreover the modeling of the vertical structure would largely correspond and therefore a two-dimensional model could provide useful information for a possible development of a three-dimensional one.

For the three kinds of applications mentioned in the introduction, it can be said that a two-dimensional side-view model can certainly serve as a tool for fundamental research, and also for the interpretation of the results from the Delft Tidal Flume. The extent to which a two-dimensional model can be predictive is in fact one of the main subjects for investigation. In an estuary with ebb and flood channels and tidal flats a side view model may not be expected to include all relevant phenomena. However, for a narrow estuary, such as the Rotterdam Waterway, where the dispersion due to lateral averaging is small, as long as large harbors and branches are schematized explicitly a branched, two-dimensional laterally averaged model to a certain extent may be predictive. Consequently, it was decided to start with the development of a two-dimensional side view model.

3. THE MATHEMATICAL MODEL

In estuaries flows are almost always turbulent. Because it is next to impossible and also of little importance to calculate the details of the fluctuating motion, the equations are averaged over a time scale which is long compared to that of the turbulent motion and small compared to the time scale of the phenomena under consideration. (For the present study this means much smaller than the tidal period.) Because of this averaging the equations no longer constitute a closed system. They contain correlation terms representing the transport of momentum (Reynolds stresses) and mass (diffusive transports) by the turbulent motion. To arrive at a closed system of equations the correlation terms must be expressed in terms of the mean quantities (turbulence model). The resulting equations no longer simulate the details of the turbulent motion but only the effect of turbulence on the mean flow behavior. When the model is laterally averaged, the basic equations contain again correlation terms (dispersion). These represent the difference between the total transport of mean momentum and mass and the transport in the laterally integrated approximation. These correlation terms also must be expressed in terms of mean quantities. In the next section the formulation of the mathematical model is presented and the consequences of certain choices are discussed.

3.1 The Basic Equations

The basic equations for the model are provided by the laterally integrated conservation laws for mass (water), momentum, and salt content. If the shallow-water approximation is made, the equation for the conservation of vertical momentum reduces to the hydrostatic pressure distribution (DHL, 1973):

$$\frac{\partial p}{\partial z} = -\rho g \tag{1}$$

in which p denotes the pressure, ρ the density, g the acceleration due to gravity, and z is the vertical coordinate (positive upward). Integration of Eq. (1) with respect to z yields the equation for the pressure

$$p = p_s + \rho_0 g(\zeta - z) + g \int_z^{\zeta} (\rho - \rho_0)\, d\theta \tag{2}$$

where p_s is the pressure at the surface, ρ_0 the fresh water density and ζ the position of the surface.

The equation for the conservation of longitudinal momentum reads

$$\frac{\partial}{\partial t}(b_c u) + \frac{\partial}{\partial x}(b_c u^2) + \frac{\partial}{\partial z}(b_c uw) + \frac{\partial}{\partial x}(b_c \tau_{xx})$$
$$+ \frac{\partial}{\partial z}(b_c \tau_{xz}) + \tau_w = -\frac{b_c}{\rho}\frac{\partial p}{\partial x} \tag{3}$$

where t is the time, x is the longitudinal coordinate, u is the longitudinal velocity, w is the vertical velocity and b_c is the current width. The modeling of the Reynolds stresses τ_{xx} and τ_{xz} will be presented in Section 3.4. The wail stress, τ_w, is computed by $\tau_w = \lambda \rho u \mid u \mid$, in which λ is the friction coefficient. The equation for the conservation of salt content and the equation of state read

$$\frac{\partial}{\partial t}(b_s c) + \frac{\partial}{\partial x}(b_c uc) + \frac{\partial}{\partial z}(b_s wc) + \frac{\partial}{\partial x}(b_s T_x) + \frac{\partial}{\partial z}(b_s T_z) = 0 \tag{4}$$

$$\rho = \rho_0 + \beta c \tag{5}$$

in which β is of the order of 0.75, c denotes the salt concentration and b_s the storage width. The modeling of the diffusive transports T_x and T_z is given in Section 3.4.

Lateral averaging yields the following correction terms in Eqs. (3) and (4) (DHL, 1974a):

(1) A correction in the pressure term, which depends primarily on the lateral surface gradient and secondly on the accuracy of the cross-sectional schematization. The first contribution depends on local circumstances and cannot be parameterized in a predictive way. Consequently, this is a limitation of the predictive capabilities. The second contribution can be reduced by a refinement of the lateral schematization and therefore is not a limitation on the predictive capabilities. However, refinement demands extra computational effort.

(2) A correction in the Reynolds stresses and the diffusive transports, due to lateral variations in the velocity and concentration distribution. The contribution to the vertical terms can be neglected in comparison with the vertical turbulent exchange. The contribution in the longitudinal direction is included as a dispersive transport, as will be discussed in Section 3.4.

The continuity equation reads

$$\frac{\partial}{\partial x}(b_c u) + \frac{\partial}{\partial z}(b_s w) = 0 \tag{6}$$

Integration of Eq. (6) over the depth and substitution of the kinematic conditions

$$u\frac{\partial z_b}{\partial x} - w = 0 \qquad (z = z_b) \tag{7a}$$

and

$$\frac{\partial \zeta}{\partial t} + u\frac{\partial \zeta}{\partial x} - w = 0 (z = zeta) \tag{7b}$$

yields the equation for the variation of the cross section from which the position of the free surface is found

$$\frac{\partial A}{\partial t} + \frac{\partial}{\partial x}\left\{\int_{z_b}^{\zeta} b_c\ u\ dz\right\} = 0 \tag{8}$$

where A denotes the cross section and z_b the position of the bottom.

3.2 The Boundary Conditions

3.2.1 Free surface conditions. At the free surface the kinematic condition holds (7b), which was used for the derivation of (8). For the momentum equation the surface stress must be specified

$$\tau_{xx}\frac{\partial \zeta}{\partial x} + \tau_{xz} = -\tau_s \tag{9}$$

τ_s can be derived from the local wind speed (u_s) 10 m above the water surface, according to the formula

$$\tau_s = \rho f u_s \mid u_s \mid \tag{10}$$

For the conservation of salt the following condition is specified

$$T_x\frac{\partial \zeta}{\partial x} + T_z = 0 \tag{11}$$

3.2.2 Bottom Conditions. The bottom is treated as an impermeable no slip boundary at which u and w are zero. For the conservation of salt a condition is specified similar to the one at the surface

$$T_x \frac{\partial z_b}{\partial x} + T_z = 0 \tag{12}$$

which means no transport through the bottom.

3.2.3 Seaward Boundary. At the seaward boundary, the surface elevation must be specified

$$\zeta\,(0,t) = \zeta_0\,(t), \tag{13}$$

and a boundary condition for the longitudinal velocity must be imposed. Because generally no velocity profiles are available during a whole tidal cycle, a weak condition is imposed

$$\frac{\partial^2 u}{\partial x^2} = 0 \tag{14}$$

Finally a condition for the conservation of salt content must be given. During flood tide the increase of the concentration is specified as a function of t and z. During ebb tide a weak condition is imposed, analogous to the condition for the longitudinal velocity.

$$u(z) > 0 : \frac{\partial c}{\partial t} = g(t,z) \qquad \text{if } c < c_{\max} \tag{15}$$

$$= 0 \qquad \text{if } c = c_{\max} \tag{16}$$

$$u(z) \leqslant 0 : \frac{\partial^2 c}{\partial x^2} = 0 \tag{17}$$

The seaward boundary condition must reflect the phenomenon at sea. This means that three dimensional phenomena outside the model should be reflected in the transition function $g(t,z)$. In the present model the following time-independent relation is adopted

$$g = c_n + c_s \frac{(H-z)}{H} c_n \tag{18}$$

The coefficients c_n and c_s fix the maximum transition time. The coefficient c_s reflects the possibility that around low water slack at sea a stratified situation occurs. The local value of $u(z)$ decides between use of Eq. (15), (16), and (17).

3.2.4 Upstream boundary. The upstream boundaries can be open, in which case the velocity and concentration have to be specified. Normally the open boundary is supposed to be upstream of the maximum salt intrusion, so that $c = 0$. An upstream boundary can also be closed, in which case $u = 0$ and we assume no longitudinal transport

$$T_x = 0 \tag{19}$$

3.3 The Nodes

Because the model is branched, nodes must also be modeled. For a node of N branches, equations expressing the requirements of no storage at a node, and uniqueness of the water level, the vertical velocity, and the concentration of salt at a node are as follows

$$\sum_N Q_i = 0 \tag{20}$$

$$\zeta_1 = \cdots = \zeta_i = \cdots = \zeta_N \tag{21}$$

$$w_1 = \cdots = w_i = \cdots = w_N \tag{22}$$

$$c_1 = \cdots = c_i = \cdots = c_N \tag{23}$$

3.4 The Turbulence Model

To close the system of equations given in (3.1) a turbulence model is necessary. In the literature turbulence models are classified according to the number of transport equations used for the turbulence quantities (Rodi, 1980). The simplest approach is the eddy viscosity/diffusivity concept which is a zero-equation method. This approach yields for the vertical turbulent exchange of momentum and mass respectively

$$\tau_{xz} = -\epsilon_z \frac{\partial u}{\partial z} \tag{24a}$$

$$T_x = -D_x \frac{\partial c}{\partial x} \tag{24b}$$

$$T_z = -D_z \frac{\partial c}{\partial z} \tag{24c}$$

For a boundary-layer type of flow, as we have here, theoretically this approach can be made plausible, even under tidal circumstances (DHL, 1974b). The main problem is whether the influence of the stratification on the turbulent exchange can be properly modeled, if possible in a predictive way. Abraham (1980) argues that a length scale, which depends on the distance to the bottom, may not correctly reflect that turbulence can be generated both externally (at the bottom) and internally (by stratification). However, this problem of modeling the influence of the stratification holds for higher order turbulence models as well and therefore, at the start, a mixing length model was chosen. Moreover, a mixing length model has the advantage over higher order models in being the most economic numerically.

For the turbulent exchange coefficients of mass and momentum, respectively D_z and ϵ_z, a mixing length approach is used (DHL, 1973)

$$D_z = \epsilon_z = l_m^2 \left| \frac{\partial u}{\partial z} \right| \tag{25}$$

The mixing length l_m is defined as

$$l_m = \kappa (z + z_0) \quad \text{for } 0 \leqslant z \leqslant 0.25H \tag{26a}$$

$$l_m = \kappa (0.25H + z_0) \quad \text{for } 0.25H \leqslant z \leqslant H \tag{26b}$$

in which z_0 is a measure for the roughness length and H denotes the water depth. For steady state, z_0 is related to the Chezy-value by $C = 18 \log (12R/33\ z_0)$, where R denotes the hydraulic radius.

In partly mixed estuaries, buoyancy effects influence the turbulence and these are accounted for in an empirical way. The eddy viscosity and diffusivity under homogeneous circumstances (Eq. (25)) are multiplied by damping factors, which are a function of the stratification parameter Ri, to give

$$\epsilon_z = l_m^2 \left| \frac{\partial u}{\partial z} \right| F(Ri) \tag{27a}$$

$$D_z = l_m^2 \left| \frac{\partial u}{\partial z} \right| G(Ri) \tag{27b}$$

where Ri denotes the local gradient Richardson number, defined by

$$Ri = -\frac{g}{\rho}\frac{(\partial\rho/\partial z)}{(\partial u/\partial z)^2} \tag{28}$$

An important difficulty in the use of the local Richardson number is that no turbulent energy is transferred where $\partial u/\partial z = 0$. In a tidal flow locally $\partial u/\partial z$ approaches zero at certain times and it seems unlikely that turbulent transport will then locally disappear. In view of the difficulties in specifying a local Ri-number, some authors (Bowden and Hamilton, 1975; Rodger, 1980; Odd and Rodger, 1978) prefer to work with an overall, respectively maximum value of Ri, based on larger scale features of the flow, accepting the loss of detailed information about the eddy viscosity and diffusivity.

In the literature (DHL, 1974b) many sets of relations for $F(Ri)$ and $G(Ri)$ can be found. Optimization of the choice of such a set is part of this study. From the calibration with tidal flume data it turned out (Section 6) that the best results were obtained with

$$F(Ri) = \exp(-4Ri)$$

$$G(Ri) = \exp(-18Ri) \tag{29}$$

These relations were taken from tidal flume measurements (Van Rees, 1975). In Delft Hydraulics Laboratory (1974b), it is shown by dimensional analysis that the influence of the stratification on the vertical turbulent exchange is a function of Ri, but the theory does not indicate what form this function should have. In the literature some suggestions can be found (Rossby and Montgomery, 1935; Kent and Pritchard, 1959), but without further verification. Consequently, all available relations are mainly empirical in nature, which probably will have a negative effect on predictive capabilities.

3.5 The Longitudinal Exchange of Mass and Momentum

The contribution of τ_{xx} could be neglected and for T_x the following expression was given

$$T_x = -D_x\frac{\partial c}{\partial x} \tag{24b}$$

T_x consists of two contributions, one from the longitudinal turbulent diffusion and another due to lateral averaging. For D_x a relation was adapted similar to the one introduced by Elder (1959),

$$D_x = d_1 b \mid \bar{u} \mid + d_2 \tag{30}$$

Although d_2 can be neglected compared to the average value of D_x, it is retained for numerical reasons (Section 4). There are indications from flume and field data that T_x should have negative values during ebb tide (DHL, 1980b). Physically this should be seen as a limit for the applicability of the dispersion concept.

4. THE NUMERICAL MODEL

4.1 The Numerical Technique

For the numerical integration of the system of differential equations, a finite difference technique was selected, in combination with a coordinate transformation (DHL, 1976b). In order to incorporate longitudinal and tidal variations in the bottom and surface topography, the physical domain is transformed in the vertical direction (Appendix I). This transformation yields a rectangular grid, which permits simple finite difference operations. A finite difference technique also offers the possibility of splitting the differential equations into parts for different spatial directions, which makes possible an efficient implicit difference technique.

For the discretization of the momentum equation after the splitting an explicit technique is used in the longitudinal direction and an implicit technique in the vertical direction. For numerical stability the term $\epsilon_x \partial^2 u / \partial x^2$ is added to the part in the longitudinal direction (see Section 4.2). For a boundary layer type of flow, as we have here, with a large ratio between the longitudinal and the vertical length scale, this combination of an explicit and an implicit technique demands no very restrictive stability conditions and offers sufficient accuracy in an economic way. For the discretization of the equation for the conservation of salt content a similar technique is used as for the momentum equation. For the continuity equation a central difference technique is used. For the computation of the cross section an explicit difference scheme is applied which yields the well known "longwave" criterion as stability condition (the difference equations are given in Appendix II). Depending on the scales of the problem the latter stability criterion may be the most restrictive. In that case it could be worthwhile to apply an implicit technique for the free surface. This possibility was already mentioned by Hamilton (1975). Roberts and Street (1975) compared the explicit and implicit approaches. The conclusion in that particular situation was that, taking the accuracy into account, no benefits could be gained from the use of an implicit method.

4.2 Stability

Numerical stability was investigated by means of the energy method (DHL, 1975a). This method yields sufficient conditions and can be applied to non-linear equations. In the present case only homogeneous boundary conditions were taken into account. The stability condition for the momentum equation reads:

$$\tau \leqslant \frac{2\epsilon_x}{\{ | u |_{\max} + \frac{2\epsilon_x}{\Delta x} \}^2} \tag{31}$$

and for the equation of conservation of salt content:

$$\tau \leqslant \frac{2D_x}{\{ | u |_{\max} + \frac{2D_x}{\Delta x} \}^2} \tag{32}$$

where τ denotes the time step and Δx the longitudinal step size. From Eq. (31) it can be seen that a positive ϵ_x is necessary for stability.

A Neumann stability analysis of the linearized momentum and cross-section equation yields the well-known "long wave" equation.

$$\tau < \frac{\Delta x}{| u \pm \sqrt{gH} |} \tag{33}$$

Finally Hirt's heuristic stability analysis was applied to analyze possible oscillations. Therefore, the modified equations were derived and also the boundary conditions were taken into account (DHL, 1975a, 1978). This resulted in two conditions:

$$\epsilon_x - | \frac{\tau}{8}\{16u^2 - 32\epsilon_x \frac{\partial u}{\partial x}\} | > 0 \tag{34}$$

$$D_x - | \frac{\tau}{4} \left\{ u^2 - D_x \frac{\partial u}{\partial x} \right\} - \frac{\Delta x^2}{2} \frac{\partial u}{\partial x} | > 0 \tag{35}$$

4.3 Accuracy

The accuracy of a difference scheme can be given in terms of the truncation error. For the present numerical model these errors read (for constant width):

For the momentum equation:

$$E_{Mx} = -\frac{\tau}{8} \frac{\partial^2 u}{\partial t^2} - \Delta x^2 \left\{ \frac{\partial u}{\partial x} \left(\frac{\partial^2 u}{\partial x^2} \right) - \frac{\epsilon_x}{12} \frac{\partial^4 u}{\partial x^4} - \frac{1}{6\rho} \frac{\partial^3 p}{\partial x^3} \right\} + 0(\tau^2, \Delta x^4) \tag{36a}$$

$$E_{Mz} = -\frac{\tau}{8}\frac{\partial^2 u}{\partial t^2} - \frac{\Delta z^2}{2}\left\{\frac{\partial}{\partial z}\left(\frac{\partial u}{\partial z}\frac{\partial w}{\partial z}\right)\right\}$$
$$+\frac{\Delta z^2}{12}\left\{\frac{\partial^2 \epsilon_z}{\partial z^2}\frac{\partial^2 u}{\partial z^2} + 2\frac{\partial \epsilon_z}{\partial z}\frac{\partial^3 u}{\partial z^3} + \epsilon_z\frac{\partial^4 u}{\partial z^4}\right\} + 0(\tau^2, \Delta z^3) \quad (36b)$$

For the equation of conservation of salt content:

$$E_{Dx} = -\frac{\tau}{8}\frac{\partial^2 c}{\partial t^2} + \frac{\Delta x^2}{12} D_x \frac{\partial^4 c}{\partial x^4} + 0(\tau^2, \Delta x^4) \quad (37a)$$

$$E_{Dz} = -\frac{\tau}{8}\frac{\partial^2 c}{\partial t^2} - \frac{\Delta z^2}{2}\left\{\frac{\partial}{\partial z}\left(\frac{\partial c}{\partial z}\frac{\partial w}{\partial z}\right)\right\}$$
$$+\frac{\Delta z^2}{12}\left\{3\frac{\partial^2 D_z}{\partial z^2}\frac{\partial^2 c}{\partial z^2} + 2\frac{\partial D_z}{\partial z}\frac{\partial^3 c}{\partial z^3} + D_z\frac{\partial^4 c}{\partial z^4}\right\} + 0(\tau^2, \Delta z^3) \quad (37b)$$

For the continuity equation:

$$E_c = \frac{\Delta z^2}{12}\frac{\partial^3 w}{\partial z^3} - \frac{\Delta x^4}{30}\frac{\partial^5 u}{\partial x^5} + 0(\Delta z^2, \Delta x^6) \quad (38)$$

However, the order of the truncation error alone is not necessarily a good indication for the size of the error. This was clearly demonstrated for the errors near the bottom of the discretized momentum equation. Without a special treatment near the bottom, large errors occur in the vertical velocity distribution (DHL, 1975b). Delft Hydraulics Laboratory (1978) describes a special approach (law of the wall) to approximate u close to the bottom (see Appendix III).

The accuracy of this special approach was tested by means of an analytical solution for u. In Figure 2 a comparison is presented between the analytical solution and the numerical solutions with and without the special approach. It can be seen that the special approach improves the accuracy considerably.

During the calibration local instabilities occurred in the vertical velocity and concentration distribution, especially during falling tide. (DHL, 1980c). Quantitatively these phenomena can be explained by the influence of the Ri-number: if locally a stratified situation arises in the concentration, large Ri-numbers arise. If the opposite happens, locally small Ri-numbers arise. In the first situation the large Ri-number leads to a strong damping of the turbulent exchange. The result is that the stratified situation is maintained. In the opposite case the small Ri-number introduces a weaker damping and the vertically homogeneous situation is maintained. As far as this behavior is physically relevant, it should be retained. However, numerical amplification should be avoided. Therefore, in the discretization of the local Ri-number an averaging and a smoothing procedure were introduced, which produced acceptable results.

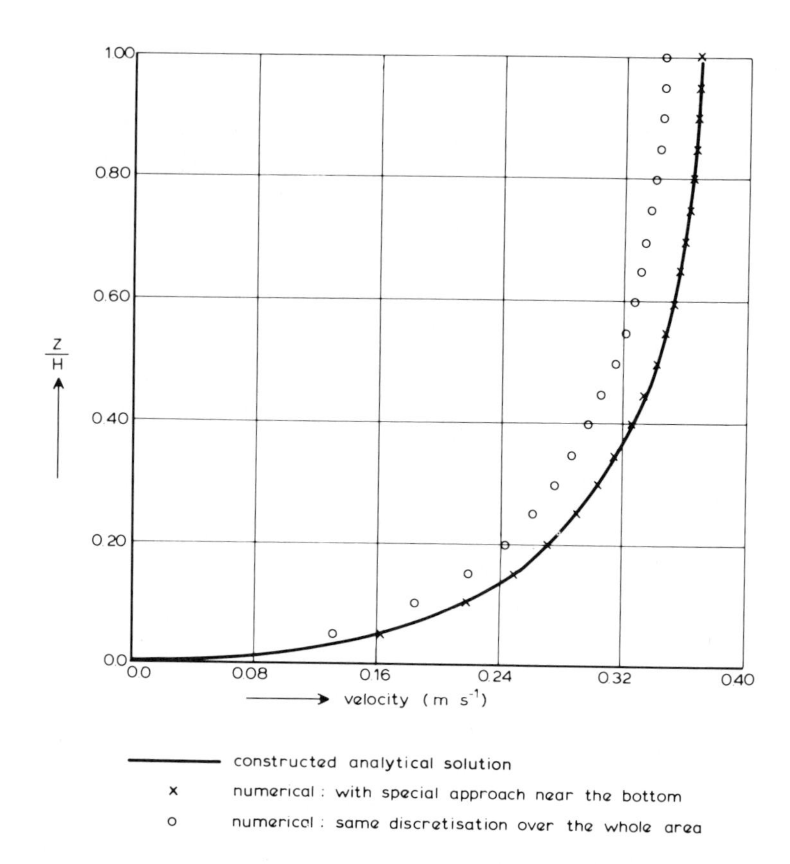

Figure 2. Comparison of analytical and numerical velocity profiles.

A kind of error which deserves special attention is the numerical viscosity and diffusivity. Equations (34) and (35) are sufficient for stability. However, if the diffusivity or the viscosity contributes significantly to the results, accuracy demands more than the conditions (34) and (35). For the tidal flume situation the diffusivity has a significant effect on the intrusion. Consequently, the numerical diffusivity should be much smaller than the physical one, which results in:

$$\left| \frac{\tau}{4} \left\{ u^2 - D_x \frac{\partial u}{\partial x} \right\} - \frac{\Delta x^2}{2} \frac{\partial u}{\partial x} \right| \ll D_x \tag{39}$$

The numerical accuracy of the whole model was tested for tidal flume circumstances (DHL, 1978). A test series was set up in which different step sizes were used, starting from a reference situation. The step sizes were varied one by one. Comparison of the successive results gave an indication about the size of the numerical error. This method has several disadvantages. The analysis has to be repeated for problems with a significantly different scale, and the analysis supposes the numerical technique to be convergent, which has not been proven rigorously. However, if no general analytical solutions are available, as in the present case, it is practically the only way to get an indication of the accuracy of the entire model.

From the results of the analysis the following conclusions concerning the numerical accuracy could be drawn: (a) The number of vertical steps, N_z, is fixed by the accuracy of the computed water level. A minimum number of steps in the vertical direction is $N_z = 12$; $N_z = 24$ yields a good accuracy. The influence on the concentration distribution is of minor importance. (b) The number of longitudinal steps, N_x, is fixed by the numerical diffusivity in the computation of the concentration (Eq. 39). For tidal flume circumstances a mininum number of steps in the longitudinal direction is $N_x = 50$, which means about 200 steps per wave length. (c) The size of the time step for the computation of the water level is fixed by stability (Eq. 33) and for the computation of the concentration by the numerical diffusivity (Eq. 39). In order to meet both conditions in an economic way different time steps were used for the computation of the water level and the concentration. This meant that for the computation of the concentration the velocities remained the same during a few time steps.

5. REVIEW OF EXISTING TWO-DIMENSIONAL SIDE VIEW MODELS

A number of two-dimensional side view, not tidally averaged models have been presented in previous reports. This section reviews five models, designated A to E in Table 1. The sixth model, F, is the subject of this report as described in Sections 3 and 4.

Table 1.
Two-dimensional Side-View Models

Model Designation	Reference to Published Description
A	Blumberg (1975, 1977, 1978)
B	Hamilton (1975)
C,	Boericke/Hogan (1977)
D	Waldrop/Farmer (1973, 1974, 1976)
E	Roberts/Street (1975)
F	Perrels/Karelse

All the models are based on the conservation laws, i.e., the momentum equation, the continuity equation for water, and the continuity equation for salt content. Some of the models neglect the longitudinal exchange of momentum; in C the convection is neglected, too. In L the longitudinal dispersion is neglected. Otherwise, they show much correspondence in the basic equations.

Some differences occur in the boundary conditions which are imposed. In A, B, D, and E wind influence is explicitly included by means of a surface stress. In E longitudinal gradients in the free surface are included in the boundary condition for the conservation of salt content. At the seaward boundary A gives a condition for conservation of salt content during ebb tide. In C, at both the downstream and the upstream end, the free surface elevation is imposed. At the bottom for the longitudinal velocity in most models a condition for the stress is imposed. At the bottom for the longitudinal velocity in most of the models a condition for the stress is specified. In E also boundary conditions for a closed boundary are specified.

Some differences also occur in the turbulence models. In B the turbulent exchange is described by a relation based on the average veocity. In E a subgrid scale model is used and for the others a mixing length model. The influence of the stratification is generally given as a function of the Richardson number. The form of the functions shows large variations. The wall stress in most of the models is based on the quadratic law. C and E utilize relations for the longitudinal dispersion, based on Elder's (1959) result.

Geometrically, D and E use a constant width; in B the width may vary in the longitudinal direction. In the other models the width may vary both vertically and longitudinally. Most of the models use a piecewise linear approximation of bottom and surface. In A and E the bottom is approximated trapezoidally. All models handle only one branch.

Concerning numerical technique, all models except D use a staggered grid with constant grid spacings. If no transformation is used, a special approach is applied near the surface and the bottom. The difficulties which arise in discretizing certain boundary conditions on a staggered grid are most clearly shown for E. The stability condition for all models is the well known "long wave" criterion. Additional conditions are given for the equation for the conservation of salt content. With A an extra iteration must be made every 50 time steps because of the phase splitting due to the use of a leap-frog scheme.

All models had been calibrated for a field situation.

6. THE CALIBRATION

For the present study calibration is defined as the adjustment of coefficients, assuming that the relations in which the coefficients appear adequately describe the phenomena under consideration, and that the coefficients may vary within a previously fixed range of physically possible values.

Before starting the actual calibration a norm was defined to allow an objective comparison of measured and computed results. Together with the numerical accuracy (4.3) and the physical accuracy (6.2.1) it offered the possibility to draw quantitative conclusions. A review of the computations and the values of the coefficients used can be found in Table 2. In Appendix IV a description of the flume and a review of the measurements is given.

6.1 Introduction of a Norm

In order to make the comparison of computations and measurements as objective as possible, a norm was defined by the variations that arise due to a variation of 15 percent in the value of z_0, which corresponds with a variation of 10 percent in the bottom roughness coefficient based on a quadratic law. The order of the variation in z_0 corresponds with the degree of uncertainty in the boundary conditions of the flume. From the results of a sensitivity analysis for homogeneous tidal flume circumstances the following norm could be derived (DHL, 1979a)

$\Delta B_0 = 0.0003$ m (mean water level)
$\Delta B_1 = 0.0007$ m (tidal amplitude of the water level)
$\Delta \phi_1 = 0.093$ rad (phase of the tidal discharge)
$\Delta A_1 = 0.0005$ m^3 s^{-1} (amplitude of the tidal discharge)
$\Delta Q_1 = 0.031$ rad (phase of the tidal discharge)

For the concentrations the norm is related to the variations in the relative depth-averaged concentrations at high water slack

$$\Delta\left(\frac{\bar{c}}{c_{\max}}\right) = 0.015$$

$\Delta L_h = 0.1$ m (intrusion length, homogeneous)

In inhomogeneous circumstances the norm for the tidal phenomena remains the same. A straightforward extension of the norm for the concentration in homogeneous circumstances to the density distribution would not be unique because of the stratification. Therefore the tolerance of the measurements in the tidal flume was taken as a norm, which is of course dependent on the instrumentation used. For the present calibration the norm reads

$\Delta \rho = 0.75$ kg m^{-3}

$\Delta L_i = 0.5$ m (intrusion length, inhomogeneous)

For the computations the order of the numerical error was made smaller than that of the norm by a proper choice of the step sizes. The order of the error in the measurements turned out to be smaller than that of the norm, too.

6.2 Results of the Homogeneous Calibration

The calibration was started for a homogeneous situation (T 22). In this situation the following coefficients were adjusted: the coefficients z_0, which is a measure for the roughness length of the bottom; the longitudinal dispersion coefficient D_x for homogeneous circumstances (see Section 3.5); the transition function for homogeneous circumstances (see Section 3.2.3); and the transfer function for homogeneous circumstances. The tidal amplitude in the tidal flume is imposed at the sea, 8 m from the seaward boundary of the flume where the boundary condition for the numerical model is imposed. The relation between the tidal amplitude at sea and at the seaward boundary is given by a transfer function on which, however, little information is available.

The infuence of separate physical coefficients was tested by comparing several numerical computations. The results of this sensitivity analysis were used for the calibration of the numerical model. The results of the homogeneous calibration are presented in Table 3 and in the Figures 3 to 5.

Figure 3 shows that the water level variations from the measurements and the computations correspond very well, with differences which are smaller than the deviations indicated by the norm. The differences in the tidal amplitude are smaller yet. When comparing the results for the tidal discharges (Table 3) it turns out that the amplitude of the measured discharge is systematically larger than the computed one. This can be explained by the fact that the flow in the flume is not exactly two-dimensional. There is a small influence of the wall, which causes lower velocities near the wall than in the middle of the flume. Therefore the discharge, which is computed from velocities measured in the center of the flume, will be higher than the laterally averaged discharge from the numerical model.

In Figure 4 a vertical velocity profile is shown at four characteristic times. Considering the confidence intervals the measured velocities are correctly reproduced. This holds for the maximum velocities at maximum flood velocity (MFV) and maximum ebb velocity (MEV), as well as for the reversal of the velocity at high water slack (HWS) and low water slack (LWS).

The relative depth-averaged Rhodamine concentrations are shown in Figure 5. Here two significant differences between the computed and measured results can be noticed: a phase lag of about 0.02 T in computed results, at $x - 3.66$m, compared to the measured ones; and, a more gradual decrease of the computed maximum concentration for increasing x. The first point can be improved by a correction of the transition function. However, the accuracy of the phase from the measurements may also be limited and therefore further research and more detailed measurements are necessary.

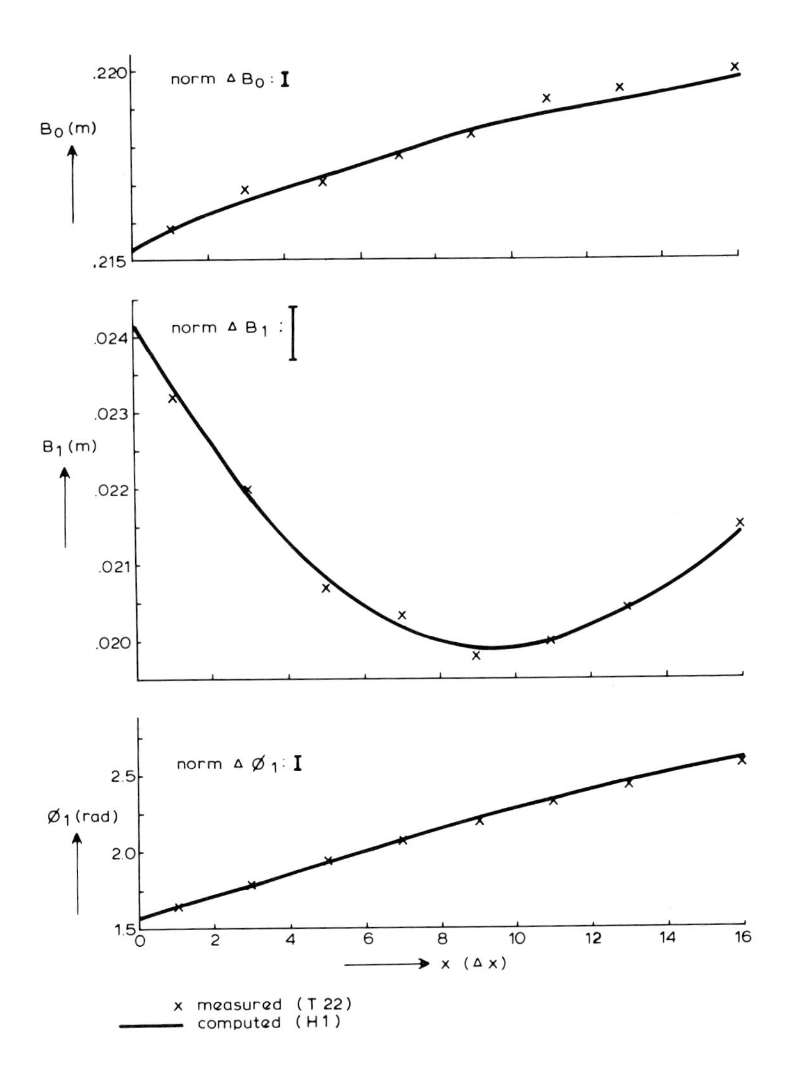

Figure 3. Results of the calibration (homogeneous). Fourier components from the water level variations.

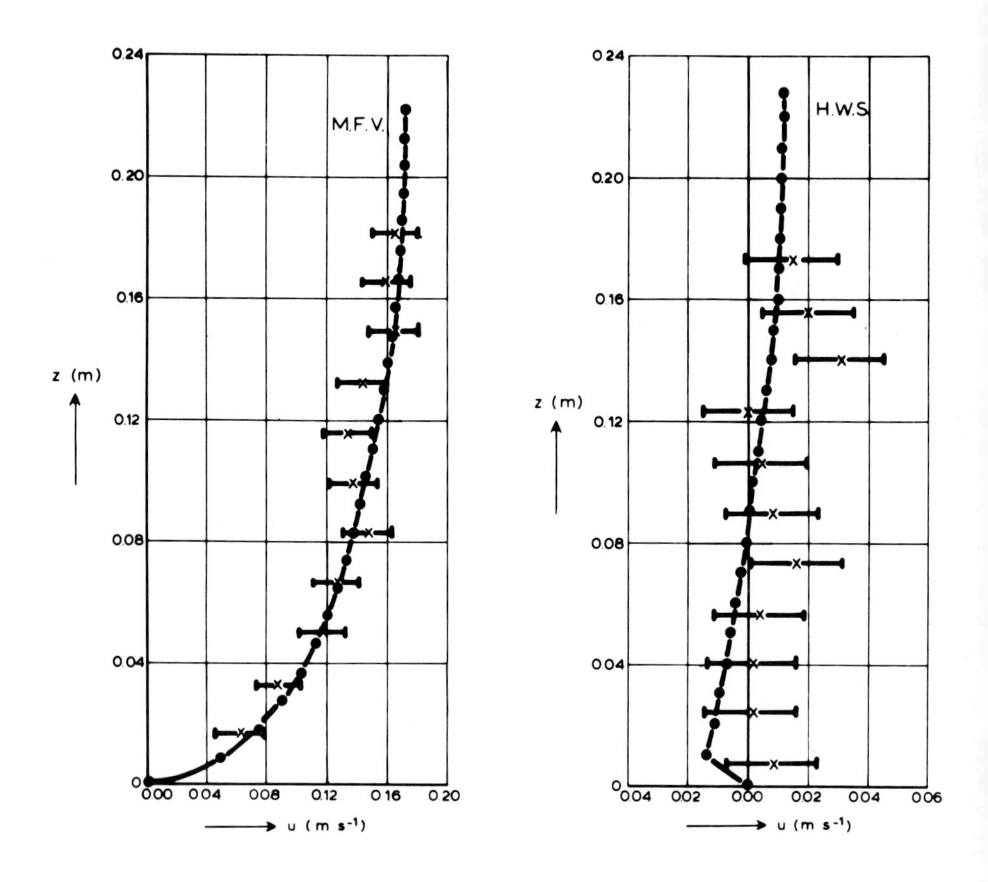

Figure 4. Results of the calibration (homogeneous). Velocity $u(x = 47.58\text{m})$ at four characteristic times.

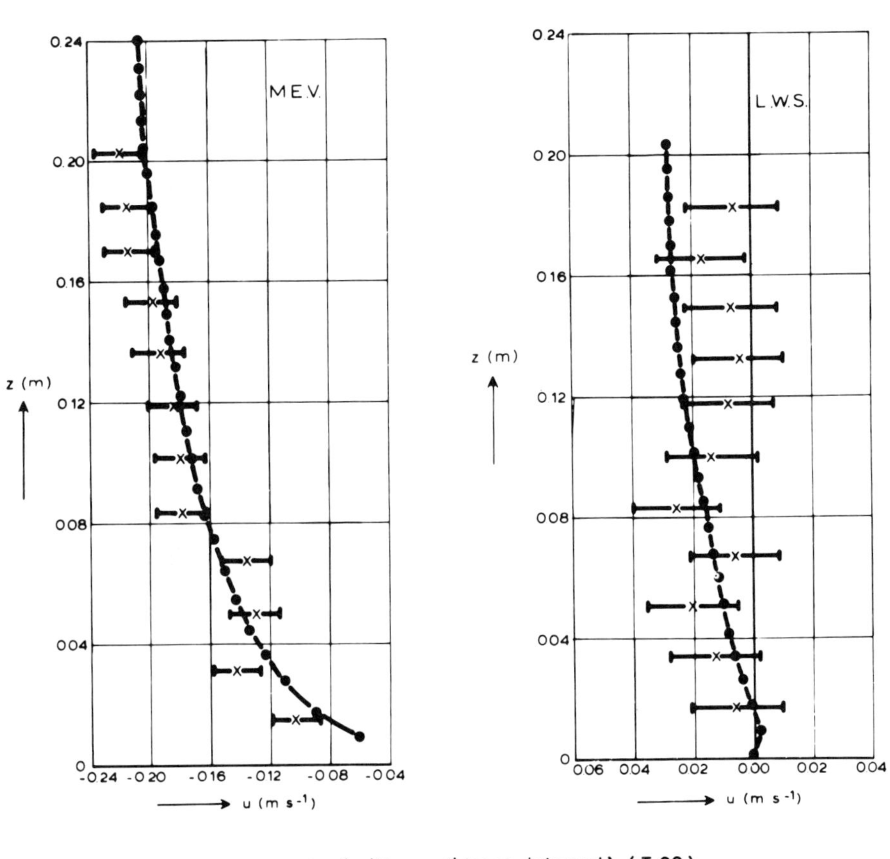

Figure 4, continued.

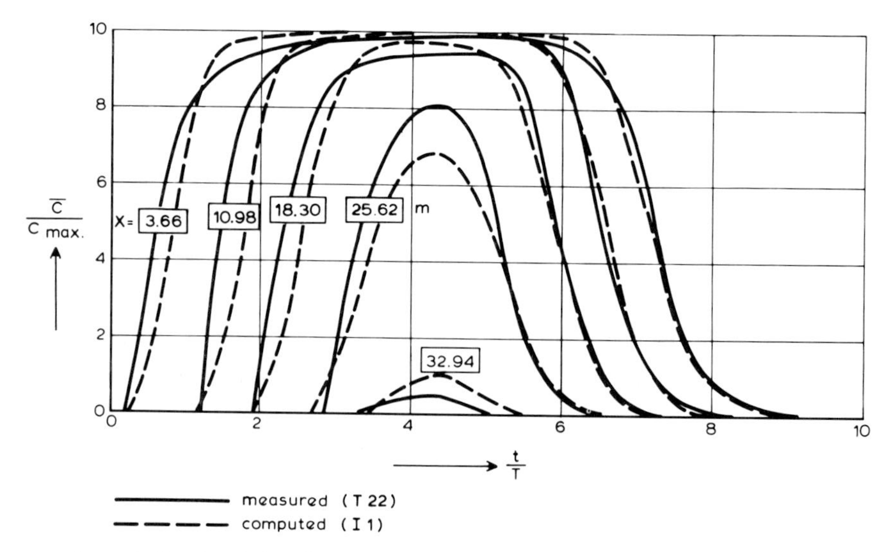

Figure 5. Results of the calibration (homogeneous). Relative depth-averaged Rhodamine concentrations.

The second point could be improved by a correction of the longitudinal dispersion. This phenomenon will return for other situations and is also a subject for further research. However, for the present case, the accuracy of the measurements did not justify further calibration.

Summarizing the results of the homogeneous calibration it can be concluded that generally the tidal phenomena are well reproduced with respect to the norm. In those cases where the norm is exceeded, the accuracy of the measurements is poor. The resistence is well reproduced by the geometric coefficient z_0, and the weak boundary condition for the velocities yields satisfactory results.

6.3 Results of the Inhomogeneous Calibration

Next the calibration was continued for an inhomogeneous situation (T 20). The following coefficients which depend on the density were adjusted: the damping functions $F(Ri)$ and $G(Ri)$, which represent the influence of the stratification on the vertical turbulent exchange of momentum and mass; the longitudinal dispersion coefficient D_x; the transition function; and the transfer function.

The main result of the inhomogeneous calibration was the important influence of the damping functions on the vertical turbulent exchange. Because the literature provides a wide range of possible functions (DHL, 1974b) the inhomogeneous calibration was started with two different sets of functions. First the relations due to Munk Anderson were used

$$F(Ri) = (1+10\ Ri)^{-0.5} \tag{40}$$

$$G(Ri) = (1+3.3\, Ri)^{-1.5} \tag{41}$$

and next a set of relations taken from the Delft Tidal flume (Van Rees, 1975 and DHL, 1976a):

$$F(Ri) = \exp\,(-4\, Ri) \tag{42}$$

$$G(Ri) = \exp(-15\, Ri) \tag{43}$$

The results of the second run were much better than the results from the first run, concerning both the tidal motion and the salt distribution (DHL, 1980a). The calibration was extended (DHL, 1980c), and it was concluded that the damping functions

$$F(Ri) = \exp(-4\, Ri) \tag{44}$$

$$G(Ri) = \exp\,(-18 Ri) \tag{45}$$

gave the best results. The results shown in this report, for both the calibration and the verification, were produced with these functions. In Figures 6 to 10 and in Table 3 the results of this inhomogeneous calibration are presented.

When the computed and measured results are compared it turns out that the water level variations are well reproduced (see Figure 6). The deviations in the tidal amplitude and the phase are within the norm. The deviations in the mean level exceed the norm at $x = 10.46$ and at $x = 18.30$m. However, this is considered to be of minor importance.

In Table 3 the Fourier components of the tidal discharges from measurements are compared with those from computations. It is striking that in this case the computed discharge tends to be larger than the measured one. This is in contradiction with what may be expected after the homogeneous results. The only possible explanation must be sought in the limited accuracy of discharges taken from measured velocities.

Figure 7 shows a measured and a computed vertical velocity distribution at four characteristic times. Apart from the stratification near the surface at MFV and at $z = H/3$ around MEV the velocities correspond well with respect to the confidence interval. It should be noticed that this also holds for the exchange flows at LWS.

In order to compare the results for the salt distribution, three aspects are distinguished: the stratification, the intrusion, and the longitudinal distribution. In Figure 8 a vertical salt distribution is shown at four characteristic times. The differences between measured and computed results are small, except at MFV and MEV, when the computed results show a stratification and the measured ones do not. This behavior corresponds with that of the velocities (Figure 7).

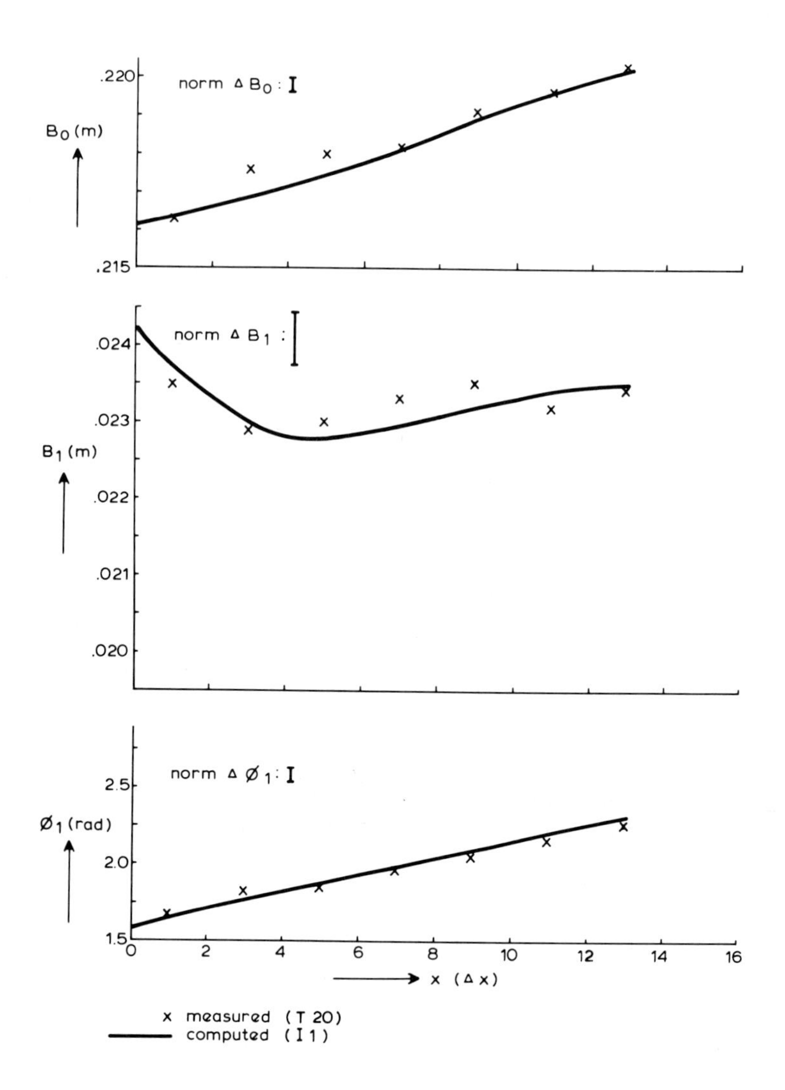

Figure 6. Results of the calibration (inhomogeneous). Fourier components from the water level variations.

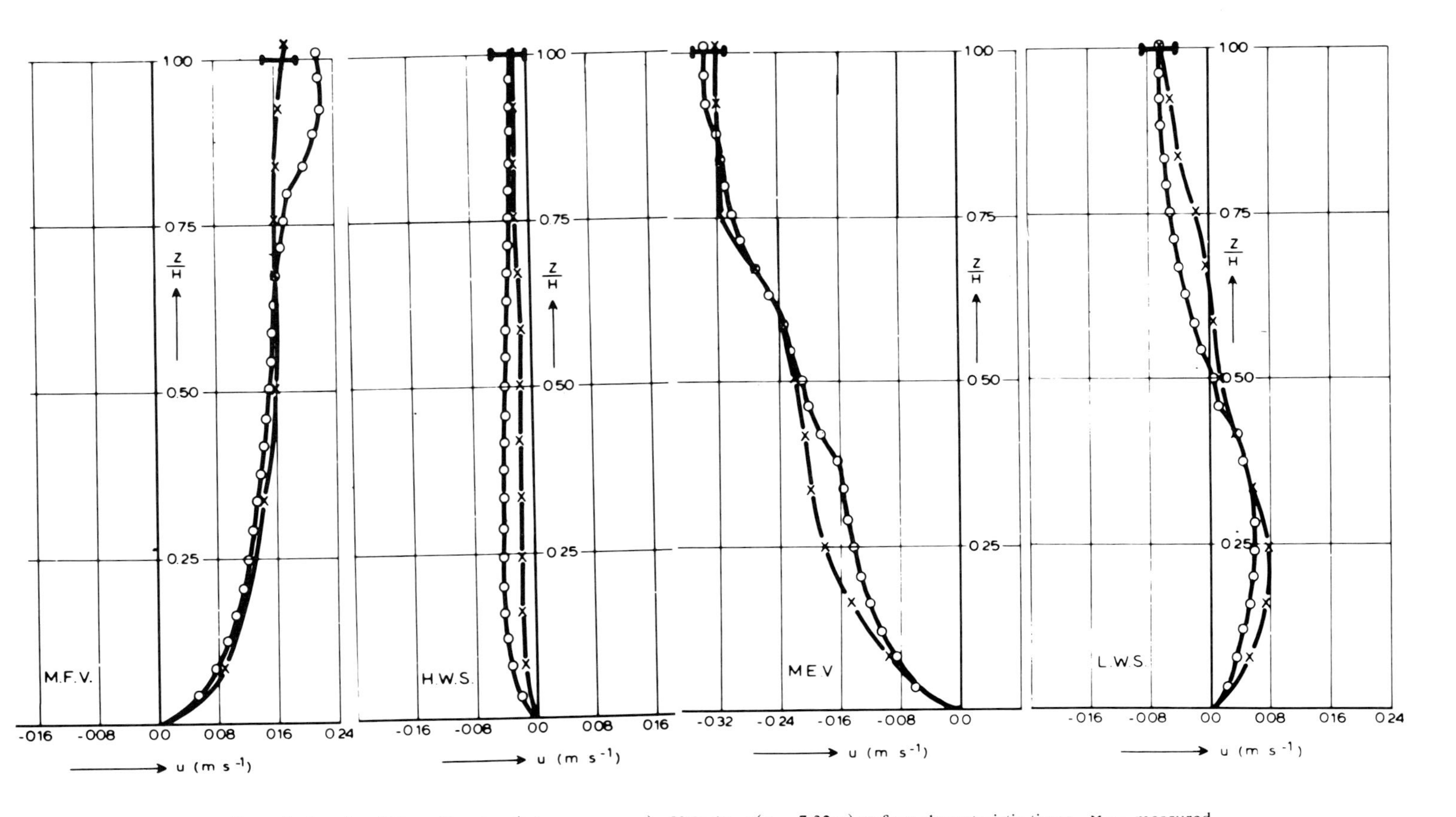

Figure 7. Results of the calibration (inhomogeneous). Velocity $u(x = 7.32\text{m})$ at four characteristic times. × = measured (T 20); ○ = computed (I 1); ⊢ = confidence interval.

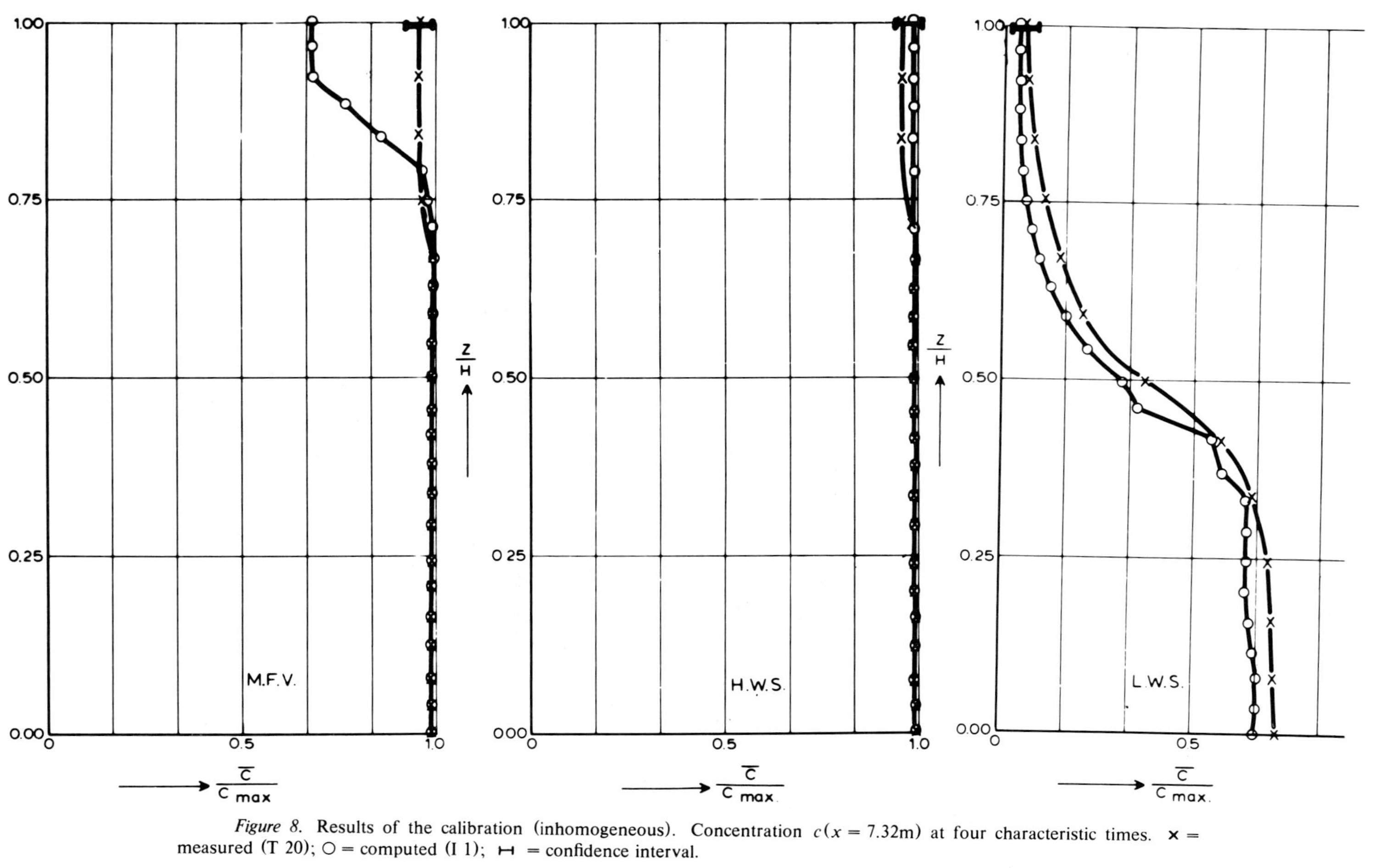

Figure 8. Results of the calibration (inhomogeneous). Concentration $c(x = 7.32\text{m})$ at four characteristic times. × = measured (T 20); ○ = computed (I 1); ⊢⊣ = confidence interval.

The longitudinal concentration distribution is shown in Figure 9, at four characteristic times. The stratification is on the average correct, but not during the whole tidal cycle. Around MEV and MFV it is correct, around LWS the stratification in the computation is too strong, and around HWS it is too weak. The intrusion tends to be too small. However, the differences are of the order of the norm. The longitudinal distribution shows significant differences. The computed densities show a more gradual course than those from measurements. This behavior can be clearly seen from the distances between the isopycnycs.

The relative depth-averaged salt concentrations are shown in Figure 10. This figure confirms the behavior of the longitudinal density distribution from Figure 9, and shows that the seaward boundary condition did fairly reproduce the salt concentration near that boundary. The largest deviations occur when the weak condition is imposed and are obviously due to differences in the stratification.

The results of the calibration lead to the following conclusions. First, the tide could be reproduced well, using the same bottom roughness coefficient z_0 for homogeneous and inhomogeneous situations. This corresponds well to the fact that z_0 reflects a geometrical property of the flume. Second, the global salt distribution was well reproduced, although in detail systematic differences could be noticed. From these differences the following conclusions with respect to the physical relations introduced in (3.4) can be drawn: (1) The mixing length concept in combination with the damping relations cannot correctly reproduce the stratification throughout a tidal cycle. As indicated in Section 3.4, apparently a limitation of the present turbulence model is met here. (2) The dispersion coefficient cannot be neglected to correctly reproduce the maximum intrusion. The dispersion coefficients used in the calibration tend to be too large (DHL, 1980b). This could partly explain the deviations in the longitudinal density distribution. The dispersion coefficient for the inhomogeneous calibration had to be taken differently from the one used in the homogeneous case. This must be credited to the influence of the stratification, but at the moment there is no knowledge about the order of the influence, nor how it could be parameterized. (3) The seaward boundary conditions for the salt conservation equation, and especially the transition function, reproduced sufficiently accurately the concentration near the seaward boundary. However, different relations had to be taken for the homogeneous and the inhomogeneous calibration. Probably here, too, the stratification has some influence, especially at sea. (4) The transfer function was kept the same and the results gave no indications for systematic improvement.

7. THE VERIFICATION

Verification is defined as the process of checking how well the calibrated model reproduces the phenomena in situations different from the one used for calibration, without changing the model. The verification presented in this chapter is a limited one, because the model is only verified for tidal flume situations. This gives an indication about the reproduction in situations which differ from the calibrated situation in the physics (e.g. stratification) but not in the geometry.

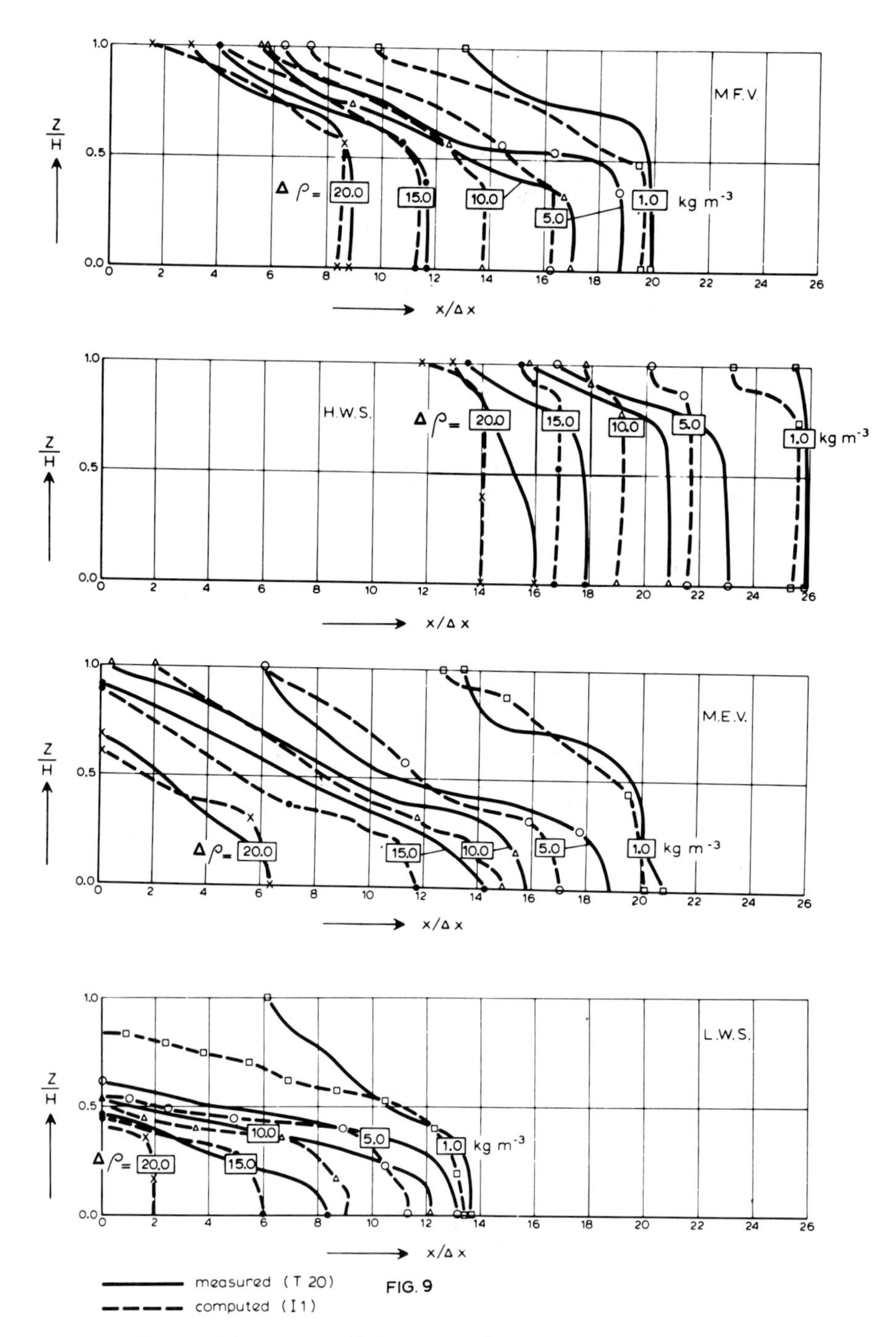

Figure 9. Results of the calibration (inhomogeneous) Longitudinal concentration distribution at four characteristic times.

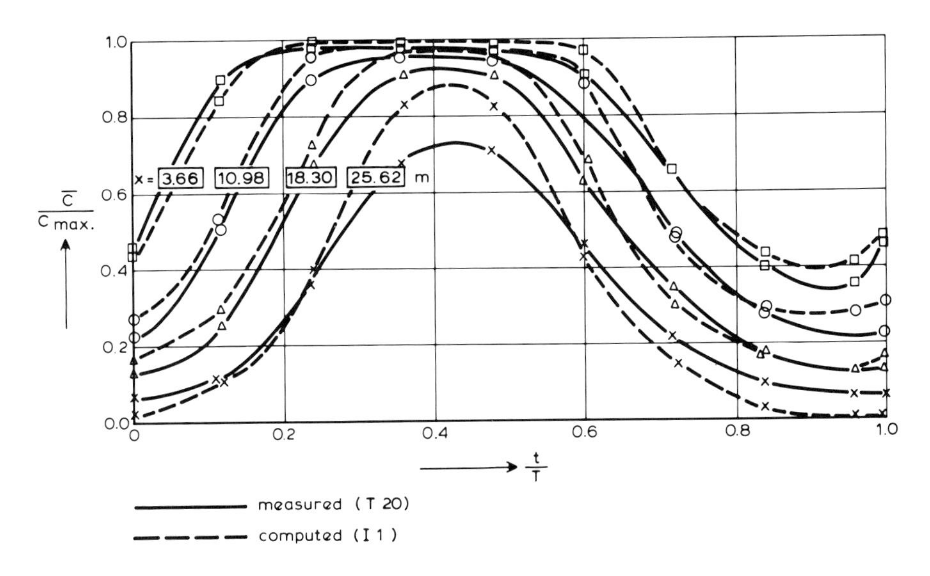

Figure 10. Results of the calibration (inhomogeneous). Relative depth averaged concentrations.

For the verification two homogeneous tests and three inhomogeneous tests were used (see Table 5). In one homogeneous and one inhomogeneous test a smaller tidal amplitude was used to obtain a more stratified situation than for the calibration. In the other tests a larger amplitude was used, which resulted in weaker stratification. For the weaker stratification two density differences were used. A description of the tests can be found in Appendix IV.

The norm used for the verification differs from that for the calibration. Here the norm is defined by the order of the deviations between measurements and computations, which occurred with the calibration. If the deviations for the verification would be of the same order or smaller, this would imply certain predictive abilities. In the opposite case the model, at least partly, fails to be predictive.

7.1 Results of the Homogeneous Verification

Figures 11 and 12 show the measured and computed water level variations for the small and large tidal amplitude. It can be seen that the mean level, B_0, and the phase, ϕ_1, are well simulated. The computed tidal amplitude, B_1, tends to be too small for smaller a_0 and too large for larger a_0. However, considering the accuracy of the measurements, the reproduction is still considered to be satisfactory.

Comparison of the computed and measured tidal dicharges, presented in Table 3, shows, given the accuracy of the measurements, no significant differences

Table 2
Review of the Computations, and the Values of the Parameters Used

Parameter	Number of Calculations						
	H1	H2	H3	I1	I2	I3	I4
L (m)	100.65	100.65	100.65	100.65	100.65	100.65	100.65
H (m)	0.216	0.216	0.216	0.216	0.216	0.216	0.216
T (s)	558.75	558.75	558.75	558.75	558.75	558.75	558.75
Q_R ($m^3 s^{-1}$)	0.0029	0.0029	0.0029	0.0029	0.0029	0.0029	0.0029
ϵ_x ($m^2 s^{-1}$)	0.37	0.37	0.37	0.37	0.37	0.37	0.37
D_x [1] ($m^2 s^{-1}$)	b	-	-	a	a	a	a
N_x	55	55	55	55	55	55	55
N_z	24	24	24	24	24	24	24
N_t	2400	2400	2400	4800	4800	4800	4800
τ_v (s)	0.93	0.93	0.93	0.93	0.93	0.93	0.93
τ_D (s)	0.23	-	-	0.23	0.23	0.23	0.23
$\Delta\rho$ ($kg\ m^{-3}$)	0.00	-	-	22.8	21.5	21.8	10.1
g(t,z) [2] ($kg\ m^{-3}s^{-1}$)	α	-	-	β	β	β	β
F(Ri) [3]	-	-	-	DTF	DTF	DTF	DTF
G(Ri) [3]	-	-	-	DTF	DTF	DTF	DTF
z_0 [4] (m)	c_1	c_1	c_1	c_1	c_1	c_1	c_1
a_0 (m)	0.025	0.0125	0.0375	0.025	0.0125	0.0375	0.0375
Damping (%)	3	3	3.5	3	3	3.5	3.5
Phase-shift (rad)	0.02	0.02	0.035	0.02	0.02	0.035	0.035
CPU [5] (s)	1165	359	363	3514	3448	3483	3445

1) The letters denote: a -- $4.5 \mid \bar{u} \mid b + .015$; b -- $2.0 \mid \bar{u} \mid + 0.005$. 2) The Greek letters denote: α -- $c_n + 7(H-z)/H\ c_n$; β -- $c_n + 3(H-z)/H\ c_n$. 3) DTF denotes: F(Ri) - exp (-4 Ri); G(Ri) - exp (-18 Ri). 4) c_1: z_0 - 0.0050 m at the first 61.3 m and z_0 = 0.0064 m at the last 34.35 m of the flume. 5) On a CDC-6600 computer.

TABLE 3

Fourier Components of the Discharges

Measurements	Computation	Position x (m)	A_0 (m^3s^{-1})	A_1 (m^3s^{-1})	Ψ_1 (rad)
T22		10.98	-0.0033	0.0290	1.189
	H1		-0.0029	0.0278	1.209
T22		47.58	-0.0031	0.0246	1.370
	H1		-0.0029	0.0237	1.352
T20		3.66	-0.0029	0.0271	1.084
	I1		-0.0029	0.0289	1.126
T20		47.58	-0.0028	0.0233	1.346
	I1		-0.0029	0.0238	1.336
T12		3.66	-0.0032	0.0182	1.077
	H2		-0.0029	0.0191	1.130
T10		3.66	-0.0035	0.0145	1.100
	I2		-0.0029	0.0197	1.078
T10		47.58	-0.0032	0.0167	1.344
	I2		-0.0029	0.0153	1.243
T32		3.66	-0.0037	0.0359	1.189
	H3		-0.00145	0.0361	1.202
T30		3.66	-0.0018	0.0374	1.148
	I3		-0.0021	0.0379	1.138
T30		47.58	-0.0025	0.0310	1.395
	I3		-0.0021	0.0306	1.349
T31		3.66	-0.0033	0.0351	1.159
	I4		-0.0021	0.0366	1.180
T31		47.58	-0.0025	0.0322	1.442
	I4		-0.0021	0.0300	1.391

between the results of the calibration and those of the verification. Because no Rhodamine concentration had been computed, a verification of the transition function and the dispersion in homogeneous circumstances is not possible. However, some remarks can be made about the transfer function. For the smaller a_0 the same transfer function was used as for the calibration (see Table 2). For the large a_0 another transfer function has to be used. Apparently the order of a_0 influences the transfer function, an influence which increases for increasing a_0.

It can be concluded that, for homogeneous circumstances, the model can predict the tidal phenomena, within the range presented in the calibration and verification. This means especially that the geometrical coefficient z_0 represents the bottom roughness correctly.

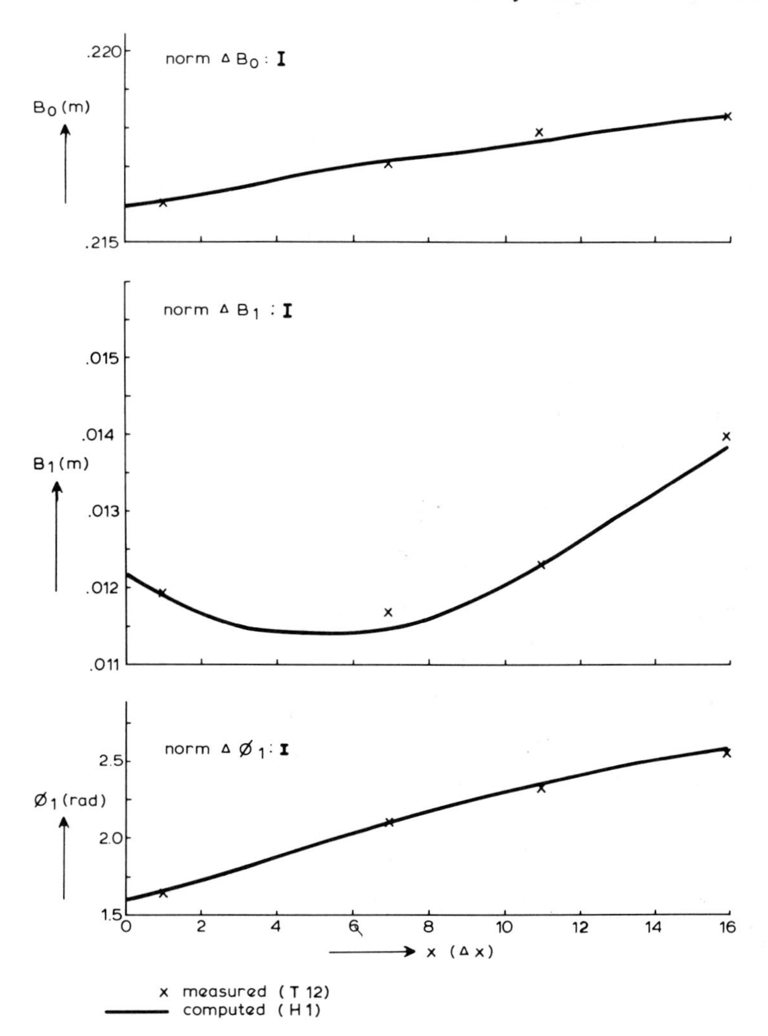

Figure 11. Results of the verification (homogeneous). Fourier components from the water level variations.

7.2 Results of the Inhomogeneous Verification

Comparison of the results for the water level variations (Figures 13 to 15) with those from the calibration (Figure 6) shows systematic differences in the mean level and the amplitude. Two reasons can be given. First, at the seaward boundary probably not the correct boundary condition has been imposed in the numerical model, as can be concluded from the large deviations at $x = \Delta x$. This means that another transfer function should have been used. Secondly, the

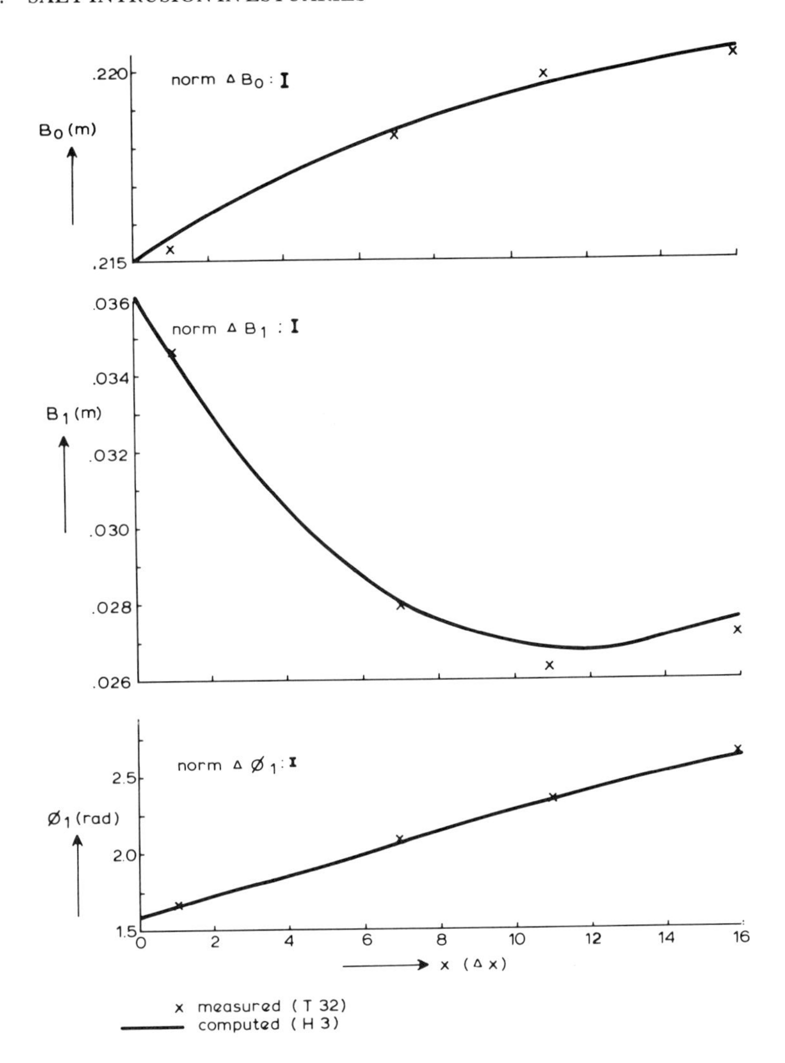

Figure 12. Results of the verification (homogeneous). Fourier components from the water level variations.

resistance of the flume is not correctly reproduced, as can be concluded from the computed and measured course of the mean level and the amplitudes. Probably this is due to the fact that the stratification could not be reproduced correctly nor, consequently, the influence of the stratification on the tide.

It should be noticed that for the small a_0 the deviations between measured and computed amplitude decrease for increasing x, while for the large a_0 the deviations increase. Obviously, the resistance for small a_0 is too strong and for large a_0 too weak. For the tidal discharges the same conclusions hold as for homogeneous circumstances: considering the accuracy no significant differences between the

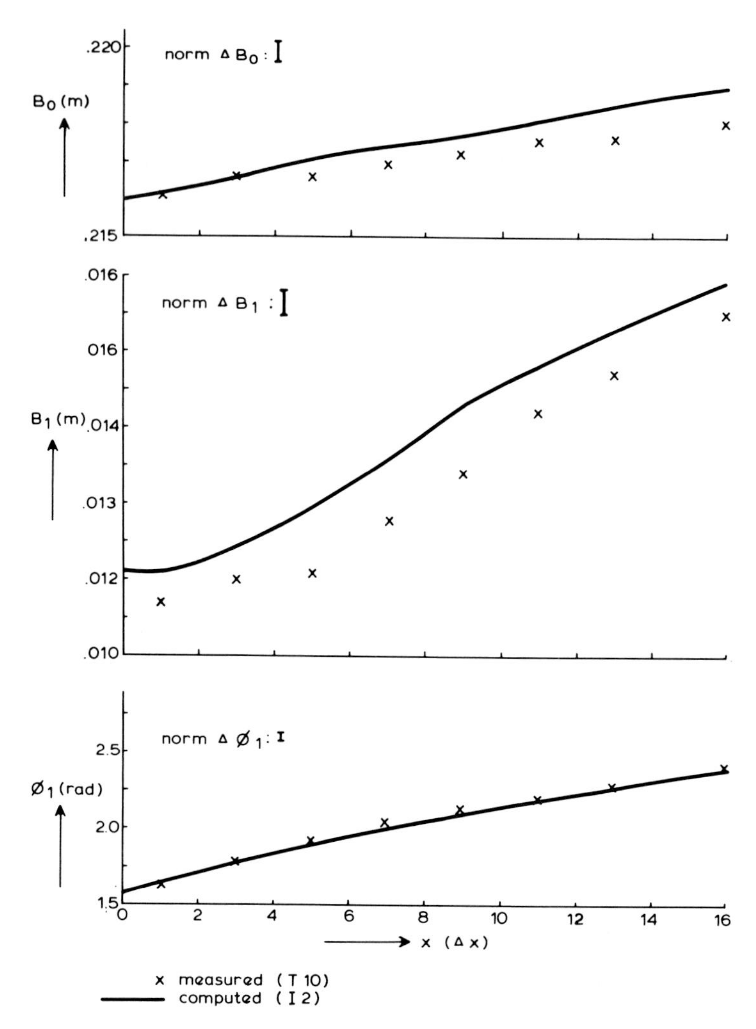

Figure 13. Results of the verification (inhomogeneous). Fourier components from the water level variations.

results of the calibration and of the verification can be seen (Table 3). In Figure 16, the longitudinal density distribution is shown at four characteristic times for the test with the small a_0. The measurements show a much more stratified situation than the computations, and the measured salt intrusion is at every moment larger than the computed one. For the isopycnyc $\Delta\rho = 1\ kg/m^3$ the difference is of the order of 4 m. The measured longitudinal density distribution near the bottom differs much from the computed one. Near the tip of the salt wedge the longitudinal gradient in the measured concentrations is larger, while downstream this gradient becomes smaller than the one in the computed results.

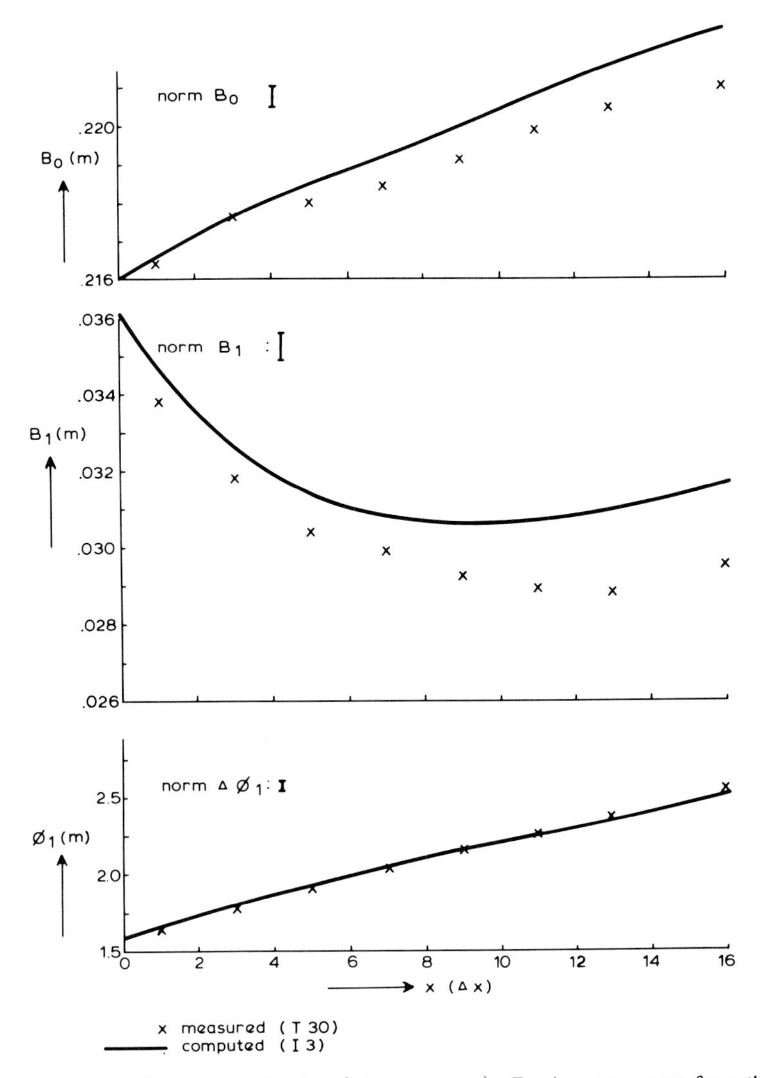

Figure 14. Results of the verification (inhomogeneous). Fourier components from the water level variations.

The longitudinal concentration distributions of the tests with a large a_0 are shown in Figures 17 and 18, respectively, for a large and a small density difference. Both figures show the reverse picture of figure 16, i.e., a computed stratification which is too strong and an intrusion length which is too large. From the results for the concentration a qualitative explanation can be given for the deviations in the water level variations. For small a_0 the computed stratification is too weak. This results in an intrusion which is too small and a resistance which is

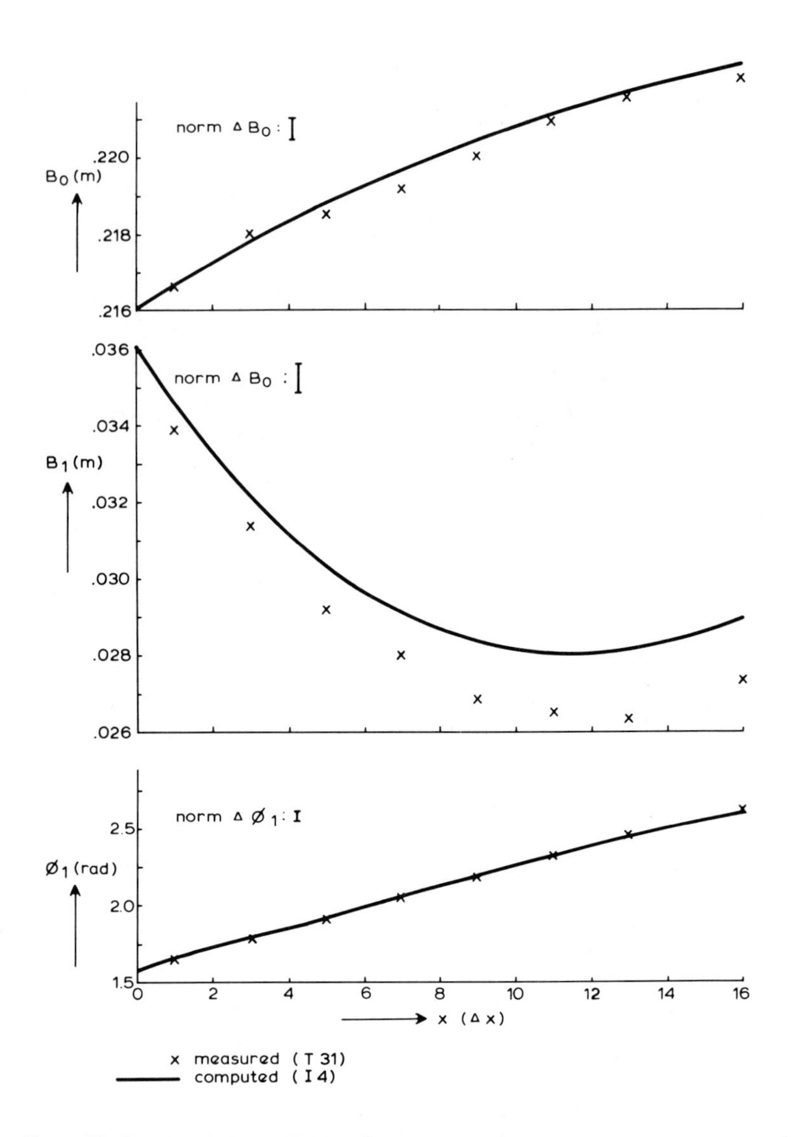

Figure 15. Results of the verification (inhomogeneous). Fourier components from the water level variations.

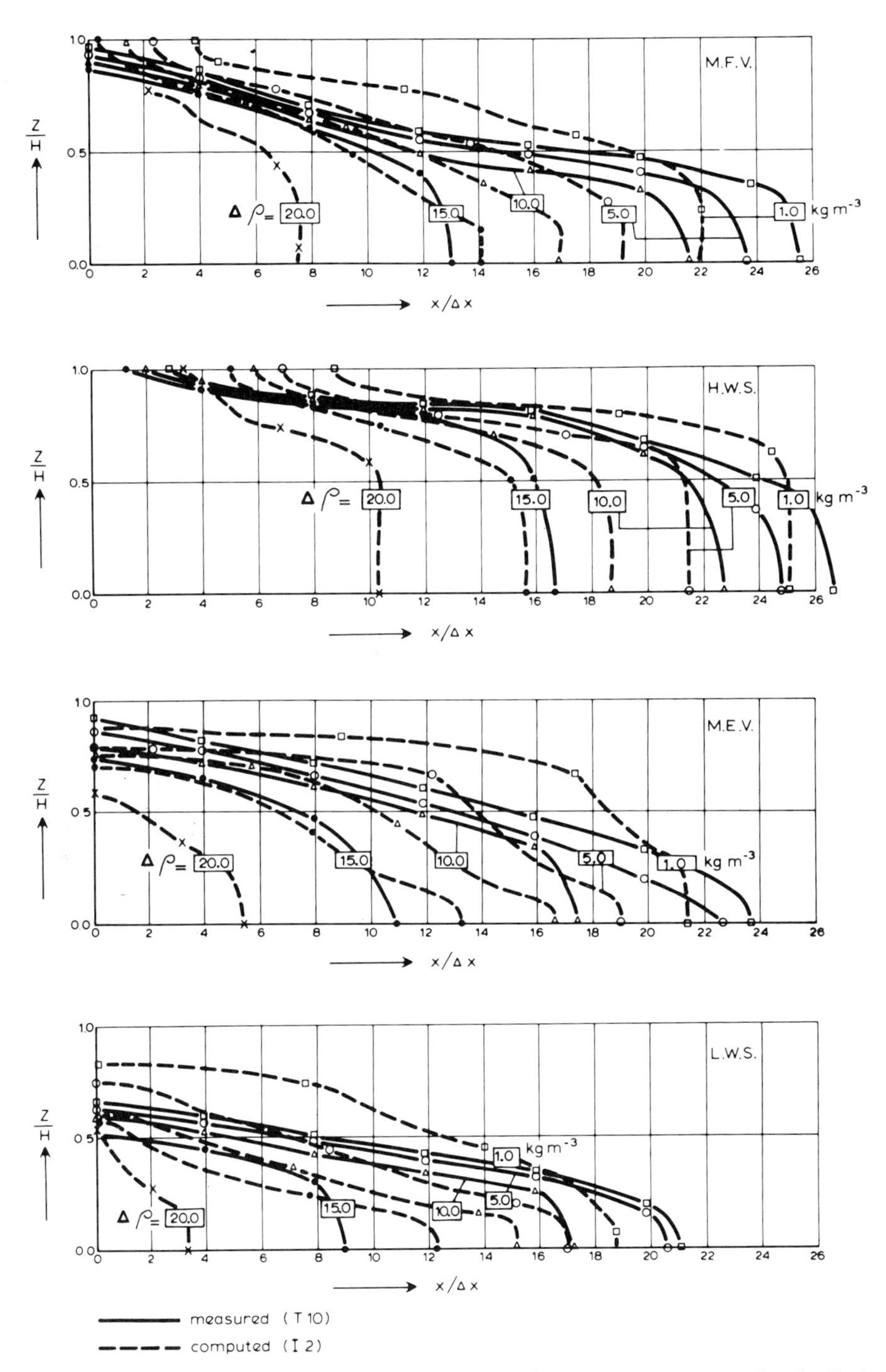

Figure 16. Results of the verification (inhomogeneous). Longitudinal concentration distribution at four characteristic times.

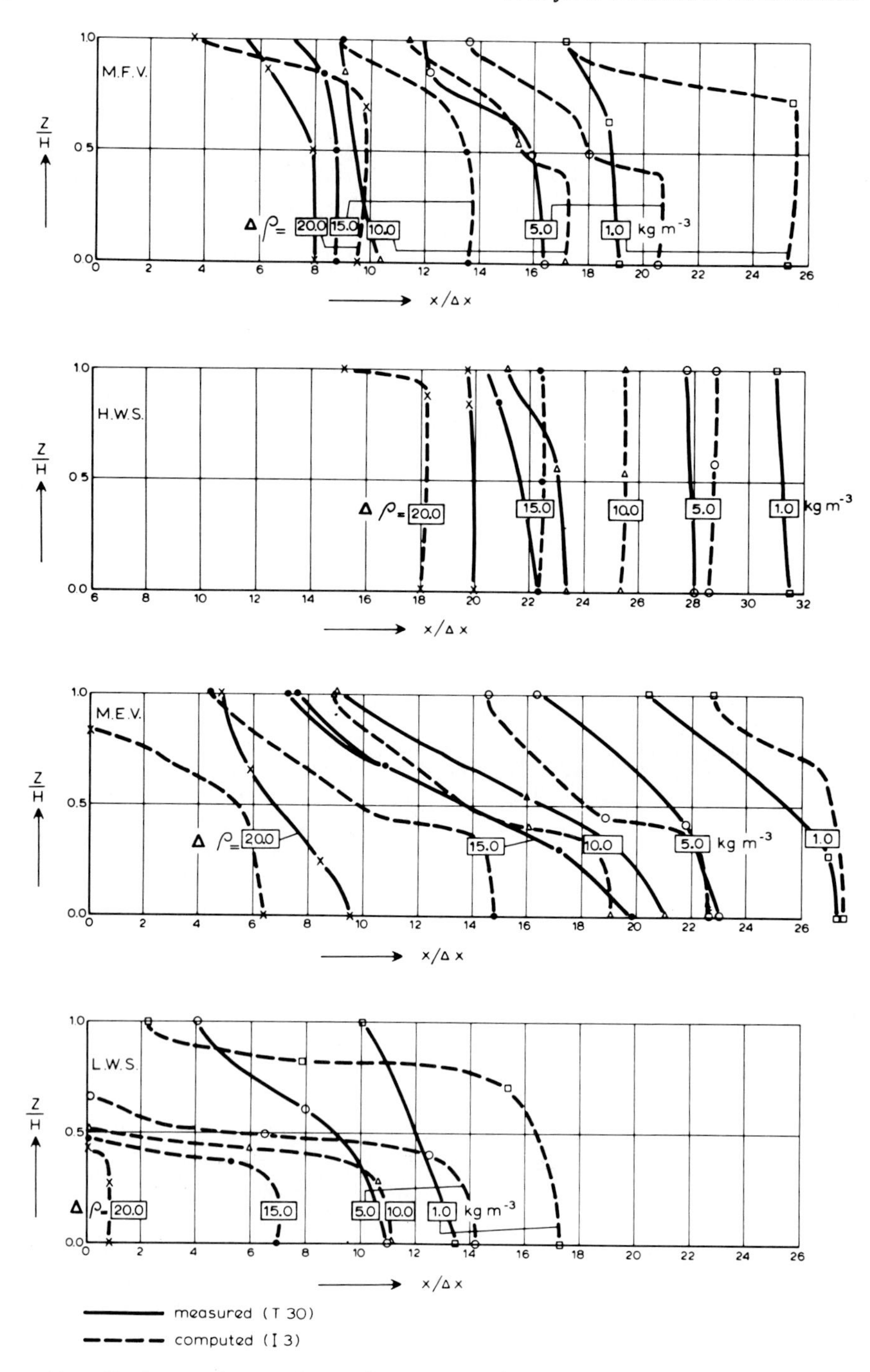

Figure 17. Results of the verification (inhomogeneous). Longitudinal concentration distribution at four characteristic times.

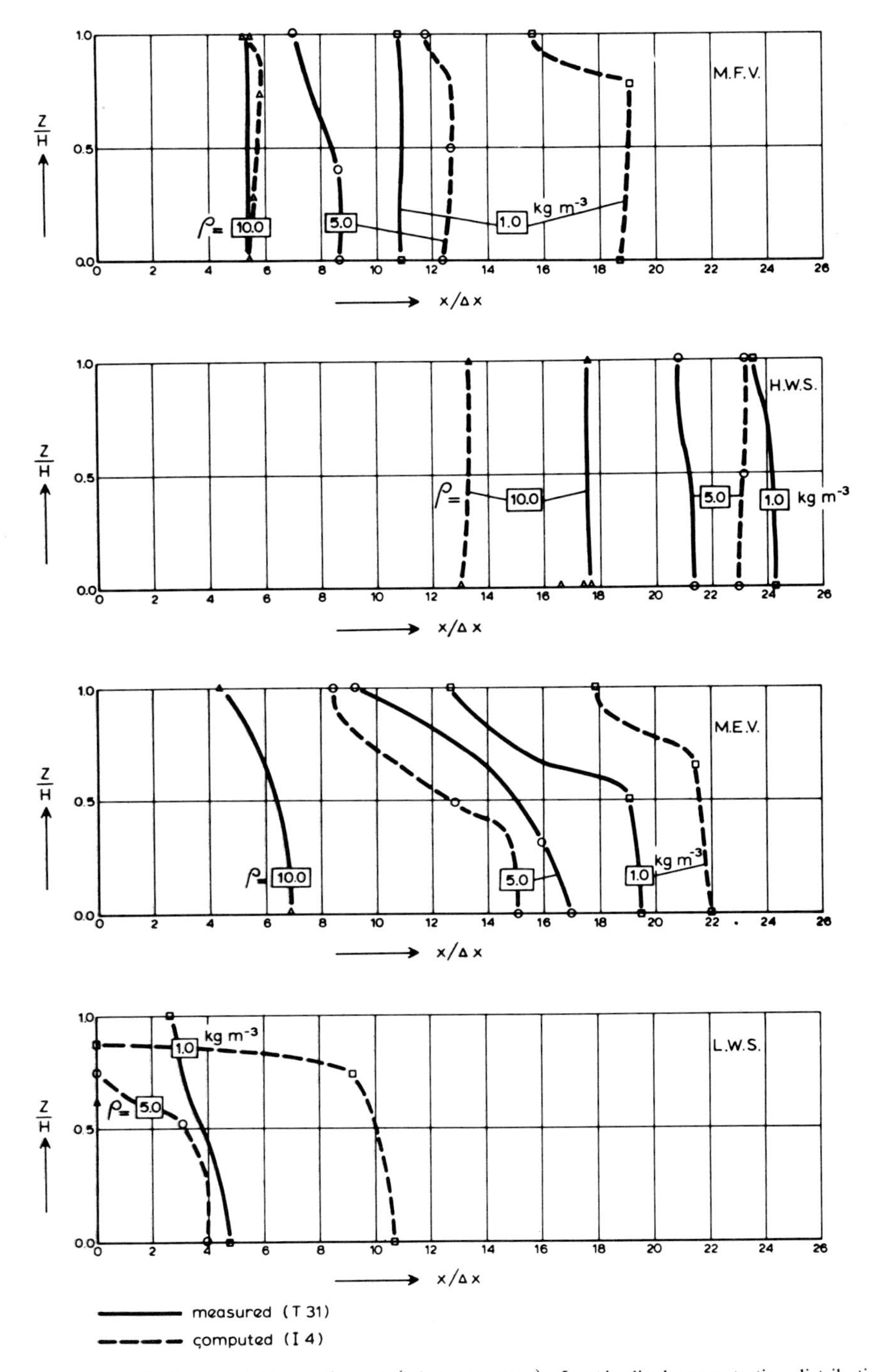

Figure 18. Results of the verification (inhomogeneous). Longitudinal concentration distribution at four characteristic times.

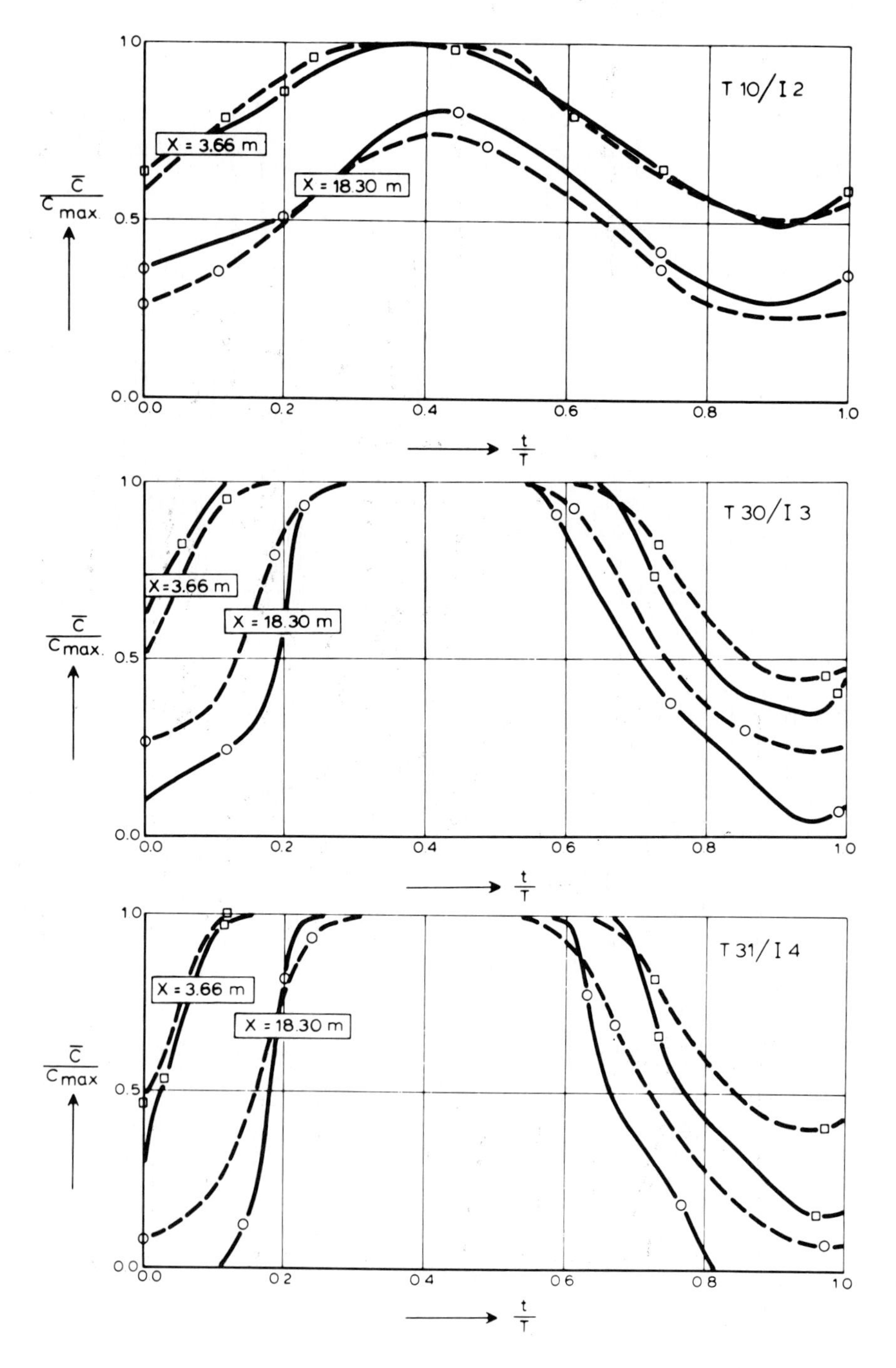

Figure 19. Results of the verificaton (inhomogeneous). Relative depth-averaged concentrations.

too strong. This corresponds well with the results already mentioned and shown in Figure 13. For large a_0 the opposite happens, as can be seen in Figures 14, 15, 17, and 18. Figure 19 shows the depth averaged concentrations near the seaward boundary of all three inhomogeneous verification tests. For these tests the same transition function was used as for the calibration. It can be seen that for small a_0, and especially in case of the small density difference, the deviations are much larger. It also should be noticed that the largest deviations occur during ebb tide, when the weak boundary conditions for the salt concentration is imposed. Obviously these results give no indication that the present transition function seriously hampers the predictive capabilities.

8. CONCLUSIONS

In this paper the development of a two-dimensional laterally averaged salt intrusion model has been presented. To test the abilities of the model, a calibration and verification in a single tidal flume have been presented. With respect to the predictive abilities the following conclusions can be drawn: (1) For a flume the tidal flow is predictable with an acceptable accuracy. For this purpose the proper bottom roughness coefficient z_0 must be found from calibration. The boundary conditions $h(t)$ at the downstream boundary and $Q(t)$ at the upstream boundary should be accurately specified. The geometrical parameter z_0 produces good results both for steady and for non-steady flow. The weak boundary condition at the downstream end yields good results, too. (2) The concentration distribution must be considered as not predictable with the present model. The present turbulence model is not able to reproduce salt distribution accurately during a whole tidal cycle, nor for a range of stratified situations. First, the knowledge about the infuence of the stratification on the vertical turbulent exchange is limited. Secondly, the relation for length scale l_m, which for the present model is a function of the distance to the bottom, is questionable. For the reproduction of the maximum salt or Rhodamine intrusion a longitudinal dispersion must be taken into account. The influence of the stratification on form and magnitude of the dispersion coefficient is hardly understood. (3) The problem of the dispersion due to lateral averaging could be avoided by the use of a three-dimensional model at the cost of much more computer effort. (4) A separate problem is the transition function. Although the relation which has been adopted in the present model could sufficiently accurately approximate the exact boundary condition, the predictive abilities are questionable, as indicated in Section 3. In the future applications, probably always some knowledge about the local situation near the seaward boundary has to be taken into account.

With respect to the three kinds of applications that are mentioned in the introduction, the development so far has proven that the model can serve as a tool for research with respect to both mathematical and numerical modeling. The calibration shows that the model is also very useful for the interpretation of measurements from the tidal flume. With respect to the use as a tool for predictions severe reservations must be made. The model has only proven to be predictive under homogeneous circumstances and a simple geometry. Consequently the predictive abilities are a major subject for further research. For stratified conditions it remains to be seen whether other turbulence closure assumptions will lead to a more detailed reproduction of the Delft Tidal Flume data than obtained with the present method.

APPENDIX I: THE TRANSFORMATION

Schematically the transformation can be presented as:

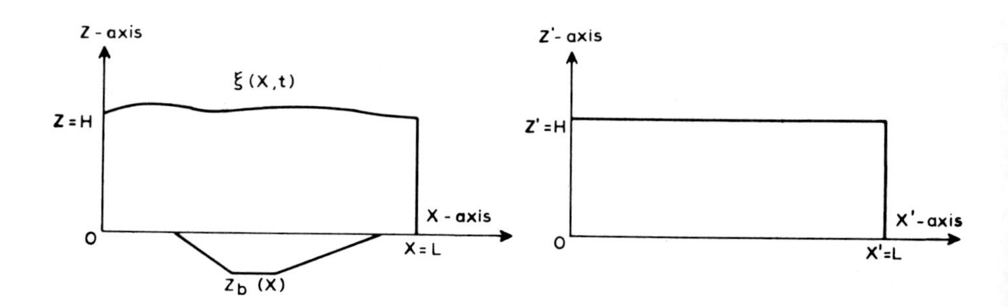

In the formula the transformation reads

$$z' = (z - z_b)\, f(x,t) \tag{A1}$$

in which

$$f(x,t) = \frac{H}{\{\zeta(x,t) - z_b(x)\}} \tag{A2}$$

From (A1) and (A2) the following transfer coefficients can be derived

$$\frac{\partial z'}{\partial t} = TF_1(t,x,z) = (z - z_b)\, H\, \frac{\left(-\dfrac{\partial \zeta}{\partial t}\right)}{(\zeta - z_b)^2} \tag{A3}$$

$$\frac{\partial z'}{\partial x} = TF_2(t,x,z) = -\frac{\partial z_b}{\partial x}\ \frac{H}{(\zeta - z_b)} + H\ (z - z_b)\ \frac{\left\{-\left(\dfrac{\partial \zeta}{\partial x} - \dfrac{\partial z_b}{\partial x}\right)\right\}}{(\zeta - z_b)^2} \tag{A4}$$

$$\frac{\partial z'}{\partial z} = TF_3(t,x) = \frac{H}{(\zeta - z_b)} \tag{A5}$$

After this transformation the equations read

Conservation of longitudinal momentum

$$\frac{\partial (b_c u)}{\partial t} + \frac{\partial (b_c u)}{\partial z'}\ TF_1 + \frac{\partial (b_c u^{2)}}{\partial x} + \frac{\partial (b_c u^{2)}}{\partial z}\ TF_2$$

$$+ \frac{\partial (b_c uw)}{\partial z'}\ TF_3 - \epsilon_x b_c \frac{\partial^2 u}{\partial x^2}$$

$$- \frac{\partial}{\partial z'}\left\{b_c \epsilon_z\ \frac{\partial u}{\partial z'}\right\} TF_3{}^2 = -\frac{b_c}{\rho}\ \frac{\partial p}{\partial x} - \frac{b_c}{\rho}\ \frac{\partial p}{\partial z}\ TF_2 \tag{A6}$$

Conservation of mass (water)

$$\frac{\partial (b_c u)}{\partial x} + \frac{\partial (b_c u)}{\partial z}\ TF_2 + \frac{\partial (b_s w)}{\partial z}\ TF_3 = 0 \tag{A7}$$

Free surface elevation

$$\frac{\partial (b_s \zeta)}{\partial t} + \frac{\partial}{\partial x}\left\{b_c \int_0^H \frac{u}{TF_3}\ dz'\right\} = 0 \tag{A8}$$

Conservation of dissolved matter (salt)

$$\frac{\partial (b_s c)}{\partial t} + \frac{\partial (b_s c)}{\partial z'}\ TF_1 + \frac{\partial (b_c uc)}{\partial x} + \frac{\partial (b_c uc)}{\partial z'}\ TF_2$$

$$+ \frac{\partial (b_s wc)}{\partial z'}\ TF_3 - \frac{\partial}{\partial x}\left\{b_s D_x \frac{\partial c}{\partial x} + b_s D_x \frac{\partial u}{\partial z'}\ \ TF_2\right\}$$

$$- \frac{\partial}{\partial z'}\left\{b_s D_x \frac{\partial c}{\partial x} + b_s D_x \frac{\partial c}{\partial z'}\ TF_2\right\} TF_2$$

$$- \frac{\partial}{\partial z'}\left\{b_s D_z \frac{\partial c}{\partial z'}\ TF_3^2\right\} = 0 \tag{A9}$$

For (A6) it was not necessary to take into account the full transformation of the longitudinal dispersion because this term is only of interest for numerical stability.

APPENDIX II: THE DIFFERENCE EQUATIONS

In this appendix the discretization is given of the differential equations (A6) to (A9) (see Appendix I).

Conservation of longitudinal momentum

$$
\begin{aligned}
\frac{b_{i,j}^{n+1}\,u_{i,j}^{n+1} - b_{i,j}^{n}\,u_{i,j}^{n}}{\tau} = & -\frac{\left\{b_{i+1}^{n}\,(u_{i+1,j}^{n})^2 - b_{i-1,j}^{n}\,(u_{i-1,j}^{n})^2\right\}}{2\Delta x} \\
& + \epsilon_x\, b_{i,j}^{n}\,\frac{\left\{u_{i+1,j}^{n} - 2u_{i,j}^{n} + u_{i-1,j}^{n}\right\}}{\Delta x^2} \\
& - \frac{b_{i,j}^{n+1}}{\rho_{i,j}^{n}}\,\frac{\left\{p_{i+1,j}^{n+1} - p_{i-1,j}^{n+1}\right\}}{2\Delta x} \\
& - \frac{b_{i,j}^{n+1}}{\rho_{i,j}^{n}}\,\frac{\left\{p_{i,j+1}^{n+1} - p_{i,j-1}^{n+1}\right\}}{2\Delta z}\,TF_{2_{i,j}} \\
& - \frac{\left\{b_{i,j+1}^{n+1}\,u_{i,j+1}^{n+1} - b_{i,j-1}^{n+1}\,u_{i,j-1}^{n+1}\right\}}{2\Delta z}\,TF_{1_{i,j}} \\
& - \frac{\left\{b_{i,j+1}^{n+1}\,u_{i,j+1}^{n}u_{i,j+1}^{n+1} - b_{i,j-1}^{n+1}\,u_{i,j-1}^{n}\,u_{i,j-1}^{n+1}\right\}}{2\Delta z}\,TF_{2_{i,j}} \\
& - \frac{\left\{b_{i,j+1}^{n+1}\,w_{i,j+1}^{n}\,u_{i,j+1}^{n+1} - b_{i,j-1}^{n+1}\,w_{i,j-1}^{n}\,u_{i,j-1}^{n+1}\right\}}{2\Delta z}\,TF_{3_i} \\
& + \left\{\frac{\left(b_{i,j+1}^{n+1} + b_{i,j}^{n+1}\right)}{2}\,\epsilon_{z_{i,j+1/2}}\,\frac{\left(u_{i,j+1}^{n+1} - u_{i,j}^{n+1}\right)}{\Delta z}\right. \\
& \left. - \frac{\left(b_{i,j}^{n+1} + b_{i,j-1}^{n+1}\right)}{2}\,\epsilon_{z_{i,j-1/2}}\,\frac{\left(u_{i,j}^{n+1} - u_{i,j-1}^{n+1}\right)}{\Delta z}\right\}\frac{TF_{3_i^z}}{\Delta z} \qquad \text{(A10)}
\end{aligned}
$$

Conservation of mass (water)

$$\frac{b_{j+1,i}\ w_{j+1,i} - b_{j,i}\ w_{j,i}}{\Delta z} = -\frac{4}{3TF_{3_i}}\left[\frac{\{b_{i+1,j+1}\ u_{i+1,j+1} - b_{i-1,j+1}\ u_{i-1,j+1}\}}{4\Delta x} + \frac{\{b_{i+1,j}\ u_{i+1,j} - b_{i-1,j}\ u_{i-1,j}\}}{4\Delta x}\right]$$
$$+ \frac{1}{3TF_3}\left[\frac{\{b_{i+2,j+1}\ u_{i+2,j+1} - b_{i-2,j+1}\ u_{i-2,j}\}}{8\Delta x} + \frac{\{b_{i+2,j}\ u_{i+2,j} - b_{i-2,j}\ u_{i-2,j}\}}{8\Delta x}\right]$$
$$- \frac{\left(TF_{2_{i,j+1}} + TF_{2_{i,j}}\right)}{2TF_{3_i}}\ \frac{\{b_{i,j+1}\ u_{i,j+1} - b_{i,j}\ u_{i,j}\}}{\Delta z} \tag{A11}$$

The free surface elevation

$$\frac{b^{n+1}_{i,N_z}\ \zeta^{n+1}_i - b^n_{i,N_z}\ \zeta^n_i}{\tau} = -\frac{4}{3}TF_{3_i}\ \frac{\left\{\sum_j [b^n_{i+1,j}\ u^n_{i+1,j} - b^n_{i-1,j}\ u^n_{i-1}]\ \Delta z\right\}}{2\Delta x} + \frac{1}{3TF_{3_i}}\frac{\left\{\sum_j [b^n_{1+2,j}\ u^n_{i+2,j} - b^n_{i-2,j}\ u^n_{i-2,j}]\Delta z\right\}}{4\Delta x} \tag{A12}$$

Conservation of dissolved matter (salt)

$$\frac{b^{n+1}_{i,j}\ c^{n+1}_{i,j} - b^n_{i,j}\ c^n_{i,j}}{\tau} = -\frac{4}{3}\ \frac{\{b^n_{i+1,j}\ u^n_{i+1,j}\ c^n_{i+1,j} - b^n_{i-1,j}\ u^n_{i-1,j}\ c^n_{i-1,j}\}}{2\Delta x} + \frac{1}{3}\frac{\{b^n_{i+2,j}u^n_{i+2,j}\ c^n_{i+2,j} - b^n_{i-2,j}u^n_{i-2,j}\ c^n_{i-2,j}\}}{4\Delta x}$$
$$+ \left\{\frac{\left(b^n_{i+1} + b^n_{i,j}\right)}{2}\ D_{x_{i+1/2,j}}\left(\frac{c^n_{i+1,j} - c^n_{i,j}}{\Delta x}\right) - \frac{\left(b^n_{i,j} + b^n_{i-1,j}\right)}{2}\ D_{x_{i-1/2,j}}\ \frac{\left(c^n_{i,j} - c^n_{i-1,j}\right)}{\Delta x}\right\}\frac{1}{\Delta x}$$
$$- \frac{\left\{b^{n+1}_{i,j+1}\ c^{n+1}_{i,j+1} - b^{n+1}_{i,j-1}\ c^{n+1}_{i,j-1}\right\}}{2\Delta z}\ TF_{1_{i,j}}$$

$$- \frac{\left\{ b_{i,j+1}^{n+1} \, u_{i,j+1}^{n+1} \, c_{i,j+1}^{n+1} - b_{i,j-1}^{n+1} \, u_{i,j-1}^{n+1} \, c_{i,j-1}^{n+1} \right\}}{2\Delta z} \; TF_{2_{i,j}}$$

$$- \frac{\left\{ b_{i,j+1}^{n+1} \, w_{i,j+1}^{n+1} \, c_{i,j+1}^{n+1} - b_{i,j-1}^{n+1} \, w_{i,j-1}^{n+1} \, c_{i,j-1}^{n+1} \right\}}{2\Delta z} \; TF_{3_i}$$

$$+ \left\{ \frac{\left(b_{i,j+1}^{n+1} + b_{i,j}^{n+1} \right)}{2} \, D_{z_{i,j+1/2}} \, \frac{\left(c_{i,j+1}^{n+1} - c_{i,j}^{n+1} \right)}{\Delta z} - \frac{\left(b_{i,j}^{n+1} + b_{i,j-1}^{n+1} \right)}{2} \, D_{z_{i,j-1/2}} \, \frac{\left(c_{i,j}^{n+1} - c_{i,j-1}^{n+1} \right)}{\Delta z} \right\} \frac{TF_{3_i}^2}{\Delta z} \qquad \text{(A13)}$$

The Richardson number

$$Ri_{i,j+1/2}{}^{*} = - \frac{8\Delta z \, g}{(\rho_{\,i,j+1} + \rho_{\,i,j})} \, \frac{\{c_{i,j+2} + c_{i,j+1} - c_{i,j} - c_{i,j-1}\}}{\{u_{i,j+2} + u_{i,j+1} - u_{i,j} - u_{i,j-1}\}^2} \qquad \text{(A14)}$$

The smoothing procedure applied to Ri^*

$$Ri_{i,j+1/2} = \tfrac{1}{4} Ri_{i,j-1/2}{}^{*} + \tfrac{1}{2} Ri_{i,j+1/2}{}^{*} + \tfrac{1}{4} Ri_{i,j+3/2}{}^{*} \qquad \text{(A15)}$$

APPENDIX III: SPECIAL DISCRETIZATION NEAR THE BOTTOM

In Delft Hydraulics Laboratory (1978), it is shown that application of the same discretization near the bottom as in the rest of the field yields inaccurate velocities. Therefore a special approximation of the Reynolds stress τ_{xz} at $z = \Delta z/2$ has to be given to replace the usual discretization.

Starting from the equation for the conservation of momentum in the x-direction (for simplicity a constant width is used here)

$$\frac{\partial u}{\partial t} + \frac{\partial u^2}{\partial x} + \frac{\partial uw}{\partial z} - \epsilon_x \frac{\partial^2 u}{\partial x^2} - \frac{\partial}{\partial z}\left(\epsilon_z \frac{\partial u}{\partial z} \right) = - \frac{1}{\rho} \frac{\partial p}{\partial x} \qquad \text{(A16)}$$

Rewriting yields approximately

$$\frac{\partial}{\partial z}\left(\epsilon_z \frac{\partial u}{\partial z} \right) = c_1 z + c_2 \qquad 0 \leqslant z \leqslant \Delta z \qquad \text{(A17)}$$

in which c_1 and c_2 are given by

$$c_1 = \left[\left\{\frac{1}{\rho}\frac{\partial p}{\partial x} + \frac{\partial u}{\partial t} + \frac{\partial u^2}{\partial x} + \frac{\partial uw}{\partial z} - \epsilon_x \frac{\partial^2 u}{\partial x^2}\right\}_{z=\Delta z} - c_2\right]/\Delta z \tag{A18}$$

$$c_2 = \left\{\frac{1}{\rho}\frac{\partial p}{\partial x} + \frac{\partial u}{\partial t} + \frac{\partial u^2}{\partial x} + \frac{\partial uw}{\partial z} - \epsilon_x \frac{\partial^2 u}{\partial x^2}\right\}_{z=0} \tag{A19}$$

This corresponds to an approximation of the Reynolds stress by a quadratic function of z

$$\epsilon_z \frac{\partial u}{\partial z} = \frac{c_1}{2} z^2 + c_2 z + c_3 \tag{A20}$$

If a mixing length approximation is applied with

$$\epsilon_z = \kappa^2 (z + z_0)^2 \left|\frac{\partial u}{\partial x}\right| \tag{A21}$$

in which z_0 is a measure for the roughness length, then substituion of Eq. (A21) into (A20) and some reordering yields

$$\frac{\partial u}{\partial z} = \left\{\frac{\sqrt{1 + \frac{c_2}{c_3} z + \frac{c_1}{2c_3} z^2}}{\kappa^2 (z+z_0)} \frac{1}{\left|\frac{\partial u}{\partial z}\right|}\right\} c_3 \frac{\sqrt{1 + \frac{c_2}{c_3} + \frac{c_1}{2c_3} z^2}}{(z + z_0)} \tag{A22}$$

with

$$\left|\frac{c_2}{c_3} z + \frac{c_1}{2c_3} z^2\right| << 1 \quad \text{for} \quad 0 \leqslant z \leqslant \Delta z \tag{A23}$$

Now the first term on the right hand side of (A22) is a constant

$$c_4 = \frac{\sqrt{1 + \frac{c_2}{c_3} z + \frac{c_1}{2c_3} z^2}}{\kappa^2 (z + z_0)} \frac{1}{\left|\frac{\partial u}{\partial z}\right|} \tag{A24}$$

Developing the square root of Eq. (A22) into a series of z yields, if higher order terms are neglected

$$\frac{\partial u}{\partial z} = c_4 \frac{\left\{\frac{c_1}{4}z^2 + \frac{c_2}{2}z + c_3\right\}}{(z + z_0)} \tag{A25}$$

which shows the correct behavior of

$$\frac{\partial u}{\partial z} \quad \text{for} \quad z \to 0$$

Integration of Eq. (A25) and substitution of the boundary conditions

$$\text{at } z = 0: \quad u = 0 \tag{A26}$$

$$\text{at } z = \Delta z: \quad u = u(\Delta z) \tag{A27}$$

yields an expression for c_3, which is linearly dependent on $u(\Delta z)$. Substitution of c_3 into Eq. (A20) gives the desired relation from which

$$\epsilon_z \left.\frac{\partial u}{\partial z}\right|_{z = \Delta z/2}$$

can be computed.

APPENDIX IV: DESCRIPTION OF THE TIDAL FLUME MEASUREMENTS

Description of the Flume

The lucite flume used for the experiments had a rectangular cross-section 0.67 m wide and 0.50 m high. Two straight sections and the bend between them had a total length of 101.5 m (Fig. 1). Downstream the flume discharges into a sea basin, 8 m long, 6 m wide and 1.1 m deeper than the flume. By means of a control value any periodic tidal movement of the water level can be generated. In the test used for the homogeneous calibration Rhodamine WT was used as tracer. The concentration (salt or Rhodamine) of the sea water was kept constant by means of a circulation system which pumps water with the desired concentration into the basin through perforated tubes on the bottom. At the upstream end of the flume there was equipment to supply separately a constant and a variable fresh water discharge. This allowed simulation of a fictitious flume which has a greater length than the present flume. The variable discharge of fresh water was programmed according to one-dimensional computations. For a detailed description of the flume see Van Rees, *et al.* (1969).

Tests Used for Calibration

The results of several tests are available with plates (2 × 2 cm) on the bottom of the flume, aranged in a diagonal pattern to obtain the desired roughness (DHL, 1979b, c). For the homogeneous calibration, test T22 was used; for the inhomogeneous one test, T20. The boundary conditions and flume parameters were kept the same, except for the density difference between river and seawater. In test T22 there was no difference, and in T20 this difference was equal to 23.4 kgm^{-3}. All boundary conditions and flow parameters are presented in Table 1.

Table 4.
Boundary Conditions and Flow Parameters

Quality	Symbol	Test T22	Test T20
Depth (averaged over T)	H	0.216 m	0.216 m
Physical length of the flume	L	100.65 m	100.65 m
Fictive length of the flume	L_F	179.34 m	179.34 m
Chézy coefficient	C	19 $m^{1/2}\ s^{-1}$	19 $m^{1/2}\ s^{-1}$
Tidal period	T	558.75 s	558.75 s
Tidal amplitude at sea	a_0	0.025 m	0.025 m
Fresh water discharge	Q_r	0.0029 $m^3\ s^{-1}$	0.0029 $m^3\ s^{-1}$
Density differences between river and sea water	$\Delta\rho$	0 kg m^{-3}	23.4 *kg* m^{-3}
Rhodamine concentration sea water	c	0.9 10^{-5} kg m^{-3}	

The tidal motion in these tests can be characterized by the following values of the estuary parameters: the flood number

$$\alpha = \frac{Q_r T}{P_t} = 0.36$$

and the internal estuary number

$$E_D = \frac{u_0^2}{\Delta\rho/\rho\ gH}\ \frac{P_t}{Q_r T} = 1.5$$

where P_t denotes the tidal prism (the volume of seawater penetrating into the flume during the flood period) and u_0 the maximum flood velocity.

An inhomogeneous estuary with these values of the estuary parameters is called partly mixed.

Tests Used for Verification

For the verification two groups of tests were used. Boundary conditions of the tests for the verification are generally the same as those used for the calibration, given in Table 4. The differences from those boundary conditions are given in Table 5.

Table 5.

Variations in the Boundary Condition for the Verification

Quantity	Dimension	T12	T10	T32	T30	T31
a_0	m	0.0125	0.0125	0.0375	0.0375	0.0375
Q_r	$m^3\ s^{-1}$	0.0029	0.0029	0.00145	0.002175	0.002175
$\Delta\rho$	$kg\ s^{-3}$	0.0	21.5	0.0	21.8	10.1

Positions of the measurements

In sixteen stations, at distances $n\Delta x$ from the downstream boundary (n varying from 1 to 16 and $\Delta x = 3.66$ m), the water level was measured. In 8 stations at respectively Δx, $3\Delta x$, $5\Delta x$, $7\Delta x$, $9\Delta x$, $11\Delta x$, $13\Delta x$, and $16\Delta x$ from the downstream boundary the velocities and Rhodamine or salt concentrations were measured at 12 positions in the vertical (with distances $\Delta z = H/13$ between each other). The Rhodamine concentration was measured at four positions, respectively $2\Delta z$, $5\Delta z$, $8\Delta x$ and $11\Delta z$ from the bottom.

Accuracy of the Measurements

Three times the standard deviation in the measurements is taken as the confidence interval. For the mean water level this yields an interval of 0.0002 m, for the tidal amplitude of the water level 0.0002 m and for the phase 0.035 rad. For the velocities the interval is 0.0225 m s^{-1}. The Rhodamine concentration has been measured by pumping water from the measuring point to the fluorimeter, which takes a travelling time of 40 s. The confidence interval of the Rhodamine measurement itself is on the order of 0.02 c_{max}. When, however, rapid variations in time occur in comparison with the traveling time (the increase from the minimum to the maximum concentration takes about 30 s) inaccuracies arise due to deviations in the phase. The confidence interval for the density measurement is 0.75 kg m^{-3}.

REFERENCES

Abraham, G. (1980). On internally generated estuarine turbulence. *Proc. 2nd Int. Symp. on Stratified Flows,* Trondheim, Paper 2.13, 344-353.

Blumberg, A. F. (1975). A numerical investigation into the dynamics of estuarine circulation. Chesapeake Bay Inst., Technical Rep. 91, The Johns Hopkins University.

Blumberg, A. F. (1977). Numerical model of estuarine circulation. *J. Hydr. Div., HY3,* 295-310.

Blumberg, A. F. (1978). The influence of density variations on estuarine tides and circulations. Estuarine and Coastal Marine Science, *6*, 209-215.

Boericke, R. R., and Hogan, J. M. (1977). An X-Z hydraulic/thermal model for estuaries. *J. of Hydr. Div., HY1,* 19-37.

Bowden, K. R., and Hamilton, P. (1975). Some experiments with a numerical model of circulation and mixing in a tidal estuary. Estuarine and Coastal Marine Science, **3**, 281-301.

Delft Hydraulics Laboratory, (1973). Computational methods for the vertical distribution of flow in shallow water. Report W 152.

Delft Hydraulics Laboratory, (1974a). Approximations in mathematical models for stratified flow. Report S 114-IV.

Delft Hydraulics Laboratory, (1974b). Momentum and mass transfer in stratified flows. Report R 880-I.

Delft Hydraulics Laboratory (1975a). Computations of density currents in estuaries, choice of the difference scheme. Report R 897-I (Dutch text).

Delft Hydraulics Laboratory (1975b). Computations of density currents in estuaries, research on weighted-residuals method. Report R 897-II (Dutch text).

Delft Hydraulics Laboratory (1976a). Two-dimensional research on turbulent diffusion coefficients. Report M 896-28 (Dutch text).

Delft Hydraulics Laboratory (1976b). Computations of density currents in estuaries, the homogeneous model. Report R 897-III (Dutch text).

Delft Hydraulics Laboratory (1978). Computation of density currents in estuaries, numerical accuracy of the model. Report R 897-IV.

Delft Hydraulics Laboratory (1979a). Computations of density currents in estuaries, calibration for homogeneous flow in a tidal flume. Report R 897-V.

Delft Hydraulics Laboratory (1979b). Investigations with bottom roughness for the verification of a two-dimensional numerical salt intrusion model, homogeneous tests. Report M 896-38A (Dutch text). Delft Hydraulics Laboratory (1979c). Investigations with bottom roughness for the verification of a two-dimensional numerical salt intrusion model, inhomogeneous tests. Report M 896-38B (Dutch text).

Delft Hydraulics Laboratory (1980a). Computation of density currents in estuaries, calibration for inhomogeneous flow in a tidal flume. Report R 897-VI.

Delft Hydraulics Laboratory (1980b). The influence of lateral variations of the diffusive transport. Report M 896-46 (Dutch text).

Delft Hydraulics Laboratory (1980c). Computation of density currents in estuaries, extended calibration for inhomogeneous flow in a tidal flume. Report R 897-VII.

Delft Hydraulics Laboratory (1980d). Verification of a two-dimensional, numerical model for inhomogeneous flows in a tidal flume. Report M 896-45 (Dutch text).

Elder, J. W. (1959). The dispersion of marked fluid in turbulent shear flow. *J. Fluid Mech., 5.*

Fischer, H. B., List, E. J., Koh, R.C.Y., Imberger, J., and Brooks, N. H., (1979). "Mixing in Inland and Coastal Waters." Academic Press, New York.

Hamilton, P. (1975), A numerical model of the verification circulation of tidal estuaries and its application to the Rotterdam Waterway. *Geophys. J. R. Astr. Soc., 40,* 1-21.

Hinwood, J. B. and Wallis, I. G. (1975a). Classification of models of tidal waters. *J. of Hydr. Div., HY10,* 1313-1331.

Hinwood, J. B. and Wallis, I. G. (1975b). Review of models of tidal waters. *J. of Hydr. Div., HY11,* 1405-1422.

Kent, R.E, and Pritchard, D. W. (1959). A test of mixing length theories in a coastal plain estuary. *J. Mar. Res.* **1**, 62-72.

Odd, N. V. M. and Rodger, J. G. (1978). Vertical mixing in stratified flows. *J. of Hydr. Div., HY3,* 337-351.

Perrels, J. A. J. and Karelse, M. (1977). A two-dimensional numerical model for salt intrusion in estuaries. Delft Hydraulics Laboratory, Publication 177.

Van Rees, A. J. (1975). Experiemental results on exchange coefficients for non-homogeneous flow. XVI IAHR Congress, Paper C36.

Van Rees, A. J. and Rigter, B. P. (1969). Flume study on salinity intrusion in estuaries. XIII IAHR Congress, Paper C33.

Roberts, R. B. and Street, R. L. (1975). Two-dimensional, hydrostatic simulation of thermally-influenced hydrodynamic flows. Stanford University, Technical Report 194.

Rodger, J. G. (1980). Simulation of stratified flows in estuaries. Proc. Int. Symp. on Stratified Flows, Paper 2.12, 338-343, Trondheim.

Rodi, W. (1980). Turbulence models and their application in hydraulics--a state of the art review. Edited by IAHR.

Rossby, C. G., and Montgomery, R. B. (1935). The layer of functional influence in wind and ocean currents. *Pap. Phys. Oceanogr.* **3**, No. 3.
Waldrop, W. R. and Farmer, R. C. (1973). Three-dimensional flow and sediment transport at river mouths. Louisiana State University, Technical Rep. 150.
Waldrop, W. R. and Farmer, R. C. (1976). A computer simulation of density currents in a flowing stream. Int. Symp. on "Unsteady flow in open channels," Newcastle-upon Tyne, England.
Waldrop, W. R., Farmer, R. C. and Bryant, P. A. (1974). Saltwater intrusion into a flowing stream. Louisiana State University, Technical Report 161.

DISCUSSION

Nicholas V. M. Odd

Hydraulics Research Station, Wallingford, United Kingdom

In my view, one of the main reasons why the model did not simulate conditions in the Delft tidal flume very accurately is the unrealistic form of the damping function, $G(Ri)$, especially as regards the value of the flux Richardson Number $Rf = D_z/\epsilon_z$. Ri should increase to a maximum value of about 0.1 as the local gradient Richardson Number increases, whereas the values prescribed by the author's damping functions tend to zero.

Local Ri	"Model Rf"	"True Rf"
0.10	0.0	0.0
0.25	7×10^{-3}	≈0.07
1.00	8×10^{-7}	≈0.10

I would expect the model to perform much better if it incorporated the mixing length damping functions proposed in Reference 1, and if some allowance was also made for viscous effects in the flume. The damped coefficient of turbulent eddy dispersion, D_z, in the flume was probably of the same order of magnitude as the coefficient of molecular diffusivity.

REFERENCES

Odd, N. V. M. and Rodger, J. G. (1978). Vertical mixing in stratified tidal flows. *J. Hydraul. Div. Proc. Am. Soc. Civ. Eng.* **104**, No. HY3, 337-351.

REPLY

The authors agree with the statement in Odd's discussion, as well as in Smith's, that the form of damping functions can be changed to a form which corresponds better to results presented in recent publications, for example that of Gartrell (1979), especially for larger Ri. However, changing the damping functions leaves the turbulence model basically unchanged. Therefore, it is questionable if such a change would give a dramatic improvement of results presented. Further we doubt very much that an approach with a less local nature, as suggested in Odd and Rodger (1978), would perform generally better than the present approach (Gartrell, 1979).

The maximum *Ri* numbers occuring in the flume were of the order of 0.6. With the present damping function *G*(*Ri*) this would imply that locally *Dz* was of the same order as the coefficient of molecular diffusivity. However, as Odd argues, the form of *G*(*Ri*) is not very realistic, especially for large *Ri*-numbers. A switch to a *G*(*Ri*) which introduces a less severe damping for high Richardson numbers would also imply that D_z stays at least one order of magnitude larger than the coefficient of molecular diffusivity.

REFERENCE

Gartrell, G. (1979). Studies on the mixing in a density-stratified shear flow. California Institute of Technology, Rep. KH-R-39.

Timothy J. Smith

Institute of Oceanographic Sciences, Crossway,
Taunton, Somerset, England

In view of the author's conclusion concerning the limited predictive ability of their model I would like to make a few remarks on how this might be improved. In a similar study, the writer developed a model for salinity intrusion in estuaries in which the eddy coefficients were determined from the local turbulence energy and a specific length scale. It was found that the calculated values of stratification and velocity shear were relatively insensitive to the vertical distribution of the length scale. This implies that scaling the eddy coefficients on the turbulence energy is sufficient to allow for the different characteristics of externally and internally generated turbulence. However, the excess stratification calculated by the authors at low water might not be due to the chosen length scale distribution but to the form of the damping functions *F*(*Ri*) and *G*(*Ri*). Recent work by the writer suggested that the exponential decay of *F*(*Ri*) and *G*(*Ri*) used by the authors is too severe at high Richardson numbers. An alternative theoretical formulation was proposed which behaves similarly to that used by the authors at low *Ri* but which decays considerably less rapidly at high *Ri*. In addition, the writer's earlier study also showed that, in laterally averaged models, the representation of the dispersion process is as important as the parameterization of the Reynolds fluxes in determining the spatial structure of the salinity distribution. Consequently, it is suggested that the authors consider the use of a more sophisticated dispersion coefficient which includes the effects of salinity gradients and also the use of different damping functions *F*(*Ri*) and *G*(*Ri*). If the predictive ability of the model is still not improved by these simple modifications then a turbulence energy closure model for the Reynolds fluxes should be used.

REPLY

It is our experience that the longitudinal dispersion coefficient *Dx* influences the stratification less than the damping functions do. The influence of *Dx* is largely confined to the longitudinal density distribution. As reported in Section 6, *Dx* is influenced by the salinity distribution. Consequently, our conclusions in Section 8 generally correspond with the suggestions of Smith.

Summarizing, in our view it still is an open question to what extent zero-order turbulence models can reproduce stratified situations. If further research shows that simple modifications will not improve the model then we are inclined to the opinion of Smith, that higher order turbulence models should be used.

Index

R

S

T

V

W